Frontiers of Combining Systems 2

STUDIES IN LOGIC AND COMPUTATION
Series Editor: **Professor Dov M. Gabbay**
King's College London, UK

7. Frontiers of Combining Systems 2
Edited by **Dov M. Gabbay** and **Maarten de Rijke**

8. Frontiers of Paraconsistent Logic*
Edited by **Diderik Batens, Chris Mortensen, Graham Priest** and **Jean Paul Van Bendegem**

* *Forthcoming*

PREVIOUS BOOKS IN THE SERIES (published by Oxford University Press)

1. Intensional Logics for Programming
Edited by **Luis Fariñas del Cerro** *and* **Martti Penttonen**

2. Substructural Logics
Edited by **Kosta Dosen** *and* **Peter Schroeder-Heister**

3. Nonstandard Queries and Nonstandard Answers
Edited by **R. Demolombe** *and* **T. Imielinski**

4. What is a Logical System?
Edited by **D. M. Gabbay**

5. Conditionals: from philosophy to computer science
Edited by **G. Crocco, L. Fariñas del Cerro** *and* **A. Herzig**

6. Logical Reasoning with Diagrams
Edited by **Gerard Allwein** *and* **Jon Barwise**

Frontiers of Combining Systems 2

Edited by

Dov M. Gabbay
King's College London, UK

Maarten de Rijke
Computational Logic Group
Institute for Logic, Language and Computation
University of Amsterdam, The Netherlands

RESEARCH STUDIES PRESS LTD.
Baldock, Hertfordshire, England

RESEARCH STUDIES PRESS LTD.
15/16 Coach House Cloisters, 10 Hitchin Street, Baldock, Hertfordshire, England, SG7 6AE

and

325 Chestnut Street, Philadelphia, PA 19106

Marketing:

Research Studies Press Ltd.
15/16 Coach House Cloisters, 10 Hitchin Street, Baldock, Hertfordshire, England, SG7 6AE

Distribution:

NORTH AMERICA
Taylor & Francis Inc.
47 Runway Road, Suite G, Levittown, PA 19057 - 4700, USA

ASIA PACIFIC
Hemisphere Publication Services
Golden Wheel Building, 41 Kallang Pudding Road #04-03, Singapore

EUROPE & REST OF THE WORLD
John Wiley & Sons Ltd.
Shripney Road, Bognor Regis, West Sussex, England, PO22 9SA

Library of Congress Cataloging-in-Publication Data

Frontiers of combining systems 2 / edited by Dov M. Gabbay, Maarten de Rijke.
p. cm. -- (Studies in logic and computation ; 7)
Includes bibliographical references and index.
ISBN 0-86380-252-4 (alk. paper)
1. Logic, Symbolic and mathematical--Congresses. 2. Computer science--Congresses.
I. Gabbay, Dov M., 1945- II. Rijke, Maarten de III. Series.

QA9.A1 .F77 1999
511.3--dc21 99-051824

British Library Cataloguing in Publication Data

A catalogue record for this book is available from the British Library.

ISBN 0 86380 252 4

Printed by Cromwell Press Ltd. in Great Britain

Preface

Dov M. Gabbay and Maarten de Rijke

This volume brings together the selected contributions to the Second International Workshop on Frontiers of Combining Systems. The main aim of the FroCoS workshop series is to bring together researchers from diverse areas who share an interest in understanding how methods or tools—both theoretical and applied—can be combined, so as to obtain general purpose systems from a collection of more restricted special purpose ones. This general theme of combining systems has attracted attention from across many disciplines, including applied and theoretical computing, logic, artificial intelligence, and natural language processing.

Let us make matters more concrete by looking at a specific example inspired by developments in artificial intelligence. Over the past decades our perception of computers and computer programs has changed several times in quite dramatic ways. With the rise of the personal computer we began to view the computer as an extension of our office desks and computer programs replaced traditional office tools such as typewriters, calculators, and filing cabinets with word processors, spreadsheets, and databases. The advent of the electronic information age is changing our view again: computers and their programs turn into ubiquitous *digital assistants.* Such assistants have become necessary due to the vast extent and scattered nature of the information landscape. In addition, today's average computer user is neither able nor willing to learn how to navigate through the information landscape with the help of more traditional tools.

Like the personal computer, digital assistants will have a substantial impact on our lives. But do they also have an impact on research in computer science? Are techniques and results already available in computer science research that could have an impact on the way digital assistants (both current and future) are developed and implemented? Combinations of logics are good candidates for a formal theory that can be helpful for the specification, development, and the execution of digital assistants. In particular, modal logics can be used for modeling both digital assistants and (aspects of) their human users. A digital assistant should have an understanding of its own abilities, knowledge, and beliefs. It should also have a representation of the knowledge, beliefs, and goals of its user and of other digital assistants with whom it might have to cooperate in order to achieve its goals. Combined modal logics seem to be perfectly suited as a representation formalism in this setting. It would be an exaggeration to claim that we have a thorough understanding of such 'combined modal formalisms.' However, a core body of notions, questions, and results for an important class of combined tools and formalisms is beginning to emerge.

As has become apparent with the publication of the proceedings of FroCoS'96 [Baader and Schulz, 1996], special issues of the *Notre Dame Journal of Formal Logic* [Blackburn and de Rijke, 1996] and of *Studia Logica* [Gabbay and Pirri, 1997], and Gabbay's book on fibring logics [Gabbay, 1998], there exists a rapidly expanding body of solutions of particular instances of the above combination problem. In addition, the issue of developing general frameworks for integrating formalisms and systems is taking on an increasingly important position on the research agenda. Despite these advances, many open problems remain, both with respect to particular instances of the combination and integration problem, and with respect to the general issue. The contributions to FroCoS'98 offer solutions on both fronts. While covering a wide spectrum of disciplines—from constraint solving to natural language semantics—, there is a clear emphasis on automated tools and logics. At the same time, there is a strong presence of algebraic methods and of term rewriting.

At FroCoS'98, the presentations were grouped together in a number of sessions under the general headings 'Theorem Proving and Rewriting,' 'Logics,' 'Systems,' and 'Constraints.' There were 6 papers in the *theorem proving and rewriting* session. The papers by Ayala-Rincón and by Giesl and Ohlebusch deal with modularity issues in term rewriting; the first paper concentrates on confluence properties of conditional rewrite systems, while the second exploits the notion of a dependency pair to study modularity of termination. Basin and Friedrich investigate the combination of weak second-order monadic logic of one successor with higher-order logic, while Benini, Nowotka and Pulley aim to provide a logical formulation for computer arithmetic, one that is suitable for coding in a theorem proving environment, along with an algorithmic engine and a decision procedure. In their invited paper, Cirstea and Kirchner give a full first-order presentation of the ρ-calculus, in which one can express first-order rewriting, λ-calculus, and non-deterministic computations, as well as combinations of these. Howe's invited contribution to the volume discusses sharing of libraries of mathematics between theorem provers through so-called shallow embeddings.

The *logics* session consisted of 8 papers. Due to reasons beyond our control, Agnes Kurucz' contribution to this session ('Weakly Associative Arrow Logics and Counting Modalities') could not be included in this volume. As to the papers that are included, Basin and Viganò use their framework for the uniform formalization of non-classical logics to analyze the complexity of the decision problems for these logics. Ghidini and Serafini define a logic for the formalization of distributed knowledge representation. In his paper, Naumann combines dynamic logic with event semantics to obtain a semantic analysis of aspectual and temporal phenomena in natural language. The combination considered by Petermann consists of reasoning components that may be either semantically or syntactically specified, and the author formulates criteria for the construction of complete calculi from such components. Two papers use category-theoretic methods for combining systems: one by Sernadas, Sernadas, Caleiro, and Mossakowski, and an invited one by Tarlecki. Whereas the former group of authors focus on a categorical analysis of the fibring construction, the latter author considers specifications in the setting of institutions linked by arrows of various kinds. Finally, Wolter and Zakharyaschev show how to construct decidable combinations of a number of expressive concept description logics and propositional temporal logics.

The emphasis of the 3 papers in the *systems* session was on general architectural issues. In their invited contribution, Apt and Schaerf demonstrate how the Alma-0 language successfully combines imperative and declarative programming. Monfroy describes the language of the **BALI** system for constraint solver collaboration, together with its operational semantics, as well as an overview of its implementation and applications. In their paper Slind and Boulton propose a general scheme for integrating theorem provers that is based on extended interaction protocols.

Finally, the session on *constraints* consisted of 4 papers. Kepser as well as Kepser and Richts are concerned with general usability aspects of combined constraint solvers. The former considers the presence of negative constraints, while the latter deals with optimization techniques for existing non-deterministic algorithms for combining constraint solvers. Prestwich provides and studies a method for combining stochastic algorithms with systematic constraint solvers, while Wang and Goldberg describe the integration of a linear arithmetic solver and a general deduction procedure for proving first-order theorems from software reasoning.

Program Committee

The program committee for FroCoS'98 consisted of the following people: Franz Baader (Aachen), David Basin (Freiburg), Jacques Calmet (Karlsruhe), Dov Gabbay (London; co-chair), Natasha Kurtonina (Philadelphia), Aart Middeldorp (Tsukuba), István Németi (Budapest), Jochen Pfalzgraf (Linz), Maarten de Rijke (Amsterdam; co-chair), Christophe Ringeissen (Nancy), Klaus Schulz (Munich), Amílcar Sernadas (Lisbon), John Slaney (Canberra), and Michael Wooldridge (London).

Out of 43 submissions, the program committee selected 17 papers for presentation at the workshop. In addition, the program committee asked Krzysztof R. Apt, Claude Kirchner, Douglas J. Howe, and Andrzej Tarlecki to give an invited talk during the meeting. We are very glad that we have been able to include their contributions in this volume.

Referees

The program committee gratefully acknowledges the help of the following referees: Slim Abdennadher, Carlos Areces, Peter Baumgartner, Patrick Blackburn, Alexander Bockmayr, Carlos Caleiro, Carlos Castro, Giuseppe De Giacomo, Makoto Hamana, Edith Hemaspaandra, Joxan Jaffar, Sara Kalvala, Maarten Marx, Christof Monz, Hans de Nivelle, Andreas Nonnengart, Robert Rodosek, Gianfranco Rossi, Ulrike Sattler, Cristina Sernadas, Frieder Stolzenburg, Georg Struth, Cesare Tinelli, Luca Viganò, Sergei Vorobyov, B. Wolff, and Hans Zantema.

Sponsors

FroCoS'98 was sponsored by the Computational Logic group at the Department of Mathematics, Computer Science, Physics and Astronomy of the University of Amsterdam; the Institute for Logic, Language and Computation (ILLC) at the University of Amsterdam; Chinaski World Wide; and the Netherlands Organization for Scientific Research (NWO). We gratefully acknowledge their generous support.

Acknowledgments

We would like to thank Carlos Areces, Christof Monz, and, especially, Marco de Vries for their extensive help as members of the local organizing committee. The founders of the FroCoS initiative, Franz Baader and Klaus U. Schulz, provided valuable input during the preparation of the event, and Mrs. Jane Spurr at King's College, London, was invaluable during the compilation of the present volume. We also want to thank Guy Robinson at Research Studies Press for his interest and support.

Dov M. Gabbay and Maarten de Rijke
(Co-chairs FroCoS'98)

REFERENCES

[Baader and Schulz, 1996] F. Baader and K.U. Schulz, editors. *Frontiers of Combining Systems.* Kluwer Academic Publishers, 1996.

[Blackburn and de Rijke, 1996] P. Blackburn and M. de Rijke, editors. *Combining Logics.* Special issue of *Notre Dame Journal of Formal Logic*, 37(2), 1996.

[Gabbay, 1998] D.M. Gabbay. *Fibring Logics.* Oxford Logic Guides no. 38. Oxford University Press, 1998.

[Gabbay and Pirri, 1997] D.M. Gabbay and F. Pirri, editors. *Combining.* Special issue of *Studia Logica*, 59(1, 2), 1997.

Table of Contents

List of Contributors

Krzysztof R. Apt, CWI, P.O. Box 94079, 1090 GB Amsterdam, The Netherlands, and Dept. of Mathematics, Computer Science, Physics & Astronomy, University of Amsterdam, The Netherlands. E-mail: apt@cwi.nl

Mauricio Ayala-Rincón, Departamento de Matemática, Universidade de Brasília, Campus Universitário - Asa Norte, 70910 900, Brasília, Brazil. E-mail: ayala@mat.unb.br

David Basin, Institut für Informatik, Universität Freiburg, Am Flughafen 17, D-79110 Freiburg, Germany. E-mail: basin@informatik.uni-freiburg.de

Marco Benini, Department of Computer Science, University of Warwick, Coventry CV4 7AL, England. E-mail: marcob@dcs.warwick.ac.uk

Richard Boulton, Division of Informatics, University of Edinburgh, 80 South Bridge, Edinburgh EH1 1HN, Scotland. E-mail: rjb@dai.ed.ac.uk

Carlos Caleiro, CMA, Departamento de Matemática, IST, Lisbon, Portugal. E-mail: css@math.ist.utl.pt

Horatiu Cirstea, LORIA and INRIA, 615, rue du Jardin Botanique, B.P. 101, 54602 Villers-lès-Nancy Cedex, France. E-mail: Horatiu.Cirstea@loria.fr

Stefan Friedrich, Institut für Informatik, Universität Freiburg, Am Flughafen 17, D-79110 Freiburg, Germany. E-mail: friedric@informatik.uni-freiburg.de

Dov M. Gabbay, Dept. of Computer Science, King's College, Strand, London WC2R 2LS, England. E-mail: dg@dcs.kcl.ac.uk

Chiara Ghidini, DISA – University of Trento, Via Inama 5, 38100 Trento, Italy. E-mail: ghidini@cs.unitn.it

Jürgen Giesl, Department of Computer Science, Darmstadt University of Technology, Alexanderstr. 10, 64283 Darmstadt, Germany. E-mail: giesl@informatik.tu-darmstadt.de

Allen Goldberg, Kestrel Institute, 3260 Hillview Avenue, Palo Alto, CA 94304, USA. E-mail: goldberg@kestrel.edu

Douglas J. Howe, Bell Labs, 700 Mountain Ave, Murray Hill, NJ 07974, USA. E-mail: howe@research.bell-labs.com

Stephan Kepser, CIS, Universität München, Oettingenstr. 67, 80538 München, Germany. E-mail: kepser@cis.uni-muenchen.de

Claude Kirchner, LORIA and INRIA, 615, rue du Jardin Botanique, B.P. 101, 54602 Villers-lès-Nancy Cedex, France. E-mail: Claude.Kirchner@loria.fr

Eric Monfroy, CWI, Kruislaan 413, 1098 SJ Amsterdam, The Netherlands. E-mail: eric@cwi.nl

Till Mossakowski, BISS, Department of Computer Science, University of Bremen, Germany. E-mail: till@informatik.uni-bremen.de

Ralf Naumann, Seminar für Allgemeine Sprachwissenschaft, University of Düsseldorf, Germany. E-mail: naumann@ling.uni-duesseldorf.de

Dirk Nowotka, Department of Computer Science, University of Warwick, Coventry CV4 7AL, England. E-mail: dirk@dcs.warwick.ac.uk

Enno Ohlebusch, Technische Fakultät, University of Bielefeld, P.O. Box 10 01 31, 33501 Bielefeld, Germany. E-mail: enno@techfak.uni-bielefeld.de

Uwe Petermann, IMN, HTWK Leipzig, 04251 Leipzig, Germany. E-mail: uwe@imn.htwk-leipzig.de

Steven Prestwich, Department of Computer Science, National University of Ireland at Cork. E-mail: s.prestwich@cs.ucc.ie

Carl Pulley, School of Computing and Mathematics, University of Huddersfield, Queensgate, Huddersfield HD1 3DH, England. E-mail: C.J.Pulley@hud.ac.uk

Jörn Richts, Theoretische Informatik, RWTH Aachen, 52056 Aachen, Germany. E-mail: richts@informatik.rwth-aachen.de

Maarten de Rijke, ILLC, University of Amsterdam, Plantage Muidergracht 24, 1018 TV Amsterdam, The Netherlands. E-mail: mdr@wins.uva.nl

Andrea Schaerf, Università di Roma "La Sapienza", Dipartimento di Informatica e Sistemistica, via Salaria 113, 00198 Roma, Italy. E-mail: aschaerf@assi.dis.uniroma1.it

Luciano Serafini, ITC – IRST, 38050 Povo, Trento, Italy. E-mail: serafini@itc.it

Amílcar Sernadas, CMA, Departamento de Matemática, IST, Lisbon, Portugal. E-mail: acs@math.ist.utl.pt

Cristina Sernadas, CMA, Departamento de Matemática, IST, Lisbon, Portugal. E-mail: `css@math.ist.utl.pt`

Konrad Slind, Computer Laboratory, University of Cambridge, New Museums Site, Pembroke St., Cambridge CB2 3QG, England. E-mail: `konrad.slind@cl.cam.ac.uk`

Andrzej Tarlecki, Institute of Informatics, Warsaw University, and Institute of Computer Science, Polish Academy of Sciences, Warsaw, Poland. E-mail: `tarlecki@mimuw.edu.pl`

Luca Viganò, Institut für Informatik, Universität Freiburg, Am Flughafen 17, D-79110 Freiburg, Germany. E-mail: `luca@informatik.uni-freiburg.de`

Tie-Chung Wang, Kestrel Institute, 3260 Hillview Avenue, Palo Alto, CA 94304, USA. E-mail: `wang@kestrel.edu`

Frank Wolter, Institut für Informatik, Universität Leipzig, Augustus-Platz 10-11, 04109 Leipzig, Germany. E-mail: `wolter@informatik.uni-leipzig.de`

Michael Zakharyaschev, Keldysh Institute for Applied Mathematics, Russian Academy of Sciences, Miusskaya Square 4, 125047 Moscow, Russia. E-mail: `mz@spp.keldysh.ru`

Programming in Alma-0, or Imperative and Declarative Programming Reconciled

Krzysztof R. Apt and Andrea Schaerf

1 INTRODUCTION

Logic programming languages, notably Prolog, rely on two important features: nondeterminism and unification. The form of nondeterminism used is usually called "don't know" nondeterminism, whereby *some* path in the computation tree should lead to a correct outcome.

There have been some efforts to incorporate this form of nondeterminism into the imperative programming paradigm. For early references see [Cohen, 1979]. More recent examples are the languages Icon of [Griswold and Griswold, 1983]) and SETL of [Schwartz *et al.*, 1986].

In [Apt *et al.*, 1998] we pursued this approach to programming by proposing another, simple, imperative language Alma-0 that supports this form of nondeterminism.

Our rationale was that almost 25 years of experience with logic programming led to an identification of the programming techniques that make it a distinct programming paradigm. The imperative programming constructs that support nondeterminism should support these programming techniques in a natural way.

And indeed, we found that a number of logic programming jewels could be reproduced in Alma-0 even though unification in the language is limited and the language offers no support for symbolic programming.

But we also found that other programs, such as the solution to the *Eight Queens* problem, could be coded in Alma-0 in a more natural way than the logic programming paradigm permits. Also, some programs, such as the solution to the *Knapsack* problem, seem to be very natural even though they use both nondeterminism and assignment.

So the hybrid programming style of Alma-0 calls for new programming techniques that need to be better understood and explored. This is the aim of this paper that can be seen as a companion article of [Apt *et al.*, 1998].

To this end we provide here a number of Alma-0 programs that show versatility of the language and provide further evidence that the constructs of the language encourage a natural style of programming. In particular, Alma-0 programs without assignment are declarative in the sense that they admit a dual reading as a logic formula.

It should be clarified that, in general, two types of nondeterminism have been considered in programming languages, "don't know" nondeterminism and "don't care" nondeterminism. According to the latter type *each* path in the computation tree should lead to a correct outcome. This form of nondeterminism is present in the guarded command language of [Dijkstra, 1975]. It leads to different issues and different considerations.

The paper is organized as follows. In Section 2 we recall the basic elements of Alma-0. In the remainder of the paper we provide selected examples of Alma-0 programs that complement those presented in [Apt *et al.*, 1998] and illustrate its use in different contexts. More specifically, in Section 3 we present two versions of a classical graph traversal problem, namely the *longest path* problem. In Section 4 we show how a typical feature of the logic programming paradigm, namely *negation as failure*, can be also profitably exploited in Alma-0. Next, in Section 5 we illustrate how executable specifications can be written in Alma-0. In Section 6 we provide a more complex example of Alma-0 programming by describing a solution to a classical scheduling problem. Finally, in Section 7 we draw some conclusions and describe the current status of the Alma project.

2 THE LANGUAGE ALMA-0

Alma-0 is an extension of a subset of Modula-2 that includes nine new features inspired by the logic programming paradigm. We briefly recall most of them here and refer to [Apt *et al.*, 1998] for a detailed presentation.

- Boolean expressions can be used as statements and vice versa. A boolean expression that is used as a statement and evaluates to `FALSE` is identified with a *failure.*

- *Choice points* can be created by the two nondeterministic statements `ORELSE` and `SOME`. The former is a dual of the statement composition and the latter is a dual of the `FOR` statement. Upon failure the control returns to the most recent choice point, possibly within a procedure body, and the computation resumes with the next branch in the state in which the previous branch was entered.

- The created choice points can be erased or iterated over by means of the `COMMIT` and `FORALL` statements. `COMMIT S END` removes the choice points created during a successful execution of `S`. `FORALL S DO T END` iterates over all choice points created by `S`. Each time `S` succeeds, `T` is executed.

- The notion of *initialized* variable is introduced and the equality test is generalized to an assignment statement in case one side is an uninitialized variable and the other side an expression with known value. The KNOWN relation is introduced to test whether a variable of a simple type is initialized.
- A new parameter passing mechanism, called *call by mixed form*, is introduced for variables of simple type. It works as follows: if the actual parameter is a variable, then it is passed by variable. If the actual parameter is an expression that is not a variable, its value is computed and assigned to a new variable v (generated by the compiler): it is v that is then passed by variable. So in this case the call by mixed form boils down to call by value.

  This parameter mechanism, denoted by MIX, is introduced to allow us to pass both values and uninitialized variables as actual parameters.

To clarify these extensions and Alma-0 programming style consider the following problem from [Gardner, 1979].

Problem 1 Ten cells numbered $0,..,9$ inscribe a 10-digit number such that each cell, say i, indicates the total number of occurrences of the digit i in this number. Find this number.

Here is a simple solution to it in Alma-0.

```
MODULE tendigit;
VAR i, j, k, l, count, sum: INTEGER;
    a: ARRAY [0..9] OF INTEGER;
BEGIN
  FORALL
    sum := 0;
    FOR i := 0 TO 9 DO
      SOME j := 0 TO 10-sum DO
        a[i] = j;
        sum := sum + j
      END;
    END;
    sum = 10;
    FOR k := 0 TO 9 DO
      count := 0;
      FOR l := 0 TO 9 DO
        IF a[l] = k THEN count := count + 1; a[k] >= count END;
      END;
      a[k] = count
    END
  DO
  FOR i := 0 TO 9 DO WRITE(a[i]) END
```

```
  END
END tendigit.
```

To better understand this program first note that any 10-digit number that is a solution to this problem has the property that the sum of its digits is 10.

Now, the first `FOR` loop nondeterministically generates 10-digit numbers, written as an array, with this property. This is done by means of a `SOME` statement. The equality `a[i] = j` is used here as an assignment, while the equality `sum = 10` is used as a test.

The second `FOR` loop tests whether a candidate array is a possible solution. The testing can be abandoned if for some `k` the count exceeds the value `a[k]`. This explains the use of the test `a[k] >= count`.

The above described code is within the `FORALL` statement, so all solutions to the problem are generated and each of them is printed. The program yields the unique solution, namely 6210001000.

The still unexplained features of Alma-0 will be discussed later.

3 GRAPH TRAVERSAL

We now illustrate by means of two examples how Alma-0 can be used in a natural way for graph-related problems.

3.1 Knight's Tour

We begin with the following well-known problem.

Problem 2 Find a knight's tour on the $n \times n$ chess board in which each field is visited exactly once.

Here is a solution in Alma-0.

```
MODULE KnightTour;
CONST
  N = 5;
TYPE
  [1..N] = [1..N];
  Board = ARRAY [1..N], [1..N] OF [1..N*N];

PROCEDURE Next(VAR row, col: INTEGER);
VAR i, j: INTEGER;
BEGIN
  EITHER i = 2;  j = 1
  ORELSE i = 1;  j = 2
  ORELSE i = -1; j = 2
  ORELSE i = -2; j = 1
  ORELSE i = -2; j = -1
```

```
  ORELSE i = -1; j = -2
  ORELSE i = 1; j = -2
  ORELSE i = 2; j = -1
  END;
  row := row + i;
  col := col + j;
  (1 <= row) AND (row <= N);
  (1 <= col) AND (col <= N)
END Next;

VAR i, j, k: INTEGER;
  x: Board;
BEGIN
  x[1,1] = 1;
  i = 1; j = 1;
  FOR k := 2 TO N*N DO
    Next(i,j);
    x[i,j] = k
  END;
  Print(x)
END KnightTour.
```

Here the `Next` procedure nondeterministically generates the coordinates of the next field, given the current one. This is done now by means of an `ORELSE` statement that explores all eight possibilities in turn.

After a call to `Next` the (implicitly) incremented value of `k` is assigned to this new field. Note that this assignment, `a[i,j] = k`, is performed by means of an equality. This is crucial, as it also prevents any field from being visited twice. Indeed, if this is the case then `a[i,j]` has already a value and the equality fails. In this case, backtracking takes place and the next, if any, candidate field is generated.

3.2 Longest Path

In the *Knight's tour* problem the $n \times n$ chess board can be viewed as a graph in which the squares are the nodes and the possible knight moves are the arcs. In this way the knight tour problem amounts to finding a simple path of maximal length. The length of this path equals n^2, the number of nodes.

Consider now a more general problem of finding the longest path in an arbitrary directed graph.

Problem 3 Given a directed graph $G = (V, E)$ and two nodes $v_1, v_2 \in V$ find the longest simple path that starts in v_1 and ends in v_2.

Recall that this decision problem is NP-complete (see [Garey and Johnson, 1979, problem ND29, page 213]).

We assume that the graph is represented by its adjacency matrix. We also employ an array for marking the visited nodes and for storing the current longest path. In what follows we use the following type declarations.

```
Graph = ARRAY [1..N],[1..N] OF BOOLEAN;
PathMark = ARRAY [1..N] OF INTEGER;
```

The basic building block that we use for traversing the graph is the following function `Successor` that upon backtracking generates all successors of a given node. The function fails if the node has no successor.

```
PROCEDURE Successor(G: Graph; X: Node): Node;
VAR i: Node;
BEGIN
  SOME i := 1 TO N DO
    G[X,i]
  END;
  RETURN i
END Successor;
```

The following procedure `LongestPath` consists of some initializations followed by a `FORALL` loop that explores all possible paths. Inside the `FORALL` loop, each path is constructed by an inner loop that searches exhaustively for unvisited successors until it gets to the requested final node.

In contrast to Problem 2, we do not know the length of the longest path in advance. Therefore we use here a `WHILE` statement rather than a `FOR` statement for constructing the path. In addition, for each generated path, we need to check its length against the currently longest one.

A node `X` is viewed as unvisited as long as `Path[X] = 0`. When `X` is visited, `Path[X]` gets the value `k` which represents the position of `X` in the path.

```
PROCEDURE LongestPath(G: Graph; InitNode, FinalNode: Node): PathMark;
VAR k, max: INTEGER;
    i: Node;
    Path, LongPath: PathMark;
BEGIN
  FOR i := 1 TO N DO Path[i] := 0 END;
  i := InitNode;
  k := 0;
  max := 0;
  FORALL
    WHILE (Path[i] = 0) AND (i <> FinalNode) DO
      k := k+1;
      Path[i] := k;
      i := Successor(G,i) (* generate a successor
                             nondeterministically *)
    END
  DO
```

```
    IF (i = FinalNode) AND (k > max)
    THEN max := k; LongPath := Path END
  END;
  RETURN LongPath
END LongestPath;
```

The longest path is delivered by means of the return value of the procedure. If no path between `InitNode` and `FinalNode` exists, then the variable `LongPath` remains uninitialized, and thus the value returned is also an uninitialized array, which can be tested within the calling procedure by using the built-in procedure `KNOWN`.

4 USE OF NEGATION

One of the important notions in logic programming is *negation by failure*. It is, in a nutshell, a meta-rule that allows us to conclude a negation of a statement from the fact that it cannot be proved (using the resolution method used in logic programming). Negation by failure is a very useful concept that allows us to write some remarkably concise Prolog programs. Also, it supports non-monotonic reasoning. Actually, the negation by failure mechanism provides a computational interpretation of the latter, a feature other main approaches to non-monotonic reasoning lack.

Negation by failure is supported in Alma-0, as well. In fact, as in logic programming, it is the mechanism used to evaluate negated statements. Consequently, we can use it in Alma-0 in the same way as in logic programming and Prolog.

In [Apt *et al.*, 1998] we already presented a number of programs that used negation. Here we show an Alma-0 solution to the proverbial *Tweety problem*, one of the classical benchmarks for non-monotonic reasoning. Let us recall it.

The problem is to reason in the presence of default assumptions. In the natural language they are often expressed by means of the qualification "usually". In what follows the "usual" situations are identified with those which are not "abnormal".

We stipulate the following assumptions.

- The birds which are not abnormal fly (i.e., birds usually fly).
- Penguins are abnormal.
- Penguins and eagles are birds.
- Tweety is a penguin and Toto is an eagle.

The problem is to deduce which of these two birds flies. Here is a solution in Alma-0, where the code for `Print` is omitted.

```
MODULE penguin;
TYPE Animal = (Tweety, Toto);

PROCEDURE penguin(MIX x: Animal);
BEGIN
  x = Tweety
END penguin;

PROCEDURE eagle(MIX x: Animal);
BEGIN
  x = Toto
END eagle;

PROCEDURE ab(MIX x: Animal);
BEGIN
  penguin(x)
END ab;

PROCEDURE bird(MIX x: Animal);
BEGIN
  EITHER penguin(x) ORELSE eagle(x) END
END bird;

PROCEDURE fly(MIX x: Animal);
BEGIN
  bird(x);
  NOT ab(x)
END fly;

VAR x: Animal;
BEGIN
  FORALL fly(x)
  DO Print(x)
  END
END penguin.
```

The use of the `MIX` parameter mechanism allows us to use each procedure both for testing and for computing, as in Prolog. In particular, the call `fly(x)` yields to a nondeterministic computing of the value of `x` using `bird(x)` and subsequent testing of it using `NOT ab(x)`.

It is instructive to compare this program with the more compact Prolog program (see, e.g., [Apt, 1997, page 303]):

```
penguin(tweety).
eagle(toto).
ab(X) :- penguin(X).
bird(X) :- penguin(X).
bird(X) :- eagle(X).
fly(X) :- not ab(X), bird(X).
```

While logically both programs amount to equivalent formulas we see that it is difficult to compete with Prolog's conciseness.

Other natural uses of negation in Alma-0 can be found in the other programs in this article.

5 EXECUTABLE SPECIFICATIONS

The next example shows that in some circumstances Alma-0 yields programs that are more intuitive than those written in Prolog.

In general, specifications can and do serve many different purposes. The issue of whether specifications should be executable or not has been for a long time a subject of a heated discussion, see, e.g. [Fuchs, 1992]. We do not wish to enter this discussion here but we will show how Alma-0 supports executable specifications in a very natural way.

As an example, consider the problem of finding the lexicographically next permutation, discussed in [Dijkstra, 1976].

To specify this problem recall that by definition a sequence $out_1, \ldots, out_N$ is a permutation of $in_1, \ldots, in_N$ if for some function π from $[1..N]$ onto itself we have

$$out_1, \ldots, out_N = in_{\pi(1)}, \ldots, in_{\pi(N)}.$$

This definition directly translates into the following Alma-0 program:

```
TYPE Sequence = ARRAY [1..N] OF INTEGER;

PROCEDURE Permutation(VAR in, out: Sequence);
VAR pi: Sequence;
    i, j: INTEGER;
BEGIN
  FOR i := 1 TO N DO
    SOME j := 1 TO N DO
      pi[j] = i
    END
  END; (* pi is a function from 1..N onto itself and ...        *)
  FOR i := 1 TO N DO
    out[i] = in[pi[i]]
  END  (* out is obtained by applying pi to the indices of in *)
END Permutation;
```

The procedure `Permutation` provides, upon backtracking, all permutations of the given input sequence.

Next, we need to define the lexicographic ordering. Let us recall the definition: the sequence $a_1, \ldots, a_N$ precedes lexicographically the sequence $b_1, \ldots, b_N$ if some i in the range $[1..N]$ exists such that for all j in the range $[1..i-1]$ we have $a_j = b_j$, and $a_i < b_i$.

In Alma-0 we write these specifications as follows:

```
PROCEDURE Lex(a,b: Sequence);
VAR i, j: INTEGER;
BEGIN
  SOME i := 1 TO N DO
    FOR j := 1 TO i-1 DO
      a[j] = b[j]
    END;
    a[i] < b[i]
  END
END Lex;
```

Now b is the lexicographically next permutation of a if

- b is a permutation of a,
- a precedes b lexicographically,
- no permutation exists that is lexicographically between a and b.

This leads us to the following procedure `Next` that uses an auxiliary procedure `Between`, which checks whether a permutation exists between `a` and `b`:

```
PROCEDURE Between(a,b: Sequence);
VAR c: Sequence;
BEGIN
  Permutation(a,c);
  Lex(a,c);
  Lex(c,b)
END Between;

PROCEDURE Next(VAR a, b: Sequence);
BEGIN
  Permutation(a,b);
  Lex(a,b);
  NOT Between(a,b)
END Next;
```

This concludes the presentation of the program. Note that it is fully declarative and it does not use any assignment. It is obviously inefficient, but still it could be used on the example given in Dijkstra's book, to compute that `1 4 6 2 9 7 3 5 8` is the lexicographically next permutation of `1 4 6 2 9 5 8 7 3`.

It is interesting to see that the above program is invertible in the sense that it can be also used to specify and compute the lexicographically previous permutation. In fact, we can use for this purpose the same procedure `Next` — it just suffices to pass now the given permutation as the second parameter of the procedure `Next`. For this purpose both parameters are passed by variable in the procedures `Next` and `Permutation`.

In this way we can compute for instance that `1 4 6 2 9 5 8 3 7` is the lexicographically previous permutation of `1 4 6 2 9 5 8 7 3`.

6 A SCHEDULING APPLICATION

We now show how Alma-0 can be employed to solve scheduling problems. In particular, we introduce a specific scheduling problem known as the *university course timetabling* problem and discuss its solution in Alma-0.

6.1 Problem Definition

The course timetabling problem consists in the weekly scheduling for all the lectures of a set of university courses in a given set of classrooms, avoiding the overlaps of lectures having common students. We consider the basic problem (which is still NP-complete). Many variants of this problem have been proposed in the literature. They involve more complex constraints and usually consider an objective function to be minimized (see [Schaerf, 1995]).

Problem 4 There are q courses $K_1, \ldots, K_q$, and each course K_i consists of k_i required lectures, and p periods $1..p$. For all $i \in 1..q$, all lectures $l \in 1..k_i$ must be assigned to a period k in such a way that the following constraints are satisfied:

Conflicts: There are c curricula $S_1, \ldots, S_c$, which are groups of courses that have common students. Lectures of courses in S_l must be all scheduled at different times, for each $l \in 1..c$.

Availabilities: There is an availability binary matrix A of size $q \times p$. If $a_{ij} = 1$ then lectures of course i cannot be scheduled at period j.

Rooms: There are r rooms available. At most r lectures can be scheduled at period k, for each $k \in 1..p$.

6.2 A Solution in Alma-0

We now provide a solution of this problem in Alma-0. We start with the constant and type definitions necessary for the program.

```
CONST
   Courses = 10;   (* p *)
   Periods = 20;   (* q *)
   Rooms = 3;      (* r *)
TYPE
   AvailabilityMatrix = ARRAY [1..Courses],[1..Periods] OF BOOLEAN;
   ConflictMatrix = ARRAY [1..Courses],[1..Courses] OF BOOLEAN;
   RequirementVector = ARRAY [1..Courses] OF INTEGER;
   TimetableMatrix = ARRAY [1..Courses],[1..Periods] OF BOOLEAN;
```

Conflicts are represented by a $q \times q$ matrix of the type `ConflictMatrix` such that the element (i, j) of the matrix is *true* if courses K_i and K_j belong simultaneously to at least one curriculum.

The solution is returned by means of a $q \times p$ boolean matrix of the type `TimetableMatrix`. Each element (i, j) of the matrix is *true* if a lecture for the course K_i is given at period j and *false* otherwise.

The procedure `Timetabling` provides the solution of this problem in Alma-0. It follows faithfully the specification of the problem and it performs an exhaustive backtracking search for a feasible solution.

For each course K_i the procedure looks for a number of periods equal to the number of lectures k_i of the course. The array `BusyRooms` counts the number of rooms already used for each period, and is used to check the room occupation constraints.

In order to avoid exploring symmetric solutions for the lectures of a course, each lecture is always scheduled later than the previously scheduled lectures of the same course. This is done by using the variable `PeriodOfPreviousLecture` which keeps track of the period of the most recently scheduled lecture.

```
PROCEDURE Timetabling(Available: AvailabilityMatrix;
                      Conflict: ConflictMatrix;
                      Requirements: RequirementVector;
                      VAR Timetable: TimetableMatrix);
VAR
   BusyRooms : ARRAY [1..Periods] OF INTEGER;
   C, C1, L, P : INTEGER;
   PeriodOfPreviousLecture : INTEGER;
BEGIN
   FOR P := 1 TO Periods DO
      BusyRooms[P] := 0;
   END;
   FOR C := 1 TO Courses DO
      PeriodOfPreviousLecture := 0;
      FOR L := 1 TO Requirements[C] DO
         SOME P := PeriodOfPreviousLecture+1 TO Periods DO
            Available[C,P];
            BusyRooms[P] < Rooms;
            FOR C1 := 1 TO C-1 DO
               NOT (Conflict[C1,C] AND Timetable[C1,P])
            END;
            Timetable[C,P] := TRUE;
            BusyRooms[P] := BusyRooms[P] + 1;
            PeriodOfPreviousLecture := P;
         END
      END
   END
END Timetabling;
```

The proposed procedure can solve only relatively small instances of the problem. For larger ones, more complex algorithms and heuristic procedures are needed (see [Schaerf, 1995]).

6.3 Additional Functionalities

If no solution to the given problem instance exists, it is, in general, necessary to relax some of the constraints. The following procedure checks whether a solution exists when one single conflict constraint is relaxed. If the solution of the relaxed instance of the problem is found, its solution is returned along with the constraint which has been relaxed. This constraint is returned by means of two courses c1 and c2 which are no longer considered in conflict.

```
PROCEDURE RelaxedTimetabling(Available: AvailabilityMatrix;
                             VAR Conflict: ConflictMatrix;
                             Requirements: RequirementVector;
                             VAR Timetable: TimetableMatrix;
                             MIX c1, c2: INTEGER);
VAR
   i, j: INTEGER;
BEGIN
   EITHER
     Timetabling(Available, Conflict, Requirements, Timetable)
   ORELSE
     SOME i := 1 TO Courses-1 DO
       SOME j := i+1 TO Courses DO
         Conflict[i,j];
         c1 = i; c2 = j;
         Conflict[i,j] := FALSE;
         Timetabling(Available, Conflict, Requirements, Timetable)
       END
     END
   END
END RelaxedTimetabling;
```

Finally, the following procedure produces all relaxed and non-relaxed solutions of the problem. The simple code for the procedures Initialize and PrintSolution is omitted.

```
PROCEDURE CreateTimetable;
VAR
  Available: AvailabilityMatrix;
  Conflict: ConflictMatrix;
  Requirements: RequirementVector;
  Timetable: TimetableMatrix;
  NbrSolutions: INTEGER;
  c1, c2: INTEGER;
BEGIN
```

```
  Initialize(Available,Conflict,Requirements,Timetable);
  NbrSolutions := 0;
  FORALL

RelaxedTimetabling(Available,Conflict,Requirements,Timetable,c1,c2)
  DO
    NbrSolutions := NbrSolutions + 1;
    WRITELN('Solution number ',NbrSolutions);
    PrintSolution(Available,Timetable);
    IF KNOWN(c1)
    THEN WRITELN('Conflict between course ', c1,' and ',c2,' relaxed')
    ELSE WRITELN('No constraint relaxed for this solution');
    END
  END;
  IF NbrSolutions > 0
  THEN WRITELN('Number of solutions : ',NbrSolutions)
  ELSE WRITELN('No solution found.');
  END;
  WRITELN
END CreateTimetable;
```

Note the use of the built-in procedure `KNOWN` that checks whether the variable `c1` is initialized or not. This test allows us to check whether a constraint has been relaxed.

Finally, note that `c1` and `c2` are passed by `MIX`. This way, not only a variable but also a constant can be supplied as an actual parameter. For example, the following call searches for a solution in which the possible relaxation involves course K_1:

```
RelaxedTimetabling(Available,Conflict,Requirements,Timetable,1,c);
```

Here `c` is an uninitialized variable.

7 CONCLUSIONS

In this paper we presented a number of programs written in Alma-0. They were chosen with the purpose of illustrating the versatility of the resulting programming style. The solutions to some other classical problems, such as α-β search, STRIPS planning, knapsack, and Eight Queens, have been already provided in [Apt *et al.*, 1998].

These programs show that imperative and logic programming can be combined in a natural and effective way. The resulting programs are in most cases shorter and more readable than their counterparts written in imperative or logic programming style.

Let us review now the work carried out on Alma-0. The implementation of the language Alma-0 is based on an abstract machine, called AAA, that combines the features of a RISC architecture and the WAM abstract machine. In the current version the AAA instructions are translated into C code.

The implementation is described in [Apt *et al.*, 1998] and explained in full detail in [Partington, 1997]. The Alma-0 compiler is available via the Web at `http://www.cwi.nl/alma`.

An executable operational specification of a large fragment of Alma-0 is provided using the ASF+SDF Meta-Environment of [Klint, 1993]. This is described in [Apt *et al.*, 1998] and comprehensively explained in [Brunekreef, 1998].

An extension of Alma-0 that integrates constraints into the language is the subject of an ongoing research project. Various issues related to such integration are highlighted in [Apt and Schaerf, 1998]: in particular, the role of logical and customary variables, the interaction between the program and the constraint store, the local and global unknowns, and the parameter passing mechanisms are considered.

Finally, in [Apt and Bezem, 1998] a computational interpretation of first-order logic based on a constructive interpretation of satisfiability with respect to a fixed but arbitrary interpretation is studied. This work provides logical underpinnings for a fragment of Alma-0 that does not include assignment and allows us to reason about Alma-0 programs written in this fragment.

REFERENCES

[Apt and Bezem, 1998] K. R. Apt and M. A. Bezem. Formulas as programs, 1998. Submitted. Available via `http://www.cwi.nl/~apt`.

[Apt and Schaerf, 1998] K. R. Apt and A. Schaerf. Integrating constraints into an imperative programming language, 1998. Submitted. Available via `http://www.cwi.nl/~apt`.

[Apt *et al.*, 1998] K. R. Apt, J. Brunekreef, V. Partington, and A. Schaerf. Alma-0: An imperative language that supports declarative programming. *ACM Toplas*, 20(5):1014–1066, 1998.

[Apt, 1997] K. R. Apt. *From Logic Programming to Prolog.* Prentice-Hall, London, U.K., 1997.

[Brunekreef, 1998] J. Brunekreef. Annotated algebraic specification of the syntax and semantics of the programming language Alma-0. Technical Report P9803, Programming Research Group, University of Amsterdam, The Netherlands, 1998. Available online at `http://www.wins.uva.nl/research/prog/reports/reports.html`.

[Cohen, 1979] J. Cohen. Non-Deterministic algorithms. *ACM Computing Surveys*, 11(2):79–94, 1979.

[Dijkstra, 1975] E. W. Dijkstra. Guarded commands, nondeterminacy and formal derivation of programs. *Communications of the ACM*, 18:453–457, 1975.

[Dijkstra, 1976] E. W. Dijkstra. *A Discipline of Programming.* Prentice-Hall, Englewood Cliffs, N.J., 1976.

[Fuchs, 1992] N. Fuchs. Specifications are (preferably) executable. *IEE Software Engineering Journal*, 7(5):323–334, 1992.

[Gardner, 1979] M. Gardner. *Mathematical Circus.* Penguin, Harmondsworth, 1979.

[Garey and Johnson, 1979] M. R. Garey and D. S. Johnson. *Computers and Intractability—A guide to NP-completeness.* W.H. Freeman and Company, San Francisco, 1979.

[Griswold and Griswold, 1983] R. E. Griswold and M. T. Griswold. *The Icon Programming Language.* Prentice-Hall, Englewood Cliffs, New Jersey, USA, 1983.

[Klint, 1993] P. Klint. A meta–environment for generating programming environments. *ACM Transactions on Software Engineering and Methodology*, 2(2):176–201, 1993.

[Partington, 1997] V. Partington. Implementation of an imperative programming language with backtracking. Technical Report P9712, Department of Mathematics, Computer Science, Physics & Astronomy, University of Amsterdam, The Netherlands, 1997. Available online at `http://www.wins.uva.nl/research/prog/reports/reports.html`.

[Schaerf, 1995] A. Schaerf. A survey of automated timetabling. Technical Report CS-R9567, CWI, Amsterdam, The Netherlands, 1995. To appear in *Artificial Intelligence Review.*

[Schwartz *et al.*, 1986] J. T. Schwartz, R. B. K. Dewar, E. Dubinsky, and E. Schonberg. *Programming with Sets — An Introduction to SETL.* Springer-Verlag, New York, 1986.

Church-Rosser Property for Conditional Rewriting Systems with Built-in Predicates as Premises

Mauricio Ayala-Rincón

1 INTRODUCTION

Rewriting techniques were applied originally only to equationally defined classes of algebras and were extended later to conditional equationally defined classes of algebras obtaining more expressiveness. For conditional (term) rewriting systems (CTRSs) the most reasonable form of interpreting the equational conditions in the premises is the *standard* one, which corresponds to intuition about conditional rewriting and allows for recursive evaluation; that is, to see whether for each equational condition its factors can be rewritten in identical terms. Confluence and termination guarantee unique normal forms, however, a restriction stronger than termination, for instance *decreasingness*, is needed in order to effectively perform computations (in conditional equationally defined classes of algebras). *Decreasingness* can be seen as a condition that ensures that the problem of validity of equational conditions is smaller than the problem of validity of the conclusions. Sets of equations and conditional equations can be considered as functional programs with rewriting as their computation mechanism. Whereas predefined operations are available in common programming languages they are not naturally incorporated into rewriting techniques. Therefore it is of great interest to combine built-in algorithms and conditional rewriting.

This paper reports generalizations of the classical proofs of Newman's diamond lemma and the Church-Rosser property for a combination of built-in predicates and CTRSs introduced in [Ayala-Rincón, 1993]. CTRSs with built-in predicates as conditions, for brevity BIS-CTRSs ("BIS" stands for "built-in standard"), are CTRSs where the premises are conjunctions of standard conditions and built-in predicates in a *basic theory*. These predicates evaluate the boolean value of terms by some procedure independent of rewriting. In order to make the process of rewriting modulo basic theories effective, they (or at least a sub theory of the basic theory) should be (effectively) decidable.

The language of the basic theory is considered as a signature over basic sorts and the remaining function symbols of the conditional specification associated with the BIS-CTRS range over extended sorts. In this way one guarantees that the whole specification is a conservative extension of the specification of the basic theory. This restriction appears to be very strong, but it allows reasonable manipulation of many important examples because it occurs often when implementing formal specifications, where new sorts are constructed from the concrete ones. This is the case in the manipulation of algebraic structures such as vectors and matrices where the access to their contents is determined by questions on their indices which are predicates in the linear arithmetic.

In [Ayala-Rincón, 1993] conditional rewriting techniques were extended in order to allow deduction in universal Horn classes with built-in predicates. Rewriting techniques were combined with built-in algorithms in such a way that deduction of Horn clauses whose conditions are purely built-in is possible. Purely rewriting is insufficient and it should be refined with a satisfiability check of the built-in conditions (matching modulo the basic theory) and *case splitting* guided by the built-in conditions. Specialized notions of termination and confluence are necessary in order to guarantee effective computation of normal forms.

This paper is organized as follows: in the second section, the combination of standard CTRSs and built-in premises originating the BIS-CTRSs and the corresponding notion of reduction are introduced; in the third section, comments on the decision procedure for universal Horn clauses (whose conditions are purely built-in) and conditions for its completeness are presented without proofs; in the fourth section, generalizations of the Newman's diamond lemma and the Church-Rosser property for BIS-CTRSs are presented; finally, conclusions and future work are given.

Notations consistent with the standard ones in the field of rewriting theory are used. Basics on rewriting can be found in [Baader and Nipkow, 1998].

Related Work. From the beginning of the application of conditional rewriting techniques in automated theorem proving and algebraic specification their inefficiency and limited expressiveness motivates the manipulation of parts of CTRSs by means of algorithms other than rewriting [O'Donnell, 1977], [Remy, 1982]. The first person to formalize a rewriting algorithm (combining satisfiability check and case splitting) for CTRSs with built-in predicates was Vorobyov [1989]. He treated CTRSs whose conditions are pure built-in predicates, without admitting standard equational conditions as BIS-CTRSs do, and developed a decision procedure which is extended for BIS-CTRSs in this work.

Built-in conditions can be seen as *constraints* which are used to represent knowledge about predefined structures converting syntactical problems of the extended specification into semantical ones of the constraint part. Kirchner et al [1990] developed an approach based on the notion of constrained rewriting, requiring the conditions to be formulated in the built-in language, whereas here standard and built-in conditions are allowed.

Dershowitz and Okada [1990] studied CTRSs with built-in predicates and standard equational conditions as premises. They briefly treated termination, confluence and a critical pair criterion for their systems. Precise proofs of their suggested results on confluence (overlay terminating and decreasing CTRSs with built-in predicates handled by rewriting, all of whose critical pairs are joinable, are confluent) can be found in [Ayala-Rincón, 1994].

Avenhaus and Becker [1992] presented a method of integrating built-in operations described by a given built-in algebra into conditional rewriting. They do not allow new sorts in the extended signature and the variables are restricted to range over basic terms. They separate the semantics of the built-in algebra from the syntax of the whole specification by introducing sort hierarchies.

Becker [1994] presented a rewrite operationalization of clausal specifications where the axioms are positive/negative conditional equations admitting predefined algebras. He studied parametric specifications with a built-in parameter theory where the extended part is not restricted as here, i.e., extended functions of basic co-domain sorts are also admitted. He gives some consistency properties which guarantee the conservation of the semantics of the built-in parameter theory and a certain class of canonical term algebras as initial models of the specification. His approach can be considered as an attempt to provide a general framework of handling specifications with predefined algebras in a semantically clean way.

The principal differences between the present approach (and Vorobyov's approach) and the previous ones is that whereas here all models of the basic theory are considered the previous proposals work with a specific model of the basic theory and they give restrictions and assumptions on the built-in part which allow an operational treatment by purely rewriting and a validity check avoiding case analysis.

Avenhaus and Becker define the rewrite relation modulo the built-in algebra in such a way that a rule can be applied only if there is an exact matching. Their notion of rewriting can essentially be seen as rewriting of equivalence classes.

Dershowitz and Okada assume that any basic term can be replaced by an equivalent normal form before rewriting is applied. This means that their built-in mechanism can also be conceived as a rewriting procedure. Under their given restrictions and assumptions, all classical results such as Newman's diamond lemma and the Church-Rosser property can be extended directly. Here, without these strong restrictions on the built-in part, it is shown how these properties are guaranteed.

2 BACKGROUND

Firstly, to obtain a subclass of universal Horn classes with built-in predicates whose deduction results are relatively easy to make effectively decidable by rewriting, appropriate restrictions on the signatures of the desired combination of conditional specifications and built-in theories are given. Secondly, basic rewrite notions for BIS-CTRSs are defined.

2.1 CTRSs with Built-in Predicates (BIS-CTRSs)

A *basic theory* T_0 over an S_0-sorted signature, Σ_0, is a many-sorted first-order Henkin[1] theory with equality. *Built-in predicates* are quantifier-free formulae of a basic theory.

The *theory of the Presburger arithmetic* (or theory of integers under addition) ($\mathcal{PAR}$), the additive number theory ($\mathcal{ANT}$) and the successor arithmetic ($\mathcal{SA}$) are examples of basic theories. The unquantified Presburger arithmetic will be taken as example of a basic theory. $\mathcal{PAR}$ can be seen as a "standard" to analyze the "absolute" expressive power of canonical rewriting systems, because arithmetic lies in the very basis of almost all formal systems of reasoning. Vorobyov [1988] has shown that the unquantified $\mathcal{PAR}$ cannot be easily covered by rewriting systems. Additionally, decision algorithms for $\mathcal{PAR}$ have been of practical relevance in order to generate appropriate induction schemata in automated theorem proving by induction using rewrite techniques [Kapur and Subramaniam, 1996].

In order to incorporate built-in predicates as conditions into the structure of universal Horn clauses (defined in an S-sorted signature Σ), built-in objects are described in a built-in language given by an S_0-sorted signature Σ_0, as mentioned above, with the following restrictions:

Restrictions on the signatures. Let Σ_0 be an S_0-sorted signature, the built-in language of a basic theory T_0. Let S_0 and S be two sets of sorts such that $S_0 \subseteq S$. Let Σ_0 and Σ be S_0- and S-sorted signatures, respectively. Suppose that $\Sigma \setminus \Sigma_0$ does not contain function symbols with co-domain sort in S_0. Terms with sort in S_0 are called *basic* and in $S \setminus S_0$ *extended terms.*

Definition 2.1 Let Σ, Σ_0 and T_0 denote signatures and basic theories satisfying the previous restriction. A universal Horn clause of the form:

$$\forall \boldsymbol{X}(\boldsymbol{t_0} = \boldsymbol{t'_0} \ \underline{\textit{if}} \ \boldsymbol{t_1} = \boldsymbol{t'_1} \wedge \ldots \wedge \boldsymbol{t_k} = \boldsymbol{t'_k} \wedge \boldsymbol{P}),$$

where, for $i = 0, \ldots, k$, t_i, t'_i are S-sorted *extended* terms of the same sort and P is a built-in predicate, is called a *universal Horn clause with built-in*

[1]I.e., for every formulae of the form $\exists x P(x)$ there is a ground term t in $T_{\Sigma_0}(\emptyset)$ such that $T_0 \models \exists x P(t) \ \underline{\textit{if}} \ P(x)$.

predicate P over the theory T_0. The built-in predicate P is often called *built-in condition.* $t_0 = t'_0$ is called the *conclusion* and $t_1 = t'_1 \wedge \ldots \wedge t_k = t'_k$ is called the *standard condition* of the clause.

Given a set H of universal Horn clauses with built-in predicates over the theory T_0. The class of all algebras over T_0 which validate all clauses in H is denoted by $Mod_{T_0}\ H$. $Mod_{T_0}\ H$ is called the *universal Horn class* of H *over the built-in theory* T_0. $Mod_{T_0}\ H$ is said to be *axiomatized* by H and the clauses in H are said to be the *axioms* of $Mod_{T_0}\ H$.

A class $\mathcal{K}$ of algebras over T_0 is said to be a *universal Horn class over the built-in theory* T_0 if $\mathcal{K} = Mod_{T_0}\ H$ for some set H of universal Horn clauses with built-in predicates over T_0.

A clause Φ is said to be a *logical consequence* of H if, for every algebra $\mathcal{A} \in Mod_{T_0} H$, $\mathcal{A} \models \Phi$. This is denoted by $H \models_{T_0} \Phi$ (or simply by $H \models \Phi$ when there is no confusion).

As usual, $\mathcal{A} \models_{T_0} H$ means that $\mathcal{A} \in Mod_{T_0} H$.

$T_0^{=}$ denotes the basic theory enriched with the uninterpreted symbols of $\Sigma \backslash \Sigma_0$, augmented with equality axioms of the form $u[x] = u[y]$ $\underline{if}$ $x = y$, (where $u[\cdot]$ is an extended context with basic argument).

The following example illustrates how rewrite deductive techniques should be modified when including a built in theory as a parameter of a conditional rewrite specification. Ending this section the formal definition of rewriting system is presented.

Example 2.1 (Arrays [Vorobyov, 1989]) Consider arrays indexed by integers which contain "elements" (for instance lists, naturals, integers, etc.) in a sort of S which will be treated by rewriting. Consider a built-in mechanism for the manipulation of integers with addition and less-equal (a decision algorithm for the $\mathcal{PAR}$ can be used).

Let $S_0 = \{int\}$, $S = S_0 \cup \{array, element\}$ be sets of sorts and Σ_0 and Σ be S_0- and S-sorted signatures, respectively, with the following declarations and rules:

$$
\begin{array}{lllll}
\Sigma_0: & \ldots, 0, 1, \ldots : \to int & & + : int \times int \to int & < : int \times int \\
 & \leq & : int \times int & > : int \times int & \geq : int \times int \\
 & =_{\mathcal{PAR}} & : int \times int & & \\
\Sigma: & \langle \_, \_, \_ \rangle & : \ array \times int \times element \to array & & \\
 & \_[\_] & : \ array \times int \to element & &
\end{array}
$$

$$
\begin{array}{lll}
1. & \langle A, i, \mathcal{L} \rangle [j] \to \mathcal{L} & \underline{if}\ i =_{\mathcal{PAR}} j \\
2. & \langle A, i, \mathcal{L} \rangle [j] \to A[j] & \underline{if}\ i \neq_{\mathcal{PAR}} j \\
3. & \langle \langle A, i, \mathcal{K} \rangle, j, \mathcal{L} \rangle \to \langle A, j, \mathcal{L} \rangle & \underline{if}\ i =_{\mathcal{PAR}} j \\
4. & \langle \langle A, i, \mathcal{K} \rangle, j, \mathcal{L} \rangle \to \langle \langle A, j, \mathcal{L} \rangle, i, \mathcal{K} \rangle & \underline{if}\ i > j \\
5. & \langle \langle A, i, \mathcal{K} \rangle, k, \mathcal{L} \rangle [j] \to \langle A, k, \mathcal{L} \rangle [j] & \underline{if}\ i \neq_{\mathcal{PAR}} j
\end{array}
$$

i, j, k are built-in variables, A is a variable of sort *array* and $\mathcal{L}, \mathcal{K}$ of sort *element*. $A[i]$ is the element in i-th position of the array A and $\langle A, i, \mathcal{K}\rangle$ is the result of replacing the object in i-th position of array A with $\mathcal{K}$.

For simplicity, to avoid confusion, $i =_{\mathcal{PAR}} j$ will be abbreviated to $i = j$.

The formula

$$\langle\langle A, i, \mathcal{L}\rangle, j, \mathcal{K}\rangle[k] = \langle\langle A, j, \mathcal{K}\rangle, i, \mathcal{L}\rangle[k] \ \underline{if} \ \{i \neq j \ \& \ i \neq k\}$$

will be proved (without using the fifth rule) by rewriting. There are two consistent cases to consider. On one side, under context $(i \neq j \ \& \ i \neq k) \ \& \ k = j$, $\langle\langle A, i, \mathcal{L}\rangle, j, \mathcal{K}\rangle[k] \rightarrow_1 \mathcal{K}$ and $\langle\langle A, j, \mathcal{K}\rangle, i, \mathcal{L}\rangle[k] \rightarrow_2 \langle A, j, \mathcal{K}\rangle[k] \rightarrow_1 \mathcal{K}$. On the other side, under context $(i \neq j \ \& \ i \neq k) \ \& \ k \neq j$, $\langle\langle A, i, \mathcal{L}\rangle, j, \mathcal{K}\rangle[k] \rightarrow_2 \langle A, i, \mathcal{L}\rangle[k] \rightarrow_2 A[k]$ and $\langle\langle A, j, \mathcal{K}\rangle, i, \mathcal{L}\rangle[k] \rightarrow_2 \langle A, j, \mathcal{K}\rangle[k] \rightarrow_2 A[k]$.

Remark. Splitting clauses by consistent cases is called *case analysis* or *case splitting*. Case splitting is considered in contextual rewriting [Zhang and Remy, 1985], but without evaluating conditions immediately because conditions are not "contexts" in a built-in theory as is the case here.

A subclass of ("operationally tractable") universal Horn classes with built-in predicates with axioms easily transformable into conditional rewrite rules is syntactically defined as follows.

Definition 2.2 Let T_0 be a basic theory and H be a set of universal Horn clauses with built-in predicates over T_0. The theory T axiomatized by H is said to be a *BIS(T_0)-theory* iff each clause of H, $t = t' \ \underline{if} \ t_1 = t'_1 \wedge \ldots \wedge t_n = t'_n \wedge P$, satisfies the variable restriction:

$$Var(P) \cup Var(t') \bigcup_{i=1}^{n} (Var(t_i) \cup Var(t'_i)) \subseteq Var(t).$$

In this case the axioms are called *BIS(T_0)-axioms*. When T_0 is clear from the context, it will be omitted, and we will simply write BIS-theory and BIS-axiom.

Example 2.2 Five BIS-axioms are obtained by replacing '$\rightarrow$' with '$=$' in example 2.1. The basic theory is the unquantified $\mathcal{PAR}$. The previous example is extended in order to obtain one with standard and built-in conditions. If there exists some predicate '*le*' (for instance less-equal) for the objects in arrays of the last example, then this can be extended to a predicate '*LE*' for arrays as follows: Let S_0, S, Σ_0, Σ be as in example 2.1 and let $S' = S \cup \{\text{BOOL}\}$ be a new set of sorts and Σ' be an S'-sorted signature with the following declarations: $\Sigma' = \Sigma \ \cup \{\text{TRUE}, \text{FALSE} \ : \rightarrow \text{BOOL}, \ \ le \ : \ element \times element \rightarrow \text{BOOL}, \ \ LE \ : \ array \times array \times int \rightarrow \text{BOOL}\}$.

BOOL is a new sort whose constants TRUE and FALSE have the obvious intended meaning. The intended meaning of $LE(A, B, n)$ is a predicate which is TRUE iff for all $0 \leq j \leq n$, $le(A[j], B[j])$ is TRUE.

6. $LE(A, A, n) = \text{TRUE}$
7. $LE(A, B, n) = LE(A, B, n-1)$ $\underline{if}$ $n > 0 \wedge le(A[n], B[n]) = \text{TRUE}$
8. $LE(A, B, n) = le(A[n], B[n])$ $\underline{if}$ $n = 0$
9. $LE(\langle A, j, \mathcal{L}\rangle, B, n) = LE(A, B, n)$ $\underline{if}$ $j > n$
10. $LE(A, \langle B, j, \mathcal{L}\rangle, n) = LE(A, B, n)$ $\underline{if}$ $j > n$
11. $LE(\langle A, j, \mathcal{L}\rangle, \langle B, j, \mathcal{K}\rangle, n) = \text{TRUE}$
    $\underline{if}$ $LE(A, B, n) = \text{TRUE} \wedge le(\mathcal{L}, \mathcal{K}) = \text{TRUE} \wedge j < n$
12. ... some sentences for the predicate *le* ...

Note that the predicates *le* and *LE* don't belong to the basic theory. The sort BOOL should not be confused with the logical semantics of the built-in theory.

Axioms which only include built-in conditions can be obtained by syntactic transformations; for example, one can change axiom 7 to $\{LE(A, B, n) = \text{TRUE} \Leftrightarrow \{LE(A, B, n-1) = \text{TRUE} \wedge le(A[n], B[n]) = \text{TRUE}\}\}$ $\underline{if}$ $n > 0$, but in the context of rewriting, by using axioms with built-in and standard conditions, we obtain more expressiveness as well as a more elegant form of specification. For axioms without built-in conditions this question is the same as to search for an equivalent unconditional system (i. e., the question of when a quasi-variety is actually a variety!). A syntactic transformation for CTRSs with built-in predicates into equivalent unconditional rewriting systems and restrictions guaranteeing preservation of the rewriting properties appear in [Ayala-Rincón, 1996].

BIS(T_0)-theories can be seen as parameterized specifications (see for example [Ehrig and Mahr, 1985]), where the built-in theory with the basic language plays the role of the formal parameter specification and the target specification is given by the S-sorted signature Σ and a set of BIS(T_0)-axioms. The target part *uses* the parameter part without modifying its logic or enriching its language. The restriction on sorts of extended functions (extended function symbols cannot be of basic sorts) guarantees automatically that the parameterized specification is a *conservative extension* of the formal parameter specification because no basic terms can be identified (or confused) through extended terms of new sorts. BIS(T_0)-axioms are appropriate to specify some structures, such as arrays, matrices, etc., where the basic objects play the role of indices. However, by the use of functions over an extended boolean sort, the usual type of examples can also be specified.

Now, a class of CTRSs closely related to $BIS(T_0)$-theories will be introduced.

Definition 2.3 Let T_0 be a built-in theory and $l = r$ $\underline{if}$ $s_1 = t_1 \wedge \ldots \wedge s_n = t_n \wedge P$ be a BIS(T_0)-axiom. The expression: $l \rightarrow r$ $\underline{if}$ $s_1 \downarrow t_1 \wedge \ldots \wedge s_n \downarrow t_n \wedge P$ is called a *BIS(T_0)-conditional rewrite rule*.

A set R of BIS(T_0)-conditional rewrite rules is called a *BIS(T_0)-conditional rewriting system (BIS-CTRS)* or a standard CTRS with built-in predicates. R_T stands for the BIS-CTRS obtained by the obvious transformation (see the following example) of the axioms H of a BIS(T_0)-theory T.

Example 2.3 (BIS-CTRS) Consider the BIS-CTRS obtained from Example 2.2 by orienting the conclusions and by replacing '=' with '$\downarrow$' in the standard conditions.

2.2 Rewriting Notions for BIS-CTRSs

Let T_0 be a fixed basic theory and R a BIS(T_0)-CTRS.

In the sequel it will be assumed that $\boldsymbol{\mathcal{R} \equiv l \to r\ \underline{if}\ s_1 \downarrow t_1 \wedge \ldots \wedge s_n \downarrow t_n \wedge P}$ is a BIS-rule in R.

Notions of contextual redex and convolution are defined before the formalization of the reduction relation.

Definition 2.4 (Contextual redex) Let u, v be extended terms such that $u|_\pi = v$ and let C be a T_0-consistent built-in predicate. If there exists a substitution σ from $Var(l)$ into $T_\Sigma(Var(v))$ such that $T_0 \models P\sigma\ \underline{if}\ C$ and $T_0^= \models l\sigma = v\ \underline{if}\ C$ hold, then v is said to be an $R(C)(\mathcal{R},\sigma)$-*contextual redex* (of u at position π). C is called the *context* of the redex.

The definition of a redex is not purely syntactic: in contrast to the standard case, it involves a semantic interpretation of terms in T_0. The possibility of redex for basic terms is avoided. The search of an $R(C)(\mathcal{R},\sigma)$-redex in a term u is also called the problem of search of a T_0-*matching* of l in u.

Definition 2.5 (Convolution) Let v be an $R(C)(\mathcal{R},\sigma)$-contextual redex of u at position π. Then the term $u' \equiv u[r\sigma]_\pi$ is called an $R(C)(v,\mathcal{R},\sigma)$-*convolution* of u.

Example 2.4 Consider the example 2.3 of arrays and let $\mathcal{R}_i$ denote its i-th rule. Let σ be the substitution: $\{i \mapsto m, j \mapsto n, A \mapsto B, \mathcal{L} \mapsto \mathcal{M}\}$. Note that the extended term $\langle B, m, \mathcal{M}\rangle[n]$ is an $R(m = n)(\mathcal{R}_1, \sigma)$-contextual redex but not an $R(m \neq n)(\mathcal{R}_1, \sigma)$-contextual redex. $\mathcal{M}$ is the $R(m = n)(\langle B, m, \mathcal{M}\rangle[n], \mathcal{R}_1, \sigma)$-convolution of $\langle B, m, \mathcal{M}\rangle[n]$.

Assumption on the Basic Theories. To make decidability of built-in predicates effective only basic theories T_0 with decidable set of *quantifier free logical consequences*: $\boldsymbol{Th^\forall(T_0)} \equiv_{def} \{P \mid T_0 \models P,\ P$ is quantifier-free$\}$ are considered. To guarantee decidability of the problem of search of R-contextual redex, universal-existential formulae[2] should be decidable too.

[2] A universal-existential formula is a formula of the form $\forall x_1 \ldots \forall x_n \exists y_1 \ldots \exists y_m P$, where P is quantifier-free.

Lemma 2.1 *Let R be a BIS(T_0)-conditional rewriting system, C be a context and u be an extended term. If the class of universal-existential formulae of the theory T_0 is decidable then there exists an algorithm to decide whether or not a position π, a rule $\mathcal{R} \in R$ and a substitution σ exist such that $u|_\pi$ is an $R(C)(\mathcal{R}, \sigma)$-contextual redex of u.*

Proof: For each rule $\mathcal{R} \equiv l \to r \underline{if} s_1 \downarrow t_1 \wedge \ldots \wedge s_n \downarrow t_n \wedge P$ in R it will be checked if there exist a position π and a substitution σ such that $u|_\pi$ is an $R(C)(\mathcal{R}, \sigma)$-contextual redex of u.

It is supposed that $\mathcal{R}$ does not contain variables in common with u and C. A redex in u occurs only in positions π such that $u|_\pi$ is an extended term with the same sort as l. For a subterm $u|_\pi$ with the same sort as l, the search of a redex can be interpreted as a question of the form: there exists σ such that $T_0^= \models P\sigma \;\&\; l\sigma = u|_\pi \underline{if} C$? that is equivalent to $T_0 \models \exists\vec{Y}\forall\vec{X}((P\&Q\&Match(l', u|_\pi)) \underline{if} C)$? where: (i) l' is obtained from l by replacing all maximal basic subterms s_j ($0 \leq j \leq k$, for some k) with new variables y_j and Q is the conjunction of all $y_j = s_j$. (ii) $Match(l', u|_\pi)$ is defined as follows: compute the match Θ from l' to $u|_\pi$ using any standard matching algorithm. If there is no match then let $Match(l', u|_\pi)$ be false. Otherwise let θ be the part of Θ assigning basic terms to basic variables $y_j \mapsto v_j$. Let us assume, $\Sigma \setminus \Sigma_0$ contains no function symbols with S_0-sorted co-domain. Therefore v_j contains only basic symbols. Now define $Match(l', u|_\pi)$ as the conjunction of all equalities $y_j = v_j$, where $y_j \mapsto v_j$ belongs to θ. (iii) $\vec{Y}$ is the union of all variables in s_j and variables y_j, for $0 \leq j \leq k$. (iv) $\vec{X}$ is the union of all variables in C and in v_j, for $0 \leq j \leq k$.

$T_0 \models \exists\vec{Y}\forall\vec{X}((P\&Q\&Match(l', u|_\pi)) \underline{if} C)$ holds exactly when its negation

$$\forall\vec{Y}\exists\vec{X}\neg((P \;\&\; Q \;\&\; Match(l', u|_\pi)) \underline{if} C)$$

is not T_0-valid. The last formula is universal-existential and, by hypothesis, can be decided.

Remark. The decision algorithm described in the preceding proof does not exhibit explicitly a substitution σ. However, for the examples of basic theories considered in this work (viz $\mathcal{PAR}$, $\mathcal{ANT}$ and $\mathcal{SA}$) one can always explicitly give a substitution[3] σ. This can also be guaranteed giving syntactical restrictions on the BIS-rules of the BIS-CTRS. In particular, if one restricts left-hand sides of the BIS(T_0) rules to contain only basic subterms which are

[3]Presburger's, Cooper's and Shostak's decision algorithms for $\mathcal{PAR}$ (see [Presburger, 1929], [Cooper, 1972] and [Shostak, 1977], respectively) search for explicit solutions of existentially quantified formulae (essentially by the method of elimination of quantifiers). For theories decided by model theoretical methods, such as $\mathcal{DNO}$ (totally and densely orders), no explicit solutions are exposed (see Rabin's chapter on decidable theories in [Barwise, 1977]).

either basic ground terms or basic variables, as is the case in the example of arrays, the search of redices is effectively computable and only decidability of the universal part of the basic theory is required. In a similar way to the case of unification in monoidal theories without constants developed by Baader and Nutt [Nutt, 1992; Baader and Nutt, 1996], it could be verified that unification problems in the theory of $\mathcal{PAR}$, for BIS-CTRSs, correspond to systems of homogeneous linear equations, because of the restriction on the sorts of the extended function symbols. A general unification algorithm modulo $\mathcal{PAR}$ for the restricted class of specifications is presented in [de Araújo and Ayala-Rincón, 1998].

It is assumed that when searching $R(\mathcal{R},\sigma)$-redex, the substitution σ can effectively be computed.

Since reduction depends on verification by rewriting of joinability of standard conditions both notions are simultaneously defined.

Definition 2.6 ($R(C)$-reduction, bi(C)-joinability) Let C be a T_0 consistent built-in predicate and let u, v be extended terms such that $u' \equiv u[r\sigma]_\pi$ is an $R(C)(v,\mathcal{R},\sigma)$-convolution of u. If, additionally, for all $1 \leq i \leq n$, $s_j\sigma, t_j\sigma$ are *bi(C)-joinable*, then it is said that u $R(C)$-reduces to the term u' and will be written

$$u \stackrel{R(C)}{\longrightarrow} u'.$$

Where $s_j\sigma, t_j\sigma$ are bi(C)-joinable, denoted by $s_j \Downarrow_C t_j$, there exist a finite index set I, a set of contexts $\{C_i\}_{i\in I}$ and sets of terms $\{s^i\}_{i\in I}$ and $\{t^i\}_{i\in I}$ such that (i) $\forall i \in I, Var(C_i) \subseteq Var(C) \cup Var(s_j) \cup Var(t_j)$; (ii) $T_0 \models C \Leftrightarrow \bigvee_{i\in I} C_i$ and $\forall i \in I$ C_i & C is T_0-consistent; (iii) $\forall i \in I$,

$$s_j \stackrel{R(C_i)^*}{\longrightarrow} s^i \text{ and } t_j \stackrel{R(C_i)^*}{\longrightarrow} t^i,$$

and (iv) $\forall i \in I, T_0^= \models s^i = t^i \ \underline{if}\ C_i$.

Second bi(C)-joinability condition corresponds to (sub)case analysis of C (which should be realized guided according to basic premises of the rules).

As usual, $\stackrel{R(C)^+}{\longrightarrow}$ and $\stackrel{R(C)^*}{\longrightarrow}$ denote the transitive and reflexive-transitive closure of the $\stackrel{R(C)}{\longrightarrow}$ relation, respectively. C is called the *context* of the reduction.

Example 2.5 Note that

$$\langle B, m+n, \mathcal{M}\rangle[n] \stackrel{R(m>0)}{\longrightarrow} B[n] \text{ and } \langle B, m+n, \mathcal{M}\rangle[n] \stackrel{R(m=0)}{\longrightarrow} \mathcal{M}.$$

Observe also that $\langle B, m+n, \mathcal{M}\rangle[n] \Downarrow_{m=n} \langle B, m+n, \mathcal{M}\rangle[n]$. For the built-in predicate

$$C \equiv m+n=l \ \&\ n>0,\ \langle\langle B,l,\mathcal{M}\rangle, m, \mathcal{N}\rangle \stackrel{R(C)}{\longrightarrow} \langle\langle B,m,\mathcal{N}\rangle, l, \mathcal{M}\rangle.$$

Remark. If C is T_0-inconsistent then for all extended terms (of the same sort) s, t, it is supposed that $s \stackrel{R(C)}{\longrightarrow} t$ (and then that $s \Downarrow_C t$). This makes sense since $T_0^= \models s = t \ \underline{if}\ C$. When $T_0 \models C$ (or there is no built-in predicate), we speak of an R-reduction. In this case, if u' is an $R(C)(v, \mathcal{R}, \sigma)$-convolution of u, then $T_0 \models P\sigma$ and $T_0^= \models (l\sigma = v)$ follow from $T_0 \models P\sigma \ \underline{if}\ C$ and $T_0^= \models (l\sigma = v) \ \underline{if}\ C$, respectively. $s \Downarrow_C t$ and $s \stackrel{R(C)}{\longrightarrow} t$ are abbreviated to $s \Downarrow t$ and $s \stackrel{R}{\longrightarrow} t$, respectively. Also it is said that u $R(\emptyset)$-reduces to the term u' ($u \stackrel{R(\emptyset)}{\longrightarrow} u'$) or simply that u R-reduces to u' ($u \stackrel{R}{\longrightarrow} u'$). Do not confuse the above abbreviations with the forthcoming notion of R-reduction.

The correctness of $R(C)$-reduction and the following basic properties are proved in [Ayala-Rincón, 1993] by using the theory of fixpoints.

Lemma 2.2 *Let C and B be T_0-consistent contexts such that $T_0 \models B \ \underline{if}\ C$ and let s, t be extended terms. Then the following implications hold:*

1. *if $s \stackrel{R(B)}{\longrightarrow} t$ then $s \stackrel{R(C)}{\longrightarrow} t$*

2. *if $s \Downarrow_B t$ then $s \Downarrow_C t$.*

The following coherence property of the BIS-CTRSs is essential in order to investigate confluence criteria.

Lemma 2.3 (Coherence) *Let R be a BIS-CTRS over T_0, u, v, v' be terms and C, B be contexts, such that $T_0^= \models u = v \ \underline{if}\ C$, $v \stackrel{R(B)}{\longrightarrow}{}^* v'$ and $T_0 \models C \ \underline{if}\ B$. Then there exists a term u' such that $u \stackrel{R(B)}{\longrightarrow}{}^* u'$ and $T_0^= \models u' = v' \ \underline{if}\ B$. This property of the BIS-CTRS is called* coherence.

Proof: It is enough to prove that if u, v, v' are terms and C, B contexts, such that $T_0^= \models u = v \ \underline{if}\ C$, $v \stackrel{R(B)}{\longrightarrow} v'$ and $T_0 \models C \ \underline{if}\ B$, then there exists a term u' such that $u \stackrel{R(B)}{\longrightarrow} u'$ and $T_0^= \models u' = v' \ \underline{if}\ B$. This means that the reduction derivation can be reproduced step by step.

Example 2.6 Consider again the example 2.3 of arrays over the $\mathcal{PAR}$ and let $B \equiv j > i \ \&\ m = i \ \&\ m = n$. Note that $\mathcal{PAR}^= \models \langle\langle A, j, \mathcal{K}\rangle, i, \mathcal{L}\rangle[m] = \langle\langle A, j, \mathcal{K}\rangle, i, \mathcal{L}\rangle[n] \ \underline{if}\ m = n$.

On one side, $\langle\langle A,\ j,\ \mathcal{K}\rangle,\ i,\ \mathcal{L}\rangle[n] \stackrel{R(B)}{\longrightarrow} \langle\langle A,\ i,\ \mathcal{L}\rangle,\ j,\ \mathcal{K}\rangle[n] \stackrel{R(B)}{\longrightarrow} \langle A,\ i,\ \mathcal{L}\rangle[n] \stackrel{R(B)}{\longrightarrow} \mathcal{L}$, by applying respectively the rules 4, 2 and 1. On the other side, by applying the same rules, one obtains $\langle\langle A, j, \mathcal{K}\rangle, i, \mathcal{L}\rangle[m] \stackrel{R(B)}{\longrightarrow} \mathcal{L}$.

Soundness of the bi-joinability is proved in [Ayala-Rincón, 1993] by induction on the *depth*[4] of bi-joinability of terms.

Lemma 2.4 *Let T be a BIS(T_0)-theory, let R be the corresponding BIS(T_0)-CTRS, let s, t be terms and let C be a built-in predicate such that $s \Downarrow_C t$. Then $T_0^= \cup T \models s = t \underline{\mathit{if}}\ C$.*

3 DECIDING CONDITIONAL EQUATIONS

In this section we give necessary notions of noetherianity and confluence and brief comments on the decision procedure for BIS-CTRSs. Formal proofs for clauses with built-in and standard conditions can be produced with the set of inference rules presented in [Ayala-Rincón, 1998]. The effective decision procedure works for restricted clauses without standard conditions, of the form $u = v\ \underline{\mathit{if}}\ C$, in $Th^{\forall}(T_0^= \cup T)$, where T is a BIS(T_0)-theory. This restriction represents a standard situation in computational problems where the conditions are in a specific theory which can be solved by built-in algorithms. However, clauses of the specification admit built-in conditions as well as standard conditions. The decision procedure is based on reduction and case-splitting guided by the conditions of the rules. In the sequel $Th^{\forall}(T_0^= \cup T)$ is restricted to formulae of this form. In the following definitions $\mathcal{R}$ denotes a BIS-rule in R of the usual form.

Notions of reduction and case-splitting are extended for restricted formulae.

Definition 3.1 (R-reduction) Let a formula $G \equiv u = v\ \underline{\mathit{if}}\ C$ and suppose that $u \stackrel{R(C)}{\longrightarrow} u'$. Then, it is said that G R-reduces to the formula G' of the form $u' = v\ \underline{\mathit{if}}\ C$, symbolically $G \stackrel{R}{\longrightarrow} G'$.

Remark. Observe that the R-reduction is a generalization of the ordinary one: consider the formula[5] $G \equiv u = u$ and suppose that $u \stackrel{R(\emptyset)}{\longrightarrow} u'$; then, for $G' \equiv u = u'$, $G \stackrel{R}{\longrightarrow} G'$ holds. It will be simply written $u \stackrel{R}{\longrightarrow} u'$. Evidently, $u \stackrel{R(\emptyset)}{\longrightarrow} u'$ and $u \stackrel{R}{\longrightarrow} u'$ coincide. By abuse of notation let $u\ \underline{\mathit{if}}\ C$ denote the term u under context[6] C. Obviously $u \stackrel{R}{\longrightarrow} u'\ \underline{\mathit{if}}\ C$ iff $u \stackrel{R(C)}{\longrightarrow} u'$.

[4] The *depth of the bi(C)-joinability* $s \Downarrow_C t$ is defined as zero if either C is T_0-inconsistent or $T_0^= \models s = t\ \underline{\mathit{if}}\ C$; otherwise it is defined as the maximum depth of the R(C_i)-reduction steps in the chains $s \stackrel{R(C_i)}{\longrightarrow}{}_* s^i$, $t \stackrel{R(C_i)}{\longrightarrow}{}_* t^i$, for $i \in I$, where the *depth of an $R(C_i)$-reduction* using the rule $l \to r\ \underline{\mathit{if}}\ l \wedge_{j=1}^m s_j \downarrow t_j \wedge P$ and substitution σ is one more than the maximum of the depth of $s_j\sigma \Downarrow_{C_i} t_j\sigma$, for $j = 1, \ldots, m$.

[5] this can be seen as $false \vee u = u$ or equivalently $u = u\ \underline{\mathit{if}}\ true$, where $true$ and $false$ are the usual logical constants.

[6] This is needed when checking equivalence of terms u and v under context C, i. e. when checking $u = v\ \underline{\mathit{if}}\ C$.

In the sequel $R(C)$-reduction and R-reduction will be identified simply by adding contexts as premises to the formulae to be reduced and by deleting context from the $R(C)$-reductions.

Definition 3.2 (Potential contextual redex) Let $u[v]_\pi$ be a term with leading symbol in $\Sigma\backslash\Sigma_0$ and let C be a T_0-consistent built-in predicate. If there exists a substitution σ from $Var(l)$ into $T_\Sigma(Var(v))$ such that both formulae $C_1 \equiv C\&P\sigma$ and $C_2 \equiv C\&\neg P\sigma$ are T_0-consistent and v is an $R(C_1)(\mathcal{R},\sigma)$-contextual redex, then v is said to be an $R(C)(\mathcal{R},\sigma)$-*potential contextual redex* (of u at position π). C is called the *context* of the potential redex.

The search for R-potential contextual redex is a problem equivalent to the search for R-contextual redex; in effect, using the notation of Lemma 2.1, answer simply whether or not $\exists\vec{Y}\forall\vec{X}((Q\&Match(l',u|_\pi))\ \underline{if}\ (C\&P)$ is T_0-valid (and then check if both formulae C_1 and C_2 are T_0-consistent).

Definition 3.3 (R-case splitting) Let G have the form $u[w]_\pi = v\ \underline{if}\ C$, where w is an $R(C)(\mathcal{R},\sigma)$-potential contextual redex of u. Let C_1, C_2 denote respectively the formulae $C\&P\sigma$ and $C\&\neg P\sigma$, as in the previous definition. If for all $i = 1,\ldots,n$, $s_i\sigma \Downarrow_{C_1} t_i\sigma$ holds then the formulae $G_1 \equiv u[r\sigma]_\pi = v\ \underline{if}\ C_1$ and $G_2 \equiv u = v\ \underline{if}\ C_2$ are said to be obtainable as the result of an $R(C)(\mathcal{R},\sigma)$-*case splitting* (for brevity R-case splitting) applied to the formula G.

Example 3.1 (Continuing Example 2.3.) The formula $\langle A,n,\mathcal{L}\rangle[m] = \mathcal{K}$ can be split with the first rule obtaining the formulae $\mathcal{L} = \mathcal{K}\ \underline{if}\ n = m$ and $\langle A,n,\mathcal{L}\rangle[m] = \mathcal{K}\ \underline{if}\ n \neq m$. Subsequently, the last formula can be rewritten with the second rule into $A[m] = \mathcal{K}\ \underline{if}\ n \neq m$.

Definition 3.4 The formula G is said to be in the *R-normal form*, iff it cannot be R-reduced and in the *R-complete normal form*, iff it is neither R-reducible nor can be split.

Let T be a BIS(T_0)-theory and R_T its corresponding BIS(T_0)-CTRS. The decision procedure for $Th^\forall(T_0^= \cup T)$ is based on the construction of a *R_T-complete reduction tree* whose root is labeled with the given formula; each internal node has either two successors labeled with formulae resulting from a case splitting or one successor labeled with a formula resulting from a convolution and all leaves are labeled with R-complete normal formulae. If one can construct an R_T-complete reduction tree for a formula its validity can be checked by deciding the validity of all formulae labeling its leaves using the decision procedure for $Th^\forall(T_0^=)$. The decision procedure includes recursive evaluation for the labels in the edges, which is the evaluation of the standard conditions of the applied rules. Evidently, one needs restrictions in order to guarantee the existence of R_T-complete reduction trees.

For BIS-CTRSs without standard conditions, Vorobyov [Vorobyov, 1989] has demonstrated that the procedure is strongly complete, provided that R_T satisfies particular properties of termination and confluence. Strong completeness means that a result of the algorithm is independent of the reduction tree constructing strategy chosen, in other words the algorithm is search-space-free.

R_T-reducibility and R_T-case splitting preserve validity of formulae in $T_0^= \cup T$. In order to show completeness of the decision procedure, it is only needed to prove that $T_0^= \cup T \models G$ implies $T_0^= \models G$ whenever G is in the R_T-complete normal form. Of course, convergence of R_T is needed. This is the usual condition for the search-space-free completeness of purely standard rewrite decision procedures. However, the notion of canonicity must be redefined substantially for BIS-CTRSs.

Definition 3.5 (c-noetherianity [Vorobyov, 1989]) A BIS-CTRS R is said to be *c-noetherian* (read it as "*complete*-noetherian"), if and only if there are no infinite sequences of contexts $C_0, C_1, \ldots$ and terms $u_0, u_1, \ldots$ such that for all $i \in \mathbb{N}$, the context C_i is T_0-consistent, $T_0 \models C_i \ \underline{if}\ C_{i+1}$ and $u_i \stackrel{R(C_i)}{\longrightarrow} u_{i+1}$.

It is known that c-noetherianity implies usual noetherianity[7] but the converse is false [Vorobyov, 1989].

As in the ordinary case, $\stackrel{R(C)}{\longrightarrow}$ calls recursively the rewrite relation for verifying the bi-joinability. This recursive definition may lead to infinite computations for just one rewrite step. To avoid this, a special notion of *decreasingness* adapted to BIS-CTRSs could be used. Originally, *decreasingness* was introduced in order to guarantee decidability of basic rewrite properties of CTRSs: termination, one-step reduction, finite reduction, joinability, reducibility, etc [Sivakumar, 1989]. Also *decreasingness* guarantees decidability of basic rewrite properties of BIS-CTRSs and is a relevant criterion to analyze their confluence [Ayala-Rincón, 1994].

Definition 3.6 (Confluence) A BIS-CTRS R is *confluent* if and only if for all extended terms u, u_1, u_2 and all contexts C, if $u \stackrel{R(C)^*}{\longrightarrow} u_1$ and $u \stackrel{R(C)^*}{\longrightarrow} u_2$ then $u_1 \Downarrow_C u_2$.

Considering the empty basic theory it can be checked that this definition of confluence as well as the definition of bi-joinability are generalizations of the standard ones.

The definition of confluence involves case splittings; in fact, the contexts of the bi-joinability are cases that occur in splittings.

[7] R is said to be noetherian, if and only if there does not exist infinite reduction chains of the form $u_0 \stackrel{R(C)}{\longrightarrow} u_1 \stackrel{R(C)}{\longrightarrow} \ldots$.

Example 3.2 Consider again the example 2.3. Let C be the context $i \neq m \,\&\, n > i$, and u be the term $\langle\langle A, n, \mathcal{L}\rangle, i, \mathcal{K}\rangle[m]$. There is a divergence: $u \stackrel{R(C)}{\longrightarrow}_2 u_1 \equiv \langle A, n, \mathcal{L}\rangle[m]$ and $u \stackrel{R(C)}{\longrightarrow}_4 u_2 \equiv \langle\langle A, i, \mathcal{K}\rangle, n, \mathcal{L}\rangle[m]$. $\mathcal{PAR}^{=} \not\models u_1 = u_2$ and u_1, u_2 are irreducible. However, if the contexts $C_1 \equiv C \,\&\, n = m$ and $C_2 \equiv C \,\&\, n \neq m$ are considered (case splitting with the rule 1), then (by abuse of notation) u_1 *if* C and u_2 *if* C splits, as follows: u_1 *if* C R-case split into u_1 *if* $(C \,\&\, n = m)$ and u_1 *if* $(C \,\&\, n \neq m)$ and u_2 *if* C R-case split into u_2 *if* $(C \,\&\, n = m)$ and u_2 *if* $(C \,\&\, n \neq m)$. Then: $u_1 \stackrel{R(C_1)}{\longrightarrow}_1 \mathcal{L}$, $u_1 \stackrel{R(C_2)}{\longrightarrow}_2 A[m]$ and $u_2 \stackrel{R(C_1)}{\longrightarrow}_1 \mathcal{L}$, $u_2 \stackrel{R(C_2)}{\longrightarrow}_2 \langle A, i, \mathcal{K}\rangle[m] \stackrel{R(C_2)}{\longrightarrow}_2 A[m]$.

A BIS-theory is said to be *canonical* or *convergent* iff its corresponding BIS-CTRS is confluent and c-noetherian. Completeness of the decision procedure, for a canonical BIS(T_0)-theory T, is based on the validity of R_T-complete normal formulae under both theories $T_0^{=} \cup T$ and $T_0^{=}$, as proved in [Ayala-Rincón, 1993].

4 EXTENSIONS OF CLASSICAL RESULTS FOR BIS-CTRSS

Newman's diamond lemma and Church-Rosser property for BIS-CTRSs are presented in order to illustrate the considerations when extending classical results on conditional rewrite theory to BIS-CTRSs. In [Avenhaus and Becker, 1992] confluence is treated for a variation of reduction in CTRSs with a built-in parameter algebra, which is essentially reduction modulo the equivalence class of the parameter algebra. An analogous approach could also be adopted for $R(C)$-reduction in BIS(T_0)-CTRSs, but by the coherence property a new notion of $R(C)$-reduction modulo $\equiv_{T_0}$ is unnecessary.

The notions of local confluence and Church-Rosser property are adapted to the BIS-CTRSs.

Definition 4.1 (Local confluence) A BIS-CTRS is *locally-confluent* if for all extended terms u, u_1, u_2 and all contexts C, if $u \stackrel{R(C)}{\longrightarrow} u_1$ and $u \stackrel{R(C)}{\longrightarrow} u_2$ then $u_1 \Downarrow_C u_2$.

Definition 4.2 (Church-Rosser Property) It is said that a BIS-CTRS satisfies the *Church-Rosser property* if and only if for all extended terms u, u_1, u_2 and all contexts C, $u \stackrel{R(C)^*}{\longleftrightarrow} v$ implies $u \Downarrow_C v$.

Newman's diamond lemma for BIS-CTRSs can be generalized.

Lemma 4.1 (Newman's diamond lemma for BIS-CTRS) *Let R be a c-noetherian BIS-CTRS. Then R is confluent if and only if R is locally-confluent.*

Proof: The proof proceeds analogously to the standard case. On the one hand, confluence implies local confluence. On the other hand, consider the following order $\succ$ on pairs of extended terms and contexts: for all extended terms u, v and contexts C, B, $(u, C) \succ (v, B)$ iff $u \overset{R(C)}{\longrightarrow} v$ and $T_0 \models C \ \underline{if}\ B$. By the basic properties (see Lemma 2.2) of $R(B)$-reduction, $(u, C) \succ (v, B)$ implies $u \overset{R(B)}{\longrightarrow} v$. Now, consider the reflexive-transitive closure of $\succ$ (by simplicity, denoted also by $\succ$). $\succ$ is a partial order and by the c-noetherian hypothesis it is noetherian. In fact, from an infinite chain $(u_1, C_1) \succ (u_2, C_2) \succ \cdots$ one can construct an infinite reduction chain: $u_1 \overset{R(C_1)}{\longrightarrow}{}^* u_2 \overset{R(C_2)}{\longrightarrow}{}^* \cdots$ contradicting c-noetherian property.

Suppose that R is locally-confluent but not confluent. Then there exists a pair (u_0, C_0) minimal with respect to $\succ$ such that $v_1 {}^*\overset{R(C_0)}{\longleftarrow} u_0 \overset{R(C_0)}{\longrightarrow}{}^* v_2$, for some extended terms v_1, v_2 which are in R-complete normal form (i. e., $v_1 = v_2 \ \underline{if}\ C_0$ is in R-complete normal form) and $T_0^= \not\models v_1 = v_2 \ \underline{if}\ C_0$. Let u_1' and u_2' be terms such that $v_1 {}^*\overset{R(C_0)}{\longleftarrow} u_1' \overset{R(C_0)}{\longleftarrow} u_0 \overset{R(C_0)}{\longrightarrow} u_2' \overset{R(C_0)}{\longrightarrow}{}^* v_2$. By local confluence $u_1' \Downarrow_{C_0} u_2'$. This means that there exist an index set I, a context set $\{B_i\}_{i \in I}$ and sets of terms $\{w_1^i\}_{i \in I}$, $\{w_2^i\}_{i \in I}$, which satisfy the definition of bi-joinability. There exists an index $j \in I$, see Figure 1, such

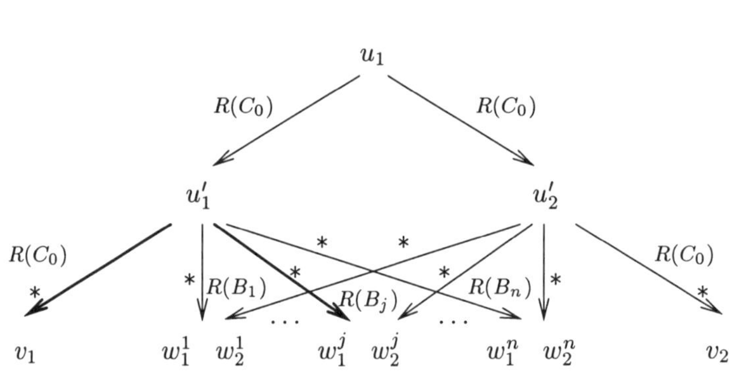

Figure 1: Local confluence and c-noetherianity imply confluence

that the context B_j and its corresponding terms w_1^j, w_2^j (which without loss of generality can be supposed to be in R-complete normal form) satisfy either $T_0^= \not\models w_1^j = v_1 \ \underline{if}\ B_j$ or $T_0^= \not\models w_2^j = v_2 \ \underline{if}\ B_j$. Contrariwise, if for all $j \in I$, $T_0^= \models w_1^j = v_1 \ \underline{if}\ B_j$ and $T_0^= \models w_2^j = v_2 \ \underline{if}\ B_j$ then for all $j \in I$, $T_0^= \models v_1 = v_2 \ \underline{if}\ B_j$, which implies that $T_0^= \models \bigwedge_{j \in I}(v_1 = v_2 \ \underline{if}\ B_j)$ or equivalently, $T_0^= \models \bigwedge_{j \in I}(\neg B_j) \vee v_1 = v_2$ which can be seen as $T_0^= \models \neg \bigvee_{j \in I}(B_j) \vee v_1 = v_2$. Thus $T_0^= \models \bigvee_{j \in I} v_1 = v_2 \ \underline{if}\ (B_j)$ and since $T_0 \models C_0 \Leftrightarrow \bigvee_{j \in I}(B_j)$, it concludes that $T_0^= \models v_1 = v_2$ *if* C_0 (giving a contradiction).

Suppose, without limitation of generality, that $T_0^= \not\models w_1^j = v_1$ *if* B_j. This contradicts the minimal assumption on (u_0, C_0) because $T_0 \models C_0$ *if* B_j and $u_0 \overset{R(B_j)}{\longrightarrow} u_1'$, i.e., $(u_0, C_0) \succ (u_1', B_j)$. Therefore R must be confluent.

Of course, the previous lemma does not hold under usual noetherianity hypothesis.

Lemma 4.2 (C-R property) *A BIS-CTRS R is Church-Rosser if and only if it is confluent.*

Proof: Evidently, the Church-Rosser property implies confluence. Conversely, suppose that R is confluent and let u, v be terms and C a context such that $u \overset{R(C)^*}{\longleftrightarrow} v$. The proof is made by induction on the length n of the derivation $u \overset{R(C)^n}{\longleftrightarrow} v$.

Cases $n = 0, 1, 2$ hold by definition of $R(C)$-reduction.

Suppose that this holds for $n = k$, i.e., $u \overset{R(C)^k}{\longleftrightarrow} v$ implies $u \Downarrow_C v$.

To show the case $n = k+1$, suppose that $u \overset{R(C)^k}{\longleftrightarrow} w \overset{R(C)}{\longleftrightarrow} v$. There are two sub-cases to consider.

First, $u \overset{R(C)^k}{\longleftrightarrow} w \overset{R(C)}{\longleftarrow} v$. By induction hypothesis $u \Downarrow_C w$, which means that there exist an index set I, a set of contexts $\{C_i\}_{i \in I}$ and sets of terms $\{u^i\}_{i \in I}, \{w^i\}_{i \in I}$ such that $T_0 \models C \Leftrightarrow \bigvee_{i \in I} C_i$, $T_0 \models u^i = w^i$ *if* C_i and $u \overset{R(C_i)^*}{\longrightarrow} u^i$, $w \overset{R(C_i)^*}{\longrightarrow} w^i$. By basic properties of $R(C)$-reduction, $v \overset{R(C_i)}{\longrightarrow} w$ since $T_0 \models C$ *if* C_i and $v \overset{R(C)}{\longrightarrow} w$. Therefore $v \overset{R(C_i)^*}{\longrightarrow} w^i$, for all $i \in I$, which concludes $u \Downarrow_C v$.

Second, (see Figure 2), $u \overset{R(C)^k}{\longleftrightarrow} w \overset{R(C)}{\longrightarrow} v$. On the one side, by induction hypothesis, there are sets I, $\{C_i\}_{i \in I}$, $\{u^i\}_{i \in I}$ and $\{w^i\}_{i \in I}$, which give the bi(C)-joinability, $u \Downarrow_C w$, as in the first case. On the other side, by confluence hypothesis, for all $i \in I$, $w^i \Downarrow_{C_i} v$, which means that for all $i \in I$ there exist an index set I_i, a set of contexts $\{B_j\}_{j \in I_i}$ and sets of terms $\{t^j\}_{j \in I_i}, \{v^j\}_{j \in I_i}$, such that $T_0 \models C_i \Leftrightarrow \bigvee_{j \in I_i} B_j$, $w^i \overset{R(B_j)^*}{\longrightarrow} t^j$, $v \overset{R(B_j)^*}{\longrightarrow} v^j$ and $T_0^= \models t^j = v^j$ *if* B_j. Since for any $j \in I_i$, $T_0 \models C_i$ *if* B_j and $u \overset{R(C_i)^*}{\longrightarrow} u^i$ then $u \overset{R(B_j)^*}{\longrightarrow} u^i$. By coherence, there exists a t_j' such that $u^i \overset{R(B_j)^*}{\longrightarrow} t_j'$ and $T_0^= \models t_j' = t^j$ *if* B_j. Note that $u \overset{R(B_j)^*}{\longrightarrow} t_j'$ and $T_0^= \models t_j' = v^j$ *if* B_j. Therefore $\bigcup_{i \in I} I_i$, $\bigcup_{i \in I} \{B_j\}_{j \in I_i}$, $\bigcup_{i \in I} \{t_j'\}_{j \in I_i}$ and $\bigcup_{i \in I} \{v^j\}_{j \in I_i}$ are respectively an index set, a context set and sets of terms which give the confluence $u \Downarrow_C v$.

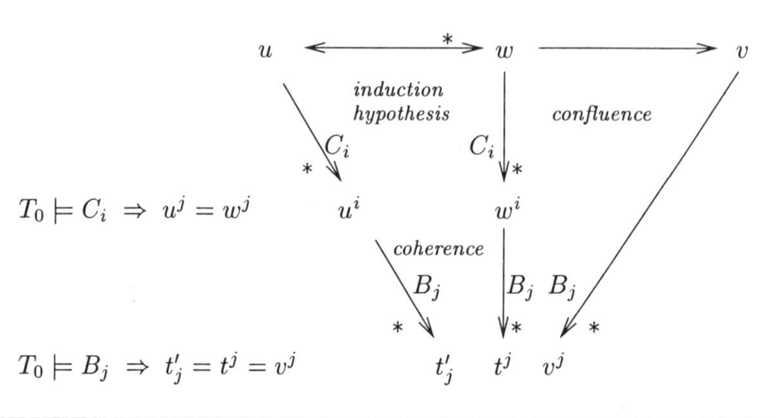

Figure 2: Church-Rosser versus confluence

Remark. In contrast with standard CTRSs, Lemma 4.2 does not imply that $u \overset{R(C)}{\longleftrightarrow}{}^* v$ <u>*if*</u> $u \Downarrow_C v$, because of possible case-splits in the bi(C)-joinability of u and v.

5 CONCLUSIONS

The significance of BIS-CTRSs becomes evident since they are a proper extension of standard CTRSs which improves their expressiveness as well as their efficiency; in fact, without built-in predicates, BIS-CTRSs reduce to standard CTRSs and their bigger expressiveness is a direct consequence of one of the universal Horn clauses with built-in predicates with respect to quasi-varieties. This class of combined systems is relevant because specifying with purely conditional equational axioms one has a strong (inherent) restriction when producing the corresponding implementation; that is the construction of a type functional implementation. Including built-in predicates in the specification, one supposes that built-in algorithms for the logical part of the specification are previously implemented too. It allows us to produce combined functional-logical implementations.

The concept of reduction for BIS-CTRSs includes a notion of contextual redex incorporating the semantics of the basic theory and case splitting guided by the conditions of the rules. Appropriate notions of termination and confluence were developed to guarantee completeness of the axiomatizations.

To illustrate the considerations when extending classical results on conditional rewrite theory, detailed proofs of Newman's diamond lemma and the Church-Rosser property were presented.

General T_0-unification and special critical pairs notion remain to be separately examined for the basic theories of interest. Also, it is of great interest to develop specialized techniques to realize inductive proofs using as a built-in parameter the theory of the Presburger arithmetic [Ayala-Rincón and Gadelha, 1997].

REFERENCES

[Avenhaus and Becker, 1992] J. Avenhaus and K. Becker. Conditional rewriting modulo a built-in algebra. SEKI-Report SR-92-11, Fachbereich Informatik, Universität Kaiserslautern, Postfach 3049, D-67653 Kaiserslautern (Germany), 1992.

[Ayala-Rincón and Gadelha, 1997] M. Ayala-Rincón and L. M. R. Gadelha. Some Applications of (Semi-)Decision Algorithms for Presburger Arithmetic in Automated Deduction based on Rewriting Techniques. *La Revista de La Sociedad Chilena de Ciencia de la Computación*, 2(1):14–23, 1997.

[Ayala-Rincón, 1993] M. Ayala-Rincón. *Expressiveness of Conditional Equational Systems with Built-in Predicates.* PhD thesis, Universität Kaiserslautern, Kaiserslautern (Germany), December 1993.

[Ayala-Rincón, 1994] M. Ayala-Rincón. Confluence of Conditional Rewriting Systems with Built-in Predicates and Standard Premises as Conditions. In *XXI Seminário Integrado de Software e Hardware, Caxambú, Brazil*, pages 507–521, August 1994.

[Ayala-Rincón, 1996] M. Ayala-Rincón. Transforming TRSs into CRSs — About Elimination of the Conditions —. In *XXII Latin American Conference on Informatics (PANEL'96), Bogotá, Colombia*, pages 322–334, June 1996.

[Ayala-Rincón, 1998] M. Ayala-Rincón. A Deductive Calculus for Conditional Equational Systems with Built-in Predicates as Premises. *Revista Colombiana de Matemáticas*, 1998. Proceedings version in *XV International Conference of the Chilean Computer Science Society*, pages 25-36, 1995.

[Baader and Nipkow, 1998] F. Baader and T. Nipkow. *Term Rewriting and All That.* Cambridge University Press, 1998.

[Baader and Nutt, 1996] F. Baader and W. Nutt. Combination Problems for Commutative/Monoidal Theories or How Algebra Can Help in Equational Unification. *Journal of Applicable Algebra in Engineering, Communication and Computing*, 7(4):309–337, 1996.

[Barwise, 1977] J. Barwise, editor. *Handbook of Mathematical Logic*, volume 90 of *Studies in Logic and the foundations of Mathematics.* North-Holland, 1977.

[Becker, 1994] K. Becker. *Rewrite Operationalization of Clausal Specifications with Predefined Structures.* PhD thesis, Universität Kaiserslautern, Kaiserslautern (Germany), April 1994.

[Cooper, 1972] D. C. Cooper. Theorem Proving in Arithmetic without Multiplication. *Machine Intelligence*, 7:91–99, 1972.

[de Araújo and Ayala-Rincón, 1998] I. E. T. de Araújo and M. Ayala-Rincón. An Algorithm for General Unification Modulo Presburger Arithmetic. 1998. Manuscript.

[Dershowitz and Okada, 1990] N. Dershowitz and M. Okada. A Rationale for Conditional Equational Programming. *Theoretical Computer Science*, 75:111–138, 1990.

[Ehrig and Mahr, 1985] H. Ehrig and B. Mahr. *Fundamentals of Algebraic Specification 1.* EATCS Monographs on Theoretical Computer Science. Springer, 1985.

[Kapur and Subramaniam, 1996] D. Kapur and M. Subramaniam. New Uses of Linear Arithmetic in Automated Theorem Proving by Induction. *Journal of Automated Reasoning*, 16(1/2), 1996.

[Kirchner *et al.*, 1990] Claude Kirchner, Hélène Kirchner, and M. Rusinowitch. Deduction with Symbolic Constraints. *Revue d'Intelligence Artificielle*, 4(3):9–52, 1990. Special issue on Automatic Deduction.

[Nutt, 1992] W. Nutt. Unification in Monoidal Theories is Solving Linear Equations over Semirings. Research Report RR-92-01, Deutsche Forschungszentrum für Künstliche Intelligenz, DFKI GmbH, Stuhlsatzenhausweg 3, D-66123 Saarbrücken, Germany, 1992.

[O'Donnell, 1977] M. J. O'Donnell. *Computing in Systems Described by Equations*, volume 58 of *LNCS*. Springer, 1977.

[Presburger, 1929] M. Presburger. Über die Vollständigkeit eines gewissen Systems der Arithmetik ganzer Zahlen, in welchem die Addition als einzige Operation hervortritt. In *1. Kongres matematyków krajow slowiańskich, Warsaw*, pages 92–101, 1929. In German.

[Remy, 1982] J.-L. Remy. *Etudes des Systemes de Reecriture Conditionnels et Applications aux Types Abstraits Algebriques.* PhD thesis, C.R.I., Nancy, 1982. In French.

[Shostak, 1977] R. E. Shostak. On the SUP-INF Method for Proving Presburger Formulas. *Journal of the Association for Computing Machinery*, 24(4):529–543, October 1977.

[Sivakumar, 1989] G. Sivakumar. *Proofs and Computations in Conditional Equational Theories.* PhD thesis, University of Illinois at Urbana-Champaign, 1989.

[Vorobyov, 1988] S. G. Vorobyov. On the Arithmetic Inexpressiveness of Term Rewriting Systems. In *Third Symp. on Logics in Computer Sciences, Edinburgh Scotland*, pages 212–217, July 1988.

[Vorobyov, 1989] S. G. Vorobyov. Conditional rewrite rule systems with Built-in Arithmetic and Induction. In N. Dershowitz, editor, *Proc. Third Int. Conf. on Rewriting techniques and Applications, Chapel-Hill, NC*, volume 355 of *LNCS*, pages 492–512. Springer, April 1989.

[Zhang and Remy, 1985] H. Zhang and J-L. Remy. Contextual rewriting. In J.-P. Jouannaud, editor, *Proc. First International conference on Rewriting Techniques and Applications, Dijon, France*, volume 202 of *LNCS*, pages 46–62. Springer, May 1985.

Combining WS1S and HOL

David Basin and Stefan Friedrich

1 INTRODUCTION

In this paper we investigate the combination of two logics and the integration of two powerful and complementary theorem proving systems for these logics:

- a decision procedure for WS1S (the weak monadic second-order logic of one successor) implemented in the MONA system, and
- an implementation of HOL (higher-order logic) in the ISABELLE system.

There are compelling reasons for investigating this combination. WS1S belongs to a class of monadic logics that are among the more expressive decidable logics known and many decision problems can be embedded in WS1S [Thomas, 1990]. The logic is also well suited for reasoning about many kinds of systems that can be modeled using automata [Basin and Klarlund, 1998]. However, as with all decidable logics, its expressiveness is limited. In contrast, HOL is a very expressive foundation for reasoning about programs and systems; however, proof construction in HOL typically requires considerable user guidance. A combination potentially offers the complementary advantages of both: the expressive logic of HOL can be used to specify problems and MONA can be used to automatically solve subproblems expressible in WS1S that arise during HOL proofs.

In this paper, we examine the problems that arise in carrying out such a combination and propose solutions to these problems. First, the combination must be sound: only valid HOL formulae should be provable. Semantic methods are required here since the decision procedure for WS1S is a semantic one; the validity of WS1S formulae is determined by translating them into automata that recognize models. We show how a semantic embedding of WS1S can be used to guarantee the correctness of the combination. Within a conservative extension of HOL with theories of sets, numbers, and arithmetic, we formalize the semantics of WS1S formulae. WS1S formulae then correspond to a subclass of HOL formulae and MONA can be used to determine the validity of problems in this class.

Second, the combination must be usable: we require a way of generating WS1S subproblems during proofs in HOL, which can be solved by MONA. In our case, this is equivalent to generating problems in the range of our semantic embedding. There cannot be a general method of doing this, since that would give us a decision procedure for HOL, which is impossible. However, we show that for two particular problem domains such methods are possible. First, we show how to solve different classes of *linear arithmetic* problems in HOL. Linear arithmetic is the theory consisting of formulae made up from numeric constants, variables, addition, arithmetic relations ($<$, $\leq$, $=$), and the standard first-order logical connectives. We present decision procedures, and their integration in HOL, based on the linear arithmetic theories of the natural numbers and integers.

The second method is rather different and more difficult to characterize crisply. The problem domain we examine is reasoning about the correctness of parametric sequential circuits, e.g. a family of n-bit counters, which is parameterized by the number of bits n and its behaviour is a function of time. Such circuits cannot be directly formalized in WS1S or in any decidable logic.[1] Such problems can, however, be formalized in HOL and we show that it is possible to eliminate one of the two unbounded parameters in cases where the goal can be reduced to a new goal involving only finitely many instances of this parameter. The reduced problem is expressible in WS1S whereby MONA can be used to establish properties that are invariant over time.

The remainder of the paper is organized as follows. In Section 2 we provide background both on ISABELLE's implementation of HOL, and on WS1S and its implementation in MONA. In Section 3 we describe the combination, its mechanization in HOL, and its correctness. Then, we consider two rather different applications: in Section 4 we show how to automate arithmetic reasoning over the natural numbers and integers, and in Section 5 we present applications for verifying parameterized sequential hardware. Finally, in Section 6, we compare with related work and draw conclusions.

2 BACKGROUND

2.1 Isabelle/HOL

We use ISABELLE's implementation of HOL for our work; integration with other theorem provers, such as the HOL system or PVS, would be similar.

ISABELLE [Paulson, 1994] is a generic theorem prover in which *object logics* are implemented in ISABELLE's *metalogic*, which is a fragment of intuitionistic higher-order logic based on the (polymorphically) typed lambda-calculus. The logical connectives of this fragment are implication (written $\Longrightarrow$), universal quantification (written $\bigwedge$), and equality (written $\equiv$). Object logics

[1] The reason for this is that the two parameters (bit-width and time) are independent. Such problems correspond, abstractly, to problems on grids, which are undecidable.

are encoded by declaring a *theory*, which is a signature and a set of axioms. Theorems are constructed interactively by applying tactics (programs that construct proofs) in the metalogic. Theorems may also be imported from external theorem provers, which can serve as oracles for particular classes of problems.

Terms in HOL are constructed from the λ-calculus and three additional operators: implication $\rightarrow$, equality $=$, and Hilbert's description-operator ϵ. All terms are typed. For example, for x of type α, $\epsilon x.\, P(x)$ is a term of type α; this term denotes some a, for which $P(a)$ holds, if such an a exists, and an arbitrary term of type α otherwise (types are non-empty). As is standard in HOL, all other logical connectives are defined using these primitives, e.g., $\exists\, x.\ P(x) \equiv P(\epsilon x.P(x))$. There is no particular symbol for bi-implication; this is expressed using equality.

The initial theory of HOL is minimal and contains only 7 axioms. ISABELLE supports building hierarchies of theories based on (definitional) extensions with new types and constants. To ensure consistency, new inference rules are *derived*, e.g., the standard introduction and elimination rules for $\exists$ are derived from the above definition. A number of extensions come distributed with HOL. Important for our work are theories of typed sets, typed finite sets, natural numbers, and integers. Finite sets and natural numbers are inductively defined. Such inductive types come with induction principles and standard functions (e.g., addition, multiplication, and proper subtraction) are defined by primitive recursion. The integers are defined as equivalence classes over pairs of natural numbers in the standard way. We will say more about these data-types and their accompanying theories as we use them below.

2.2 WS1S and MONA

Research on WS1S and related monadic theories goes back to Büchi [Büchi, 1960] and Elgot [Elgot, 1961]. We briefly review syntax, semantics, and decidability.

Definition 1 Let x and X range over disjoint sets V_1 and V_2 of (first and second-order) variables. The language of WS1S is described by the following grammar.

$$\begin{aligned} t \; &::= \; x \mid 0 \mid \mathsf{s}(t) \\ \phi \; &::= \; t = t \mid t \in X \mid \phi \wedge \phi \mid \neg\phi \mid \exists x : \phi \mid \exists X : \phi \end{aligned}$$

Hence terms are built from first-order variables, the constant 0, and the successor symbol. Atomic formulae are built from the equality and membership predicates and formulae consist of atomic formulae and are closed under conjunction, negation, and quantification over first and second-order variables. Other connectives and quantifiers can be defined using standard classical equivalences, e.g., $\forall x : \phi \equiv \neg\exists x : \neg\phi$; the symbol $\{\!\!\}$ represents the empty set. Also, it is possible to represent propositional variables and quantification over them (e.g. the propositional variable P can be represented by the atomic formula $0 \in X_P$, where X_P is a new second-order variable).

In the semantics of WS1S, formulae are interpreted in $\mathbb{N}$, where $\mathsf{0}$ and s are zero and the successor function respectively, $=$ is equality over numbers, and $\in$ is membership. The logic is 'weak' because second-order variables are interpreted over *finite* subsets of the domain. The following embedding makes this interpretation clear. Here nat, finnat, and bool are the HOL types of naturals, finite sets of naturals, and truth-values, respectively (i.e., relations are formalized as boolean-valued functions) and Suc denotes the successor function on natural numbers. In the sequel we shall omit typing constraints if types are known from the context.

Definition 2 The *embedding function* $\lceil \_ \rceil$ from WS1S formulae to HOL formulae is recursively defined as follows:

$$
\begin{aligned}
\lceil x \rceil &= x :: \mathsf{nat} && \text{if } x \in \mathrm{V}_1\\
\lceil X \rceil &= X :: \mathsf{finnat} && \text{if } X \in \mathrm{V}_2\\
\lceil \mathsf{0} \rceil &= \mathsf{0} :: \mathsf{nat}\\
\lceil \mathsf{s}(t) \rceil &= (\mathsf{Suc} :: \mathsf{nat} \Rightarrow \mathsf{nat})(\lceil t \rceil)\\
\lceil t_1 = t_2 \rceil &= (= :: [\mathsf{nat}, \mathsf{nat}] \Rightarrow \mathsf{bool})(\lceil t_1 \rceil)(\lceil t_2 \rceil)\\
\lceil t \in X \rceil &= (\in :: [\mathsf{nat}, \mathsf{finnat}] \Rightarrow \mathsf{bool})(\lceil t \rceil)(\lceil X \rceil)\\
\lceil \phi_1 \wedge \phi_2 \rceil &= (\wedge :: [\mathsf{bool}, \mathsf{bool}] \Rightarrow \mathsf{bool})(\lceil \phi_1 \rceil)(\lceil \phi_2 \rceil)\\
\lceil \neg \phi \rceil &= (\neg :: \mathsf{bool} \Rightarrow \mathsf{bool})(\lceil \phi \rceil)\\
\lceil \exists x :: \phi \rceil &= (\exists :: (\mathsf{nat} \Rightarrow \mathsf{bool}) \Rightarrow \mathsf{bool})(\lambda x :: \mathsf{nat}. \lceil \phi \rceil)\\
\lceil \exists X : \phi \rceil &= (\exists :: (\mathsf{finnat} \Rightarrow \mathsf{bool}) \Rightarrow \mathsf{bool})(\lambda X : \mathsf{finnat}. \lceil \phi \rceil)
\end{aligned}
$$

The problem of determining if WS1S sentences are true under the above interpretation is decidable (see, e.g. [Thatcher and Wright, 1967; Thomas, 1990]). The decision procedure is semantically based; it translates a WS1S formula ϕ to an automaton A_ϕ that, essentially, recognizes valuations for the free variables under which ϕ is true. It is easy to determine from the resulting automaton A_ϕ whether a sentence is true in the above structure.

The MONA system implements this decision procedure. Input to MONA is a script consisting of a sequence of definitions followed by a formula ϕ to be proven. MONA computes A_ϕ and, depending on the result, declares ϕ to be valid or delivers a counter-example. MONA is implemented to be as efficient as possible [Henriksen et al, 1995] and works well in practice on a large range of problems despite the non-elementary worst-case complexity of WS1S; empirical evidence of this and an analysis of why this is the case is given in [Basin and Klarlund, 1998].

3 COMBINING MONA AND ISABELLE/HOL

We formalize the semantics of WS1S in ISABELLE's HOL. We create a theory that formalizes the semantic domains of WS1S (natural numbers and finite sets of natural numbers) as HOL-types. The embedding function (Definition 2) maps WS1S formulae to formulae of this theory and the image of this mapping characterizes what we may call *the WS1S-subset of HOL*. We explain below how we implement the inverse of the embedding so that formulae in this subset of HOL can be submitted to MONA and the results can be incorporated in HOL proofs.

3.1 Logical Basis and Correctness

Hooking an 'oracle' to a theorem prover is risky business. The oracle could be buggy or there could be a mismatch between the semantics of the oracle and semantics of the logic used by the theorem prover. The only way to avoid a buggy oracle is to reconstruct a proof in the theorem prover based on output from the oracle, or perhaps formally verify the oracle itself. For a semantics based decision procedure, proof reconstruction is not a realistic option: one would have to formalize the entire automata-theoretic machinery within HOL and this would amount to verifying the oracle.

By taking a semantic based translation approach (sometimes called semantic embedding in the literature) we cannot avoid the problem of a buggy oracle, but we can formally (although not inside of HOL) show that there is no semantics mismatch and that we do not compromise the consistency of HOL by accepting theorems proved by MONA.

We define in HOL a theory Finset in which we model the semantic domains of WS1S. The set $\mathbb{N}$ is represented by the type nat, which is defined inductively. The Peano axioms can be derived from this definition, including a second-order induction axiom. It follows that every standard model [Andrews, 1986; Gordon and Melham, 1993] of the theory must interpret the type nat (up to isomorphism) as the set $\mathbb{N}$. A similar statement holds for our inductive definition of the type finnat, which is defined as the set of finite subsets of nat. The inductive definition characterizes the finite powerset of $\mathbb{N}$, $\mathcal{F}(\mathbb{N})$, uniquely and every model of the theory must interpret the type finnat (up to isomorphism) as the set $\mathcal{F}(\mathbb{N})$.

Given the above, we can show by induction on the structure of terms and formulae of WS1S that the terms and formulae of WS1S are interpreted in the same way as their image under the embedding function of Definition 2; consequently for every valid WS1S-formula ϕ, the image $\lceil \phi \rceil$ is valid in every standard model of HOL. By soundness of ISABELLE/HOL's deductive system (see [Regensburger, 1994]), we conclude that $\neg \lceil \phi \rceil$ is not derivable in the theory Finset and therefore we may accept $\lceil \phi \rceil$ consistently as a theorem of the theory Finset. To summarize:

Theorem 3 *If an WS1S-formula ϕ is valid in WS1S we can add $\lceil \phi \rceil$ consistently to the theorems of the theory* Finset.

3.2 Mechanization

Our goal is to link up two different systems. Mechanically, this entails identifying formulae of an ISABELLE/HOL proof-goal that are in the WS1S-subset of HOL, translating them to WS1S, invoking MONA, and updating the proof when MONA succeeds. These formulae are identified syntactically; they contain only terms of types nat and finnat and operations corresponding to those in WS1S. (see Definition 2). Polymorphic operators such as equality or quantification can be distinguished by their type instance. This allows us to strictly control which formulae are translated.

Typically, users work by adding definitions to theories, thereby extending the language of HOL with new constants, e.g., the constant IF defined by $\mathsf{IF}\, e\, x\, y \equiv (e \to x) \wedge (\neg e \to y)$. When such definitions are also expressible in WS1S, we can include them in our translations. We do this by allowing the user to define corresponding definitions in MONA, and extend the embedding function, provided that the right-hand side of the definition contains only terms that are already in the image of the embedding function. The constants defined in this way are recorded together with their definitions in a particular data structure called a *definition set.* When translating a formula to WS1S, every constant is replaced by a call of the respective macro and the macro definition is added to the declaration part of the generated MONA input-file.

To connect ISABELLE and MONA, we use the general 'oracle interface' provided by ISABELLE. Oracles are declared in ISABELLE's theory definitions by specifying a top-level ML-function that takes arbitrary data, e.g. a proof goal, and returns a term of ISABELLE's meta-level type of truth-values. This function may be an implementation of a decision procedure, a model-checker, or it may call, as in our case, an external reasoning tool.

We have written a tactic that uses the oracle to solve goals arising in HOL proofs. This tactic takes a goal and a definition set, and invokes the oracle. If MONA can prove the resulting formula, then that formula is promoted by the oracle interface to a theorem, which we use to solve the goal; otherwise the tactic fails.

The translation of a subgoal in which one tries to prove a conclusion C from assumptions $A_1, \ldots, A_n$ is organized into several steps. First, we attempt to translate separately each assumption and the conclusion. If this fails for some assumption A_i, then we replace this assumption by true; effectively, we try then to prove the subgoal without A_i. If the translation fails for the conclusion we replace it by false and try to prove a contradiction from the assumptions. This yields the assumptions $A'_1, \ldots, A'_n$ and the conclusion C' in the language of MONA. We join them together as a nested implication from which we take the universal closure and thus obtain a MONA-formula of the form

$$\forall X_1 \ldots \forall X_k\, \forall x_1 \ldots \forall x_l,\colon\ A'_1 \to \ldots \to A'_n \to C'$$

4 APPLICATION: LINEAR ARITHMETIC

In the previous section we showed how to soundly integrate MONA with HOL by formalizing the semantics of WS1S as a theory-extension of HOL. The resulting coding though is rather 'low level'. It is easier for users to encode problems using higher level concepts like numbers and arithmetic functions as opposed to finite sets and WS1S definable relations on them.

In this section we show how to link these levels in the case of arithmetic. We build a kind of interpreter that allows us to translate arithmetic as expressed in standard ISABELLE theories to the (embedded) WS1S theory. This interpretation is done completely within HOL using formally derived rewrite rules; hence correctness is guaranteed.

We will consider several arithmetic theories in this section. For example, *Presburger arithmetic* [Presburger, 1929] is the first-order theory of $(\mathbb{N}, +)$. This theory can easily be extended with inequalities as well as multiplication and division by constants. It also can be extended to a theory of integers by considering pairs of numbers. In [Büchi, 1960] and [Elgot, 1961] Büchi and Elgot note that decidability of WS1S implies the decidability of Presburger arithmetic. After first presenting arithmetic over number representations from finnat, we will consider the first-order theories of $(\mathbb{N}, +)$ and $(\mathbb{Z}, +)$.

4.1 Arithmetic over finnat

We use a binary encoding to represent a natural number n as an element N of type finnat. For all i, $i \in N$ holds if and only if the ith digit in the binary representation of n is 1, e.g., the number 5 is represented by the set $\{\!|0,2|\!\}$ since $5 = 2^0 + 2^2$. Thus the empty set $\{\!||\!\}$ represents zero and the successor relation is defined as

$$\begin{aligned} \mathsf{SUCC}\, M\, N \equiv\ \exists p.\, \forall i. \quad & (i < p \to i \notin M \wedge i \in N) \\ \wedge \quad & (i = p \to i \in M \wedge i \notin N) \\ \wedge \quad & (p < i \to (i \in M) = (i \in N))\,, \end{aligned}$$

i.e., M represents the successor of the number represented by N. To show that this is a proper encoding for the natural numbers, we prove that an analogue of the Peano axioms hold for it; that is we prove the following five formulae.

1) $\{\!||\!\} :: \mathsf{finnat}$
2) $\forall M.\, \exists N.\, \mathsf{SUCC}\ N\ M$
3) $\forall N.\, \neg\mathsf{SUCC}\ \{\!||\!\}\ M$
4) $\forall K\, L\, M\, N.\, \mathsf{SUCC}\, L\ K\ \wedge\ \mathsf{SUCC}\ N\ M\ \wedge\ L = N\ \to\ K = M$
5) $\forall P.\ (P\, \{\!||\!\})\ \wedge\ (\forall K\, L.\, (P\, K)\ \wedge (\mathsf{SUCC}\, L\, K) \to (P\, L)) \to (\forall N.\, (P\, N))$

The first holds trivially, since $\{\!||\!\}$ is an element of type finnat. We prove the next three automatically using MONA as an oracle. This was a pleasant surprise: our oracle was often useful in proving theorems required to extend its scope. The fifth theorem formalizes an induction principle: all elements of type finnat can be reached from $\{\|\}$ via the successor relation. This formula cannot be proven by MONA since it involves quantification over predicates. However, it can be derived in ISABELLE/HOL using the induction principle associated with the inductively defined finite-set type and in this derivation we can use MONA to discharge different proof obligations.

To use MONA as an arithmetic decision procedure requires defining arithmetic operations for addition, equality, and inequality for numbers encoded as finite sets. The following is an example of encoding addition, i.e. the number represented by S is the sum of the numbers represented by A and B.

$$\begin{aligned}\mathsf{ADD}\, S\, A\, B \equiv\ & \exists C.\, 0 \notin C \wedge \\ & \forall p. \quad \mathsf{mod\_two}\ (p \in A)\ (p \in B)\ (p \in C)\ (p \in S) \\ & \quad \wedge\ \mathsf{at\_least\_two}\ (p \in A)\ (p \in B)\ (p \in C)\ ((\mathsf{Suc}\, p) \in C)\end{aligned}$$

This implements the standard addition algorithm (over binary representations of numbers) where the pth bit of the result S is the sum of the pth bits of A, B, and the carry (here C) mod 2 and the pth carry is set if at least two of the previous inputs and the carry are set. The auxiliary predicates mod_two and at_least_two are defined as follows.

$$\begin{aligned}\mathsf{mod\_two}\, a\, b\, c\, s &\equiv a = b = c = s \\ \mathsf{at\_least\_two}\, a\, b\, c\, d &\equiv d = (a \wedge b) \vee (b \wedge c) \vee (a \wedge c)\end{aligned}$$

4.2 Arithmetic over nat

The above definitions allow us to extend our oracle's definition set and solve arithmetic problems expressed using relations over finnat, i.e., arithmetic problems where numbers are formalized as finite sets representing their binary encodings. This has limited use in practice. Users of HOL usually perform arithmetic reasoning directly within the theory nat. Formally, we can define a translation between these two theories: we recursively define mappings val and bin between the types nat and finnat and prove the following properties, which characterize them as isomorphisms (we have named the properties for subsequent use).

$$\begin{array}{ll}
(\mathit{val_Empty}) & \mathsf{val}\ \{\|\} = 0 \\
(\mathit{val_SUCC}) & \mathsf{SUCC}\ M\ N \implies \mathsf{val}\ M = \mathsf{Suc}(\mathsf{val}\ N) \\
(\mathit{bin_0}) & \mathsf{bin}\ 0 = \{\|\} \\
(\mathit{bin_Suc}) & \mathsf{bin}\ (\mathsf{Suc}\ n) = (\epsilon M.\ \mathsf{SUCC}\ M\ (\mathsf{bin}\ n)) \\
(\mathit{val_inverse}) & \mathsf{val}(\mathsf{bin}\ n) = n \\
(\mathit{bin_inverse}) & \mathsf{bin}(\mathsf{val}\ N) = N \\
(\mathit{ADD}) & \mathsf{ADD}\ (\mathsf{bin}\ s)\ (\mathsf{bin}\ a)\ (\mathsf{bin}\ b) = (s = a + b)
\end{array}$$

Next, we automate the use of this encoding, effectively hiding it from the user; we derive rewrite rules as HOL theorems that can be used by ISABELLE's simplifier in order to automatically translate arithmetic operations on numbers to formulae built from predicates over sets. Our translation rules are of three kinds. First, rules that replace relations between numbers by relations between sets, e.g.

$$(eq\_c) \qquad (n = m) \;=\; (\mathsf{bin}\ n = \mathsf{bin}\ m)$$

Second, rules that propagate applications of bin downwards through the term structure and replace numeric functions by the corresponding predicates over sets. There is a rule for each operation; e.g. for addition we have:

$$(add\_c) \qquad P(\mathsf{bin}(a+b)) \;=\; (\exists S.\ \mathsf{ADD}\ S\ (\mathsf{bin}\ a)\ (\mathsf{bin}\ b) \wedge P(S))$$

Third, rules for quantifiers that translate quantification over numbers to quantification over sets, e.g.

$$(all\_c) \qquad (\forall n :: \mathsf{nat}.\ P(n)) \;=\; (\forall N :: \mathsf{finnat}.\ P(\mathsf{val}\ N))$$

As previously mentioned, these rules are formally derived in HOL.

A simple example should clarify how the rules together define the translation. Let us consider how rewriting transforms the statement that addition is commutative to a formulae that can be proved by MONA. Rewriting performs the following steps:

$$\forall a\, b.\ a + b = b + a$$

$$\xrightarrow{(eq\_c)} \quad \forall a\, b.\ \mathsf{bin}(a+b) \;=\; \mathsf{bin}(b+a)$$

$$\xrightarrow{(add\_c)} \quad \forall a\, b.\ \exists S.\ \mathsf{ADD}\ S\ (\mathsf{bin}\ a)\ (\mathsf{bin}\ b) \wedge (S = \mathsf{bin}(b+a))$$

$$\xrightarrow{(add\_c)} \quad \begin{array}{l}\forall a\, b.\ \exists S.\ \mathsf{ADD}\ S\ (\mathsf{bin}\ a)\ (\mathsf{bin}\ b)\\ \quad \wedge\ (\ \exists T.\ \mathsf{ADD}\ T\ (\mathsf{bin}\ b)\ (\mathsf{bin}\ a) \wedge (S = T))\end{array}$$

$$\xrightarrow{(all\_c)}{}^{2} \quad \begin{array}{l}\forall A\, B.\ \exists S.\ \mathsf{ADD}\ S\ (\mathsf{bin}\ (\mathsf{val}\ A))\ (\mathsf{bin}\ (\mathsf{val}\ B))\\ \quad \wedge\ (\ \exists T.\ \mathsf{ADD}\ T\ (\mathsf{bin}\ (\mathsf{val}\ B))\ (\mathsf{bin}\ (\mathsf{val}\ A)) \wedge (S = T))\end{array}$$

$$\xrightarrow{(bin\_inverse)}{}^{4} \quad \begin{array}{l}\forall A\, B.\ \exists S.\ \mathsf{ADD}\ S\ A\ B\\ \quad \wedge\ (\ \exists T.\ \mathsf{ADD}\ T\ B\ A) \wedge (S = T))\end{array}$$

Note that a and b are of type nat whereas A and B are of type finnat. This rewriting is performed by a tactic that is applied before invoking the oracle. Hence the translation is performed automatically and it serves to extend the scope of the oracle to function not just over set based relations like ADD, bin or val, but also to numbers and standard functions and operators on them, which is what users usually require.

4.3 Arithmetic over int

We now turn our attention to the integers and extending our interface to them. Our solution is based on providing a bijection zenc between int and nat. This function maps negative integers to odd natural numbers and positive integers to even natural numbers in the following way:

$$\mathsf{zenc}(z) = \begin{cases} 2 \cdot z & \text{if } 0 \leq z \\ -(2 \cdot z + 1) & \text{if } z < 0 \end{cases}$$

In our binary representation, this corresponds to shifting the mantissa one digit to the left and using the vacated 0-th bit as a sign bit. The inverse of zenc is called zdec. The bijection zbin and it inverse zval between finnat and int are functional compositions of bin with zenc and, respectively, of zdec with val.

$$\begin{aligned} \mathsf{zbin}(z) &\equiv \mathsf{bin}(\mathsf{zenc}(z)) \\ \mathsf{zval}(Z) &\equiv \mathsf{zdec}(\mathsf{val}(Z)) \end{aligned}$$

Addition for this encoding of the integers can be represented by a predicate ZADD, which is defined similarly to ADD.

Completing the extension is analogous to the extension to nat: we derive rules that characterize zbin as an isomorphism and after that we extend the set of rewriting rules. The extension automatically translates formulae involving integer arithmetic to our encoding of finite sets.

4.4 Experience

Arithmetic reasoning plays an important role in many theorem proving applications such as program verification. Our tactics can be used to automatically discharge first-order arithmetic problems (in the language of 0, +, and inequalities) that arise when using the theories nat and int. It is easy also to extend these procedures too as required. For example, we can add WS1S definable relations like proper subtraction (over the natural numbers), subtraction (over the integers), division or remainder by 2, and the like.

Our tactics operate automatically. Trivial examples like commutativity or associativity of addition are proved in less than two seconds on a Sparc Station-20. As a less trivial example, consider the following verification problem that arises in showing the correctness of our encoding of the integers described in the previous section.

$$\mathsf{ZADD}\ s\ a\ b = (\mathsf{zval}\ s = \mathsf{zval}\ a + \mathsf{zval}\ b)$$

Expanding the definitions of zval and zdec yields the following goal.

$$
\begin{array}{l}
(\mathsf{val}\ s \bmod 2 = 0 \rightarrow \\
\quad (\mathsf{val}\ a \bmod 2 = 0 \rightarrow \\
\qquad\quad (\mathsf{val}\ b \bmod 2 = 0 \rightarrow \mathsf{ZADD}\ s\ a\ b = (\mathsf{val}\ s\ \mathsf{div}\ 2 = \mathsf{val}\ a\ \mathsf{div}\ 2 + \mathsf{val}\ b\ \mathsf{div}\ 2)) \\
\qquad \wedge\ (\mathsf{val}\ b \bmod 2 \neq 0 \rightarrow \mathsf{ZADD}\ s\ a\ b = (\mathsf{Suc}(\mathsf{val}\ s\ \mathsf{div}\ 2 + \mathsf{val}\ b\ \mathsf{div}\ 2) = \mathsf{val}\ a\ \mathsf{div}\ 2))) \\
\quad \wedge\ (\mathsf{val}\ a \bmod 2 \neq 0 \rightarrow \\
\qquad\qquad (\mathsf{val}\ b \bmod 2 = 0 \rightarrow \mathsf{ZADD}\ s\ a\ b = (\mathsf{Suc}(\mathsf{val}\ s\ \mathsf{div}\ 2 + \mathsf{val}\ a\ \mathsf{div}\ 2) = \mathsf{val}\ b\ \mathsf{div}\ 2)) \\
\qquad\quad \wedge\ (\mathsf{val}\ b \bmod 2 \neq 0 \rightarrow \mathsf{ZADD}\ s\ a\ b = \mathsf{False}))) \\
\wedge\ (\mathsf{val}\ s \bmod 2 \neq 0 \rightarrow \\
\qquad (\mathsf{val}\ a \bmod 2 = 0 \rightarrow \\
\qquad\qquad (\mathsf{val}\ b \bmod 2 = 0 \rightarrow \mathsf{ZADD}\ s\ a\ b = \mathsf{False}) \\
\qquad\quad \wedge\ (\mathsf{val}\ b \bmod 2 \neq 0 \rightarrow \mathsf{ZADD}\ s\ a\ b = (\mathsf{val}\ b\ \mathsf{div}\ 2 = \mathsf{val}\ a\ \mathsf{div}\ 2 + \mathsf{val}\ s\ \mathsf{div}\ 2))) \\
\quad \wedge\ (\mathsf{val}\ a \bmod 2 \neq 0 \rightarrow \\
\qquad\qquad (\mathsf{val}\ b \bmod 2 = 0 \rightarrow \mathsf{ZADD}\ s\ a\ b = (\mathsf{val}\ a\ \mathsf{div}\ 2 = \mathsf{val}\ b\ \mathsf{div}\ 2 + \mathsf{val}\ s\ \mathsf{div}\ 2)) \\
\qquad\quad \wedge\ (\mathsf{val}\ b \bmod 2 \neq 0 \rightarrow \mathsf{ZADD}\ s\ a\ b = (\mathsf{Suc}(\mathsf{val}\ a\ \mathsf{div}\ 2 + \mathsf{val}\ b\ \mathsf{div}\ 2) = \mathsf{val}\ s\ \mathsf{div}\ 2))))
\end{array}
$$

Such a goal would be rather tedious for a human to prove interactively. It is solved by our tactic in 43.6 seconds. Interestingly though (and a possible area for improvement), almost all the time is spent on rewriting; MONA requires only 0.13 seconds to verify the translated formula.

5 APPLICATION: CIRCUIT VERIFICATION

To illustrate the generality and flexibility of our combination, we present a second example from a rather different domain: formal reasoning about parametric circuit descriptions. As previously noted, systems with multiple independent parameters fall out of the scope of WS1S but can be formalized in HOL. Our verification strategy is to reduce the number of parameters to a single parameter, at which point the result may be WS1S formalizable.

Here we consider an example of a circuit with two parameters: time, and bit-width. The key idea is that we can eliminate the time parameter (eliminating bit-width is more difficult, but also possible) by reducing the correctness problem to show that an invariant holds over consecutive time-instances. We demonstrate this with the verification of a parameterized family of n-bit counters. Due to space limitations we only sketch the main ideas; full formalization and proof details are given in [Friedrich, 1998].

In hardware verification, like software verification, one establishes that an implementation fulfills its specification. A common approach is to formalize both the implementation and the specification as relations between the circuit's inputs and outputs [Camilleri *et al.*, 1986]. Circuits are built from primitive relations corresponding to transistors, gates, and the like, are combined by conjunction, and 'wired together' using shared variables, hidden by existential quantifiers. For sequential circuits, signals are modelled as functions from time (discrete in our case) to port-values, which are truth values. It is often desirable to establish the correctness of parametric families of circuits. That is, rather than establish that an n-bit counter is correct for particular values of n, we show that the entire family is correct for all $n \in \mathsf{nat}$. Note that parametric circuits require parametric signals (buses); in our work we model

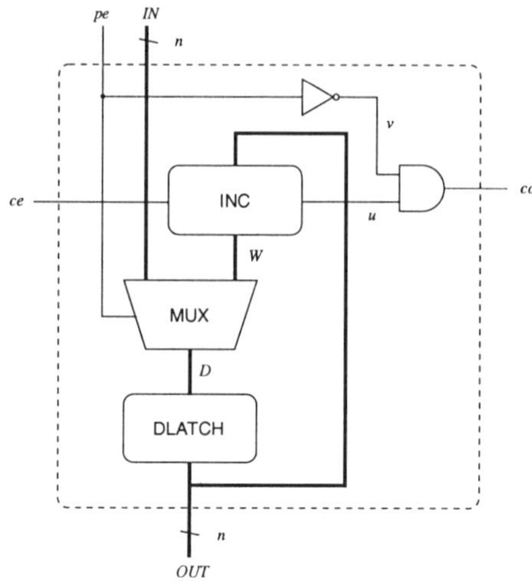

Figure 1: Implementation of the n-bit counter

these as functions from time to elements of type finnat with the convention that the i-th bit of a bus B is set at time t if and only if $i \in B(t)$.

Figure 1 contains a diagram of a schematic n-bit counter. It takes an n-bit bus *IN* as input and yields an n-bit output *OUT* and a carry-out bit *co*. In addition, it takes other input signals, namely *pe* (parallel enable) and *ce* (count enable), which control the function performed by the counter. The components INC, MUX and DLATCH are parameterized implementations of an n-bit incrementer, multiplexor and dlatch, respectively. Assuming that these have been modelled in HOL, we model the implementation of the n-bit counter by the following formulae.

$$\begin{aligned}
&\mathsf{COUNTER}\ n\ IN\ pe\ ce\ OUT\ co\ \equiv \\
&\qquad \exists\, D\ W\ u\ v.\ \mathsf{INC}\ n\ OUT\ ce\ W\ u\ \wedge\ \mathsf{MUX}\ n\ pe\ IN\ W\ D \\
&\qquad\qquad \wedge\ \mathsf{DLATCH}\ n\ D\ OUT\ \wedge\ \mathsf{NOT}\ pe\ v\ \wedge\ \mathsf{AND}\ v\ u\ co
\end{aligned}$$

This circuit is explicitly parameterized by the bit-width n. The time parameter is implicitly part of the specification as all inputs are signals, which are functions of time.

The values of *pe* and *ce* determine the counter's action. For any time instance t, if *pe* holds at time t (written by applying the function *pe* to the argument t) then the value output by the counter at time $t+1$ is the same as that at time t. Otherwise, if *ce* is set, the value at time $t+1$ is one greater than at time t, unless there is an overflow, in which case the output value is

0. The following HOL formula makes this precise.

$$
\begin{array}{l}
\mathsf{COUNTER\_SPEC}\ n\ t\ IN\ pe\ ce\ OUT\ co \equiv \\
\quad \text{if } pe\ t \\
\quad \text{then } \mathsf{valn}\ n\ (OUT(\mathsf{Suc}\ t)) = \mathsf{valn}\ n\ (IN\ t)\ \wedge\ \neg\ co\ t \\
\quad \text{else if } ce\ t \\
\qquad \text{then if } \mathsf{MAX}\ n\ (OUT\ t) \\
\qquad\quad \text{then } \mathsf{valn}\ n\ (OUT(\mathsf{Suc}\ t)) = 0\ \wedge\ co\ t \\
\qquad\quad \text{else } \mathsf{valn}\ n\ (OUT(\mathsf{Suc}\ t)) = \mathsf{Suc}(\mathsf{valn}\ n(OUT\ t))\ \wedge\ \neg\ co\ t \\
\qquad \text{else } \mathsf{valn}\ n\ (OUT(\mathsf{Suc}\ t)) = \mathsf{valn}\ n\ (OUT\ t)\ \wedge\ \neg\ co\ t
\end{array}
$$

We can now state what it means for the counter to be correctly implemented as an equivalence between the implementation and the specification.

$$\mathsf{COUNTER}\ n\ IN\ pe\ ce\ OUT\ co = \forall t.\ \mathsf{COUNTER\_SPEC}\ n\ t\ IN\ pe\ ce\ OUT\ co$$

All free variables are implicitly universally quantified, i.e., we show the equivalence for all possible port-values. As previously noted, the time parameter on the left-hand side is implicit in our use of signals. This parameter is explicit, however, in the specifications of the components INC_SPEC, MUX_SPEC, etc.

The counter is proved correct in three steps. First we expand its definition and replace the implementations of the components by their specifications (these subcomponents are first proved correct with respect to their specifications). This reveals the implicit time parameter on the left-hand side and results in the following goal:

$$
\begin{array}{rl}
(\exists\ D\ W\ u\ v. & (\forall t.\ \mathsf{INC\_SPEC}\ t\ n\ OUT\ ce\ W\ u) \\
& \wedge\ (\forall t.\ \mathsf{MUX\_SPEC}\ t\ n\ pe\ IN\ W\ D) \\
& \wedge\ (\forall t.\ \mathsf{DLATCH\_SPEC}\ t\ n\ D\ OUT) \\
& \wedge\ (\forall t.\ \mathsf{NOT\_SPEC}\ t\ pe\ v) \\
& \wedge\ (\forall t.\ \mathsf{AND\_SPEC}\ t\ v\ u\ co)) \\
= & (\forall t.\ \mathsf{COUNTER\_SPEC}\ n\ t\ IN\ pe\ ce\ OUT\ co)
\end{array}
$$

In the second step we eliminate the time parameter. To do this we pull the universal quantifier outside the equivalence using derived inference rules for quantification. Pulling the universal quantifier over an existential quantifier can be seen as a kind of reverse Skolemization where the (internal) signals D and W, which are functions of type nat $\Rightarrow$ finnat, are replaced by port values D' and W' of type finnat and where the signals u and v, which are of type nat $\Rightarrow$ bool, are replaced by port values of type bool, respectively. Moreover, pulling the time quantifier over the equivalence corresponds to choosing a fixed (but arbitrary) time point at which implementation and specification are compared. This step results in the following proof obligation:

$$\begin{array}{l} \forall t.(\exists\, D'\ W'\ u'\ v'.\ \mathsf{INC\_SPEC}'\ t\ n\ OUT\ ce\ W'\ u' \\ \qquad \wedge\ \mathsf{MUX\_SPEC}'\ t\ n\ pe\ IN\ W'\ D' \\ \qquad \wedge\ \mathsf{DLATCH\_SPEC}'\ t\ n\ D'\ OUT \\ \qquad \wedge\ \mathsf{NOT\_SPEC}'\ t\ pe\ v' \\ \qquad \wedge\ \mathsf{AND\_SPEC}'\ t\ v'\ u'\ co) \\ \quad = (\mathsf{COUNTER\_SPEC}\ n\ t\ IN\ pe\ ce\ OUT\ co) \end{array}$$

After this, we can simplify matters a bit by removing universal quantifiers and replacing all remaining applications of signals to time points by fresh variables that represent the values of the signals at these time points. For example, we replace all occurrences of $OUT\ t$ with the variable $OUT0$ and all occurrences of $OUT(\mathsf{Suc}\ t)$ with the variable $OUT1$.

This leaves us with a new proof goal that no longer contains the time parameter t. (Here, we have expanded the definitions of $\mathsf{INC\_SPEC}'$, $\mathsf{MUX\_SPEC}'$ etc.)

$$\begin{array}{l} (\exists\, D'\ W'\ u'\ v'.(\ \mathsf{if}\ ce0 \\ \qquad \mathsf{then}\ \ \mathsf{if}\ \mathsf{MAX}\ n\ OUT0 \\ \qquad\qquad \mathsf{then}\ \mathsf{valn}\ n\ W' = 0\ \wedge\ u' \\ \qquad\qquad \mathsf{else}\ \mathsf{valn}\ n\ W' = \mathsf{Suc}(\mathsf{valn}\ n\ OUT0)\ \wedge\ \neg u' \\ \qquad \mathsf{else}\ \mathsf{valn}\ n\ W' = \mathsf{valn}\ n\ OUT0\ \wedge\ \neg u') \\ \quad \wedge\ (\mathsf{if}\ pe0\ \mathsf{then}\ \mathsf{EQ}\ n\ D'\ IN0\ \mathsf{else}\ \mathsf{EQ}\ n\ D'\ W') \\ \quad \wedge\ \mathsf{EQ}\ n\ OUT1\ D' \\ \quad \wedge\ v' = (\neg pe0) \\ \quad \wedge\ co0 = (v'\ \wedge\ u')) \\ = (\ \mathsf{if}\ pe0 \\ \quad \mathsf{then}\ \mathsf{valn}\ n\ OUT1 = \mathsf{valn}\ n\ IN0\ \wedge\ \neg co0 \\ \quad \mathsf{else}\ \ \mathsf{if}\ ce0 \\ \qquad \mathsf{then}\ \ \mathsf{if}\ \mathsf{MAX}\ n\ OUT0 \\ \qquad\qquad \mathsf{then}\ \mathsf{valn}\ n\ OUT1 = 0\ \wedge\ co0 \\ \qquad\qquad \mathsf{else}\ \mathsf{valn}\ n\ OUT1 = \mathsf{Suc}(\mathsf{valn}\ n\ OUT0)\ \wedge\ \neg co0 \\ \qquad \mathsf{else}\ \mathsf{valn}\ n\ OUT1 = \mathsf{valn}\ n\ OUT0\ \wedge\ \neg co0) \end{array}$$

This is WS1S formalizable. Invoking our oracle is the third step and this completes the proof in 20 seconds.

6 RELATED WORK AND CONCLUSIONS

6.1 Related Work

The combination of logics and related provers is an active area of research. Various groups have investigated decision procedures for arithmetic reasoning and their integration with theorem provers. The theory of linear integer arithmetic was shown to be decidable by Presburger in 1929. The application of WS1S to arithmetic was documented by Büchi in [Büchi, 1960]. Boyer and Moore have integrated a procedure for the universal fragment of linear natural arithmetic in their prover NQTHM [Boyer and Moore, 1988]. The PVS system, based on higher-order logic, implements a solver for linear equations [Owre *et al.*, 1992], and the STEP system combines decision procedures for partial orders and linear integer arithmetic with a semi-decision procedure for first-order logic [Manna *et al.*, 1994].

In each of the above combinations, the decision procedure is part of the prover and is 'hardwired' to operate over specific data-types and syntactic formulae classes. Such close coupling avoids translation and linkage to an external prover, but it introduces inflexibility. As documented by Boyer and Moore [Boyer and Moore, 1988], in practice few subgoals fit the requirements of such decision procedures; their successful use requires closely integrating the procedure with other theorem proving activities. Our formalization, which is based on an open coupling between WS1S and HOL, offers advantages in this regard. WS1S is a decision procedure for a general theory of finite sets of natural numbers in which arithmetic can be encoded. The encoding is made by, and transparent to, the user and can easily be modified to different HOL datatypes (e.g., natural numbers or integers). Additional relations in HOL can be user defined and incorporated into the translation, provided they are WS1S expressible.

Relevant to our second application is research on using model checkers to reason about systems with infinite or parameterized state spaces, which cannot directly be model-checked. Kurshan and Lamport [Kurshan and Lamport, 1993] have presented a similar connection between a proof checker for TLA and the COSPAN model-checker; they present a parameterized n-bit multiplier verified in TLA based on induction and the verification of an 8-bit multiplier by COSPAN. The interaction between the two systems is very loose: both systems were used independently leaving the users to communicate between them and to insure correctness of the interaction.

Our parameter elimination technique is closely related to work of Wolper and Lovinfosse [Wolper and Lovinfosse, 1989] and Kurshan and McMillan [Kurshan and McMillan, 1989]. They present techniques where model checking is used to verify an invariant that formalizes an induction step. Our work differs from theirs in that we use a general purpose decision procedure (as opposed to a model checker) and the parameter elimination steps are completely formalized within HOL. This kind of formalized reduction is also found in work at SRI where a decision procedure for the mu-calculus is implemented in PVS and used to reason about infinite state systems, which can be reduced to finite state systems using abstraction techniques [Rajan *et al.*, 1995] (see also [Müller and Nipkow, 1995]).

6.2 Conclusions

Our experience with the combination is positive. WS1S has a simple semantics that leads to a simple embedding whose correctness is easily established. Despite this simplicity, WS1S is expressive, has a wide range of applications, and one can easily embed or combine other decision procedures with WS1S. The rather different nature of our two examples is evidence of this flexibility.

Our work is just a starting point, and there are many interesting open problems. Complexity is one of them. WS1S is non-elementary in worst case; however, MONA can solve many complex problems quickly in practice. Theoretical reasons for this and elementary bounds for certain sub-problems are given in [Basin and Klarlund, 1998]. Theoretical and practical comparisons with other decision procedures for arithmetic constitute future work. Another question concerns the scope and applicability of our embedding. In the case of arithmetic, we can precisely state when our decision procedure is applicable. For other applications, like verification of parameterized sequential systems, the situation is not so clear: although a reduction from an undecidable class of problems to a decidable one cannot always succeed, there may be useful characterizations of when it can work.

REFERENCES

[Andrews, 1986] Peter B. Andrews. *An Introduction to Mathematical Logic and Type Theory: To Truth Through Proof.* Academic Press, 1986.

[Basin and Klarlund, 1998] David Basin and Nils Klarlund. Automata based symbolic reasoning in hardware verification. *The Journal of Formal Methods in Systems Design*, 13(3):255–288, 1998.

[Boyer and Moore, 1988] R. S. Boyer and J. S. Moore. Integrating decision procedures into heuristic theorem provers: A case study with linear arithmetic. *Machine Intelligence*, 11:83–124, 1988.

[Büchi, 1960] J. R. Büchi. Weak second order arithmetic and finite automata. *Zeitschrift für mathematische Logik und Grundlagen der Mathematik*, 6:66–92, 1960.

[Camilleri *et al.*, 1986] A. J. Camilleri, M. J. C. Gordon, and T. F. Melham. Hardware verification using higher-order logic. In D. Borrione, editor, *From HDL Descriptions to Guaranteed Correct Circuit Designs.* North Holland, September 1986.

[Elgot, 1961] C. C. Elgot. Decision problems of finite automata design and related arithmetics. *Transactions of the AMS*, 98:21–52, 1961.

[Friedrich, 1998] Stefan Friedrich. Integration of a Decision Procedure for Second-Order Monadic Logic in a Higher-Order Logic Theorem Proving Environment. Master's thesis, Universität des Saarlandes, 1998.

[Gordon and Melham, 1993] Mike J. C. Gordon and Tom F. Melham. *Introduction to HOL.* Cambridge University Press, 1993.

[Henriksen et al, 1995] Jesper G. Henriksen et al. Mona: Monadic second-order logic in practice. In Ed Brinksma et al, editor, *Tools and Algorithms for the Construction and Analysis of Systems, First International Workshop, TACAS'95*, volume 1019 of *Lecture Notes in Computer Science*, pages 89–110, Heidelberg, May 1995. Springer-Verlag.

[Kurshan and Lamport, 1993] Robert Kurshan and Leslie Lamport. Verification of a multiplier: 64 bits and beyond. In Costas Courcoubetis, editor, *Proceedings of the Conference on Computer-Aided Verification*, volume 697 of *Lecture Notes in Computer Science*, pages 166–179, Heidelberg, 1993. Springer-Verlag.

[Kurshan and McMillan, 1989] Robert Kurshan and Ken McMillan. A structural induction theorem for processes. In *Proceedings of the 8th Annual ACM Symposium on Principles of Distributed Computing*, pages 239–247. ACM Press, 1989.

[Manna *et al.*, 1994] Zohar Manna, Anuchit Anuchitanukul, Nikolaj Bjorner, Anca Browne, Edward Chang, Michael Colon, Luca de Alfaro, Harish Devarajan, Henny Sipma, and Tomas Uribe. STeP: The stanford temporal prover. Technical Report CS-TR-94-1518, Stanford University, Computer Science Department, June 1994.

[Müller and Nipkow, 1995] Olaf Müller and Tobias Nipkow. Combining model checking and deduction for I/O-automata. In Ed Brinksma et al, editor, *Tools and Algorithms for the Construction and Analysis of Systems, First International Workshop, TACAS'95*, volume 1019 of *Lecture Notes in Computer Science*, pages 1–16, Heidelberg, May 1995. Springer-Verlag.

[Owre *et al.*, 1992] S. Owre, J. M. Rushby, and N. Shankar. PVS: A prototype verification system. In Deepak Kapur, editor, *Proc. of the 11th Intern. Conf. on Autom. Deduction (CADE)*, volume 607 of *Lecture Notes in Artificial Intelligence*, pages 748–752, Heidelberg, 1992. Springer-Verlag.

[Paulson, 1994] Lawrence C. Paulson. *Isabelle : a generic theorem prover; with contributions by Tobias Nipkow*, volume 828 of *Lecture Notes in Computer Science.* Springer-Verlag, Heidelberg, 1994.

[Presburger, 1929] M. Presburger. Über die Vollständigkeit eines gewissen Systems der Arithmetik ganzer Zahlen, in welchen, die Addition als einzige Operation hervortritt. In *Comptes Rendus du Premier Congrès des Mathématicienes des Pays Slaves*, pages 92–101, 395, Warsaw, 1929.

[Rajan *et al.*, 1995] S. Rajan, N. Shankar, and M.K. Srivas. An integration of model-checking with automated proof checking. In Pierre Wolper, editor,

Computer-Aided Verification, CAV '95, volume 939 of *Lecture Notes in Computer Science*, pages 84–97, Heidelberg, June 1995. Springer-Verlag.

[Regensburger, 1994] Franz Regensburger. *HOLCF: Eine konservative Erweiterung von HOL um LCF*. PhD thesis, Technische Universität München, November 1994.

[Thatcher and Wright, 1967] J. W. Thatcher and J. B. Wright. Generalized finite automata theory with an application to a decision problem of second-order logic. *Mathematical Systems Theory*, 2(1):57–81, August 1967.

[Thomas, 1990] Wolfgang Thomas. Automata on infinite objects. In Jan van Leeuwen, editor, *Handbook of Theoretical Computer Science*, volume B, chapter 4. MIT Press/Elsevier, 1990.

[Wolper and Lovinfosse, 1989] P. Wolper and V. Lovinfosse. Verifying properties of large sets of processes with network invariants. In *Proceedings of the International Workshop on Automatic Verification Methods for Finite State Systems*, volume 407 of *Lecture Notes in Computer Science*, pages 68–80, Heidelberg, June 1989. Springer-Verlag.

A Recipe for the Complexity Analysis of Non-Classical Logics

David Basin and Luca Viganò

1 INTRODUCTION

Background. In previous work [Basin *et al.*, 1998], we introduced a framework for the uniform formalization of large families of non-classical logics with Kripke semantics, such as modal, relevance and intuitionistic logics. We formalized a family of related logics in terms of two interacting parts, each a natural deduction system: a base logic of labelled formulas, and a relational theory characterizing properties of the accessibility relation of the Kripke models. While the base logic stays fixed for the whole family, we generated particular logics in the family by appropriately instantiating the relational theory. The result is a modular approach for presenting both partial and complete fragments of families of non-classical logics, that also supports uniform proofs of soundness and completeness and proof normalization.

In [Basin *et al.*, 1997], we used our framework to give decision procedures for the propositional modal logics K and T. This work exploits the connection between our normalizing labelled natural deduction systems and cut-free labelled sequent systems. We analyze the resulting sequent systems for these two logics and show how to restrict their structural rules. This, combined with an analysis of the accessibility relation of the corresponding Kripke models, yields decision procedures with bounded space requirements, i.e. $O(n \log n)$ space, for both logics.

Contribution. In this paper, we generalize the above method and provide a recipe for establishing upper-bounds on the space complexity of the decision problem for propositional non-classical logics. One of the key ideas is that our framework is restricted to those logics whose relational theories are Horn clause axiomatizable. Although this decision limits the class of logics we can formalize and hence analyze, it has the advantage of yielding normalizing natural deduction systems and corresponding cut-free sequent systems. This is a first step towards decidability. However, even in this restricted setting, not all logics formalizable are decidable (e.g. the relevance logics R and E, as shown in [Anderson *et al.*, 1992]), and there can be no recursive procedure that yields nontrivial (i.e. recursive) complexity bounds for decidable instances. Hence additional steps are required. We show how to use our framework to factor the remaining analysis into two independent subproblems: bounding the application of structural rules and bounding the complexity of reasoning about the accessibility relation. If the user can provide these bounds, our recipe yields a decision procedure with bounded space requirements. As examples, we use our framework to give PSPACE decision procedures for the modal logics K4 and S4, and for the implication-conjunction fragment $B[\rightarrow, \wedge]$ of the basic relevance logic B. The upper-bounds that we establish for these logics are new or equal to the best upper-bounds given in the literature.

Organization. In Section 2 we introduce our framework for formalizing non-classical propositional logics using labelled sequent systems. In Section 3 we give our recipe for constructing decision procedures and establishing their complexity. We present examples in Section 4, and discuss related work in Section 5.

Due to space restrictions, proofs have been omitted or considerably shortened; details can be found in [Viganò, 1997; Viganò, 1998a; Viganò, 1998b].

2 CUT-FREE LABELLED SEQUENT SYSTEMS

In this section we provide a general framework for formalizing non-classical logics as cut-free labelled sequent systems, and give examples of both modal and relevance logics. We assume the reader is familiar with non-classical logics, natural deduction and sequent systems [Anderson *et al.*, 1992; Hughes and Cresswell, 1996; Prawitz, 1965; Troelstra and Schwichtenberg, 1996].

2.1 Framework

We base our labelled sequent systems for non-classical logics on a generalized notion of non-classical (or *non-local*) connectives associated with relations. Let W be a denumerable set of *labels* $x, y, z, a, b, \ldots$, and R an $n+1$-ary relation over W. We factor our presentations into two parts, a fixed *base logic* and a varying *relational theory*. The former reasons about labelled well-formed formulas (*lwffs*), written $x{:}A$, which pair a label x and a non-classical formula A. The latter reasons about the labels and relations between them, and manipulates relational well-formed formulas (*rwffs*), written $R\,x\,x_1 \ldots x_n$. The lwff $x{:}A$ expresses that the formula A holds at world x, the rwff $R\,x\,x_1 \ldots x_n$ expresses that $(x, x_1, \ldots, x_n)$ is in the relation R.

Formulas are built from connectives, which are partitioned into two families: *local* and *non-local*. Typical local connectives are conjunction ($\wedge$) and material implication ($\supset$). Local connectives are evaluated locally, at the same world, e.g. $\models^{\mathfrak{M}} x{:}A \supset B$ iff $\models^{\mathfrak{M}} x{:}A$ implies $\models^{\mathfrak{M}} x{:}B$, where $\models^{\mathfrak{M}} x{:}A$ holds iff A is true at world x in the Kripke model $\mathfrak{M}$.

Let Γ and Δ (possibly annotated) vary over finite multisets of lwffs and rwffs, respectively; sequents in our systems have the form $\Gamma, \Delta \vdash \Gamma'$ or $\Delta \vdash R\,x\,x_1 \ldots x_n$. To reason about local connectives, we adapt traditional sequent rules [Prawitz, 1965; Troelstra and Schwichtenberg, 1996] by adding labels and parametric rwffs, e.g.

$$\frac{\Gamma, \Delta \vdash \Gamma', x{:}A_1 \quad x{:}A_2, \Gamma, \Delta \vdash \Gamma'}{x{:}A_1 \supset A_2, \Gamma, \Delta \vdash \Gamma'} \supset\!\mathrm{L} \qquad \frac{x{:}A_1, \Gamma, \Delta \vdash \Gamma', x{:}A_2}{\Gamma, \Delta \vdash \Gamma', x{:}A_1 \supset A_2} \supset\!\mathrm{R}$$

$$\frac{x{:}A, x{:}B, \Gamma, \Delta \vdash \Gamma'}{x{:}A \wedge B, \Gamma, \Delta \vdash \Gamma'} \wedge\mathrm{L} \qquad \frac{\Gamma, \Delta \vdash \Gamma', x{:}A \quad \Gamma, \Delta \vdash \Gamma', x{:}B}{\Gamma, \Delta \vdash \Gamma', x{:}A \wedge B} \wedge\mathrm{R}$$

The meaning of a non-local connective at some world is defined in terms of conditions at other worlds. A non-local connective $\mathcal{M}$ of arity n is associated with an $n+1$-ary relation R on worlds, and the truth of $x{:}\mathcal{M}A_1 \ldots A_n$ is evaluated non-locally at the worlds R-accessible from x; i.e. in terms of the truth of $x_1{:}A_1, \ldots, x_n{:}A_n$ where $R\,x\,x_1 \ldots x_n$. That is, we extend the relation $\models^{\mathfrak{M}}$ to express truth for rwffs in a Kripke model $\mathfrak{M}$ with an $n+1$-ary relation $\mathfrak{R}$: $\models^{\mathfrak{M}} R\,x\,x_1 \ldots x_n$ iff $(x, x_1, \ldots, x_n) \in \mathfrak{R}$. Then we can give the following evaluation clause for $\mathcal{M}$:

(1) $\models^{\mathfrak{M}} x{:}\mathcal{M}A_1 \ldots A_n$ iff for all $x_1, \ldots, x_n((\models^{\mathfrak{M}} R\,x\,x_1 \ldots x_n$ and $\models^{\mathfrak{M}} x_1{:}A_1$ and...and $\models^{\mathfrak{M}} x_{n-1}{:}A_{n-1})$ imply $\models^{\mathfrak{M}} x_n{:}A_n)$.

This semantics is captured by the following rules for $\mathcal{M}$, where, in $\mathcal{M}$R, $x_1, \ldots, x_n$ are distinct and do not occur in $\Gamma, \Delta \vdash \Gamma', x{:}\mathcal{M}A_1 \ldots A_n$.

$$\frac{\Delta \vdash Rxx_1 \ldots x_{n-1}x_n \quad \Gamma, \Delta \vdash \Gamma', x_1{:}A_1 \ \cdots\ \Gamma, \Delta \vdash \Gamma', x_{n-1}{:}A_{n-1} \quad x_n{:}A_n, \Gamma, \Delta \vdash \Gamma'}{x{:}\mathcal{M}A_1 \ldots A_n, \Gamma, \Delta \vdash \Gamma'} \mathcal{M}\mathrm{L}$$

$$\frac{x_1{:}A_1, \ldots, x_{n-1}{:}A_{n-1}, \Gamma, \Delta, R\,x\,x_1 \ldots x_{n-1}\,x_n \vdash \Gamma', x_n{:}A_n}{\Gamma, \Delta \vdash \Gamma', x{:}\mathcal{M}A_1 \ldots A_n} \mathcal{M}\mathrm{R}$$

The evaluation clauses of modal $\Box$ and relevant implication $\rightarrow$ are special cases of (1), in that $\Box$ is interpreted in terms of a binary (accessibility) relation $R^{\Box}$ [Hughes and Cresswell, 1996], and $\rightarrow$ is interpreted using a ternary (compossibility) relation $R^{\rightarrow}$ [Anderson *et al.*, 1992]. Thus, as instances of $\mathcal{M}$L and $\mathcal{M}$R, we have the rules

$$\frac{\Delta \vdash x\,R^{\Box}\,y \quad y{:}A, \Gamma, \Delta \vdash \Gamma'}{x{:}\Box A, \Gamma, \Delta \vdash \Gamma'}\ \Box\mathrm{L} \qquad \frac{\Gamma, \Delta, x\,R^{\Box}\,y \vdash \Gamma', y{:}A}{\Gamma, \Delta \vdash \Gamma', x{:}\Box A}\ \Box\mathrm{R}$$

$$\frac{\Delta \vdash R^{\rightarrow}\,x\,y\,z \quad \Gamma, \Delta \vdash \Gamma', y{:}A \quad z{:}B, \Gamma, \Delta \vdash \Gamma'}{x{:}A \rightarrow B, \Gamma, \Delta \vdash \Gamma'}\ {\rightarrow}\mathrm{L}$$

$$\frac{y{:}A, \Gamma, \Delta, R^{\rightarrow}\,x\,y\,z \vdash \Gamma', z{:}B}{\Gamma, \Delta \vdash \Gamma', x{:}A \rightarrow B}\ {\rightarrow}\mathrm{R}$$

where, for simplicity, we write $R^{\Box}$ in infix notation. $\Box$R has the side condition that the label y does not occur in $\Gamma, \Delta \vdash \Gamma', x{:}\Box A$. In $\rightarrow$R, the labels y and z are distinct and do not occur in $\Gamma, \Delta \vdash \Gamma', x{:}A \rightarrow B$.

The rules for local and non-local connectives are called *logical rules* in that they define the behavior of the logical connectives. To define sequent systems for non-classical logics, we also need the following *axioms* (AXl, $\bot$L, AXr), and *structural rules* (weakening and contraction) for lwffs and rwffs.

$$\frac{}{x{:}A \vdash x{:}A}\ \mathrm{AXl} \qquad \frac{}{y{:}\bot \vdash x{:}A}\ \bot\mathrm{L} \qquad \frac{}{R\,x\,x_1 \ldots x_n \vdash R\,x\,x_1 \ldots x_n}\ \mathrm{AXr}$$

$$\frac{\Gamma, \Delta \vdash \Gamma'}{x{:}A, \Gamma, \Delta \vdash \Gamma'}\ \mathrm{WlL} \qquad \frac{\Gamma, \Delta \vdash \Gamma'}{\Gamma, \Delta \vdash \Gamma', x{:}A}\ \mathrm{WlR}$$

$$\frac{x{:}A, x{:}A, \Gamma, \Delta \vdash \Gamma'}{x{:}A, \Gamma, \Delta \vdash \Gamma'}\ \mathrm{ClL} \qquad \frac{\Gamma, \Delta \vdash \Gamma', x{:}A, x{:}A}{\Gamma, \Delta \vdash \Gamma', x{:}A}\ \mathrm{ClR}$$

$$\frac{\Gamma, \Delta \vdash \Gamma'}{\Gamma, \Delta, R\,x\,x_1 \ldots x_n \vdash \Gamma'}\ \mathrm{WrL} \qquad \frac{\Delta, R\,a\,a_1 \ldots a_n, R\,a\,a_1 \ldots a_n \vdash R\,x\,x_1 \ldots x_n}{\Delta, R\,a\,a_1 \ldots a_n \vdash R\,x\,x_1 \ldots x_n}\ \mathrm{CrL}$$

Note that for rwffs we do not need rules for weakening and contraction on the right, since our relational sequents $\Delta \vdash R\,x\,x_1 \ldots x_n$ are single-conclusioned.

Other connectives can be defined in the usual manner, e.g. $\neg A$ as $A \supset \bot$ and $\Diamond A$ as $\neg\Box\neg A$, and their corresponding rules derived.[1]

2.2 Examples

We start by defining the sequent system S(K) for the base modal logic K.

[1]We here consider only 'universal' non-local connectives such as $\Box$ and $\rightarrow$. In terms of these, we can define 'existential' non-local connectives such as $\Diamond$, which have an evaluation clause in which the metalevel quantification is existential and the body of the clause is a conjunction. Our account here can also be extended to treat 'non-local' negation, see [Basin *et al.*, 1998].

Definition 2.1 The labelled sequent system S(K) consists of the axioms AXl, ⊥L and AXr, the logical rules ⊃L, ⊃R, □L and □R, and the structural rules WlL, WlR, ClL, ClR, WrL and CrL, where we instantiate R to $R^{\Box}$.

Given a base logic, a family of related logics is generated by varying the behavior of the relation. In [Basin *et al.*, 1998] we show how to exploit this for logics where the behavior of the relation is Horn clause axiomatizable[2]; in such cases we can uniformly prove soundness and completeness with respect to the corresponding Kripke semantics, and normalization results for the resulting systems. As examples, we present extensions of S(K) by giving *relational rules* that formalize properties of $R^{\Box}$. For example, we formalize reflexivity and transitivity of $R^{\Box}$ in terms of the rules

$$\frac{}{\Delta \vdash x\,R^{\Box}\,x}\;refl \qquad \text{and} \qquad \frac{\Delta \vdash x\,R^{\Box}\,y \quad \Delta \vdash y\,R^{\Box}\,z}{\Delta \vdash x\,R^{\Box}\,z}\;trans$$

Definition 2.2 The labelled sequent systems S(T) and S(K4) are obtained by extending S(K) with *refl* and *trans*, respectively. S(S4) is the extension of S(K) with both relational rules.

Relevance logics are formalized analogously, i.e. by extending a fixed base system with relational rules formalizing properties of the compossibility relation $R^{\rightarrow}$. However, relevance logics require slightly more machinery than modal logics, for reasons we shall now explain by means of an example: the implication-conjunction fragment $B[\rightarrow, \wedge]$ of the base relevance logic B.

To begin with, in $B[\rightarrow, \wedge]$, and thus in B, the relation $R^{\rightarrow}$ is postulated to satisfy the *identity property*, $(0, x, x) \in R^{\rightarrow}$ for all x, where 0 denotes the *actual world*. We formalize the identity of $R^{\rightarrow}$ with the rule

$$\frac{}{\Delta \vdash R^{\rightarrow}\,0\,x\,x}\;iden$$

Thus, the relational theory of $S(B[\rightarrow, \wedge])$ is not empty, unlike that of S(K).

[2]Most, but not all, non-classical logics with first-order axiomatizable accessibility relations can be formalized using Horn clauses, e.g. the modal logics in the Geach hierarchy including K, T, K4, S4, S5, etc. By the restriction to Horn clauses we gain proof normalization (proofs can be transformed into a normal form similar to that defined by Prawitz [1965]), while retaining completeness; in the full first-order case, on the other hand, we lose either normalization or completeness [Basin *et al.*, 1998]. In our sequent systems in this paper, Horn clause axioms are formalized without logical connectives by using rules; for example, $(R_1 \wedge \ldots \wedge R_n) \supset R$ is formalized as

$$\frac{\Delta \vdash R_1 \quad \ldots \quad \Delta \vdash R_n}{\Delta \vdash R}$$

In addition, many non-classical logics, e.g. intuitionistic and relevance logics, require truth to be monotonic. To express this, we define a partial order $\sqsubseteq$ on worlds, where, for intuitionistic logic, $\sqsubseteq$ coincides with the accessibility relation, while for relevance logics it can be defined in terms of $R^{\rightarrow}$ and the actual world 0, i.e. $x \sqsubseteq y$ iff $R^{\rightarrow}\, 0\, x\, y$. (Modal logics are also monotonic, but in a trivial way where $\sqsubseteq$ reduces to equality.) Then we can postulate the *atomic monotony property*, i.e. for any worlds x and y, and for any propositional variable p, if $\models^{\mathfrak{M}} x \sqsubseteq y$ and $\models^{\mathfrak{M}} x{:}p$, then $\models^{\mathfrak{M}} y{:}p$. For the most common non-classical logics, such as intuitionistic and relevance logics, this property immediately generalizes to arbitrary formulas [Basin *et al.*, 1998]; that is, by induction on the structure of A it follows that

(2) $$\text{if } \models^{\mathfrak{M}} x \sqsubseteq y \text{ and } \models^{\mathfrak{M}} x{:}A, \text{ then } \models^{\mathfrak{M}} y{:}A\,.$$

Monotony is postulated also for rwffs, and we formalize the monotony properties in terms of the following rules, where $j < n$ in the schematic relational monotony rule *monr*(j).

$$\frac{\Delta \vdash x \sqsubseteq y \quad \Gamma, \Delta \vdash \Gamma', x{:}A}{\Gamma, \Delta \vdash \Gamma', y{:}A}\ monl \qquad \frac{\Delta \vdash y_n \sqsubseteq x \quad \Delta \vdash R\, y_0 \ldots y_{n-1}\, y_n}{\Delta \vdash R\, y_0 \ldots y_{n-1}\, x}\ monr(n)$$

$$\frac{\Delta \vdash x \sqsubseteq y_j \quad \Delta \vdash R\, y_0 \ldots y_{j-1}\, y_j\, y_{j+1} \ldots y_n}{\Delta \vdash R\, y_0 \ldots y_{j-1}\, x\, y_{j+1} \ldots y_n}\ monr(j)$$

S(B[$\rightarrow, \wedge$]) contains these rules, i.e. we define:

Definition 2.3 The labelled sequent system S(B[$\rightarrow, \wedge$]) consists of AXl, AXr, $\wedge$L, $\wedge$R, $\rightarrow$L, $\rightarrow$R, WlL, WlR, ClL, ClR, WrL, CrL, *monl*, *monr*(0), *monr*(1), *monr*(2) and *iden*, where R is $R^{\rightarrow}$ and $x \sqsubseteq y$ is $R^{\rightarrow}\, 0\, x\, y$.

We now give two example proofs in our labelled sequent systems. The following is a proof of $\vdash 0{:}(A \wedge B) \rightarrow B$ in S(B[$\rightarrow, \wedge$]).

$$\dfrac{\dfrac{\dfrac{\dfrac{}{R^{\rightarrow}\, 0\, x\, y \vdash R^{\rightarrow}\, 0\, x\, y}\ \text{AXr} \quad \dfrac{\dfrac{\dfrac{}{x{:}B \vdash x{:}B}\ \text{AXl}}{x{:}A, x{:}B \vdash x{:}B}\ \text{WlL}}{x{:}A, x{:}B, R^{\rightarrow}\, 0\, x\, y \vdash x{:}B}\ \text{WrL}}{x{:}A, x{:}B, R^{\rightarrow}\, 0\, x\, y \vdash y{:}B}\ monl}{x{:}A \wedge B, R^{\rightarrow}\, 0\, x\, y \vdash y{:}B}\ \wedge\text{L}}{\vdash 0{:}(A \wedge B) \rightarrow B}\ \rightarrow\text{R}$$

To exhibit the use of relational rules and contraction, we prove $\vdash x{:}\Box\neg\Box B \supset \Box\neg\Box\Box B$ in S(K4), i.e.

(3)

$$
\dfrac{\dfrac{}{x\,R^{\Box}\,y \vdash x\,R^{\Box}\,y}\,\text{AXr} \qquad \dfrac{\dfrac{\dfrac{\Pi \atop x\,R^{\Box}\,y, y\,R^{\Box}\,z \vdash x\,R^{\Box}\,z \qquad \dfrac{\dfrac{\dfrac{\dfrac{}{y\,R^{\Box}\,z \vdash y\,R^{\Box}\,z}\,\text{AXr}}{x\,R^{\Box}\,y, y\,R^{\Box}\,z \vdash y\,R^{\Box}\,z}\,\text{WrL} \qquad \Pi'}{y{:}\Box\Box B, x\,R^{\Box}\,y, y\,R^{\Box}\,z \vdash z{:}B, z{:}\Box B}\,\Box\text{L}}{z{:}\neg\Box B, y{:}\Box\Box B, x\,R^{\Box}\,y, y\,R^{\Box}\,z \vdash z{:}B}\,\neg\text{L}}{x{:}\Box\neg\Box B, y{:}\Box\Box B, x\,R^{\Box}\,y, y\,R^{\Box}\,z \vdash z{:}B}\,\Box\text{L}}{x{:}\Box\neg\Box B, y{:}\Box\Box B, x\,R^{\Box}\,y \vdash y{:}\Box B}\,\Box\text{R}}{y{:}\neg\Box B, x{:}\Box\neg\Box B, y{:}\Box\Box B, x\,R^{\Box}\,y \vdash}\,\neg\text{L}}{\dfrac{\dfrac{\dfrac{\dfrac{x{:}\Box\neg\Box B, x{:}\Box\neg\Box B, y{:}\Box\Box B, x\,R^{\Box}\,y \vdash}{x{:}\Box\neg\Box B, y{:}\Box\Box B, x\,R^{\Box}\,y \vdash}\,\text{ClL}}{x{:}\Box\neg\Box B, x\,R^{\Box}\,y \vdash y{:}\neg\Box\Box B}\,\neg\text{R}}{x{:}\Box\neg\Box B \vdash x{:}\Box\neg\Box\Box B}\,\Box\text{R}}{\vdash x{:}\Box\neg\Box B \supset \Box\neg\Box\Box B}\,\supset\text{R}}\,\Box\text{L}
$$

where, for brevity, we use the derived rules for $\neg$, and where Π stands for

$$
\dfrac{\dfrac{\dfrac{}{x\,R^{\Box}\,y \vdash x\,R^{\Box}\,y}\,\text{AXr}}{x\,R^{\Box}\,y, y\,R^{\Box}\,z \vdash x\,R^{\Box}\,y}\,\text{WrL} \qquad \dfrac{\dfrac{}{y\,R^{\Box}\,z \vdash y\,R^{\Box}\,z}\,\text{AXr}}{x\,R^{\Box}\,y, y\,R^{\Box}\,z \vdash y\,R^{\Box}\,z}\,\text{WrL}}{x\,R^{\Box}\,y, y\,R^{\Box}\,z \vdash x\,R^{\Box}\,z}\,\textit{trans}
$$

and Π' for

$$
\dfrac{\dfrac{\dfrac{\dfrac{}{z{:}\Box B \vdash z{:}\Box B}\,\text{AXl}}{z{:}\Box B, y\,R^{\Box}\,z \vdash z{:}\Box B}\,\text{WrL}}{z{:}\Box B, x\,R^{\Box}\,y, y\,R^{\Box}\,z \vdash z{:}\Box B}\,\text{WrL}}{z{:}\Box B, x\,R^{\Box}\,y, y\,R^{\Box}\,z \vdash z{:}B, z{:}\Box B}\,\text{WlR}
$$

Note that the contraction in (3) is indispensable, i.e. $\vdash x{:}\Box\neg\Box B \supset \Box\neg\Box\Box B$ cannot be proved in S(K4) without (at least) one application of ClL; cf. § 4.

2.3 Metatheoretic Properties

Normalizing labelled natural deduction systems and labelled sequent systems without cut are intertranslatable; this follows by a straightforward adaptation and extension of the proof given by Prawitz [1965, App. A], and the reader is referred to [Basin *et al.*, 1997; Viganò, 1997] for details. From this and previous results in [Basin *et al.*, 1998], it follows that:

Theorem 2.4 *Our cut-free labelled sequent systems are sound and complete with respect to the corresponding Kripke semantics.*

3 A RECIPE FOR COMPLEXITY ANALYSIS

Our recipe for the complexity analysis of non-classical logics is based on bounding which formulas can appear in proofs and with how many repetitions. We begin by observing that the absence of cut implies that our sequent systems have a subformula property: in any proof of a sequent $\Gamma, \Delta \vdash \Gamma'$, only labelled subformulas of Γ and Γ' occur. Let the *grade* of an lwff $x{:}A$ be the number of connectives in A. If we apply the rules *backwards* to build proofs 'top-down', starting with the end-sequent and working towards the axioms, then we can partition rules into two types:

(i) rules that 'simplify' sequents, in the sense that we obtain lwffs of smaller grade, or we just weaken or delete formulas;
(ii) rules that do not simplify sequents.

Examples of (ii) are the contraction rules and *monl*, which duplicate or repeat existing formulas, and relational rules such as *trans* and the relational monotony rules *monr*(i). All the other rules introduced above, in particular the logical rules, are of type (i).

Since the subformula property bounds the number of different formulas arising in proofs, bounding proof search reduces to the problem of bounding applications of the rules of type (ii). That is, we must bound how often lwffs appear and applications of non-simplifying relational rules. Our recipe is based on the user providing both these bounds and in the following we give guidelines for (and examples of) establishing them. To begin with though, we simplify matters by showing that for all our sequent systems certain kinds of applications of rules of type (ii) can always be eliminated or bounded.

3.1 Logic Independent Bounds

We begin by observing that by induction on the structure of lwffs we have:

Lemma 3.1 *We can restrict applications of monl to atomic lwffs, i.e. to lwffs $x{:}A$ where A is atomic.*

As shown below, this lemma often allows us to bound applications of *monl* provided that we can bound relational reasoning.[3] Independent of relational reasoning, we have the following three results. The first result is:

Lemma 3.2 *The rule* CrL *is always eliminable.*

[3]The lemma holds for the most common non-classical logics, i.e. those for which (2) holds, but may fail for some 'artificial' logics in which A is required to possess a particular form for (2) to hold in the first place; examples of such logics are discussed in [Basin *et al.*, 1998].

This follows because the principal rwff of a relational rule is introduced in the succedent of the sequent, so that contracted rwffs can be introduced in the antecedent of a sequent only by applications of WrL. Thus, we can delete both CrL and the corresponding WrL. For example, we transform

$$\cfrac{\begin{array}{c}\cfrac{\begin{array}{c}\Pi_1\\ \Delta_1 \vdash R\,x\,x_1\ldots x_n\end{array}}{\Delta_1, R\,a\,a_1\ldots a_n \vdash R\,x\,x_1\ldots x_n}\,\text{WrL}\\ \Pi_2\\ \Delta_2, R\,a\,a_1\ldots a_n, R\,a\,a_1\ldots a_n \vdash R\,x\,x_1\ldots x_n\end{array}}{\Delta_2, R\,a\,a_1\ldots a_n \vdash R\,x\,x_1\ldots x_n}\,\text{CrL}$$

to

$$\begin{array}{c}\Pi_1\\ \Delta_1 \vdash R\,x\,x_1\ldots x_n\\ \Pi_2^{\dagger}\\ \Delta_2, R\,a\,a_1\ldots a_n \vdash R\,x\,x_1\ldots x_n\end{array}$$

where we use '†' to denote that in the derivation $\Pi_i^{\dagger}$ we apply the same sequence of rules applied in the derivation Π_i; in other words, Π_i and $\Pi_i^{\dagger}$ differ only in their parametric formulas.

Let us call an *lwff-rule* a rule that has an lwff as its principal formula. Suppose that a sequent S results from sequents $S_1, \ldots, S_n$ by first applying the lwff-rule ρ_1 and then applying the lwff-rule ρ_2, where (i) each of the premises of ρ_2 results from an application of ρ_1, (ii) each application of ρ_1 introduces or contracts the same lwff, and (iii) the principal formula of ρ_1 is parametric in the application of ρ_2. We then say that ρ_1 *permutes over* ρ_2 if the original inference can be replaced by one in which the sequent S is derived from $S_1, \ldots, S_n$ by applying first ρ_2 and then ρ_1. By inspection of the rules, we have the second result:

Lemma 3.3 *Every lwff-rule permutes over any other lwff-rule, with the exception of* $\mathcal{M}$L, *which permutes over every lwff-rule other than* $\mathcal{M}$R.

The reason that □L does not, in general, permute over □R is that both rules may have the same active rwff $x\,R^{\Box}\,y$, e.g.

$$\cfrac{\cfrac{\Delta, x\,R^{\Box}\,y \vdash x\,R^{\Box}\,y \quad y{:}A, \Gamma, \Delta, x\,R^{\Box}\,y \vdash \Gamma', y{:}B}{x{:}\Box A, \Gamma, \Delta, x\,R^{\Box}\,y \vdash \Gamma', y{:}B}\,\Box\text{L}}{x{:}\Box A, \Gamma, \Delta \vdash \Gamma', x{:}\Box B}\,\Box\text{R}$$

Similarly, $\rightarrow$ L does not permute over $\rightarrow$ R when both rules have the same active rwff $R\,x\,x_1\ldots x_n$.

By adapting and extending the proof for classical propositional logic given by Zeman [1973], and by Lemma 3.2 and Lemma 3.3, we then have the third result:

Lemma 3.4 *Let* Π *be a proof of a theorem* $\vdash d{:}D$ *in one of our sequent systems. Then we can transform* Π *so that it does not contain contractions, except for contractions of lwffs of the form* $x{:}\mathcal{M}A_1 \ldots A_n$.

In other words, we can always eliminate contractions of lwffs whose principal connective is local, and need only retain contractions of lwffs with non-local connectives. This follows by three nested inductions: on the number of contractions in Π, on the grade of the contracted lwff, and on the *rank* of the contraction (the largest number of steps immediately preceding the contraction and containing at least one of the two contracted lwffs); see [Viganò, 1997; Viganò, 1998a; Viganò, 1998b] for details. Note that this lemma cannot generally be extended to non-local connectives. For example, we cannot always reduce the rank of a left contraction of $x{:}\Box A$ by permuting an application of $\Box$L over an application of $\Box$R following it. In the case of ClR, there are modal and relevant theorems whose proofs require a sequent containing two copies of $x{:}\Box A$ or $x{:}A \to B$ in its succedent.[4]

Thus, given a particular base logic, we only need to analyze contractions of $x{:}\mathcal{M}A_1 \ldots A_n$. Moreover, the analysis is modular in the sense that extensions with relational rules only require us to consider the additional cases that arise for $\mathcal{M}$L. (This is because relational rules only affect applications of $\mathcal{M}$L, for which they provide new possible active rwffs.)

3.2 The Recipe

We have thus reduced the problems with rules of type (ii) to the task of bounding applications of contractions of $x{:}\mathcal{M}A_1 \ldots A_n$, *monl* with atomic lwffs, and relational rules such as *trans* and the monotony rules. This is a logic dependent task: for a cut-free labelled sequent system S($\mathcal{L}$), formalizing an arbitrary non-classical logic $\mathcal{L}$, we may not always succeed in this task. Moreover, the bounds on the rules are not independent since the bounds on contractions and *monl* explicitly depend on bounding relational reasoning.

Guidelines and examples of this task are provided in § 3.4, § 3.5 and § 4. To structure our account, we now anticipate some details and give our full recipe for the logic $\mathcal{L}$ in Figure 1; note that we write the three recipe steps in italics, while commentary is written in Times Roman type face. It will also simplify matters to postpone discussion of step *(2)* until after discussion of step *(3)*; we proceed by giving our general decision procedure and analyzing its space requirements.

---

[4]For an example of the need of ClR, try to prove $\vdash x{:}\Box\Box\neg B \supset (\Box B \supset \Box A)$ in S(K5), the system obtained by extending S(K) with the relational rule

$$\frac{\Delta \vdash x R^{\Box} y \quad \Delta \vdash x R^{\Box} z}{\Delta \vdash y R^{\Box} z}$$

(1) Give a cut-free labelled sequent system for S($\mathcal{L}$), based on a generalized notion of non-classical connectives associated with relations.

(i) By Theorem 2.4, S($\mathcal{L}$) is sound and complete.

(ii) S($\mathcal{L}$) satisfies a subformula property. This tells us which formulas may appear in proofs, and based on such knowledge rules are partitioned into two types: (i) simplifying rules, and (ii) non-simplifying rules.

(iii) By our logic independent bounds in § 3.1, the rules of type (ii) are: contractions of $x{:}\mathcal{M}A_1 \ldots A_n$, applications of *monl* with atomic lwffs, and relational rules such as *trans* and *monr*(*i*).

(2) Provide bounds for the remaining non-simplifying rules of S($\mathcal{L}$), if any, by following the guidelines and examples in § 3.4, § 3.5 and § 4, possibly bringing in 'relational oracles' to decide $\Delta \vdash R\,x\,x_1 \ldots x_n$.

(3) Compute the space requirements of the decision procedure of § 3.3, based on the results of step *(2)* and the guidelines of § 3.4 and § 3.5.

Figure 1: The recipe for an arbitrary non-classical logic $\mathcal{L}$

3.3 Decision Procedure

Given a non-classical logic $\mathcal{L}$, we apply the rules of the associated system S($\mathcal{L}$) backwards to build a proof 'top-down' starting with the end-sequent and working towards the axioms. Assume that step *(2)* in the recipe has been completed successfully, i.e. we have established bounds on the non-simplifying rules of S($\mathcal{L}$), so that no infinite branches arise and proof search terminates. To bound space requirements we begin by distinguishing between two kinds of branching that arise in the search space for proofs.

Conjunctive branching: applying rules with multiple premises builds a branching tree, where all branches must be proven.

Disjunctive branching: more than one rule may be applicable and a given rule may be applicable in different ways.

Conjunctive branching, caused by rules like $\supset$L, $\wedge$R and $\mathcal{M}$L, leads to proofs that, if stored in their entirety, are exponential in size. Disjunctive branching arises when more than one rule may be applied to a sequent, or when a single rule can be applied to more than one formula, e.g. weakening or contraction.

We now adapt and extend standard techniques for developing search algorithms with good space bounds (e.g. [Hudelmaier, 1993; Basin *et al.*, 1997]). Rather than storing entire proofs, we store a sequent and a stack that maintains information sufficient to reconstruct branching points. Each stack entry is a triple: the name of the rule applied, the principal formula of the sequent, and an index. This allows us to reconstruct the sequent associated with a branching point by replaying the stack entries to that point. The index records sufficient information such that on return to the branching point we can generate the remaining branches. For example, for a conjunctive branching point associated with $\supset$L, the index is a bit indicating the first or second premise; for a disjunctive branching point associated with weakening, the index is a pointer indicating which formula is weakened.

A proof begins with the end-sequent $S = \vdash d{:}D$ and the empty stack. Each rule application generates a new sequent and appropriately extends the stack. (This is a disjunctive branching point; we assume rules are ordered and we apply them in order.) If the generated sequent is an axiom and the stack contains no conjunctive branching points that still need to be explored, then S is provable. Otherwise we pop entries off the stack until we find a conjunctive branching point that must be further explored and then generate the next branch (first incrementing the index on the stack to record this). Alternatively, if we arrive at a sequent that is not an axiom and no rule applies, then we pop stack entries and continue at the first available disjunctive branching point with the choices that remain (e.g. apply a different weakening, determined by the index, or apply the next available rule). If no such branching point remains, then S is not provable.

Observe that for any given sequent $S = \vdash d{:}D$ and sequent system S($\mathcal{L}$), there are only finitely many conjunctive and disjunctive branching points. Provided there are no branches of infinite length, i.e. that step *(2)* in the recipe establishes bounds on applications of the non-simplifying rules of S($\mathcal{L}$), then application of the above procedure will terminate and find a proof whenever one exists. We now analyze the space requirements for this procedure.

3.4 Space Complexity Bounds

The above procedure trades time for space: it is not time-efficient but it minimizes space usage by storing the minimum information required to check provability. The actual space required is a function of three things: the length $\mathtt{l}$ of the stack, the size $\mathtt{e}$ of a stack entry, and the size $\mathtt{s}$ required to store any single sequent that could arise in the proof. The overall space requirement is then $O((\mathtt{l}\,\mathtt{e}) + \mathtt{s})$. A useful trick to reduce both $\mathtt{e}$ and $\mathtt{s}$ is to observe that if we save a copy of the end-sequent S, then we can represent any subformula by an index into S, which requires only $\log n$ bits (where $n = |S|$, the number of symbols in the string representation of S).

Let us give a very simple example of this: the cut-free labelled sequent system S(PL) for classical propositional logic PL, consisting of the axioms and rules AXl, $\bot$L, $\supset$L, $\supset$R, WlL and WlR, operating over sequents that contain only lwffs (i.e. Δ is always empty). By Lemma 3.4, in S(PL) we can eliminate the contraction rules; thus we can measure the S(PL)-size of an lwff $x{:}A$ to be $|A|$, and the S(PL)-size of a sequent to be the sum of the sizes of its lwffs; each rule application is measure decreasing. (The S($\mathcal{L}$)-size of an lwff or sequent α is the size of α in S($\mathcal{L}$).) Hence, for an end-sequent S, $\mathtt{l}$ is bounded by the S(PL)-size of S, which is in turn bounded by $n = |S|$. Using the above trick, we see that $\mathtt{e}$ is $\log n$ plus a few extra bits, and $\mathtt{s} = O(n \log n)$ since $O(n)$ possible subformulas can occur in any sequent (labels are irrelevant in S(PL)), each requiring an index of size $\log n$. Overall we have an upper-bound on the space complexity of $O(n \log n)$. Note that this is not optimal, cf. the discussion in § 4.

The same analysis can be applied to establish the space complexity of the decision procedure for an arbitrary non-classical logic $\mathcal{L}$, but things are more complicated since in S($\mathcal{L}$) we must also deal with labels, rwffs and non-simplifying rules. We have assumed that applications of these rules have been bounded in step *(2)*. This means that every application of a rule of S($\mathcal{L}$), other than *monl* or non-simplifying relational rules, decreases some measure $\mathtt{m}$ on sequents, defined in terms of the S($\mathcal{L}$)-size of S.

Let us consider this in more detail. As observed above, contraction cannot always be eliminated. However, in many logics its application can be bounded, i.e. there exists a bounding function $f(S)$ that, given a sequent $S = \vdash d{:}D$, tells us the maximum number of contractions of $x{:}\mathcal{M}A_1 \dots A_n$ that are needed to prove S in S($\mathcal{L}$). We incorporate this into our proof procedure by explicitly annotating lwffs or sequents with *contraction indices* and adopting contraction rules to account for these (see § 3.5). This additional annotation will not, however, change our overall space complexity bounds.

In its most general form (see the examples below), $f(S)$ explicitly depends on the relational rules of S($\mathcal{L}$). For relational reasoning, it is helpful in practice (although sometimes not needed, e.g. for S(K), S(T) and S(B[$\rightarrow, \wedge$]) in § 4) to allow the user to bring in non-proof based algorithms with known space-requirements. For example, if we are analyzing the complexity of a modal logic like K4 where the accessibility relation $R^{\Box}$ is transitive, then determining the status of $\Delta \vdash x\, R^{\Box}\, y$ is equivalent to determining whether $x\, R^{\Box}\, y$ is in the transitive closure of Δ; this can be determined by depth-first search in linear time (and hence space). Our recipe allows the user to provide *relational oracles*, which can be used in the above proof procedure to decide relational queries, including those arising at applications of relational monotony rules *monr*(i).

An application of *monl* is also a non-simplifying rule application, and it constitutes a case of disjunctive branching. But Lemma 3.1 tells us that *monl* need only be applied to atomic formulas, so that it can be applied at most once for each world in a branch, and application is again bounded by the number of worlds since $\sqsubseteq$ is a partial order. This means that for S($\mathcal{L}$) the length $\mathtt{l}$ of the stack is bounded by the measure $\mathtt{m}$, and the overall space requirement is $O((\mathtt{m}\,\mathtt{e}) + \mathtt{s} + \mathtt{r})$ where $\mathtt{r}$ is the space requirement of the oracle needed to decide any relational query that arises in the proof of S. To conclude the analysis, we must therefore show how to compute $\mathtt{e}$ and $\mathtt{s}$ for S($\mathcal{L}$).

Unlike for S(PL), for S($\mathcal{L}$) we must take labels and rwffs seriously. However, a bound on the length of proofs in S($\mathcal{L}$) can serve as an upper-bound on the number of rwffs generated: rwffs can only be generated by applications of $\mathcal{M}$L, $\mathcal{M}$R or relational rules, and the number of these applications is linear in the length $\mathtt{l}$ of the proof itself. Not all of these rules, on the other hand, generate new labels: labels can only be generated by applications of $\mathcal{M}$R, and thus each S($\mathcal{L}$)-proof of S contains at most $O(\mathtt{m})$ labels. Moreover, given our bounds on contractions and *monl*, the number of lwffs in a generated sequent is also bounded by $O(\mathtt{m})$, like that of rwffs. As for S(PL), we can represent the $O(\mathtt{m})$ labels using $O(\log \mathtt{m})$ bits. Similarly, we can represent a subformula by an index into the end-sequent S, which requires $O(\log \mathtt{m})$ bits. Thus, using the above trick again, we see that $\mathtt{e}$ is $\log \mathtt{m}$ plus a few extra bits and $\mathtt{s} = O(\mathtt{m} \log \mathtt{m})$. Summarizing, our procedure for deciding $S = \vdash d{:}D$ in S($\mathcal{L}$) has an overall space requirement of $O(\mathtt{m} \log \mathtt{m} + \mathtt{r})$.

If the user provides the bounds of step *(2)*, then she will be able to compute $\mathtt{m}$ and $\mathtt{r}$, and thus the overall space required by our decision procedure. Note that the space requirement of S(PL) is an instance of this. We will give more examples in § 4, but first we give some guidelines for establishing the logic dependent bounds on non-simplifying rules.

3.5 Logic Dependent Bounds

Above we argued how bounds on relational reasoning, provided by an oracle or simply by applying relational rules, allow us to bound applications of *monl* and contractions. We now explain how to modify contractions to account for the bounding function $f(S)$ and define the measure $\mathtt{m}$ based on it.

We can annotate sequents or lwffs with contraction indices that track how many contractions are allowed, and use rules that propagate the indices appropriately. For this task, we consider two different sets of contraction rules. Our first set contains the following two rules, and covers the case when $f(S)$ provides contraction indices annotating sequents.

$$\frac{x{:}\mathcal{M}A_1 \ldots A_n, x{:}\mathcal{M}A_1 \ldots A_n, \Gamma, \Delta \vdash^{p-1} \Gamma'}{x{:}\mathcal{M}A_1 \ldots A_n, \Gamma, \Delta \vdash^{p} \Gamma'}\ \mathrm{ClL}^p$$

$$\frac{\Gamma, \Delta \vdash^{p-1} \Gamma', x{:}\mathcal{M}A_1 \ldots A_n, x{:}\mathcal{M}A_1 \ldots A_n}{\Gamma, \Delta \vdash^{p} \Gamma', x{:}\mathcal{M}A_1 \ldots A_n}\ \mathrm{ClR}^p$$

Both rules have the side condition that $p > 0$. We have annotated sequents with a *sequent contraction index* p, set to $p = f(S)$ at the start of the backwards proof, which states how many contractions we may perform in each branch of the proof from this point (i.e. sequent) upwards. It is decremented at every contraction, and is imported in the premises of branching rules, e.g.

$$\frac{\Gamma, \Delta \vdash^{p} \Gamma', x{:}B_1 \quad \Gamma, \Delta \vdash^{p} \Gamma', x{:}B_2}{\Gamma, \Delta \vdash^{p} \Gamma', x{:}B_1 \wedge B_2}\ \wedge\mathrm{R} \qquad \frac{\Delta \vdash x R^{\Box} y \quad y{:}B, \Gamma, \Delta \vdash^{p} \Gamma'}{x{:}\Box B, \Gamma, \Delta \vdash^{p} \Gamma'}\ \Box\mathrm{L}$$

Note that we do not import p into the left premise $\Delta \vdash x R^{\Box} y$ of $\Box$L since we can eliminate contractions of rwffs by Lemma 3.2.

Given these bounded contraction rules, we define the S($\mathcal{L}$)-size of an lwff $x{:}A$ to be $2 \times |A|$, and we define the S($\mathcal{L}$)-size of a sequent S with contraction index p lexicographically as the pair (p, Σ), where Σ is the sum of the S($\mathcal{L}$)-sizes of the lwffs of S plus the number of rwffs in S. With the exception of *monl* and non-simplifying relational rules, every rule, including the contraction rules ClL^p and ClR^p, reduces this (lexicographically ordered) measure. Thus the S($\mathcal{L}$)-size of S is the measure $\mathtt{m}$ required by our recipe.

There are, however, sequent systems, e.g. S(T) in § 4, in which we (or 'the user') can give an alternative measure that can yield better complexity bounds. This happens when the analysis of contraction tells us that we can replace the sequent contraction index with indices on individual lwffs. In other words, instead of controlling contraction over the whole sequent, we shift the control to its lwffs by directly indexing them; this occurs when the analysis yields, instead of $f(S)$, a function g on lwffs telling us how many times we are allowed to contract each of them (in each branch of the proof). In this case we adopt the following set of contraction rules, where we annotate lwffs with an *lwff contraction index* q, set to $q = g(x{:}\mathcal{M}A_1 \ldots A_n)$ at the first appearance of $x{:}\mathcal{M}A_1 \ldots A_n$ in the backwards proof.

$$\frac{(x{:}\mathcal{M}A_1 \ldots A_n)^0, (x{:}\mathcal{M}A_1 \ldots A_n)^{q-1}, \Gamma, \Delta \vdash \Gamma'}{(x{:}\mathcal{M}A_1 \ldots A_n)^q, \Gamma, \Delta \vdash \Gamma'}\ \mathrm{ClL}^q$$

$$\frac{\Gamma, \Delta \vdash \Gamma', (x{:}\mathcal{M}A_1 \ldots A_n)^0, (x{:}\mathcal{M}A_1 \ldots A_n)^{q-1}}{\Gamma, \Delta \vdash \Gamma', (x{:}\mathcal{M}A_1 \ldots A_n)^q}\ \mathrm{ClR}^q$$

Both rules have the side condition that $q > 0$. After a contraction, one of the contracted lwffs has a contraction index of 0, so that it cannot be further contracted, while the other contracted lwff may be contracted $q - 1$ times. As with p, we do not decrement q when we apply branching rules, e.g.

$$\frac{\Delta \vdash R^{\rightarrow} a\,b\,c \qquad \begin{array}{c}(x{:}A_1 \rightarrow A_2)^{q_j}, \\ \Gamma, \Delta \vdash \Gamma', b{:}B\end{array} \qquad \begin{array}{c}(c{:}C \rightarrow D)^{g(c:C \rightarrow D)}, \\ (x{:}A_1 \rightarrow A_2)^{q_j}, \Gamma, \Delta \vdash \Gamma'\end{array}}{(a{:}B \rightarrow (C \rightarrow D))^{q_i}, (x{:}A_1 \rightarrow A_2)^{q_j}, \Gamma, \Delta \vdash \Gamma'}\ {\rightarrow}\mathrm{L}$$

Note that we have set the index of $c{:}C \rightarrow D$ to its initial value $g(c{:}C \rightarrow D)$.

For these 'improved' bounded contraction rules, we can refine the sizes of lwffs and sequents. We define the S($\mathcal{L}$)-size of an lwff $(x{:}A)^q$ to be $2 \times |A|$, if $q = 0$; and $4 \times |A| \times (q + 1)$, if $q > 0$. By defining the S($\mathcal{L}$)-size of a sequent S to be the sum of the S($\mathcal{L}$)-sizes of the lwffs in S plus the number of rwffs in S, we obtain a measure $\mathtt{m}$ that, like the one above, is reduced by every rule except *monl* and non-simplifying relational rules. This new measure may be smaller than the first one; for example, in the case of S(T), we can replace a sequent contraction index that is a linear function of the end-sequent with constant (0 or 1) lwff contraction indices.

4 EXAMPLES

As examples, we briefly show that our recipe yields PSPACE decision procedures for the modal logics K, T, K4 and S4, intuitionistic logic J, and B$[\rightarrow, \wedge]$. Detailed accounts can be found in the references given at the end of this paper.

The cut-free labelled sequent systems for these logics all satisfy a common property, which we call the $\mathcal{M}$-*disjunction property*:

(4) if $\Gamma, \Delta \vdash \Gamma', x{:}\mathcal{M}A_1 \ldots A_n, x{:}\mathcal{M}B_1 \ldots B_n$ is provable in $\mathrm{S}(\mathcal{L})$,
then so is either $\Gamma, \Delta \vdash \Gamma', x{:}\mathcal{M}A_1 \ldots A_n$ or $\Gamma, \Delta \vdash \Gamma', x{:}\mathcal{M}B_1 \ldots B_n$

where $\mathcal{L}$ is one of the above logics. The semantic intuition behind this is that in the Kripke models for $\mathcal{L}$ we only need to follow one 'chain' of worlds.[5] An immediate corollary of (4) is the eliminability of ClR in $\mathrm{S}(\mathcal{L})$ (set $B_i = A_i$ for each $1 \leq i \leq n$). This, combined with Lemma 3.4, tells us that in $\mathrm{S}(\mathcal{L})$ we only need to consider left contractions of $x{:}\mathcal{M}A_1 \ldots A_n$, together with possible applications of *monl* with atomic lwffs and non-simplifying relational rules. Bounding these rules is a logic dependent task.

Modal logics. In [Basin *et al.*, 1997] we gave $O(n \log n)$ space bounds for both K and T. This fits in our recipe as follows. Lemma 3.4 extends to eliminate all contractions in S(K). ClL is not eliminable in S(T) (try to prove $\vdash x{:}\neg\Box\neg(B \supset \Box B)$), but it suffices to contract each $x{:}\Box A$ at most once in each branch; thus in S(T) we only require the contraction rule ClL^g where $g(x{:}\Box A) = 1$. Since $\Delta \vdash x\, R^{\Box}\, y$ is provable in S(K) iff $x\, R^{\Box}\, y \in \Delta$, and in S(T) iff $x\, R^{\Box}\, y \in \Delta$ or $x = y$, $\mathtt{m}$ and $\mathtt{r}$ are both bounded by $n = |S|$ for both logics, resulting in $O(n \log n)$ space bounds.

S(K4) is an example of a system requiring the contraction rule ClL^p, since ClL can be neither eliminated nor bounded by a constant (try to prove $\vdash x{:}\Box\neg\Box B \supset \Box\neg\Box^i B$, where $\Box^i$ means a series of i consecutive $\Box$'s; cf. the proof (3), where $i = 2$). The analysis of ClL in S(K4) is considerably more complicated than that in S(K) or S(T), but in [Viganò, 1998a] we show that to prove $S = \vdash d{:}D$ in S(K4) we need at most $n = |S|$ contractions of $x{:}\Box A$, i.e. that $f(S) = n$. Thus, given our lexicographic ordering, $\mathtt{m}$ is $O(n^2)$ since p

[5] Similar to the more common intuitionistic disjunction property, (4) holds provided that Γ and Γ' satisfy certain logic dependent conditions; informally, that Γ and Γ' do not contain lwffs that are labelled with x or with a world preceding x, and that are the principal formulas of applications of branching rules in the proof of $\Gamma, \Delta \vdash \Gamma', x{:}\mathcal{M}A_1 \ldots A_n, x{:}\mathcal{M}B_1 \ldots B_n$, since such rules implicitly contract lwffs.

and Σ are both bounded by n. Since $\mathtt{r}$ is bounded by $O(n)$ (we require linear space to decide whether $x\,R^{\Box}\,y$ is in the transitive closure of Δ), we then have an overall upper-bound of $O(n^2 \log n)$.

The bounds that we established for S(T) and S(K4) are independent, in the sense that we can immediately combine them to give an $O(n^2 \log n)$ space upper-bound for S(S4). Note that we can then exploit the standard [Troelstra and Schwichtenberg, 1996] linear translation of intuitionistic logic J into S4 to establish an $O(n^2 \log n)$ upper-bound for provability in J. This bound is bigger than the one given by Hudelmaier [1993], which, however, relies on a specialized analysis.

Relevance logic B[$\rightarrow, \wedge$]**.** To simplify the analysis of the complexity of B[$\rightarrow, \wedge$], it is helpful to prove some preliminary results [Viganò, 1998b]: We can eliminate applications of *iden*, *monr*(1) and *monr*(2) in S(B[$\rightarrow, \wedge$])-proofs of $S = \vdash 0{:}D$. Then it follows that, as with S(K), we can eliminate contractions in S(B[$\rightarrow, \wedge$]). Hence, since $\Delta \vdash R^{\rightarrow}\,x\,y\,z$ is provable iff $R^{\rightarrow}\,x\,y\,z \in \Delta$ or it follows by an application of *monr*(0), both $\mathtt{m}$ and $\mathtt{r}$ are bounded by $n = |S|$. This results again in an $O(n \log n)$ space bound.

Concluding remarks. Our recipe structures complexity arguments and can ease analysis since various bounds on proof search are established once and for all in a logic independent way. It is, of course, not a panacea: establishing the remaining bounds may require considerable user analysis and the quality of the overall results depends on the quality of this analysis. Moreover, in some cases, as in classical or intuitionistic propositional logic, it may not be possible to obtain optimal bounds with our recipe. However, for other logics, such as K, T, K4 and S4, we have established bounds equal to the best given in the literature [Cerrito and Cialdea Mayer, 1997; Hudelmaier, 1996], which in turn improved the bounds given by Ladner [1977]. In the case of B[$\rightarrow, \wedge$], following the recipe helped us to establish new results for a basic relevance logic.

5 RELATED WORK

The decidability of non-classical logics is often established semantically, by proving that they possess the finite model property. Although effective in establishing decidability, this often yields poor complexity bounds. For example, for the PSPACE decidable modal logics K, T and S4, there are classes of satisfiable formulas in which every satisfying structure contains exponentially many worlds [Halpern and Moses, 1992].

To get more refined upper-bounds, logics are typically recast as cut-free sequent systems in which rules are measure decreasing. The systems of Hudelmaier [1996] for the modal logics K, T and S4, and of Hudelmaier [1993] and Dyckhoff [1992] for intuitionistic logic are typical in this regard. These systems are quite specialized and difficult to invent. Our work is close to that of Fitting [1983], who gives decision procedures for many prefixed modal tableaux systems. Although our labelled sequent systems share characteristics with Fitting's systems, the analysis of decidability relies on different mechanisms. In particular, our recipe is based on explicit contractions bounds, which are needed to yield good complexity bounds. Fitting's analysis, where termination of the decision procedure is shown using König's Lemma, leads instead to implementations based on loop-checking, as opposed to measure-decreasing rule applications as in our work. Moreover, although it establishes decidability, his analysis does not provide good complexity bounds.

Perhaps the work closest to ours in spirit is that of Cerrito and Cialdea Mayer [1997], who avoid loop-checking by eliminating contractions in standard (unlabelled) sequent systems for K, T, K4 and S4. However, their analysis yields larger complexity bounds.

The analysis of decidability and complexity for relevance logics is a more subtle issue, and positive results have been presented alongside open problems and several negative results (including the undecidability of all principal relevance logics); cf. [Anderson *et al.*, 1992; Restall, 1998]. Proof theoretic decision procedures have been given for B and other relevance logics, in particular based on Belnap's *display logic*, but we are not aware of subrecursive bounds for B or its fragments. We plan to use our recipe to investigate the complexity of other relevance logics.

Acknowledgments. We thank Seán Matthews, Nicola Olivetti and the anonymous referees for discussions and comments about this work.

REFERENCES

[Anderson *et al.*, 1992] A. R. Anderson, N. D. Belnap, Jr., and J. M. Dunn. *Entailment, The Logic of Relevance and Necessity*, volume 2. Princeton University Press, Princeton, New Jersey, 1992.

[Basin *et al.*, 1997] D. Basin, S. Matthews, and L. Viganò. A new method for bounding the complexity of modal logics. In G. Gottlob, A. Leitsch, and D. Mundici, editors, *Proceedings of KGC'97*, LNCS 1289, pp. 89–102. Springer, Berlin, 1997.

[Basin *et al.*, 1998] D. Basin, S. Matthews, and L. Viganò. Natural deduction for non-classical logics. *Studia Logica*, 60(1):119–160, 1998.

[Cerrito and Cialdea Mayer, 1997] S. Cerrito and M. Cialdea Mayer. A polynomial translation of S4 into T and contraction-free tableaux for S4. *Logic Journal of the IGPL*, 5(2):287–300, 1997.

[Dyckhoff, 1992] R. Dyckhoff. Contraction-free sequent calculi for intuitionistic logic. *Journal of Symbolic Logic*, 57(3):795–807, 1992.

[Fitting, 1983] M. Fitting. *Proof methods for modal and intuitionistic logics.* Kluwer, Dordrecht, 1983.

[Halpern and Moses, 1992] J. Y. Halpern and Y. O. Moses. A guide to completeness and complexity for modal logics of knowledge and belief. *Artificial Intelligence*, 54(3):319–379, 1992.

[Hudelmaier, 1993] J. Hudelmaier. A $O(n \; log \; n)$-space decision procedure for intuitionistic propositional logic. *Journal of Logic and Computation*, 3(1):63–75, 1993.

[Hudelmaier, 1996] J. Hudelmaier. Improved decision procedures for the modal logics K, T and S4. In H. Kleine Büning, editor, *Proceedings of CSL'95*, LNCS 1092, pp. 320–334. Springer, Berlin, 1996.

[Hughes and Cresswell, 1996] G. E. Hughes and M. J. Cresswell. *A new introduction to modal logic.* Routledge, London, 1996.

[Ladner, 1977] R. E. Ladner. The computational complexity of provability in systems of modal propositional logics. *SIAM Journal of Computing*, 6(3):46–65, 1977.

[Prawitz, 1965] D. Prawitz. *Natural deduction.* Almqvist and Wiksell, Stockholm, 1965.

[Restall, 1998] G. Restall. Displaying and deciding substructural logics 1: Logics with contraposition. *Journal of Philosophical Logic*, 27(2):179–216, 1998.

[Troelstra and Schwichtenberg, 1996] A. S. Troelstra and H. Schwichtenberg. *Basic proof theory.* Cambridge University Press, Cambridge, UK, 1996.

[Viganò, 1997] L. Viganò. *A framework for non-classical logics.* PhD thesis, Universität des Saarlandes, Saarbrücken, Germany, Sept. 1997.

[Viganò, 1998a] L. Viganò. Bounding the complexity of the transitive modal logics K4 and S4. Manuscript, 1998.

[Viganò, 1998b] L. Viganò. The computational complexity of the relevance logic $B[\rightarrow, \wedge]$. Manuscript, 1998.

[Zeman, 1973] J. J. Zeman. *Modal logic: the Lewis-modal systems.* Oxford University Press, Oxford, 1973.

Computer Arithmetic: Logic, Calculation, and Rewriting

Marco Benini, Dirk Nowotka, and Carl Pulley

1 INTRODUCTION

Computer arithmetic is the mathematical theory which underlies the way calculational machines operate on integer numbers.

Computers manipulate *integer numbers* of a finite, fixed precision, internally represented as strings of bits of fixed length. A processor's hardware [Braun, 1963] is built to perform additions, multiplications, and other standard arithmetical operations along with *logical* operations like "or", "and", "not", "exclusive or", and so on.

The distinguishing features of computer arithmetic are:

- *logical* operations, i.e., the ability to calculate bit per bit the conjunction, disjunction and negation of *integers*. For clarity, since we deal with a logical theory, we will refer to these operations with the adjective *bitwise*;

- *fixed precision*, which means that every representable number lies in a fixed, well-defined range of values and every operation must signal exceptions when unable to provide a result which fits into that range, usually by means of carry and/or overflows bits.

The goal of this paper is to provide a logical formulation for computer arithmetic, suitable for coding in a theorem proving environment, along with an algorithmic, as opposed to a logically declarative, calculational engine based on equational rewriting techniques, which forms the natural counterpart for the logical representation, and a decision procedure providing efficiency in computation, without compromising validity of results.

Even if it may appear to be a very specialized topic, an integrated decision procedure and a calculational engine for computer arithmetic is very general and very useful as well.

Dropping the constraint of finite precision, computer arithmetic is actually an extension of the standard theory of integer numbers [Smorynski, 1991], i.e., it provides the usual operations on integers as well as bitwise operations.

From the point of view of formal verification [Gordon, 1986], computer arithmetic is essential, since it is the theory we are really dealing with when we execute a program, or when we design a digital circuit.

What we are about to show is a logical theory plus a rewriting engine to perform calculations and an automatic procedure for deciding (in)equalities. Our discussion will be general, but we will constantly refer, for the applicative aspects, to our particular implementation which uses Isabelle's higher-order logic theory (Isabelle/HOL) [Paulson, 1997a] and Standard ML [Paulson, 1996].

Our tool is particularly designed to approach the formal verification of object code, since it was developed as part of a bigger project, Holly [Benini *et al.*, 1998], on this subject. Despite its origin, the designing purposes do not affect the generality of the ideas, nor the applicability of the implementation to other theories and problems.

The outline of the paper is as follows: Section 2 is devoted to the precise description of what computer arithmetic is; in Section 3, we specify each component of the system to deal with computer arithmetic; Section 4 explains how the combination of the previous parts allows us to tackle non trivial problems and how the interaction of components is the key for success in this application; in Section 5 we briefly discuss what we have achieved.

2 COMPUTER ARITHMETIC

We said in the introduction that computer arithmetic is the logical theory which models the way computers treat numbers. In fact, we developed an abstract model for the arithmetic implemented by computers.

Every computer works by using bits, bytes, words and so on. These are the elements (interpreted as *numbers*) the Arithmetical Logical Unit (ALU, the part of a processor which is devoted to perform calculations [Braun, 1963]) is able to deal with. There are significant differences between processors in the *size* of numerical types (i.e., the precision) and in their *bounds* (i.e., the range of values an element can represent). Nowadays most processors are able to deal with operations on bytes, double bytes and quadruple bytes; the usual convention they use for the sign of a number is the two's complement representation, and they are able to perform additions, subtractions, multiplications, divisions, bitwise conjunction, bitwise disjunction and bitwise negation on the numerical types they provide.

We wish to produce a general theory, which closely resembles the way ALUs perform calculations, but we do not want to choose a particular set of numerical types for a specific processor. Since we aim to be very general, we adopt a simplifying decision: we specify only one numerical type, integers (formally represented as the type **int**), with no bounds and infinite precision, and we represent them in the two's complement notation, with the full set of standard operations. In section 3.1, we will provide the technical details.

We can define the computer numerical types by *subtyping* [Andrews, 1986; Paulson, 1997c]:

$$\begin{array}{ll}\textbf{unsigned\_bytes} & = \{x : \textbf{int} \mid 0 \leq x < 256\} \\ \textbf{signed\_bytes} & = \{x : \textbf{int} \mid -128 \leq x < 128\}\end{array} \quad ,$$

or by *quotienting* with an equivalence relation [Paulson, 1997a]:

$$\textbf{byte} = \textbf{int}/\{\langle x, y\rangle \mid x \bmod 256 = y \bmod 256\} \ .$$

If we do not consider the bitwise operations, **int** is isomorphic to $\mathbb{Z}$ (the ring of integers numbers). In this way our system can be effectively used as an arithmetical oracle on integers.

We can even use our tool as an oracle for standard arithmetic, that is the theory of natural numbers: they could be represented by subtyping

$$\textbf{nat} = \{x : \textbf{int} \mid x \geq 0\} \ ,$$

or by quotienting

$$\textbf{nat} = \textbf{int}/\{\langle x, y\rangle \mid x = y \vee x = -y\} \ .$$

Both of these axiomatizations have some difficulties: the former does not respect subtraction, e.g., $3 - 7 = -4$ on integers, but $3 - 7$ should be 0 on naturals; the latter does not respect the ordering relation, e.g., $-3 < 2$ on integers, but $3 \not< 2$ on naturals.

For this reason, and because Isabelle provides a good implementation for the theory of natural numbers (terms of type *nat*), our tool is able to treat them directly.

Our system is developed as a package for the higher-order logic theory of the Isabelle theorem prover, so it seems worthwhile to compare it with similar theories/packages for Isabelle and other theorem provers.

Isabelle has no specialized decision procedures for the theory of integer numbers: its formalization of natural number arithmetic is quite advanced and a powerful simplification procedure is able to decide most elementary propositions in a reasonably efficient way. But, for the moment, the theory of integer numbers is quite undeveloped; for instance, it lacks the division and modulus operations, and, without modifying the simplifier, it fails to prove simple goals like

$$x + 1 - x = 1 \ .$$

The comparison with other, similar, theorem provers is more problematic, since their designs are different and it is not simple to identify a subsystem which could be properly compared with our tool. From a very general point of view, the HOL theorem prover [Gordon and Melham, 1993] has a decision procedure (coded up as a tactic) which could be roughly equivalent to the calculation engine (Section 3.2), while the logical theory is an essential part of the logic the prover provides. There are no specialized decision procedures for arithmetic in HOL, but only clever instances of the tactics for the general logic.

The situation for PVS [Owre *et al.*, 1996] is different: this theorem prover provides a decision procedure for integer arithmetic, which is based on the same algorithm we use (see Section 3.3); the logical representation for integers is part of the standard logic, and an equivalent of the calculational engine is embedded into the simplification tactics. But, in contrast with our tool, PVS does not provide bitwise operations on pure integers, but it develops a specialized theory (called *bitvector*) to reason about binary represented numbers.

3 THE COMPONENTS

The concept of computer arithmetic is implemented by three major parts. A *logical theory* for arithmetic is developed in section 3.1. It comprises a definition of integer numbers and operations on them using Isabelle/HOL [Paulson, 1997b] as an example environment. Section 3.2 presents the *calculation engine* used to do calculations and simplifications on arithmetical expressions efficiently. Last but not least, a *decision procedure* is introduced in section 3.3. We have chosen SHOSTAK'S SUP-INF method [Shostak, 1979] for proving formulas of an extension of quantifier-free Presburger arithmetic.

3.1 The Logical Development

Our first goal is to devise a logical theory which describes what we intend to call integer numbers, the definitions of operations and their basic properties. This theory is an instrument to reason, in a formal way, e.g., using a theorem prover, about properties of integers which could be modeled inside computer arithmetic.

As remarked in Section 2, computer arithmetic is based on a particular representation of numbers: every number is represented as a string of binary digits. Formally the type for integer numbers is defined as follows[1]:

$$\begin{aligned}\texttt{datatype}\ \mathit{int} = {} & \texttt{PlusSign} \\ \mid {} & \texttt{MinusSign} \\ \mid {} & \texttt{Bcons}\ \mathit{int}\ \mathit{bool}\end{aligned}$$

The value `PlusSign` stands for 0 while the value `MinusSign` stands for -1; looking at the binary representation, `PlusSign` is the infinite string where

[1]This representation is borrowed from the `Bin` theory of Isabelle/HOL.

every bit is 0, while `MinusSign` is the infinite string where every bit is 1. The `Bcons` constructor works by appending a bit (*False* for 0, and *True* for 1) to a string of bits.

Intuitively, 6 (binary 110) will be represented as

$$\texttt{Bcons (Bcons (Bcons PlusSign}\ True)\ True)\ False$$

and -6 (binary $\ldots 1010$) will be represented as

$$\texttt{Bcons (Bcons (Bcons MinusSign}\ False)\ True)\ False$$

The mapping from `int` to integers is straightforward:

$$\begin{array}{ll} \texttt{PlusSign} & \stackrel{\text{map}}{\longmapsto} 0 \\ \texttt{MinusSign} & \stackrel{\text{map}}{\longmapsto} -1 \\ \texttt{Bcons}\ x\ y & \stackrel{\text{map}}{\longmapsto} 2(\text{map}\ x) + (\text{if}\ y\ \text{then}\ 1\ \text{else}\ 0) \end{array} \quad ,$$

and vice versa

$$\begin{array}{lll} 0 & \stackrel{\text{map}'}{\longmapsto} \texttt{PlusSign} & \\ -1 & \stackrel{\text{map}'}{\longmapsto} \texttt{MinusSign} & \\ 2x & \stackrel{\text{map}'}{\longmapsto} \texttt{Bcons}\ (\text{map}'\ x)\ False & \text{if}\ x \neq 0 \\ 2x+1 & \stackrel{\text{map}'}{\longmapsto} \texttt{Bcons}\ (\text{map}'\ x)\ True & \text{if}\ x \neq -1 \end{array} \quad .$$

We will use, for clarity, the usual numerical representation[2], or a binary representation, where it is convenient. Both notations are unambiguous.

An important point about this representation lies in the fact that it is not fully determined: the same integer could be represented in different, but equivalent, ways. For example 5 could be represented by $\ldots 0101$, or by $\ldots 00000101$. The difference between two equivalent representations is in the trailing bits: formally,

$$\texttt{Bcons PlusSign}\ False = \texttt{PlusSign}$$

and

$$\texttt{Bcons MinusSign}\ True = \texttt{MinusSign}\ .$$

Propagating these equations through the inductive structure of the definition of *int*, we eventually get a normal form for our representation. In this way, an easy definition of equality is given: two representations are equal if and only if their normal forms are syntactically identical.

The definition of the successor and predecessor functions, as well as of addition, subtraction and multiplication is straightforward, and it is summarized in Figure 1.

[2] So `PlusSign` becomes 0, `MinusSign` becomes -1, `Bcons` x *True* is written as $2x+1$ and `Bcons` x *False* is written as $2x$.

Successor function

$$\begin{aligned} \mathrm{Succ}(0) &= 1 \\ \mathrm{Succ}(-1) &= 0 \\ \mathrm{Succ}(2x+1) &= 2\,\mathrm{Succ}(x) \\ \mathrm{Succ}(2x) &= 2x+1 \end{aligned}$$

Predecessor function

$$\begin{aligned} \mathrm{Pred}(0) &= -1 \\ \mathrm{Pred}(-1) &= -2 \\ \mathrm{Pred}(2x+1) &= 2x \\ \mathrm{Pred}(2x) &= 2\,\mathrm{Pred}(x)+1 \end{aligned}$$

Addition

$$\begin{aligned} 0+x &= x \\ -1+x &= \mathrm{Pred}(x) \\ x+0 &= x \\ x+-1 &= \mathrm{Pred}(x) \\ (2x+1)+(2y+1) &= 2\,\mathrm{Succ}(x+y) \\ (2x+1)+2y &= 2(x+y)+1 \\ 2x+(2y+1) &= 2(x+y)+1 \\ 2x+2y &= 2(x+y) \end{aligned}$$

Numeric Complement

$$\begin{aligned} -0 &= 0 \\ -(-1) &= 1 \\ -(2x+1) &= \mathrm{Pred}(-2x) \\ -2x &= 2\,(-x) \end{aligned}$$

Subtraction

$$x-y = x+(-y)$$

Multiplication

$$\begin{aligned} 0\cdot x &= 0 \\ -1\cdot x &= -x \\ (2x+1)\cdot y &= 2(x\cdot y)+y \\ 2x\cdot y &= 2(x\cdot y) \end{aligned}$$

Figure 1: Definitions for arithmetical operations

Formalizing division is more complex, since it is a partial operation, i.e., it is not defined when the divisor is 0. For the same reason, the remainder operation (mod) is awkward to define in a direct way. We define these pair of operators as the ones satisfying the basic equation:

$$x/y + x \bmod y = x \qquad \text{when } y \neq 0$$

under the additional constraint

$$0 \leq x \bmod y < y$$

Bitwise operations are easily defined by induction on the structure of integer representation: the precise statements are given in Figure 2.

Conjunction

$$\begin{array}{ll}
0 \mathbin{\underline{\wedge}} x & = 0 \\
-1 \mathbin{\underline{\wedge}} x & = x \\
x \mathbin{\underline{\wedge}} 0 & = 0 \\
x \mathbin{\underline{\wedge}} -1 & = x \\
(2x+1) \mathbin{\underline{\wedge}} (2y+1) & = 2(x \mathbin{\underline{\wedge}} y) + 1 \\
(2x+1) \mathbin{\underline{\wedge}} 2y & = 2(x \mathbin{\underline{\wedge}} y) \\
2x \mathbin{\underline{\wedge}} (2y+1) & = 2(x \mathbin{\underline{\wedge}} y) \\
2x \mathbin{\underline{\wedge}} 2y & = 2(x \mathbin{\underline{\wedge}} y)
\end{array}$$

Negation

$$\begin{array}{ll}
\underline{\neg} 0 & = -1 \\
\underline{\neg} -1 & = 0 \\
\underline{\neg}(2x+1) & = 2(\underline{\neg} x) \\
\underline{\neg}(2x) & = 2(\underline{\neg} x) + 1
\end{array}$$

Disjunction

$$\begin{array}{ll}
0 \mathbin{\underline{\vee}} x & = x \\
-1 \mathbin{\underline{\vee}} x & = -1 \\
x \mathbin{\underline{\vee}} 0 & = x \\
x \mathbin{\underline{\vee}} -1 & = -1 \\
(2x+1) \mathbin{\underline{\vee}} (2y+1) & = 2(x \mathbin{\underline{\vee}} y) + 1 \\
(2x+1) \mathbin{\underline{\vee}} 2y & = 2(x \mathbin{\underline{\vee}} y) + 1 \\
2x \mathbin{\underline{\vee}} (2y+1) & = 2(x \mathbin{\underline{\vee}} y) + 1 \\
2x \mathbin{\underline{\vee}} 2y & = 2(x \mathbin{\underline{\vee}} y)
\end{array}$$

Figure 2: Definitions for bitwise operations

In order to reason about inequalities we need to define the ordering relation. This is done as a two-step process: first, we reduce '*less than*' to '0 *is less than*':

$$x < y \equiv \exists z.0 < z \wedge x + z = y \ ,$$

then, we state the conditions under which an integer is positive

$$\begin{aligned} 0 &\not< 0 \\ 0 &\not< -1 \\ 0 < 2x &\equiv 0 < x \\ 0 < 2x+1 &\equiv 0 < x \vee 0 = x \quad . \end{aligned}$$

In the usual way we define $x \leq y \equiv x < y \vee x = y$, $x \geq y \equiv y \leq x$ and $x > y \equiv y < x$.

In order to be able to deduce interesting properties about numbers, we need some induction principles: we provide the ones summarized in Figure 3. They encode induction over the structure of the binary representation, and two variants of the standard induction over natural numbers. These instruments are used to prove a wide set of lemmas which provides a (partial) validation for the other components, specifically for the rewrite rules of the calculation engine and the axioms of the decision procedure.

$$\frac{P(k) \qquad \begin{array}{c} P(x), x \leq k \\ \vdots \\ P(x+1) \end{array}}{\forall x \leq k . P(x)} \qquad \frac{P(k) \qquad \begin{array}{c} P(x), x \geq k \\ \vdots \\ P(x+1) \end{array}}{\forall x \geq k . P(x)}$$

$$\frac{P(0) \quad P(-1) \quad \begin{array}{c} P(x) \\ \vdots \\ P(2x) \end{array} \quad \begin{array}{c} P(x) \\ \vdots \\ P(2x+1) \end{array}}{\forall x . P(x)}$$

Figure 3: Induction principles

3.2 The Calculational Engine

A theory of integer numbers is developed in a logical framework; see section 3.1. The purpose of the calculational engine is to do reductions on integer expressions outside the logic.

A calculational engine works in a syntactical way by term rewriting and in a semantical way by doing actual calculations on integer numbers using an implementation programming language. It has been implemented in a functional programming language, namely *Standard ML* [Paulson, 1996]. This engine is defined by a function that takes the representation of an integer expression and returns a term that represents an equivalent, but reduced integer expression. We clarify next what we understand by *representation* of an integer expression, by *equivalent*, and by *reduced.*

The syntax of an internal language, in which the calculations and rewritings are done, is fixed by a data type. The corresponding construct `IntTerm` in ML is shown in Figure 4. It defines inductively a set of terms, called **Int** for the rest of this section.

```
datatype IntTerm = IntConstant of term * IntTerm list
                 | IntValue of int
                 | IntSucc of IntTerm
                 | IntPred of IntTerm
                 | IntComp of IntTerm
                 | IntPlus of IntTerm * IntTerm
                 | IntMinus of IntTerm * IntTerm
                 | IntTimes of IntTerm * IntTerm
                 | IntDivide of IntTerm * IntTerm
                 | IntModulus of IntTerm * IntTerm;
```

Figure 4: The syntax of the internal language.

Since the calculational engine is used to reason about integer expressions, that data type closely corresponds to the syntax for integer expressions used in section 3.1. In fact, to allow a smooth combination of deduction system and calculational engine, there is an obvious mapping from integer expressions to their representations in the internal language. That mapping is a bijection between the quotient of integer expressions formed by the equivalence relation, that relates different representations of the same integer number, and the representation of integer expressions in ML. See section 4 for the combination of reasoners.

We call expressions, formed by `IntTerm`, integer expressions for the rest of this section, bearing in mind that these are just a *representation* of the expressions developed in 3.1.

The manipulation of integer expressions is done by a set of ML functions, called *rewrite functions.* Figure 5 shows such a function. The pattern matching feature of ML is used here to rewrite terms that have a certain structure and leave others unchanged.

```
fun Axiom1 (IntPlus (IntValue 0, x):  IntTerm) =
      x
  | Axiom1 default = default;
```

Figure 5: A rewrite function.

A *rewrite rule* is gained by taking a lemma about the equivalence of integer expressions deduced from the definition of integers on the logical level and directing the equation, say from left to right and then translating the left- and the right-hand side to **Int**, using the obvious mapping where integer variables are taken to ML variables to allow pattern matching. A rewrite function is made from such rules in the obvious way. The following shows the preparation steps to implement the rewrite function of figure 5. We take a lemma given by the definition of integers in section 3.1.

$$x + 0 = x$$

This lemma translates to the rewrite rule

$$\texttt{IntPlus (}x\texttt{, IntValue }0\texttt{)} \rightarrow x$$

where x is a ML variable over `IntTerm`.

This tight relation between rewrite functions in ML and lemmas derived on the logical level gives a good justification for the claim that an integer expression is rewritten into an expression which is equivalent by means of the logical definition of integers. A formal treatment of such rewriting is beyond the scope of this paper, textbooks on term rewriting or overview articles such as [Dershowitz and Jounnaud, 1990] may be referred to for the standard proofs necessary.

A crucial matter for the practical use of the calculational engine is to guarantee termination. Since the rewrite procedure stops only if no more rules can be applied to any subterm of a given expression, it is therefore obvious that one cannot choose arbitrary lemmas to make them to rewrite rules. For instance, the unconditional application of

$$\texttt{IntPlus (}x\texttt{, }y\texttt{)} \rightarrow \texttt{IntPlus (}y\texttt{, }x\texttt{)}$$

is always possible if the argument expression contains addition.

The rewriting stops if every rewrite rule produces a "reduced" version of its argument, if applied. To express this reduction a well-founded ordering over **Int** has to be defined, and then one has to show that every rewrite rule in the system, if applied, gives a result that is strictly smaller than the argument. Provided that holds, the rewrite process must terminate, because the reduction cannot go on forever.

An ordering over terms can be conveniently defined by a homomorphism from the ground terms of **Int** to an algebra $\mathcal{A}$ (of the same signature) which already has a well founded ordering. Let $\succ$ be such a well founded ordering on $\mathcal{A}$, then the monotonicity condition:

$$f_{\mathcal{A}}(\ldots x \ldots) \succ f_{\mathcal{A}}(\ldots y \ldots) \quad \text{if} \quad x \succ y$$

for all operations $f_{\mathcal{A}}$ and all x, y in $\mathcal{A}$ has to hold, as well.

The mapping τ_1 : **Int** $\rightarrow \{2, 3, \ldots\}$ explained in Figure 6 is a good starting point for an overall termination argument, where $\succ$ is the usual well-ordering on natural numbers.

$$
\begin{array}{llll}
\texttt{IntConstant}_{\mathcal{A}}(a,\ b) = 2 & & \texttt{IntPlus}_{\mathcal{A}}(a,\ b) & = a+b+1 \\
\texttt{IntValue}_{\mathcal{A}}\ a & = 2 & \texttt{IntMinus}_{\mathcal{A}}(a,\ b) & = a+2^b+1 \\
\texttt{IntSucc}_{\mathcal{A}}\ a & = a+4 & \texttt{IntTimes}_{\mathcal{A}}(a,\ b) & = a\cdot b \\
\texttt{IntPred}_{\mathcal{A}}\ a & = a+4 & \texttt{IntDivide}_{\mathcal{A}}(a,\ b) & = a\cdot b \\
\texttt{IntComp}_{\mathcal{A}}\ a & = 2^a & \texttt{IntModulus}_{\mathcal{A}}(a,\ b) & = a\cdot b
\end{array}
$$

Figure 6: A reduction mapping.

A reduction ordering $\succ_{\tau_1}$ over **Int** is now given by:

$$s \succ_{\tau_1} t \quad \text{iff} \quad \tau_1(s) \succ \tau_1(t)$$

with s, $t \in$ **Int**. However, we need further reduction relations for coping with associative and commutative rewrite rules. Such measures are given by the mappings τ_2 and τ_3 sketched in Figure 7 and Figure 8, respectively, where a suitable ordering relation $<$ has to be defined.

$$
\begin{array}{l}
\texttt{IntPlus}_{\mathcal{A}}\ (a,\ \texttt{IntPlus}_{\mathcal{A}}\ (b,\ c)) = 2 \\
\texttt{IntPlus}_{\mathcal{A}}\ (\texttt{IntPlus}_{\mathcal{A}}\ (a,\ b),\ c) = 3 \\
\ldots
\end{array}
$$

Figure 7: A reduction mapping.

$$
\begin{array}{l}
\texttt{IntPlus}_{\mathcal{A}}\ (a,\ b) = 2 \quad \text{if} \quad a < b \\
\texttt{IntPlus}_{\mathcal{A}}\ (a,\ b) = 3 \quad \text{if} \quad b < a \\
\ldots
\end{array}
$$

Figure 8: A reduction mapping.

The reduction ordering $\succ_{\tau}$, which we use is built from τ_1, τ_2, and τ_3 in the standard way of composing orderings by the lexicographical ordering of the cross-product. That is, given s and $t \in$ textbfInt, then $s \succ_{\tau} t$ iff $s \succ_{\tau_1} t$, or in case $s = t$, then $s \succ_{\tau_2} t$, or in case $s = t$, then $s \succ_{\tau_3} t$. The termination of the calculational engine can be easily shown with $\succ_{\tau}$.

The following example appeared when an actual correctness proof of a program was performed by the authors. It shows the effectiveness of reasoning with this rewrite engine in practice. We do not use the syntax of `IntTerm` here to increase readability. The expression:

$$c + ((3 + (f(x) + (-4))) + (-c))$$

is subsequently reduced to

$$c + ((f(x) + (3 + (-4))) + (-c)),$$
$$c + ((f(x) + (-1)) + (-c)),$$
$$c + (f(x) + ((-1) + (-c))),$$
$$c + (f(x) + ((-c) + (-1))),$$
$$c + ((-c) + (f(x) + (-1))),$$
$$0 + (f(x) + (-1)),$$

and finally

$$f(x) + (-1).$$

The treatment of rewriting bitwise operations is done in the same way as the integer expressions. For reasons of space and readability, we do not go further into this here.

3.3 The Decision Procedure

The third part of our representation of Computer Arithmetic consists of a decision procedure for integer arithmetic with function symbols. We use SHOSTAK'S SUP-INF algorithm [Shostak, 1977; Shostak, 1979] for solving unquantified Presburger formulas which might also contain an unlimited number of function and predicate symbols.

Intuitively, Presburger formulas are those that can be built up from integers and variables over integers, addition, equality and inequality relations, and first-order logical connectives. The decision procedure we use here, see [Shostak, 1979], operates on an extension of the quantifier-free version of Presburger arithmetic. This extension allows an unlimited number of n-ary function symbols $f^{(n)} : \mathbb{Z}^n \to \mathbb{Z}$, and n-ary predicate symbols $P^{(n)} : \mathbb{Z}^n$, with $0 \leq n$, in each formula. These symbols are treated as undefined, i.e., they are not interpreted. A small example of this kind of formula is the following:

$$x < f(y) \wedge f(y) \leq (x + 1) \rightarrow (P(x) \leftrightarrow P(f(y) + (-1))) .$$

More generally, SHOSTAK'S algorithm works on unquantified formulas of first-order logic which contain any function and predicate symbols over the set of integers. Those symbols are not interpreted except the function $+$ and the predicates $<$, $\geq$, $>$, and $=$. A more general treatment of decision procedures is given in [Shostak, 1984] and [Cyrluk *et al.*, 1996].

The decision procedure is linked to the theory of integers in Isabelle/HOL by a straightforward translation from logical formulas to an internal language similar to the way described in section 3.2. In addition to that, we allow multiplication with constants, for they can be rewritten as a finite sum, and also subtraction, since that can be rewritten as addition by complementing the second addend.

The decision procedure rejects terms which cannot be translated into the internal language, i.e., which are not recognized as formulas of the described extension of unquantified Presburger arithmetic. An exception is signaled in that case and the calling procedure, e.g., a proof tactic, has to cope with it.

If a formula was read in, the procedure gives a *true* or *false* as answer, depending on which conclusion it has reached.

4 THE COMBINED SYSTEM

The strength of our approach to computer arithmetic is the combination of systems that have different fields of application but which are chosen and implemented to work together in order to solve a demanding problem. Efficient interaction and a reasonable "division of labour" was a major design goal.

Stemming from the demand to reason about computer arithmetic, a theory was developed which fits these needs and overcomes limitations of existing solutions by extending them, i.e., providing the division and modulus operation and integrating bitwise operations on integers.

In order to make use of this theory we need a way to reason about it. The most basic way to do that is to use basic rewriting of goals by rules gained from the equational theory induced by the definition of integers. Isabelle provides a powerful rewrite engine, called *simplifier*, that supports the proving procedure in a sophisticated way. Nevertheless, its generality prevents the simplifier from being as efficient and useful as a specialized rewrite engine could be. Many arithmetical goals could not be easily proved with that tool alone. Apart from rewriting, actual calculations like division cannot be done in a reasonable way in Isabelle. The calculational engine is used to remedy such problems.

Simplifying arithmetical expressions by calculations and simple rewriting is a basic way of reasoning but is too "primitive" to provide efficient tactics. SHOSTAK'S decision procedure is therefore used to solve more sophisticated goals when reasoning about computer arithmetic. It should be remarked that the computational engine is used to aid the decision procedure directly by rewriting an integer expression into a proper format such that the decision procedure can reason about it.

Technically, the integration of the calculational engine into Isabelle is done as a simple tactic that calls an *oracle* which maps an Isabelle term to the internal language of the calculational engine. Given the conclusion of the current subgoal, the oracle then provides the result of the calculations as a theorem which states that the argument and the reduced expression are equivalent. The subgoal and that theorem are then resolved.

Alternatively, the calculational engine is linked to the simplifier in Isabelle. By setting up a simplification procedure the external reasoner can aid the simplifier when rewriting complex subgoals which involve arithmetical expressions.

The decision procedure is linked to the theorem prover in a similar way as the calculational engine.

Finally, we can say that a logical theory of integers and bitwise operations provides a suitable basis for modeling computer arithmetic. The reasoning in that theory is made feasible by the tight coupling of a specialized rewrite engine, combined with calculation capabilities, and a powerful decision procedure.

5 CONCLUSIONS

A combined system for reasoning in computer arithmetic has been shown in this contribution. An integrated approach of logical, calculational and rewriting techniques makes it feasible to deduce complex theorems in computer arithmetic.

For the future, we plan to generalize the decision procedure in the way discussed in [Shostak, 1984] and [Cyrluk *et al.*, 1996].

Acknowledgment. We would like to acknowledge the help of Dr. Sara Kalvala in improving the ideas and the form of this article.

This work has been sponsored by EPSRC under grant GR/K52447.

A EXAMPLES

The following examples have been performed in Isabelle94-8 using version 1.0 beta 2 of CAT[3] (Computer Arithmetic Toolkit), developed by the authors.

The examples show the formulas that are reasoned about and the corresponding (slightly modified) dialog in the Isabelle/HOL environment.

Example 1

$$P(z) \wedge z = 1 \wedge g(y) = z + 4 \rightarrow f(g(y)) = f(3 + 2z) \vee \neg P(1)$$

```
> goal thy "((P(z) & z = #1) & g(y::int) = z + #4) --> \
\          ((f(g(y))::int) = f(#3 + #2 * z) | ~(P(#1)))";
Level 0
...
> by (supinf_tac 1);
Level 1
(P z & z = #1) & g y = z + #4 --> f (g y) = f (#3 + #2 * z) | ~ P #1
No subgoals!
```

[3]See http://www.dcs.warwick.ac.uk/holly/CAT

Example 2

$\forall x.\forall y.(x < y - 4) \vee (x = y - 4) \vee (x = y - 3) \vee (x = y - 2) \leftrightarrow (x < y - 1)$

```
> goal ExtBin.thy "! x y. (x < y - #4 | \
\                          x = y - #4 | \
\                          x = y - #3 | \
\                          x = y - #2) =  (x < y - #1)";
Level 0
...
> by (supinf_tac 1);
Level 1
! x y. (x < y-#4 | x = y-#4 | x = y-#3 | x = y-#2) = (x < y-#1)
No subgoals!
```

The examples 3 and 4 are taken from [Shostak, 1977].

Example 3

The formula $\forall x.\forall y.2x + 3y \neq 1$ is not valid; proof level 0 and 1 are the same.

```
> goal ExtBin.thy "! x y. ~(#2 * x + #3 * y = #1)";
Level 0
! x y. #2 * x + #3 * y ~= #1
 1. ! x y. #2 * x + #3 * y ~= #1
...
> by (supinf_tac 1);
Warning: same as previous level
Level 1
! x y. #2 * x + #3 * y ~= #1
 1. ! x y. #2 * x + #3 * y ~= #1
```

Example 4

The formula $\forall x.\forall y.2x+2y \neq 1$ is valid but SUP-INF cannot prove its validity. This example shows the incompleteness of the decision procedure.

```
> goal ExtBin.thy "! x y. ~(#2 * x + #2 * y = #1)";
Level 0
! x y. #2 * x + #2 * y ~= #1
 1. ! x y. #2 * x + #2 * y ~= #1
...
> by (supinf_tac 1);
Warning: same as previous level
Level 1
! x y. #2 * x + #2 * y ~= #1
 1. ! x y. #2 * x + #2 * y ~= #1
```

The following examples show the use of the calculational engine in CAT.

Example 5

The formula $(x + y)^2 = x^2 + 2xy + y^2$ is solved by reflexivity of "=" after it was rewritten.

```
> goal ExtBin.thy "(x + y) zexp #2 = x * x + #2 * x * y + y * y";
Level 0
...
> by (extbin_tac 1);
Level 1
(x + y) zexp #2 = x * x + #2 * x * y + y * y
No subgoals!
```

Example 6

The formula $x - 1 = y + (2 + (x + -3) + -y)$ is solved by reflexivity of "=" after it was rewritten.

```
> goal ExtBin.thy "x - #1 = y + (#2 + (x + $~ #3) + $~ y)";
Level 0
...
> by (extbin_tac 1);
Level 1
x - #1 = y + (#2 + (x + $~ #3) + $~ y)
No subgoals!
```

REFERENCES

[Andrews, 1986] P. Andrews. *An Introduction to Higher-Order Logic: to Truth through Proof.* Academic Press, New York, 1986.

[Benini *et al.*, 1998] Marco Benini, Sara Kalvala, and Dirk Nowotka. Program abstraction in a higher-order logic framework. In J. Rundy and M. Newwey, *Proceedings of Theorem Proving in Higher-Order Logic '98 International Conference.* LNCS 1479, pages 12–26, 1998.

[Braun, 1963] Edward L. Braun. *Digital Computer Design.* Academic Press, New York and London, 1963.

[Cyrluk *et al.*, 1996] David Cyrluk, Patrick Lincoln, and N. Shankar. On Shostak's decision procedure for combinations of theories. In M. A. McRobbie and J. K. Slaney, editors, *Automated Deduction—CADE-13*, number 1104 in Lecture Notes in Artificial Intelligence, pages 463–477, New Brunswick, NJ, July/August 1996. Springer-Verlag.

[Dershowitz and Jounnaud, 1990] Nachum Dershowitz and Jean-Pierre Jounnaud. Rewrite systems. In J. van Leeuwen, editor, *Handbook of Theoretical Computer Science*, pages 244–320. Elsevier Science Publishers, 1990.

[Gordon and Melham, 1993] Michael J. C. Gordon and Tom F. Melham. *Introduction to HOL: A Theorem Proving Environment for Higher Order Logic*. Cambridge University Press, 1993.

[Gordon, 1986] Michael J. Gordon. Why higher-order logic is a good formalism for specifying and verifying hardware. In G. Milne and P. Subrahmanyan, editors, *Formal Aspects of VLSI Design: Proceedings of the 1985 Edinburgh Workshop on VLSI*, Amsterdam, 1986. North Holland.

[Owre *et al.*, 1996] S. Owre, S. Rajan, J.M. Rushby, N. Shankar, and M.K. Srivas. PVS: Combining specification, proof checking, and model checking. In Rajeev Alur and Thomas A. Henzinger, editors, *Computer-Aided Verification, CAV '96*, number 1102 in Lecture Notes in Computer Science, pages 411–414, New Brunswick, NJ, July/August 1996. Springer-Verlag.

[Paulson, 1996] Lawrence C. Paulson. *ML for the Working Programmer*. Cambridge University Press, 2nd edition, 1996.

[Paulson, 1997a] Lawrence C. Paulson. *Isabelle's Object-logics*, chapter Higher-Order Logic, pages 59–99. In [1997b], May 1997.

[Paulson, 1997b] Lawrence C. Paulson. Isabelle's object-logics. Technical Report 286, Computer Laboratory, University of Cambridge, May 1997.

[Paulson, 1997c] Lawrence C. Paulson. Isabelle's reference manual. Technical Report 283, Computer Laboratory, University of Cambridge, May 1997.

[Shostak, 1977] Robert E. Shostak. On the SUP-INF method for proving presburger formulas. *JACM*, 24(4):529–543, October 1977.

[Shostak, 1979] Robert E. Shostak. A practical decision procedure for arithmetic with function symbols. *JACM*, 26(2):351–360, April 1979.

[Shostak, 1984] Robert E. Shostak. Deciding combinations of theories. *JACM*, 31(1):1–12, January 1984.

[Smorynski, 1991] Craig Smorynski. *Logical Number Theory*, volume I. Springer-Verlag, 1991.

Combining Higher-Order and First-Order Computation Using ρ-calculus: Towards a Semantics of ELAN

Horatiu Cirstea and Claude Kirchner

1 INTRODUCTION

The ρ-calculus is a calculus of explicit rewrite rule application that uniformly integrates the first-order and higher-order computation paradigms.

Because these two paradigms are individually extremely attractive and have useful complementary features, a lot of attention has been devoted to the study of their combination, especially in the last decade. This has been addressed either by enriching first-order rewriting with higher-order capabilities or by adding algebraic features to λ-calculus allowing one to deal with equality in an efficient way. In the first case, we cite the works on CRS [Klop *et al.*, 1993] and other higher-order rewriting systems [Nipkow and Prehofer, 1998], in the second case the works on combination of λ-calculus with term rewriting [Okada, 1989; Breazu-Tannen, 1988; Gallier and Breazu-Tannen, 1989; Jouannaud and Okada, 1997] to mention only a few.

We come up with another point of view arising from of our previous works on the control of term rewriting [Kirchner *et al.*, 1995; Vittek, 1994; Borovanský *et al.*, 1998]. Indeed we realized that the tool needed to control rewriting should be made explicit and could be itself naturally described using rewriting. This leads to the ρ-calculus; the rewriting calculus which is a calculus of explicit rule application.

In this calculus the application of a rule (say $a \to b$) to the top level of a term (like the constant a) is represented as the object of the calculus $[a \to b](a)$, which evaluates to the singleton $\{b\}$. When the application of a rule fails, like in $[a \to b](c)$, because a does not match the constant c, the expression evaluates to the empty set $\emptyset$. Of course variables may appear in rewrite rules, like $[f(x) \to x](f(a))$ that evaluates to $\{a\}$. In fact when evaluating this expression, the variable x is bound to a by the matching mechanism and we have shown in [Borovanský *et al.*, 1998] how a rewrite rule can be considered as a function with matching as the parameter passing mechanism.

A λ-expression $\lambda x.t$ could be represented as the rule $x \rightarrow t$ and the β-reduction of the redex $(\lambda x.t\, u)$ is exactly the reduction of the ρ-term $[x \rightarrow t](u)$ into the ρ-term $\{t\{x/u\}\}$. Of course, the substitution mechanism should preserve the correct variable bindings via the appropriate α-conversion. In order to make this point clear and as initiated in [Dowek *et al.*, 1995], we make a strong distinction between substitution (which takes care of variable binding) and grafting (that performs replacement directly). So basic ρ-calculus objects are built from the signature, the abstraction operator $\rightarrow$ and the application operator $[\,](\,)$.

Another important feature of the ρ-calculus is its ability to handle non-determinism in the sense of sets of results. This is achieved via the explicit handling of reduction result sets, including the empty set that records failure of rule application. It allows us to make use in an explicit and direct way of non-terminating or non-confluent (equational) rewrite systems. For example if the symbol $+$ is assumed to be commutative then applying the rule $x + y \rightarrow x$ to the term $a + b$ results in $\{a, b\}$ since there are two different ways to apply this rule modulo commutativity.

After introducing the ρ-calculus and its primary relevance in the combination of first and higher-order computations, the purpose of this paper is to define and study an explicit substitution calculus for the ρ-calculus and to provide a full yet simple semantics for the rewrite rules used in the ELAN language.

In Section 2 we first introduce the syntax of the basic ρ-calculus and we give examples of its expressiveness. The deduction rules are then presented together with considerations on the matching mechanism and on the application of the substitutions. At the end of this section we explain the strategy guiding the deduction rules that should be used in order to obtain a confluent calculus. In Section 3, we define and study a calculus of explicit substitutions, $\rho\sigma$, that generalizes the standard $\lambda\sigma$ calculus. The syntax and the deduction rules are presented in this new context and the same confluence results as in the non-explicit case are found. In Section 4 we detail how to use the ρ-calculus to give an operational semantics to the ELAN language. This permits in particular to give a full account of ELAN's strategy objects and their use. We conclude by providing research directions that are of significant interest in the context of ELAN and more generally of rewrite based languages like ASF+SDF [van den Brand *et al.*, 1997], ML, Maude [Clavel *et al.*, 1998] or CafeOBJ [Futatsugi and Nakagawa, 1996].

This paper does not contain all the technical details and several proofs that can be found in [Cirstea and Kirchner, 1998].

2 THE ρ-CALCULUS

The ρ-calculus is defined by five components:

- First its syntax that makes precise the formation of the terms manipulated by the calculus as well as the substitutions that are used during the evaluation mechanism. In the case of ρ-calculus, the core of the term formation relies on rewrite rules and rule application.
- The description of the substitution application on terms. This description is often given at the meta-level, except for explicit substitution frameworks like the one we describe in this paper. In the case of ρ-calculus, substitutions are higher-order substitutions and not grafting.
- The matching algorithm used to bind variables to their actual arguments. In the case of ρ-calculus, this is in general higher-order matching and in practical cases pattern [Miller, 1991], equational [Jouannaud and Kirchner, 1991] or syntactic matching.
- The rules describing the way the calculus operates. It is the glue between the previous components and the simplicity and clarity of these rules are fundamental for the calculus usability.
- The strategy guiding the application of the rules. Depending on the strategy employed we can obtain different properties for the calculus.

This section makes explicit all these components for the ρ-calculus.

2.1 Syntax, Substitution and Matching

Definition 2.1 We consider a set $\mathcal{F} = \bigcup_m \mathcal{F}_m$ of ranked function symbols, where $\mathcal{F}_m$ is the subset of function symbols of arity m, and $\mathcal{X}$ a set of variables. We denote by $\mathcal{T}(\mathcal{F}, \mathcal{X})$ the set of first-order terms built on $\mathcal{F}$ using the variables in $\mathcal{X}$. The set of basic ρ-terms, denoted $\mathcal{F}_\rho$, can be inductively defined by the following grammar:

$$\text{terms} \quad t ::= x \mid \{t, \ldots, t\} \mid f(t, \ldots, t) \mid [t](t) \mid t \to t$$

where $x \in \mathcal{X}$ and $f \in \mathcal{F}$.

We adopt a very general discipline for the rule formation, and we do not enforce any of the standard restrictions used when defining the rewriting relation like non-variable left-hand-sides or occurrence of the right-hand-side variables in the left-hand-side. We also allow rules containing rules as well as rule application. We consider that the symbols $\{\}$ and $\emptyset$ both represent the empty set.

The main intuition behind this syntax is that a rewrite rule is an abstractor, the left-hand-side of which determines the bound variables and some contextual information. Having new variables in the right-hand-side is just the ability to have free variables in the calculus. We will come back to this later but, in order to support the intuition of the reader, let us mention that the λ-term $\lambda x.(y\ x)$ corresponds to the ρ-term $x \to [y](x)$ and that standard first-order rewrite rules [Dershowitz and Jouannaud, 1990; Baader and Nipkow, 1998] are clearly embedded in the calculus.

Example 2.1 Some more examples of ρ-terms are:

- $[x \to y](a)$; as expected, we will see below why the result is $\{y\}$.
- $[x \to (x \to x)]([x \to y](a))$; a more complicated ρ-term similar to the λ-term $((\lambda x.\lambda x.x)\ (\lambda x.y\ a))$.
- $[(a \to b) \to c](a)$; a ρ-term without corresponding λ-term.

As in λ-calculus, α-conversion should be used in order to obtain a correct substitution calculus and the first order substitution (called here grafting) is not directly suitable for our calculus.

Computing the matching substitutions from a ρ-term t to a ρ-term t' is an important parameter in the evaluation rules of the ρ-calculus. For a theory T a *T-match-equation* is a formula of the form $t \ll^?_T t'$, where t and t' are ρ-terms. A substitution σ is a solution of the T-match-equation $t \ll^?_T t'$ if $T \models \sigma(t) = t'$. A *T-matching system* is a conjunction of T-match-equations. A substitution is a solution of a T-matching system P if it is a solution of all the T-match-equations in P. We denote by $\mathbf{F}$ a T-matching system without solution.

Since in general we consider higher-order terms as well as arbitrary equational theories, T-matching is in general undecidable, even when restricted to first-order equational theories [Jouannaud and Kirchner, 1991].

A T-matching system is called *trivial* when all substitutions are a solution of it. We define the function *Solution* on a T-matching system $\mathcal{S}$ as returning the set of all T-matches of $\mathcal{S}$ when $\mathcal{S}$ is not trivial and $\{\mathbb{ID}\}$ where $\mathbb{ID}$ is the identity substitution when $\mathcal{S}$ is trivial. When the matching algorithm fails *Solution* returns the empty set ($\emptyset$).

One may consider using constraints as in constrained higher-order resolution [Huet, 1973] or constrained deduction [Kirchner *et al.*, 1990] if one wants to use undecidable matching theories. But we are primarily interested here in the decidable cases and we restrict ourselves later in this paper to just considering syntactic matching. In this case we assume that the left members of matching equations are composed only of first-order terms (i.e. not containing sets, arrows or applications). Notice that the last ρ-term presented in Example 2.1 does not satisfy this restriction.

2.2 Deduction Rules

Given a rewrite rule $l \to r$, its application to a term t (denoted $[l \to r](t)$) consists of replacing the term t by σr, where σ is the substitution obtained by T-matching l on t and σr represents the application of the substitution on the term r as detailed above. The semantics of the application of a rewrite rule is given by the rules in Figure 1.

Fire	$[l \to r](p)$
$\Longrightarrow$	$r\langle\!\langle Solution(l \ll^?_T p)\rangle\!\rangle$
Congruence	$[f(u_1,\ldots,u_n)](f(v_1,\ldots,v_n))$
$\Longrightarrow$	$\{f([u_1](v_1),\ldots,[u_n](v_n))\}$
Congruence_fail	$[f(u_1,\ldots,u_n)](g(v_1,\ldots,v_m))$
$\Longrightarrow$	$\emptyset$
Distrib	$[\{u_1,\ldots,u_n\}](v)$
$\Longrightarrow$	$\{[u_1](v),\ldots,[u_n](v)\}$
Batch	$[v](\{u_1,\ldots,u_n\})$
$\Longrightarrow$	$\{[v](u_1),\ldots,[v](u_n)\}$
Switch$_{\mathbf{L}}$	$\{u_1,\ldots,u_n\} \to v$
$\Longrightarrow$	$\{u_1 \to v,\ldots,u_n \to v\}$
Switch$_{\mathbf{R}}$	$u \to \{v_1,\ldots,v_n\}$
$\Longrightarrow$	$\{u \to v_n,\ldots,u \to v_n\}$
OpOnSet	$f(v_1,\ldots,\{u_1,\ldots,u_m\},\ldots,v_n)$
$\Longrightarrow$	$\{f(v_1,\ldots,u_1,\ldots,v_n),\ldots,f(v_1,\ldots,u_m,\ldots,v_m)\}$
Flat	$\{u_1,\ldots,\{v_1,\ldots,v_n\},\ldots,u_m\}$
$\Longrightarrow$	$\{u_1,\ldots,v_1\ldots,v_n,\ldots,u_m\}$

Figure 1: Basic ρ_T-calculus

The *Fire* rule *starts* the reduction process by T-matching the left-hand side of the rule on the argument of the rule. This generalizes the β-rule of λ-calculus. As we have seen before, the matching problem $l \ll^?_T t$ is solved in the theory T and, depending on the theory T, we obtain various rewriting calculi of interest.

Since rule application occurs at the top level of terms, in order to push rule application deeper into terms, we introduce two *Congruence* rules that deal with the application of a term of the form $f(t_1, \ldots, t_n)$ (where $f \in \mathcal{F}_n$) to another term of the same form. When we have the same head symbol for the two terms of the application $[u](v)$ the arguments of the term u are applied on those of the term v argumentwise. If the head symbols are not similar an empty set is obtained.

The $Distrib, Batch$ and $Switch_L, Switch_R$ rules describe the propagation of the sets on the application and abstraction operators. The *OpOnSet* rule describes the semantics of a function with set arguments. The *Flat* rule is used to flatten the sets of sets.

The behavior of the "$\langle\!\langle\rangle\!\rangle$" operator is described by the rule *Propagate* in Figure 2. Note that this rule is applied at the meta-level and is not made explicit in the ρ-calculus. In the next section we give a version of the ρ-calculus with explicit substitutions in the spirit of [Abadi *et al.*, 1991].

$$\textbf{Propagate} \quad r\langle\!\langle\{\sigma_1, \ldots, \sigma_n\}\rangle\!\rangle \quad \rightsquigarrow \quad \{\sigma_1\, r, \ldots, \sigma_n\, r\}$$

Figure 2: Substitution rules

The *Propagate* rule yields as a result the union of the partial results obtained by applying each of the substitutions to the respective term. When the matching problem has no solution the result is the empty set.

Example 2.2 The reduction for a ρ-term representing the application of a rewrite rule on a term:

$$[f(x,y) \to g(x,y)](f(a,b)) \stackrel{Fire}{\Longrightarrow} g(x,y)\langle\!\langle Solution(f(x,y) \ll^? f(a,b))\rangle\!\rangle$$
$$\Longrightarrow g(x,y)\langle\!\langle\{(x \mapsto a, y \mapsto b)\}\rangle\!\rangle \stackrel{Propagate}{\rightsquigarrow} \{g(a,b)\}$$

We note that the syntax and the deduction rules of the ρ-calculus can be restricted to some simpler calculi such as λ-calculus and term rewriting. A λ-term of the form $\lambda x.t$ is represented by a ρ-term of type $x \to t$ and a rewrite rule is also a ρ-rewrite-rule (see Section 4.2).

2.3 Properties of the ρ-calculus

The general ρ-calculus is obviously non-confluent. The main reasons for the non-confluence are undesired matching failures involving uninstantiated variables and unreduced terms and the set representation for the results of reductions.

The first main reason comes from the fact that a matching system containing free variables or unreduced terms that lead to a failure can become successful when the variables are instantiated or the terms reduced. Example 2.3

illustrates the case of a matching failure due to an uninstantiated variable. For the case of unreduced terms we can consider the term $[a \to b]([a \to a](a))$ that reduces to $\{b\}$ or $\emptyset$ following the position where the rule *Fire* is applied first. Similarly, we can obtain different derivations for terms containing non-left linear rules ($[f(x,x) \to b](f(a,[a \to a](a)))$) or when a rule is applied on a set ($[f(x) \to b](\{f(a)\})$).

Example 2.3

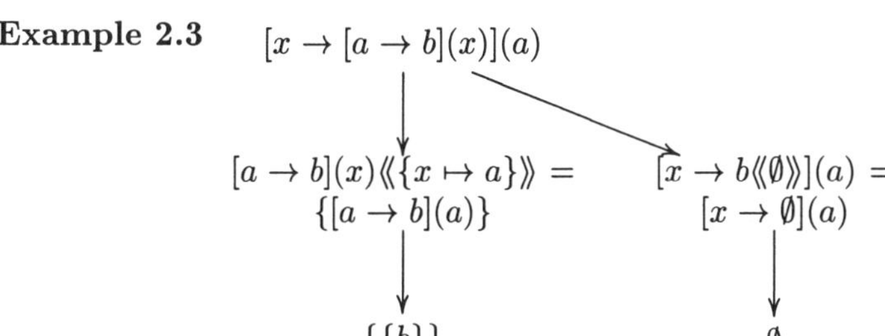

As far as it concerns the handling of the non-determinism (sets of results) we have two types of problems. On one hand, we can have empty sets representing a matching failure that are eventually not propagated properly (for instance $[x \to b](\emptyset)$ reduces to $\{b\}$ or $\emptyset$ according to the rule applied) and on the other hand, we have sets with more than one element that can lead to undesirable results in a non-linear context (Example 2.4).

Example 2.4

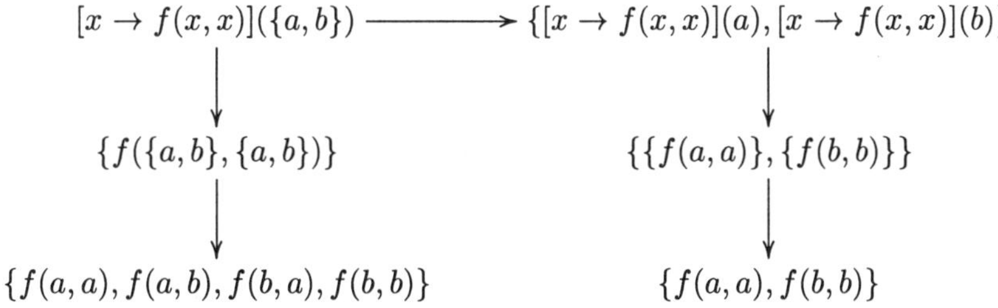

Every time a ρ-term is reduced using one of the rules *Fire*, *Congruence* and *Congruence_fail* of the ρ-calculus, a set is generated. These rules are the ones that describe the application of a rule at the top level or deeper in a term. The set obtained when applying one of the above rules can trigger the application of the other rules of the calculus. These rules deal with the (propagation of) sets and compute a "set-normal form" for the ρ-terms by pushing out the set braces and flattening the imbricated sets.

Therefore, we consider that the rules of the ρ-calculus consist of a set of *deduction* rules (*Fire*, *Congruence*, *Congruence_fail*) and a set of *computation* rules (*Distrib*, *Batch*, $Switch_L$, $Switch_R$, *OpOnSet*, *Flat*) and that the reduction behaves as a deduction modulo ([Dowek *et al.*, 1998]).

Let us now define the relations induced by the rules of the ρ-calculus and present their main properties.

Definition 2.2 We call *Set* the reduction relation over $\mathcal{F}_\rho$ induced by the rules *Distrib*, *Batch*, $Switch_L$, $Switch_R$, *OpOnSet* and *Flat*.

The following relations are induced by the relation *Set*:

- $\longrightarrow_S$ is the one step *Set*-reduction (the compatible closure of *Set*),
- $\stackrel{*}{\longrightarrow}_S$ is the *Set*-reduction (the transitive-reflexive closure of $\longrightarrow_S$),
- $=_S$ is the equivalence relation generated by $\stackrel{*}{\longrightarrow}_S$.

Definition 2.3 We call *Fire* the reduction relation over $\mathcal{F}_\rho$ induced by the rules *Fire*, *Congruence* and *Congruence_fail*. Starting from this relation, we consider the following relations induced by the relations *Fire* and *Set*:

- $\longrightarrow_F$ is the compatible closure of the relation *Fire*,
- $\longrightarrow_{F/S}$ is the relation $\longrightarrow_F$ modulo the relation $=_S$ defined as usual ([Aho *et al.*, 1972]): given two ρ-terms u, v we have $u \longrightarrow_{F/S} v$ when there exist two terms u', v' such that $u =_S u'$ and $u' \longrightarrow_F v'$ and $u =_S u'$,
- $\stackrel{*}{\longrightarrow}_{F/S}$ is the transitive-reflexive closure of the relation $\longrightarrow_{F/S}$.

Note that the relation $\stackrel{*}{\longrightarrow}_{F/S}$ is not the transitive closure of the relation $\longrightarrow_F$ ($\stackrel{*}{\longrightarrow}_F$) modulo the relation $=_S$.

Since we can consider the reduction in the ρ-calculus as the relation induced by the rules of the calculus ($\longrightarrow_F \cup \longrightarrow_S$) or as a rewriting modulo ($\longrightarrow_{F/S}$), we can prove the confluence of the two relations.

As already said the calculus is not confluent if no strategy is used for guiding the application of the reduction rules but the confluence can be obtained if a reduction strategy is imposed.

Definition 2.4 We call *ConfStrat* the strategy whose reductions are denoted by $\longrightarrow_{CS}$ and that satisfies the next conditions, where the notation $u\lceil t \rceil_p$ denotes the term u having t at occurrence p.

- If $u\lceil t \rceil_p \stackrel{*}{\longrightarrow}_{CS} u\lceil \emptyset \rceil_p$ then for all term v such that $u\lceil t \rceil_p \stackrel{*}{\longrightarrow}_{CS} v$, we have $v \stackrel{*}{\longrightarrow}_{CS} \emptyset$ (i.e. "error" propagation is strict),
- if $[l \to r](p) \stackrel{Fire}{\longrightarrow}_{CS} r'\lceil p \rceil_i \lceil p \rceil_j$ then for all term v such that $p \stackrel{*}{\longrightarrow}_{CS} v$, there exists a ρ-term t such that $v \stackrel{*}{\longrightarrow}_{CS} \{t\}$ or $v \stackrel{*}{\longrightarrow}_{CS} \emptyset$ (i.e. v cannot be reduced to a set with more than one element).

From now on we consider that the rules of the calculus are guided by the strategy *ConfStrat*. Of course, more operational strategies can be given for obtaining the confluence of the calculus.

For proving the confluence of $\longrightarrow_{F/S}$ we try to find a relation δ that has the diamond property and whose transitive closure is the relation $\xrightarrow{*}_{F/S}$. In this case, we can conclude that the relation $\xrightarrow{*}_{F/S}$ has the diamond property and thus, $\longrightarrow_{F/S}$ is confluent.

We can choose for δ the reflexive closure of $\longrightarrow_{F/S}$. Unfortunately, it can be easily checked that this relation does not have the diamond property. We shall take an approach similar to the "parallel reduction" due to *Tait and Martin-Löf* and prove that $\longrightarrow_{F\|/S}$ is confluent and its transitive closure is $\xrightarrow{*}_{F/S}$.

Lemma 2.1 *If* $\longrightarrow_{F\|}$ *is the parallelization of the relation* $\longrightarrow_F$ *then* $\longrightarrow_{F\|}$ *is strongly confluent.*

Proof: By induction on the structure of the ρ-terms and using a "substitution lemma". □

Lemma 2.2 *The relation* $\longrightarrow_S$ *is confluent and terminating.*

Proof: We prove the termination by using a polynomial interpretation for the rewrite rules. The local confluence is obtained by proving the convergence of all critical pairs. We can conclude the confluence of the relation $\longrightarrow_S$. □

Proposition 2.1 *The relation* $\longrightarrow_{F/S}$ *is confluent.*

Proof: First, we show that if $t \longrightarrow_S t'$ and $t \longrightarrow_{F\|} s$, then there exists s' such that $t' \xrightarrow{*}_S \longrightarrow_{F\|} \xrightarrow{*}_S s'$ and $s \xrightarrow{*}_S s'$. The confluence of the relation $\longrightarrow_S$ and the strong confluence of the relation $\longrightarrow_{F\|}$ allow us to prove that $\longrightarrow_{F\|/S}$ has the diamond property.

It is then easy to prove that the relation $\xrightarrow{*}_{F/S}$ is the transitive closure of the relation $\longrightarrow_{F\|/S}$ and, thus, we conclude that $\longrightarrow_{F/S}$ is confluent. □

It is worth mentioning that the confluence of the relation induced by the rules of the calculus is similar to the approach proposed in [Curien *et al.*, 1996] for proving the confluence of $\lambda_{\Uparrow}$.

Proposition 2.2 *The relation* $\xrightarrow{*}_S \longrightarrow_F \xrightarrow{*}_S$ *is confluent.*

Proof: The Lemma of Yokouchi ([Curien *et al.*, 1996]) is used to show the strong confluence of $\xrightarrow{*}_S \longrightarrow_{F\|} \xrightarrow{*}_S$. Since $\longrightarrow_F \subset \longrightarrow_{F\|} \subset \xrightarrow{*}_F$ we obtain the confluence of $\xrightarrow{*}_S \longrightarrow_F \xrightarrow{*}_S$. □

3 THE ρ_σ-CALCULUS

Up to now the substitution application was not part of the calculus. In order to make explicit the role of substitutions we define an explicit version of ρ-calculus. We start by giving the syntax and the basic deduction rules for the ρ_σ-calculus with explicit substitutions and we present the confluence results.

Comparing the ρ_σ-calculus to the $\lambda_\Uparrow$-calculus one should notice the ability to have several variables bound by the ρ-abstractor ($\rightarrow$) and the possibility to use contextual information handled by the matching mechanism.

3.1 Syntax

In what follows we concentrate on a presentation that relies on the de Bruijn's numbering for variables [de Bruijn, 1972]. This way the renaming of variables is handled via incrementation or decrementation of the integers representing variables.

If we consider a set $\mathcal{F} = \bigcup_m \mathcal{F}_m$ of ranked function symbols, where $\mathcal{F}_m$ is the subset of functions of arity m then the $\rho\sigma$-terms are formed according to the following rules:

$$\textsf{terms} \quad t ::= n \mid \emptyset \mid \{t, \ldots, t\} \mid f(t, \ldots, t) \mid [t](t) \mid t \rightarrow_n t \mid t\langle s \rangle$$

$$\textsf{substitutions} \quad s ::= \mathbb{ID} \mid \uparrow \mid \Uparrow(s) \mid t.s \mid s \circ s$$

where $n \in \mathbf{N}^*$ and $f \in \mathcal{F}$.

A rule of the form $(l \rightarrow_n r)$ represents a rule with n bound variables by its left-hand side, that is, l contains n free variables (e.g. $1 + 1 + 2 \rightarrow_2 1 + 2$).

3.2 Deduction Rules

The semantics of the application of a rewrite rule is given by the deduction rules of the $\rho\sigma_T$-calculus (Figure 3) and by the rules describing the application of a substitution on a term (Figure 4).

The deduction rules are similar to the ones of the ρ-calculus but this time we deal with indices instead of variable names and the application of substitutions is explicit in the calculus.

The rule *Fire* starts the reduction process by matching the left-hand side of the rewrite rule on the initial input term and applying the obtained substitution(s) on the right-hand side of the rule. The implicit $\langle\!\langle\rangle\!\rangle$ substitution mechanism becomes explicit in the new form of the rule *Fire*. The sub-system σ_ρ applies explicitly the obtained substitutions on the right-hand side of the rule. All the other rules are very similar to their non-explicit counterparts.

We consider a matching algorithm that returns a result of the form

$$nt_1. \ldots .nt_n.\mathbb{ID}$$

Fire	$[l \rightarrow_n r](p)$ $\Longrightarrow$ $\{r\langle\sigma_1\rangle, \ldots, r\langle\sigma_n\rangle\}$ where $\sigma_i \in \mathcal{S}olution(l \ll^?_T p)$
Congruence	$[f(u_1, \ldots, u_n)](f(t_1, \ldots, t_n))$ $\Longrightarrow$ $\{f([u_1](t_1), \ldots, [u_n](t_n))\}$
Congruence_fail	$[f(u_1, \ldots, u_n)](g(t_1, \ldots, t_m))$ $\Longrightarrow$ $\emptyset$ if $f \neq g$
Distrib	$[\bigcup_{i=1}^n \{u_i\}](t)$ $\Longrightarrow$ $\bigcup_{i=1}^n \{[u_i](t)\}$
Batch	$[u](\bigcup_{i=1}^n \{t_i\})$ $\Longrightarrow$ $\bigcup_{i=1}^n \{[u](t_i)\}$
Switch$_\mathbf{L}$	$\bigcup_{i=1}^n \{u_i\} \rightarrow_k v$ $\Longrightarrow$ $\bigcup_{i=1}^n \{u_i \rightarrow_k v\}$
Switch$_\mathbf{R}$	$u \rightarrow_k \bigcup_{i=1}^n \{v_i\}$ $\Longrightarrow$ $\bigcup_{i=1}^n \{u \rightarrow_k v_i\}$
OpOnSet	$f(u, \ldots, \bigcup_{i=1}^n \{t_i\}, \ldots, v)$ $\Longrightarrow$ $\bigcup_{i=1}^n \{f(u, \ldots, t_i, \ldots, v)\}$
Flat	$\{u_1, \ldots, \{v_1, \ldots, v_n\}, \ldots, u_m\}$ $\Longrightarrow$ $\{u_1, \ldots, v_1, \ldots, v_n, \ldots, u_m\}$

Figure 3: Basic $\rho\sigma_T$-calculus

if the classical matching algorithm [Jouannaud and Kirchner, 1991] returns the result $\{x_i \mapsto t_i\}_{i=1,\ldots,n}$ for the same problem and nt_i are the de Bruijn representation of the terms t_i in the corresponding referential. The transformation of a term in the de Bruijn representation is described below. As in section 2.1, we define the function *Solution* on a T-matching system $\mathcal{S}$ as returning the set of all T-matches of $\mathcal{S}$ when $\mathcal{S}$ is not trivial, $\{\mathbb{ID}\}$ when $\mathcal{S}$ is trivial and the empty set ($\emptyset$) when the matching algorithm fails.

The rewriting rules of the reduction relation σ_ρ are very similar to the ones described in [Curien *et al.*, 1996] and [Pagano, 1998] for the relations σ and $\sigma_\Uparrow$ respectively.

We abbreviate by $\uparrow^n$ the composition of n symbols $\uparrow$ (i.e. $\uparrow \circ \ldots \circ \uparrow$) and by $\Uparrow^n (s)$ the application n times of $\Uparrow$ (i.e. $\Uparrow (\ldots(\Uparrow (s)\ldots)))$.

The rules in σ_ρ applied on a term $t\langle s\rangle$ yield a normal form that does not contain the application of a substitution if the substitution s does not represent a matching system without solution ($\emptyset$) and provide $\emptyset$ as solution in this latter case. The application of a substitution on an abstraction, application, set and function is described by the rules *lam*, *app*, *set* and $f_\Uparrow$ respectively.

In the $\lambda\sigma$-calculus the term λa binds the variable 1 in the term a and we can say that λ has one (implicit) argument that is bound and that the binding arity of λ is (1). For a more detailed discussion about binding arities one should refer to [Pagano, 1998]. The unary symbol "λ" from the σ system is replaced in our calculus by the binary symbol "$\rightarrow_n$" of a binding arity (n, n) (each of the two arguments of $\rightarrow_n$ contains n bound variables). All the first-order symbols "f" have a binding arity (0,...,0).

Definition 3.1 The calculus consisting of the deduction rules in Figure 3 and the subsystem σ_ρ (Figure 4) is called the ρ_σ-calculus.

The transformation of a term with real names from the ρ-calculus in a term using natural indices is similar to the transformation of a λ term in a λ_{DB} term except the rule dealing with the rewrite rules.

We consider a list of bound variables (e.g. $x.y.z.Nil$) that is called a referential. Having done a referential $\mathcal{R}$ we define recursively the translation of a term t, written $tr(t, \mathcal{R})$.

1. $tr(x, \mathcal{R}) = j$, where j is the first position of x in $\mathcal{R}$,
2. $tr(f(t_1, \ldots, t_n, \mathcal{R})) = f(tr(t_1, \mathcal{R}), \ldots, tr(t_n, \mathcal{R}))$,
3. $tr([a](b), \mathcal{R}) = [tr(a, \mathcal{R})](tr(b, \mathcal{R}))$,
4. $tr(l \rightarrow r, \mathcal{R}) = tr(l, var(l).\mathcal{R}) \rightarrow_{\|var(l)\|} tr(r, var(l).\mathcal{R})$,

where $var(l)$ represents the list of free variables of the term l and $\|var(l)\|$ its length.

Example 3.1 If we consider the term $[f(x, y) \rightarrow g(x, y, z)](f(a, b))$ its transformation is $[f(1, 2) \rightarrow_2 g(1, 2, 3)](f(a, b))$ in the referential $z.Nil$.

lam	$(u \to_n v)\langle s\rangle$	$\Rightarrow$	$u\langle \Uparrow^n (s)\rangle \to_n v\langle \Uparrow^n (s)\rangle$
app	$[u](v)\langle s\rangle$	$\Rightarrow$	$[u\langle s\rangle](v\langle s\rangle)$
clos	$u\langle s\rangle\langle t\rangle$	$\Rightarrow$	$u\langle s \circ t\rangle$
vs1	$n\langle \uparrow\rangle$	$\Rightarrow$	$(n+1)$
vs2	$n\langle \uparrow \circ s\rangle$	$\Rightarrow$	$(n+1)\langle s\rangle$
fvc	$1\langle u.s\rangle$	$\Rightarrow$	u
fvl1	$1\langle \Uparrow (s)\rangle$	$\Rightarrow$	1
fvl2	$1\langle \Uparrow (s) \circ t\rangle$	$\Rightarrow$	$1\langle t\rangle$
rvc	$(n+1)\langle u.s\rangle$	$\Rightarrow$	$n\langle s\rangle$
rvl1	$(n+1)\langle \Uparrow (s)\rangle$	$\Rightarrow$	$n\langle s \circ \uparrow\rangle$
rvl2	$(n+1)\langle \Uparrow (s) \circ t\rangle$	$\Rightarrow$	$n\langle s \circ (\uparrow \circ t)\rangle$
id	$u\langle \mathbb{ID}\rangle$	$\Rightarrow$	u
set	$\bigcup_{i=1}^{n}\{u_i\}\langle s\rangle$	$\Rightarrow$	$\bigcup_{i=1}^{n}\{u_i\langle s\rangle\}$
ass	$(s \circ t) \circ v$	$\Rightarrow$	$s \circ (t \circ v)$
map	$(u.s) \circ t$	$\Rightarrow$	$u\langle t\rangle.(s \circ t)$
sc	$\uparrow \circ (u.s)$	$\Rightarrow$	s
sl1	$\uparrow \circ \Uparrow (s)$	$\Rightarrow$	$s \circ \uparrow$
sl2	$\uparrow \circ \Uparrow (s) \circ t$	$\Rightarrow$	$s \circ (\uparrow \circ t)$
l1	$\Uparrow (s) \circ \Uparrow (t)$	$\Rightarrow$	$\Uparrow (s \circ t)$
l2	$\Uparrow (s) \circ (\Uparrow (t) \circ v)$	$\Rightarrow$	$\Uparrow (s \circ t) \circ v$
le	$\Uparrow (s) \circ (u.t)$	$\Rightarrow$	$u.(s \circ t)$
il	$\mathbb{ID} \circ s$	$\Rightarrow$	s
ir	$s \circ \mathbb{ID}$	$\Rightarrow$	s
li	$\Uparrow (\mathbb{ID})$	$\Rightarrow$	$\mathbb{ID}$
$\mathbf{f}_{\Uparrow}$	$f(u_1, \ldots, u_n)\langle s\rangle$	$\Rightarrow$	$f(u_1\langle s\rangle, \ldots, u_n\langle s\rangle)$

Figure 4: The σ_ρ calculus

3.3 Properties of the ρ_σ-calculus

The system applying substitutions is confluent and terminating as stated below:

Lemma 3.1 *σ_ρ is locally confluent and terminating.*

Proof: We use the methods proposed in [Curien *et al.*, 1996] and [Pagano, 1998] for the relations σ and $\sigma_\Uparrow$ respectively. □

In order to prove now that the ρ_σ-calculus is itself confluent, we can use the notion of Explicit Reduction Systems(XRS) as defined by [Pagano, 1998] or a direct proof as done for $\lambda_\Uparrow$ in [Curien *et al.*, 1996].

The first approach consists in expressing the ρ_σ-calculus as an XRS satisfying the confluence conditions stated in [Pagano, 1998]. The most restrictive condition is the orthogonality between the high-order rules ($Fire$ in our case) and the first-order rules (all the other rules presented in Figure 3). When an appropriate strategy for guiding the application of the rule $Fire$ is used, this condition is satisfied and the calculus is confluent:

Proposition 3.1 *We suppose that the $Fire$ rule is applied on a sub-term $[l \to_n r](p)$ of a term t only when r and p contain no redexes and no sets and p contains no free variables that are bound in t. Following this strategy, the ρ_σ-calculus is confluent.*

The definitions from Section 2.3 are extended for the relations induced by the rules of the ρ_σ-calculus and the same conditions on the application of the rule $Fire$ are considered. For the direct proof of the confluence of the calculus, when the strategy from the Definition 2.4 is used, we consider two appropriate relations and we show that the conditions for the Lemma of Yokouchi ([Curien *et al.*, 1996]) are satisfied. The first relation is $\longrightarrow_{F\|}$ and the second one is $(\overset{*}{\longrightarrow}_\sigma \longrightarrow_S \overset{*}{\longrightarrow}_\sigma)$ and, for simplicity, we denote this last one by $\longrightarrow_{\sigma^* S \sigma^*}$.

Lemma 3.2 *The relation $\longrightarrow_{\sigma^* S \sigma^*}$ is confluent and terminating.*

Proof: The termination of $\longrightarrow_{\sigma^* S \sigma^*}$ is proved by giving a polynomial interpretation to the operators of the calculus.

First, we show the "compatibility" between $\longrightarrow_S$ and $\longrightarrow_\sigma$: if $t \longrightarrow_S t'$ and $t \longrightarrow_\sigma s$ then there exists s' such that $t' \overset{*}{\longrightarrow}_\sigma s'$ and

$$s \overset{*}{\longrightarrow}_\sigma \longrightarrow_S \overset{*}{\longrightarrow}_\sigma s'.$$

Using the compatibility between the two relations and their confluence we obtain the confluence of $\longrightarrow_{\sigma^* S \sigma^*}$ by induction on the number of $\longrightarrow_\sigma$ reductions. □

Proposition 3.2 *The ρ_σ-calculus is confluent.*

Proof: We show that the relations $\longrightarrow_{F\|}$ and $\longrightarrow_{\sigma^* S\sigma^*}$ are compatible in the sense explained in the previous lemma and we apply the lemma of Yokouchi [Curien *et al.*, 1996] for the relations $\longrightarrow_{F\|}$ and $\longrightarrow_{\sigma^* S\sigma^*}$. □

As one would expect, we have a strong relationship between the ρ-calculus and the ρ_σ-calculus:

Proposition 3.3 *Given t and t' two ρ-terms. Then $t \stackrel{\rho}{\Longrightarrow} t'$ iff $tr(t,\mathcal{R}) \stackrel{\rho\sigma}{\Longrightarrow} tr(t',\mathcal{R})$ for any $\mathcal{R}$. If t and t' are α-equivalent then $tr(t,\mathcal{R})$ and $tr(t',\mathcal{R})$ are equal for any $\mathcal{R}$.*

3.4 Primal Strategies

The basic ρ-calculus is already quite powerful because of the matching and substitution mechanisms it relies on, but we would like to enrich it further to allow an easy specification of choice operators. Indeed the whole calculus is intended to facilitate the specification and execution of non-deterministic, possibly non-confluent, rewrite systems as needed for concurrent systems as well as general deduction systems like theorem provers or constraint solvers.

The basic calculus could be enriched in two ways. The first one is to consider a fixpoint operator; the second one consists of an extension of the syntax and of the inference rules of the calculus in order to integrate directly the desirable operators. We have chosen the second approach as it is more direct and close to what is done in ELAN, as we will discuss below.

Considering the ρ-terms as functions that apply rewrite rules, we called them ***strategies***, and since we are building an extension of the existing syntax of terms, we call ***primal strategies*** basic terms to which we add a sequence operator in order to denote the successive application of two strategies.

To this end, the syntax from section 2.1 is extended with the operators **id** and **fail** that correspond respectively to the identity rule ($x \to x$) and the failure rule whose application fails on any term ($x \to \emptyset$):

terms $\quad t ::= id \mid fail$

and the deduction rules are:

Identity	id	$\Longrightarrow$	$x \to x$
Fail	$fail$	$\Longrightarrow$	$x \to \emptyset$

The sequential application of two strategies is achieved by using the concatenation operator ";":

terms $\quad t ::= t; t$

The behavior of the concatenation operator is expressed by the following deduction rule:

$$\textbf{Compose} \quad [s_1; s_2](t) \quad \Longrightarrow \quad [s_2]([s_1](t))$$

Definition 3.2 *Primal strategies* are built starting from rewrite rules (i.e. ρ-terms of the form $l \to r$) and the two operators id and $fail$ and using concatenation. Their application is defined by the rules in Figure 1, *Identity*, *Fail* and *Compose*.

One can notice that the operators introduced in this section are just shortcuts for some particular ρ-terms and, thus, the properties of the ρ-calculus are still valid for the calculus extended with the above aliases.

Example 3.2 If we consider the primal strategy

$$g(f(f(x) \to x)); g(f(x) \to x); g(a) \to a$$

then the only term on which the application of the strategy yields as a result a non-empty set is $g(f(f(a)))$.

If the term is of the form $g(f(t))$ with t not of the form $f(t')$ then the *Congruence* rule can be applied but the matching would fail and the result is the empty set. If t is of the form $f(t')$ with t' different from a then the application of the strategy $g(f(f(x) \to x)); g(f(x) \to x)$ to the term would yield as a result $g(t')$ and we obtain once again the empty set due to a matching failure when applying the last rule. In the other cases the rule *Congruence_fail* would lead immediately to an empty set result.

3.5 Elementary Strategies

The strategy language is now enriched by operators, allowing us to describe the computation space in a convenient way.

The added operators select the final result(s) from a list of results and choose the strategy to be applied from a list of strategies. For the first type of selection we introduce the unary operators **one** and **all** that operate on strategies and whose semantics are described by the following rules:

$$\begin{array}{llll} \textbf{One} & [one(s)](t) & \Longrightarrow & \{t'\} \\ & & & \text{where } t' \in [s](t) \\ \textbf{All} & [all(s)](t) & \Longrightarrow & [s](t) \end{array}$$

Then we can also introduce selection operators **select-one**, **select-first**, **select-all** on strategy tuples ([Borovanský *et al.*, 1998]). Their semantics is defined by the rules:

$$\begin{array}{lll} \textbf{Select\_first} & [select\_first(s_1, \ldots, s_n)](t) \Longrightarrow [s_j](t) \\ & \qquad \text{if } \bigcup_{i=1}^{j-1}[s_i](t) = \emptyset \text{ and } [s_j](t) \neq \emptyset \\ \textbf{Select\_one} & [select\_one(s_1, \ldots, s_n)](t) \Longrightarrow [s_i](t) \\ & \qquad \text{if } [s_i](t) \neq \emptyset \\ \textbf{Select\_all} & [select\_all(s_1, \ldots, s_n)](t) \Longrightarrow \bigcup_{i=1}^{n}[s_i](t) \end{array}$$

The operator **select-first** returns the result of the first successful application, **select-one** selects one of the successful applications and **select-all** provides the union of all (successful) applications. These selection operators allow us to describe the operators **first**, **dc**(dont care) and **dk**(dont know) by the rules:

Dk	$[dk(s_1, \ldots, s_n)](t)$	$\Longrightarrow$	$[select\_all(all(s_1), \ldots, all(s_n))](t)$
Dc	$[dc(s_1, \ldots, s_n)](t)$	$\Longrightarrow$	$[select\_one(all(s_1), \ldots, all(s_n))](t)$
First	$[first(s_1, \ldots, s_n)](t)$	$\Longrightarrow$	$[select\_first(all(s_1), \ldots, all(s_n))](t)$

Intuitively, the selectors are used to select the strategy and the **one** and **all** operators determine the number of results to be returned when applying the respective strategy to the input term.

Some other strategies can be constructed starting from the above operators. For example, an elementary strategy that chooses one result in each set of results provided by a sequence of strategies is :

Dc_One $[dc\_one(s_1, \ldots, s_n)](t) \Longrightarrow$
$[select\_one(one(s_1), \ldots, one(s_n))](t)$

Note also that the result selector can be made more precise when a different representation for results is chosen. For example, when using lists, the **one** operator could be completed with "nth" ones.

It is easy to show that the **dk** operator can be expressed in the basic ρ-calculus. It is an open problem whether this is the case for the **dc** and **first** operators.

3.6 Defined Strategies

Using this kind of operators and primal strategies we can define more elaborate strategies. For example, a function **map** can be defined in an explicit way by:

Map1	$[map(s)](nil)$	$\Longrightarrow$	nil
Map2	$[map(s)](a.l)$	$\Longrightarrow$	$[s](a).[map(s)](l)$

or in an implicit (and more compact) way:

Map $map(s) \Longrightarrow first(nil,\ s.map(s))$

Other basic strategy definitions, such as **iterate** or **repeat**, can be given in a similar way.

More complicated control structures can be constructed if we extend the rule *Congruence* with new cases that deal with special head symbols in the strategy application:

Congruence_M $[\Phi(s_1, \ldots, s_n)](f(t_1, \ldots, t_n)) \Longrightarrow$
$f([s_1](t_1), \ldots, [s_n](t_n))$

Congruence_S $[\Psi(s)](f(t_1, \ldots, t_n)) \Longrightarrow$
$f([s](t_1), \ldots, [s](t_n))$

where Φ, Ψ are operators of the language ($\Phi, \Psi \notin \mathcal{F}$) acting as wildcards for the head symbol of a strategy application. These operators are similar to the ones presented in [Visser and el Abidine Benaissa, 1998].

A strategy operator applying a strategy in an *innermost* way is then easy to define:

$$\textbf{Im} \quad im(s) \quad \Longrightarrow \quad \Psi(im(s)); s$$

while the *outermost* strategy has a similar description:

$$\textbf{Om} \quad om(s) \quad \Longrightarrow \quad s; \Psi(om(s))$$

Depending on the order used in evaluating the arguments of the function in the right-hand side of the rule *Congruence_S* the strategy *im* is a *leftmost innermost, rightmost innermost* or *random innermost* strategy.

4 THE ELAN ENVIRONMENT AND ITS ρ-CALCULUS SEMANTICS

4.1 ELAN's Rewrite Rules

ELAN is an environment for specifying and prototyping deduction systems in a language based on labeled conditional rewrite rules and strategies to control rule application. The ELAN system offers a compiler and an interpreter of the language. The ELAN language allows us to describe in a natural and elegant way various deduction systems and to experiment with, for example, the combination of theorem provers, constraint solvers and decision procedures [Vittek, 1994; Kirchner *et al.*, 1995; Borovanský *et al.*, 1996]. It has been experimented on several non-trivial applications ranging from constraint solvers to logic programming and automated theorem proving[1].

ELAN provides a kernel that implements the leftmost innermost standard rewriting strategy, the elementary strategies, and which allows the iteration of the construction on defined strategies as presented in the previous section.

A partial semantics could be given to an ELAN program using rewriting logic [Meseguer, 1992], but more conveniently ELAN's rules can be expressed using the ρ-calculus and thus an ELAN program is just a list of ρ-terms. A rule with no conditions and no local assignments $l \Rightarrow r$ is represented by $l \to r$. The local assignments are let-like constructions that allow applications of strategies on some terms. The general syntax of an ELAN rule is:

$$[\ell] \quad l \Rightarrow r \qquad [\ \textbf{if}\ \ cond \ \mid \ \textbf{where}\ \ y := (S)u\ ]^*$$

We should notice that the square brackets ([ ]) in ELAN are used to indicate the label of the rule and should be distinguished from the square brackets of the ρ-calculus that represent the application of a rule (strategy).

[1]See `http://www.loria.fr/ELAN/` for more details

Example 4.1 An example of an ELAN rule describing one of the deduction rules of the sequent calculus is:

```
[AND] H |- (P && Q) => True
                          where S1 := (dedstrat) H |- Q
                          if S1 =  True
                          where S2 := (dedstrat) H |- P
                          if S2 = True   end
```

The strategy used for evaluating the conditions is a leftmost innermost standard rewriting strategy.

4.2 The ρ-calculus Representation of ELAN Rules

As noted before, term rewriting can be described using the ρ-calculus. Conditional rewrite rules can also be represented in a simple extension of the ρ-calculus.

A conditional rewrite rule of the form

$$l \to r \ \ if \ \ c$$

is represented by the ρ-term

$$l \to [\{True \to r, False \to \emptyset\}]([normalize](c))$$

or even the simpler (but maybe less suggestive) one:

$$l \to [True \to r]([normalize](c))$$

where *normalize* represents the strategy used to evaluate the condition.

Example 4.2 We consider the rewrite rule $f(x) \to g(x) \ \ if(x \geq 1)$ applied to the term $f(2)$.

$$[f(x) \to [True \to g(x)]([normalize](x \geq 1))](f(2)) \Longrightarrow$$
$$\{[True \to g(2)]([normalize](2 \geq 1))\} \Longrightarrow \{[True \to g(2)](\{True\})\}$$
$$\Longrightarrow \{\{[True \to g(2)](True)\}\} \Longrightarrow \{\{\{g(2)\}\}\} \Longrightarrow \{g(2)\}$$

Indeed, an appropriate ρ-term represents any derivation:

Proposition 4.1 *Given t and t' two terms in $\mathcal{T}(\mathcal{F},\mathcal{X})$ and $\mathcal{R}$ a first order conditional term rewrite system, if $t \stackrel{*}{\longrightarrow}_{\mathcal{R}} t'$ then there exists a ρ-term u constructed using the primal strategies of $\mathcal{R}$ such that $[u](t) \stackrel{*}{\longrightarrow}_{\rho} \{t'\}$.*

The rules of the system ELAN can be expressed using the ρ-calculus. A rule with no conditions and no local assignments $l \Rightarrow r$ is represented by $l \to r$ and a conditional rule is expressed as above. The ELAN rewrite rules with local assignments can be given as well a ρ-calculus representation as shown in the following example:

Example 4.3 ELAN's rule

```
[deriveSum] p_1 + p_2 => p_1' + p_2'
                         where p_1' := (derive)p_1
                         where p_2' := (derive)p_2
```

can be represented by one of the following two ρ-terms

$$p_1 + p_2 \to [derive](p_1) + [derive](p_2)$$
$$p_1 + p_2 \to [p_1' \to [p_2' \to p_1' + p_2']([derive](p_2))]([derive](p_1))$$

If we consider more general ELAN rules containing local assignments as well as conditions on the local variables, the combination of the previous two methods should be done carefully. If we had used the former representation for this kind of rules we would have obtained some incorrect results as in Example 4.4. In order to simplify the terms considered we omit the strategies used for reducing the conditions and simply denote $[normalize](c)$ by c.

Example 4.4 We consider the following ELAN rule:

```
[] x => y
       where y := (dk(1 => 2,1 => 3)) x
       if y >= 3 end
```

that would be represented by the ρ-term

$$x \to [True \to [dk(1 \to 2, 1 \to 3)](x)]([dk(1 \to 2, 1 \to 3)](x) >= 3)$$

which, when applied to the term 1, yields the following derivation:

$$\begin{aligned}
&[x \to [True \to [dk(1 \to 2, 1 \to 3)](x)]([dk(1 \to 2, 1 \to 3)](x) >= 3)](1) \\
&\Longrightarrow \{[True \to [dk(1 \to 2, 1 \to 3)](1)]([dk(1 \to 2, 1 \to 3)](1) >= 3)\} \\
&\Longrightarrow \{[True \to \{2,3\}](\{2,3\} >= 3)\} \\
&\Longrightarrow \{[True \to \{2,3\}](2 >= 3) \cup [True \to \{2,3\}](3 >= 3)\} \\
&\Longrightarrow \{[True \to \{2,3\}](False) \cup [True \to \{2,3\}](True)\} \\
&\Longrightarrow \{\emptyset \cup [True \to \{2,3\}](True)\} \\
&\Longrightarrow \{2,3\}
\end{aligned}$$

while the correct result is: $\{3\}$

The problem in Example 4.4 is the double evaluation of the local variable y: once in the condition and once in the right-hand side of the rule. If the local variable is evaluated to a set of results this set is returned if one of its elements satisfies the condition while the appropriate result would be the subset of elements satisfying the condition. Thus, we need a mechanism that evaluates only once all the local assignments in a rule.

We denote by $t\lceil s_1 \rceil_{p_1} \cdots \lceil s_n \rceil_{p_n}$ the term t containing the sub-terms s_1, $\ldots$, s_n at the positions $p_1, \ldots, p_n$ respectively.

Without loss of generality, we consider an ELAN rule which has the following form:

$$\begin{array}{lll} [\text{label}] & l & \Rightarrow \; r\lceil x_1 \rceil_{q_1} \lceil x_2 \rceil_{q_2} \\ & & \quad \text{where } x_1 := (s_1)t_1 \\ & & \quad \text{if } C_1 \lceil x_1 \rceil_{p_1} \\ & & \quad \text{where } x_2 := (s_2)t_2 \\ & & \quad \text{if } C_2 \lceil x_1 \rceil_{r_1} \lceil x_2 \rceil_{r_2} \end{array}$$

where t_{i+1} can depend on x_i.

We denote by *normalize* the ρ-term corresponding to the leftmost innermost strategy of ELAN. If we use an approach similar to the latter ρ-representation of conditional rules, the ELAN rule above is expressed as the ρ-term:

$$\begin{array}{l} l \to \; [x_1 \to \; [\{True \to \; [x_2 \to [\{ \; True \to r\lceil x_1 \rceil_{q_1} \lceil x_2 \rceil_{q_2}, \\ \qquad\qquad\qquad\qquad\qquad\qquad False \to \emptyset\}](C_2 \lceil x_1 \rceil_{r_1} \lceil x_2 \rceil_{r_2}) \\ \qquad\qquad\qquad\qquad]([s_2](t_2)), \\ \qquad\qquad False \to \emptyset\}](C_1 \lceil x_1 \rceil_{p_1}) \\ \qquad]([s_1](t_1)) \end{array}$$

or the simpler one:

$$\begin{array}{l} l \to \; [x_1 \to \; [True \to \; [x_2 \to \; [True \to r\lceil x_1 \rceil_{q_1} \lceil x_2 \rceil_{q_2}] \\ \qquad\qquad\qquad\qquad\qquad\qquad (C_2 \lceil x_1 \rceil_{r_1} \lceil x_2 \rceil_{r_2}) \\ \qquad\qquad\qquad\qquad]([s_2](t_2))] \\ \qquad\qquad (C_1 \lceil x_1 \rceil_{p_1}) \\ \qquad]([s_1](t_1)) \end{array}$$

We notice that, when evaluating the application of such a rule, the ρ-rule $Fire$ should be applied on top before applying it on the right-hand side of the rule, that is exactly the condition requested to get a confluent ρ-calculus.

The way this transformation applies on a rewrite rule and the corresponding reduction of the obtained ρ-term is illustrated in Example 4.5.

Example 4.5 For the following rewrite rule

```
[]x => h(z,l(y))
       where y:= (s1) x+1
       where z:= (s2) x+1
       if y > z
```

the transformed term is:

$$x \to [y \to [z \to [True \to h(z,l(y))](y > z)]([s_2](x{+}1))]([s_1](x{+}1))$$

This term applied to the term 1 with $s_1 = s_2 = dk(2 \to 3, 2 \to 4)$ yields the following derivation:

$$\begin{array}{l} [x \to [y \to [z \to [True \to h(z,l(y))](y > z)]([s_2](x+1))]([s_1](x+1))](1) \\ \Rightarrow \; \{[y \to [z \to [True \to h(z,l(y))](y > z)]([s_2](1+1))]([s_1](1+1))\} \\ \Rightarrow \; \{[y \to [z \to [True \to h(z,l(y))](y > z)]([s_2](1+1))](\{3,4\})\} \\ \ldots \end{array}$$

$$
\begin{aligned}
&\Rightarrow \{\{\{\emptyset\} \cup \{\emptyset\}\} \cup \{\{\{h(3, l(4))\}\} \cup \{\emptyset\}\}\} \\
&\Rightarrow \{\{\{\{h(3, l(4))\}\}\}\} \\
&\Rightarrow \{h(3, l(4))\}
\end{aligned}
$$

The semantics of the primal and elementary strategies of ELAN are reflected by the corresponding strategies in the ρ-calculus and the user defined strategies are similar to the ρ defined strategies.

5 CONCLUSION

We have presented the general properties of the ρ-calculus and given a first-order presentation of the calculus using explicit substitutions.

The ρ-calculus is both conceptually simple as well as very expressive. This allowed us to show how the ρ-calculus can be used to give a semantics to ELAN rules. More generally, the applications to many other frameworks have to be investigated, including rewrite based languages like ASF+SDF, ML, Maude or CafeOBJ but also non-deterministic transition systems.

Among the many research directions concerned by the use of the ρ-calculus for the combination of first-order and higher-order paradigms we are now deepening the relationship of ρ-calculus with rewriting logic [Meseguer, 1992] and we plan to investigate the relationship of this calculus with higher-order rewrite concepts like CRS and HOR [Oostrom and Raamsdonk, 1993].

REFERENCES

[Abadi *et al.*, 1991] M. Abadi, L. Cardelli, P.-L. Curien, and J.-J. Lévy. Explicit substitutions. *Journal of Functional Programming*, 1(4):375–416, 1991.

[Aho *et al.*, 1972] A. V. Aho, R. Sethi, and J. D. Ullman. Code optimization and finite Church-Rosser systems. In *Proceedings of Courant Computer Science*, pages 89–105. Prentice Hall, 1972.

[Baader and Nipkow, 1998] Franz Baader and Tobias Nipkow. *Term Rewriting and all That*. Cambridge University Press, 1998.

[Borovanský *et al.*, 1996] Peter Borovanský, Claude Kirchner, Hélène Kirchner, Pierre-Etienne Moreau, and Marian Vittek. ELAN: A logical framework based on computational systems. In José Meseguer, editor, *Proceedings of the first international workshop on rewriting logic*, volume 4 of *Electronic Notes in TCS*, Asilomar (California), September 1996.

[Borovanský *et al.*, 1998] Peter Borovanský, Claude Kirchner, and Hélène Kirchner. A functional view of rewriting and strategies for a semantics of ELAN. In Masahiko Sato and Yoshihito Toyama, editors, *The Third*

Fuji International Symposium on Functional and Logic Programming, pages 143–167, Kyoto, April 1998. World Scientific. Also report LORIA 98-R-165.

[Breazu-Tannen, 1988] V. Breazu-Tannen. Combining algebra and higher-order types. In *Proceedings 3rd IEEE Symposium on Logic in Computer Science, Edinburgh (UK)*, pages 82–90, 1988.

[Cirstea and Kirchner, 1998] Horatiu Cirstea and Claude Kirchner. *Combining Higher-Order and First-Order Computation Using ρ-calculus: Towards a semantics of* ELAN. *Full Paper.* http://www.loria.fr/~cirstea/Papers/RhoCaculus.ps, 1998.

[Clavel *et al.*, 1998] M. Clavel, F. Durán, S. Eker, P. Lincoln, and J. Meseguer. An Introduction to Maude (Beta Version). Technical report, SRI International, Computer Science Laboratory, Menlo Park, (CA, USA), March 1998.

[Curien *et al.*, 1996] P.-L. Curien, Th. Hardin, and J.-J. Lévy. Confluence properties of weak and strong calculi of explicit substitutions. *Journal of the ACM*, 43(2):362–397, 1996.

[de Bruijn, 1972] N. G. de Bruijn. Lambda calculus with nameless dummies, a tool for automatic formula manipulation, with application to the Church-Rosser theorem. *Proc. Koninkl. Nederl. Akademie van Wetenschappen*, 75(5):381–392, 1972.

[Dershowitz and Jouannaud, 1990] N. Dershowitz and J.-P. Jouannaud. Rewrite Systems. In J. van Leeuwen, editor, *Handbook of Theoretical Computer Science*, chapter 6, pages 244–320. Elsevier Science Publishers B. V. (North-Holland), 1990.

[Dowek *et al.*, 1995] Gilles Dowek, Thérèse Hardin, and Claude Kirchner. Higher-order unification via explicit substitutions, extended abstract. In Dexter Kozen, editor, *Proceedings of LICS'95*, pages 366–374, San Diego, June 1995.

[Dowek *et al.*, 1998] Gilles Dowek, Thérèse Hardin, and Claude Kirchner. Theorem proving modulo. Rapport de Recherche 3400, Institut National de Recherche en Informatique et en Automatique, April 1998. `http://pauillac.inria.fr/~dowek/RR-3400.ps.gz`.

[Futatsugi and Nakagawa, 1996] K. Futatsugi and A. Nakagawa. An Overview of Cafe Project. In *Proceedings of First CafeOBJ Workshop*, Yokohama (Japan), August 1996.

[Gallier and Breazu-Tannen, 1989] J. Gallier and V. Breazu-Tannen. Polymorphic rewriting conserves algebraic strong normalization and confluence. In *16th Colloquium Automata, Languages and Programming*, volume 372 of *Lecture Notes in Computer Science*, pages 137–150. Springer-Verlag, 1989.

[Huet, 1973] G. Huet. A mechanization of type theory. In *Proceeding of the third international joint conference on artificial intelligence*, pages 139–146, 1973.

[Jouannaud and Kirchner, 1991] J.-P. Jouannaud and Claude Kirchner. Solving equations in abstract algebras: a rule-based survey of unification. In Jean-Louis Lassez and G. Plotkin, editors, *Computational Logic. Essays in honor of Alan Robinson*, chapter 8, pages 257–321. The MIT press, Cambridge (MA, USA), 1991.

[Jouannaud and Okada, 1997] Jean-Pierre Jouannaud and Mitsuhiro Okada. Abstract data type systems. *Theoretical Computer Science*, 173(2):349–391, 28 February 1997.

[Kirchner *et al.*, 1990] Claude Kirchner, Hélène Kirchner, and M. Rusinowitch. Deduction with symbolic constraints. *Revue d'Intelligence Artificielle*, 4(3):9–52, 1990. Special issue on Automatic Deduction.

[Kirchner *et al.*, 1995] Claude Kirchner, Hélène Kirchner, and Marian Vittek. Designing constraint logic programming languages using computational systems. In P. Van Hentenryck and V. Saraswat, editors, *Principles and Practice of Constraint Programming. The Newport Papers.*, chapter 8, pages 131–158. The MIT press, 1995.

[Klop *et al.*, 1993] J.W. Klop, V. van Oostrom, and F. van Raamsdonk. Combinatory reduction systems: introduction and survey. *Theoretical Computer Science*, 121:279–308, 1993.

[Meseguer, 1992] José Meseguer. Conditional rewriting logic as a unified model of concurrency. *Theoretical Computer Science*, 96:73–155, 1992.

[Miller, 1991] Dale Miller. A logic programming language with lambda-abstraction, function variables, and simple unification. In Peter Schroeder-Heister, editor, *Extensions of Logic Programming: International Workshop, Tübingen, Germany, December 1989*, volume 475 of *Lecture Notes in Computer Science*, pages 253–281. Springer-Verlag, 1991.

[Nipkow and Prehofer, 1998] Tobias Nipkow and Christian Prehofer. Higher-order rewriting and equational reasoning. In W. Bibel and P. Schmitt, editors, *Automated Deduction — A Basis for Applications. Volume I: Foundations.* Kluwer, 1998.

[Okada, 1989] Mitsuhiro Okada. Strong normalizability for the combined system of the typed λ calculus and an arbitrary convergent term rewrite system. In Gaston H. Gonnet, editor, *Proceedings of the ACM-SIGSAM 1989 International Symposium on Symbolic and Algebraic Computation: ISSAC '89 / July 17–19, 1989, Portland, Oregon*, pages 357–363, New York, NY 10036, USA, 1989. ACM Press.

[Oostrom and Raamsdonk, 1993] Vincent van Oostrom and Femke fan Raamsdonk. Comparing combinatory reduction systems and higher-order rewrite systems. In *HOA'93*, volume 816 of *Lecture Notes in Computer Science*, pages 276–304. Springer-Verlag, 1993.

[Pagano, 1998] Bruno Pagano. X.R.S : Explicit Reduction Systems - A First-Order Calculus for Higher-Order Calculi. In Claude Kirchner and Hélène Kirchner, editors, *15th International Conference on Automated Deduction*, LNAI 1421, pages 72–87, Lindau, Germany, July 5–July 10, 1998. Springer-Verlag.

[van den Brand *et al.*, 1997] M. van den Brand, P. Olivier, L. Moonen, and T. Kuipers. Implementation of a Prototype for the New ASF Meta-environment. In *Proceedings of International Workshop on Theory and Practice of Algebraic Specifications ASF+SDF 97, Amsterdam (The Nederlands)*, Workshops in Computing. Springer-Verlag, September 1997.

[Visser and el Abidine Benaissa, 1998] Eelco Visser and Zine el Abidine Benaissa. A core language for rewriting. In Claude Kirchner and Hélène Kirchner, editors, *Proceedings of the second International Workshop on Rewriting Logic and Applications*, volume 15, `http://www.elsevier.nl/locate/entcs/volume16.html`, Pont-à-Mousson (France), September 1998. Electronic Notes in Theoretical Computer Science.

[Vittek, 1994] Marian Vittek. ELAN*: Un cadre logique pour le prototypage de langages de programmation avec contraintes.* Thèse de Doctorat d'Université, Université Henri Poincaré – Nancy 1, October 1994.

Distributed First Order Logics

Chiara Ghidini and Luciano Serafini

1 INTRODUCTION

In distributed knowledge representation systems, knowledge is not organized in a monolithic and homogeneous system, it is rather composed of a set of heterogeneous subsystems, each of which represents a certain subset of the whole knowledge. Analogously distributed reasoning is not a single process which involves the whole knowledge, it is rather a combination of reasoning processes on different subsets of the global knowledge.

Some formalisms have been proposed for the representation and integration of distributed knowledge and reasoning systems. Well known approaches are based on Multi Modal Logics [Fagin *et al.*, 1995], Labelled Deductive Systems [Gabbay, 1990; Gabbay, 1994] and Labelled Deductive Systems for Modal Logics [D'Agostino, 1992; Basin *et al.*, 1998], Annotated Logics [Subrahmanian, 1994], and finally Cooperative Information Systems [Catarci and Lenzerini, 1993]. The underlying idea of these formalisms is that truth values of certain formulae in different knowledge subsystems are related to one another. In modal logics truth values of different formulae in different worlds are related by axioms connecting different modal operators and constraints between different accessibility relations. LDS inference machine and fibered semantics relate the truth values of formulae with different labels. In annotated logics, annotated clauses enable one to relate truth values of formulae with different labels. Finally, in cooperative information systems an inclusion between concepts of different databases, which can be considered an implication between first order formulae [Borgida, 1996], is specified by interschema assertions.

In many cases, stating relations between truth values of different formulae is not enough and we need a way to represent and reason about relations between terms of different languages with different semantics. Consider an example from the Electronic Commerce scenario: a buyer agent a_b asks another agent a_v, who sells fruits, the price list of fruits. a_v will answer with a list of pairs $\langle$*fruit, price*$\rangle$. This list might contain fruits which are "unknown" to a_b, or fruits known with different names, or different items which are mapped into a single item by a_b, and so on (the reader can imagine other possible relations). In this example we need to represent and reason about the relations existing between the terms for fruits in the languages of a_v and the terms for fruits in the languages of a_b.

The problem of relating the meaning of different terms in different languages is a challenging issues in the development of distributed systems. The state of the art solution to this problem is to provide a set of common ontologies which are used to interpret the terms of different languages. A logical system for reasoning about relations among objects (terms) in different domains would be of great impact in the development of consistent and theoretically well founded ontologies.

The goal of this paper is (i) defining a formal logic, called *Distributed First Order Logic (DFOL)*, which formalizes relations among objects as well as relations among formulae of different subsystems of the same knowledge representation system, and (ii) providing a calculus for DFOL. FDOL is based on the work on contextual reasoning [Giunchiglia, 1993]. DFOL syntax is composed of a family of first order languages, each language describing a piece of the global knowledge. DFOL semantics is an extension of Local Models Semantics defined in [Giunchiglia and Ghidini, 1998] and it is defined in terms of two components: the first one is a family of sets of first order interpretations. Each set of first order interpretations, that we call *local interpretation*, contains interpretations for a first order language, all defined on a single domain. The second component is a relation, that we call *domain relation*, between domains of different sets of interpretations. DFOL calculus is based on ML systems [Giunchiglia and Serafini, 1994]. It is composed of a set of first order natural deduction systems, connected by a set of inference rules called *bridge rules.* These rules export theorems among different theories, each theory representing a piece of the global knowledge contained in a subsystem. We restrict ourselves to First Order Languages as they are powerful enough to represent most of the knowledge representation formalisms currently used. But the same ideas can be extended to other logical languages.

This paper is organized as follows. Section 2 gives a list of requirements for DFOL. In Section 3 we introduce the languages and the semantics of DFOL and the logical consequence defined on this semantics. In Section 4 we show that DFOL enjoys the requirements defined in Section 2. In Section 5 we define a calculus for DFOL. This calculus is sound and complete with respect to DFOL logical consequence (Section 6). We compare DFOL with similar approaches such as Multi Modal Logics, LDS, Annotated Logics, and Cooperative Information Systems (Section 7). We conclude in Section 8.

2 REQUIREMENTS

Before starting with the definition of DFOL let us state our requirements for a logic for distributed knowledge and reasoning systems.

Representational requirements These requirements concern the ability of a formal logic to represent the following set of aspects which are relevant in distributed knowledge representation systems.

- *different logical languages.* This is necessary as different subsystems might represent knowledge using different languages.
- *different interpretation domains.* This is necessary as different subsystems might represent knowledge about different domains at different level of abstraction.
- different *local semantics* for the symbols of each language, i.e. the semantics of a symbol in a language should not depend on anything else outside that language. This is necessary as representation languages have their own autonomous semantics.
- representing *relations among objects of different domains*, i.e. the logic must represent the fact that an object in a domain of a subsystem corresponds to one or more objects in the domain of another subsystem. This is necessary, e.g., to represent the relations between fruits in the electronic commerce example described above.
- representing *directional* relations among different domains, i.e. the relation from a domain **dom** of a subsystem S into a different domain **dom**$'$ of a subsystem S' is not the inverse of the relation from **dom**$'$ into **dom**. This represents the fact that the point of view of S about the domain **dom**$'$ of S' might be different from the S''s point of view about the domain **dom** of S.
- representing *directional* relations among the truth values of different formulae in different languages. This means that the way the truth value of a formula in the language L_i affects the truth value of a formula in another language L_j does not depend on the way the truth value of a formula in L_j affects the truth value of a formula in L_i. This is necessary since in distributed knowledge representation systems the information flow from a subsystem S to S' does not depend on the flow in the opposite direction.
- representing *partial information*, i.e. the logic must not force a formula to be either true or false in a local interpretation. This is necessary as subsystems (e.g. partial databases) might not contain complete information.
- representing *local inconsistency*, i.e. the logic must formalize the fact that an inconsistent set of formulae in a language, corresponding to an inconsistent set of facts in a subsystem, might not imply the inconsistency in other subsystems. This is necessary since inconsistent information between subsystems is often not propagated.

- preventing *hypothetical reasoning across subsystems*, i.e. the logical consequence relation $\models$ between formulae corresponding to facts in different subsystems is not closed under the rule

$$\frac{\phi_1 \models \psi \quad \phi_2 \models \psi}{\phi_1 \vee \phi_2 \models \psi}$$

  This is necessary as, in general, this kind of reasoning between subsystems is not allowed. Consider for instance the connection between the knowledge bases of two agents *John* and *Sue*. If John communicates either $phone(Mary, 1234)$ or $phone(Mary, 1235)$, then *Sue* can infer $knows(John, phone(Mary))$. However, the same fact cannot be inferred by *Sue* if *John* communicates $phone(Mary, 1234) \vee phone(Mary, 1235)$.

Structural requirements These are requirements on *how* a logic represents the features of distributed knowledge representation systems.

- *Modularity* and *compositionality.* Modularity means that each representational property described above is formalized by a formal component of the logic (an axiom schema, an inference rule, an applicability condition, etc.). Compositionality means that the formalization of a distributed knowledge representation system with a specific subset of the representational properties must be obtained by specifying the logic using the corresponding formal components. Compositionality makes the logic flexible enough to formalize a wide range of distributed knowledge representation systems.

- *Incrementality* with regard to the formal logics for the representation of non-distributed reasoning systems. This means that the logic must be based on well established and well studied logics. This structural requirement is necessary because in a logic for the representation and integration of distributed knowledge and reasoning systems we want to exploit all the previous formal results, especially the results obtained in the area of automated reasoning.

3 THE LOGIC

The goal of this section is to define the main components of a DFOL. These components are the family of languages, the interpretation, the satisfiability relation, possible constraints on the interpretations, and logical consequence.

3.1 Languages and Semantics

Let $\{L_i\}_{i \in I}$ (in the following $\{L_i\}$) be a family of first order languages defined over a set I of indexes. Each language L_i partially describes the world from a certain perspective (i's point of view). For instance in the formalization of multiagent propositional attitudes described in [Giunchiglia and Serafini, 1994] I is a set of names for agents and each L_i is the language adopted by agent i to express its beliefs about the world. In the formalization of federated databases described in [Serafini and Ghidini, 1997] I contains the indexes associated to the databases of a federated database and each L_i is the language which describes the logical schema of the i-th database in the federation.

Languages L_i and L_j are not necessarily disjoint and the same formula ϕ may occur in different languages. As L_i and L_j describe the world from different points of view, ϕ can have different meanings in the two languages. Suppose for instance that two agents i and j "live" in different places. Then the formula *it-is-raining* in the language of agent i means that it is raining where i lives, whereas the same formula in the language of agent j means that it is raining where j lives. Similarly, the same symbol, occurring in two different databases in a federation, may have different semantic interpretations in each database. In order to distinguish occurrences of the same formula in different languages we annotate in which language a formula is to be considered. A *labeled formula* is a pair $i : \phi$[1]. It denotes the formula ϕ and the fact that ϕ is a formula of the language L_i. When no ambiguity arises, labeled formulae are called formulae. Given a set of labeled formulae Γ, Γ_j denotes the set of formulae $\{\gamma \mid j : \gamma \in \Gamma\}$. From now on, a formula ϕ of the language L_i is called *i-formula*.

The semantics for $\{L_i\}$ is an extension of Local Models Semantics defined in [Giunchiglia and Ghidini, 1998]. The languages $\{L_i\}$ are interpreted in a structure which consists of two components: (i) the local interpretation for each L_i, and (ii) a binary relation between the interpretation domains of each pair of languages L_i and L_j.

Let M_i be the set of all the first order models of L_i on a given domain **dom** (see [Chang and Keisler, 1973]). $m \in M_i$ is a *local model* (of L_i). Each first order model m is a pair $\langle \mathbf{dom}, \mathcal{I} \rangle$ where **dom** is the *domain* of interpretation (or *universe*) of L_i and $\mathcal{I}$ is the the *interpretation* function. The *local satisfiability relation* is the usual satisfiability relation $\models$ between first order models and first order formulae with an assignment to the free variables. The set M_i of local models and the local satisfiability relation define the *local semantics* of L_i. The fact that the i-th subsystem contains partial knowledge about facts in L_i is represented by associating each L_i to a set of *possible models* $S_i \subseteq M_i$ on the domain $\mathbf{dom}_i$. The satisfiability of a formula in a set of possible models is defined as follows:

Definition 3.1 (Satisfiability of a formula) Let $\{S_i\}$ be a family of set of possible models for $\{L_i\}$; let an assignment a be a family $\{a_i\}$ of assignments

[1] Similar notations have been introduced by [Gabbay, 1994; Subrahmanian, 1994; Dinsmore, 1991; Masini, 1992].

a_i to the variables of L_i. A formula $i : \phi$ is *satisfied* in $\{S_i\}$ by an assignment a, in symbols $\{S_i\} \models i : \phi[a]$, if for all $m \in S_i$ $m \models \phi[a_i]$. For any set of i-formulae Γ_i, $S_i \models \Gamma_i[a]$ means that $S_i \models \gamma[a]$ for all $\gamma \in \Gamma_i$.

The second component of a model for $\{L_i\}$ is a family of *domain relations.* A domain relation r_{ij} from $\mathbf{dom}_i$ to $\mathbf{dom}_j$ is a subset of $\mathbf{dom}_i \times \mathbf{dom}_j$ ($\mathbf{dom}_i$ and $\mathbf{dom}_j$ are the domains of interpretation of two different languages L_i and L_j respectively), and represents the capability of subsystem j to map the objects of $\mathbf{dom}_i$ into its domain $\mathbf{dom}_j$. A pair $\langle d, d' \rangle$ in r_{ij} means that, from the point of view of j, d in $\mathbf{dom}_i$ is the representation of d' in $\mathbf{dom}_j$.

Notice that r_{ij} formalizes j's subjective point of view of the relation between domains and not an absolute objective point of view. Therefore $\langle d, d' \rangle \in r_{ij}$ must not be read as if d and d' represent the same object in a domain shared by i and j. This fact would have to be formalized from a point of view which is external to both i and j.

The definition of a domain relation among languages gives us flexibility in specifying the different kind of relations between domains associated to different languages. For clarifying the concept, suppose that two observers, Chiara and Luciano, represent a "piece of world" at two different approximation levels. Suppose that Chiara describes the habit of people living in a village called Poggio Rusco, whereas Luciano describes the habit of families in Italy. The domain of the language used to describe Chiara's knowledge contains objects which stand for the people living in Poggio Rusco, whereas the domain of the language used to describe Luciano's knowledge contains objects which stand for families. From the point of view of Luciano all the objects (people) in the domain of Chiara who live in the same family A and the object in its own domain representing the family A represent the same real world object. Formally, let $d_1, \ldots, d_n$ be the objects in the domain $\mathbf{dom}_c$ of Chiara representing the people living in a family A and let d be the object in the domain $\mathbf{dom}_l$ of Luciano representing family A. The fact that all the pairs $\langle d_1, d \rangle, \ldots \langle d_n, d \rangle$ are in the domain relation $r_{cl} \subseteq \mathbf{dom}_c \times \mathbf{dom}_l$ formalizes that, from the point of view of Luciano, all $d_1, \ldots, d_n$ in the domain of Chiara and the object d in his domain represent the same real world object. Suppose also that Chiara does not have any information about families. Therefore, from her point of view, there is no relation between people she describes and families described by Luciano. Therefore $r_{lc} \subseteq \mathbf{dom}_l \times \mathbf{dom}_c$ is empty. Notice that we are not imposing that the point of view of i is always different from the point of view of j. In many cases it is plausible that r_{ij} and r_{ji} are such that $\langle d, d' \rangle \in r_{ij}$ if and only if $\langle d', d \rangle \in r_{ji}$.

Definition 3.2 (Model) A *model* $\mathcal{M}$ (for $\{L_i\}$) is a pair $\langle \{S_i\}, \{r_{ij}\} \rangle$ where each $S_i \subseteq M_i$ is a set of possible models over the same domain of interpretation $\mathbf{dom}_i$; each r_{ij} is a domain relation from $\mathbf{dom}_i$ to $\mathbf{dom}_j$.

3.2 Compatibility Constraints

One of the main purposes of DFOL is to enable a formal reasoning among different first order theories. To do this we need a way of specifying constraints on DFOL models. Being a DFOL model composed of a family of possible models and a family of domain relations, the constraints on DFOL models we consider in this paper are (a) constraints on the domain relation, and (b) constraints on the possible models. We call the former *domain constraints* and the latter *interpretation constraints.*

Definition 3.3 (Domain Constraint) Let L_i and L_j be two first order languages. A *domain constraint from* L_i *to* L_j is an expression of the form T or S.

Intuitively T from L_i to L_j captures the fact that, from j perspective, $\mathbf{dom}_i$ is contained in $\mathbf{dom}_j$. Conversely S from L_i to L_j captures the fact that, from j perspective, $\mathbf{dom}_j$ is contained in $\mathbf{dom}_i$.

Consider the example of a mediator M [Wiederhold, 1992], i.e., a database which integrates the information of a set containing n databases $1, \ldots, n$. The mediator collects all the information about the individuals of each database. The fact that the domain $\mathbf{dom}_i$ of each database must be embedded by M into its domain $\mathbf{dom}_{\text{M}}$ can be formalized by the domain constraint T from each database in $1, \ldots, n$ to the mediator M.

The set of domain constraints from L_i to L_j is denoted by DC_{ij}.

Definition 3.4 (Interpretation Constraint) Let L_i and L_j be two first order languages. An *interpretation constraint from* L_i *to* L_j is an expression of the form

$$i : \phi(x_1, \ldots, x_n) \to j : \psi(x_1, \ldots, x_n)$$

where $\phi(x_1, \ldots, x_n)$ and $\psi(x_1, \ldots, x_n)$ are formulae of L_i and L_j, respectively.[2]

Intuitively $i : \phi(x_1, \ldots, x_n) \to j : \psi(x_1, \ldots, x_n)$ captures the fact that, from the point of view of j, the set of tuples of objects of $\mathbf{dom}_i$ which satisfy $\phi(x_1, \ldots, x_n)$ in L_i corresponds (via r_{ij}) to a set of tuples which satisfy $\psi(x_1, \ldots, x_n)$ in L_j. Let us consider the example of the mediator given above. Suppose that the i-th database contains data about the telephone numbers of a town (say Roma) and a relation $phone(x, y)$ stating that x has telephone number y. Suppose that $phone(x, y)$ is mapped by the mediator into its relation $telno(x, 39, 6, y)$, with the addition of international and area code. This mapping is represented by the interpretation constraint:

(1) $$i : phone(x, y) \to \text{M} : telno(x, 39, 6, y)$$

[2] For the sake of simplicity we define interpretation constraints as pairs of formulae with the same set of free variables. Interpretation constraints can be easily generalized by dropping this requirement.

The set of interpretation constraints from L_i to L_j are denoted by IC_{ij}. A *compatibility constraint* C_{ij} *from* L_i *to* L_j is a pair $C_{ij} = \langle DC_{ij}, IC_{ij}\rangle$.

Definition 3.5 (Satisfiability of Domain Constraint) Let $\mathbf{dom}_i$ and $\mathbf{dom}_j$ be the domains of interpretation of L_i and L_j, respectively. The domain relation $r_{ij} \subseteq \mathbf{dom}_i \times \mathbf{dom}_j$ satisfies the domain constraint T if for any $d \in \mathbf{dom}_i$ there is a $d' \in \mathbf{dom}_j$ such that $\langle d, d'\rangle \in r_{ij}$. Analogously r_{ij} satisfies the domain constraint S if for any $d \in \mathbf{dom}_j$ there is a $d' \in \mathbf{dom}_i$ such that $\langle d', d\rangle \in r_{ij}$.

Definition 3.6 (Satisfiability of Interpretation Constraint) Let S_i and S_j be two sets of first order models of L_i and L_j over the domains $\mathbf{dom}_i$ and $\mathbf{dom}_j$ respectively. Let r_{ij} be a domain relation from $\mathbf{dom}_i$ to $\mathbf{dom}_j$. The tuple $\langle S_i, S_j, r_{ij}\rangle$ satisfies the interpretation constraint $i : \phi(x_1, \ldots, x_n) \rightarrow j : \psi(x_1, \ldots, x_n)$ if for any $\langle d_k, d'_k\rangle \in r_{ij}$ $(1 \leq k \leq n)$, $S_i \models \phi(d_1, \ldots, d_n)$ implies that $S_j \models \psi(d'_1, \ldots, d'_n)$.

Definition 3.7 (Satisfiability of Compatibility Constraints) Given a family of compatibility constraints $C = \{C_{ij}\}$, a model $\langle\{S_i\}, \{r_{ij}\}\rangle$ is said to *satisfy* C if for each $i, j \in I$, the domain relation r_{ij} satisfies all the domain constraints in DC_{ij} and $\langle S_i, S_j, r_{ij}\rangle$ satisfies all the interpretation constraints in IC_{ij}.

3.3 Logical Consequence

Compatibility constraints imply that certain facts in a language are necessarily true as a consequence of other facts being true in, possibly distinct, languages. The formal characterization of such a relation is crucial as it allows us the detection of inconsistencies in the languages and to understand how information propagates through languages. In this section we formalize this relation among facts by means of the notion of *logical consequence with regard to a set of compatibility constraints* (or more simply logical consequence).

To define logical consequence we extend the set of variables of each L_i to a set of what we call *extended variables.* For each $j \in I$ and each variable x in L_i, $x^{j\rightarrow}$ and $x^{\rightarrow j}$ are variables of L_i. Intuitively a variable x (without indexes) occurring in $i : \phi$ is a placeholder for a generic element of $\mathbf{dom}_i$; the extended variable $x^{j\rightarrow}$ occurring in $i : \phi$ is a placeholder for an element of $\mathbf{dom}_i$ which is an image, via the domain relation r_{ji}, of the element of $\mathbf{dom}_j$ denoted by x; analogously $x^{\rightarrow j}$ occurring in $i : \phi$ is a placeholder for an element of $\mathbf{dom}_i$ which is a pre-image, via r_{ij}, of the element of $\mathbf{dom}_j$ denoted by x.

We also extend the assignment a to all extended variables. a is an *admissible assignment* if for any variable x and any variable $x^{i\rightarrow}$ and $x^{\rightarrow i}$

1. if $\mathsf{T} \in DC_{ij}$, then $\langle a_i(x), a_j(x^{i\rightarrow})\rangle \in r_{ij}$
2. if $\mathsf{S} \in DC_{ji}$, then $\langle a_j(x^{\rightarrow i}), a_i(x)\rangle \in r_{ji}$.

For any set of i-formulae Γ, ϕ, $\Gamma[a] \models_{S_i} \phi[a]$ if and only if for all $m \in S_i$, $m \models \Gamma[a]$ implies $m \models \phi[a]$.

Definition 3.8 (Logical Consequence) Let Γ be a set of formulae. A formula $i : \phi$ is a *logical consequence* of Γ with regard to a set of compatibility constraints C, in symbols $\Gamma \models_C i : \phi$, if for all the models $\langle \{S_i\}, \{r_{ij}\} \rangle$ satisfying the compatibility constraints C and for all the admissible assignments a, if for all $j \neq i$, $S_j \models \Gamma_j[a]$, then $\Gamma_i[a] \models_{S_i} \phi[a]$.

4 FULFILLMENT OF THE REQUIREMENTS

In this section we show that DFOL satisfies the representational and structural properties introduced in Section 1. Proofs of theorems are reported in [Ghidini, 1998].

The capability of DFOL to cope with different languages, interpretation domains, and semantics is a consequence of how we define DFOL syntax and semantics.

Relations between objects in different domains are represented using the domain relation. The basic relations of totality and surjectivity between different domains are formalized by imposing that the domain relation satisfies domain constraints T and S. Frequently occurring relations between domains, such as containment, abstraction, and embedding can be expressed using domain constraints and interpretation constraints on the equality predicate.

Domain containment: From the point of view of j each element of the domain of i corresponds at least to an element in its domain. This is easily captured by imposing domain constraint T from L_i to L_j.

Embedded domains: From the point of view of j the domain of i is isomorphic to a subset of its domain. This can be represented by imposing domain constraint T from L_i to L_j and interpretation constraint

$$i : x \neq y \rightarrow j : x \neq y$$

The set of domain relations from $\mathbf{dom}_i$ to $\mathbf{dom}_j$ satisfying these constraints is the set of total injective relations from $\mathbf{dom}_i$ to $\mathbf{dom}_j$. Under these constraints it can be proved that for any $n \geq 1$

$$(2) \qquad i : \exists x_1, \ldots x_n \bigwedge_{i \neq j} x_i \neq x_j \models_C j : \exists x_1, \ldots x_n \bigwedge_{i \neq j} x_i \neq x_j$$

(2) means that $\mathbf{dom}_j$ contains at least as many elements as $\mathbf{dom}_i$.

Abstracted domains: From the point of view of j each object of its domain is an abstraction of a set of objects of the domain of i. This can be represented by imposing the domain constraint S from i to j and the interpretation constraint

$$i : x = y \rightarrow j : x = y$$

The domain relations that satisfy these constraints are the surjective functions from $\mathbf{dom}_i$ to $\mathbf{dom}_j$. Under these constraints (3) it can be proved that for any $n \geq 1$

$$(3) \qquad i : \forall x_1, \ldots x_n \bigvee_{i \neq j} x_i = x_j \models_C j : \forall x_1, \ldots x_n \bigvee_{i \neq j} x_i = x_j$$

(3) means that $\mathbf{dom}_j$ contains at most as many objects as $\mathbf{dom}_i$.

Isomorphic domains: From the point of view of j, its domain is isomorphic to the domain of i. This is obtained by combining embedding and abstraction. The domain relations that satisfy both the constraints of embedding and abstraction are the isomorphisms from $\mathbf{dom}_i$ to $\mathbf{dom}_j$.

DFOL allows for representing directional relations between domains since domain relation and domain constraints from L_i to L_j and from L_j to L_i are independent. Analogously, an interpretation constraint from L_i to L_j never affects the logical consequence in the opposite direction. Indeed the following theorem holds:

Theorem 4.1 (Directionality) *Let $i : \phi$ and $j : \psi$ be closed formulae such that ψ is classically satisfiable. There is a set of compatibility constraints C such that $i : \phi \models_C j : \psi$ and $j : \neg\psi \not\models_C i : \neg\phi$.*

DFOL represents partiality. Indeed the model associates to each language L_i a set S_i of possible models. We represent complete distributed knowledge defining a model for $\{L_i\}$ as $\langle\{m_i\}\{r_{ij}\}\rangle$ where m_i is a (single) local model for the language L_i.

DFOL represents local inconsistency. Indeed, $i : \bot \models_{IC} j : \bot$ does not hold unless the interpretation constraint $i : \bot \rightarrow j : \bot$ is added to C.

DFOL prevents hypothetical reasoning across subsystems. This property is a consequence of the following theorem.

Theorem 4.2 (Hypothetical reasoning) *Let $i : \phi_1$, $i : \phi_2$, and $j : \psi$ be classically satisfiable closed formulae. There is a set of compatibility constraints C such that $i : \phi_1 \models_C j : \psi$, $i : \phi_2 \models_C j : \psi$, and $i : \phi_1 \vee \phi_2 \not\models_C j : \psi$.*

5 THE CALCULUS

The goal of this section is to define a sound and complete calculus for the logical consequence in DFOL.

The calculus for DFOL is based on Multilanguage systems (ML systems) [Giunchiglia and Serafini, 1994], which are a generalization of Prawitz classical Natural Deduction (ND) [Prawitz, 1965]. An ML system is a triple $\langle\{L_i\}, \{\Omega_i\}, \Delta\rangle$ where $\{L_i\}$ is a family of languages, $\{\Omega_i\}$ is a family of sets of axioms, and Δ is the deductive machinery. Δ contains two kinds of inference rules: *i rules*, i.e., inference rules with premises and conclusions in the same language, and *bridge rules*, i.e., inference rules with premises and conclusions belonging to different languages. Notationally, we write inference rules with, e.g., a single premise, as follows:

$$\frac{i : \psi}{i : \phi}\, ir \qquad\qquad \frac{i : \psi}{j : \phi}\, br$$

ir is an i rule while *br* is a bridge rule. Derivability in a ML system is defined in [Giunchiglia and Serafini, 1994]; roughly speaking it is a generalization of the notion of deduction in natural deduction.

An ML system for a DFOL with languages $\{L_i\}$ and compatibility constraints C is composed of the same family of languages, the sets of axioms corresponding to the basic facts for each L_i, the i rules for local reasoning, and the bridge rules which formalize the compatibility constraints.

Definition 5.1 (First Order ML system) The ML system MS for the DFOL with languages $\{L_i\}$, basic facts $\{\Omega_i \subseteq L_i\}$, and compatibility constraints $C = \{\langle DC_{ij}, IC_{ij}\rangle\}$ is the tuple MS $= \langle\{L_i\}, \{\Omega_i\}, \Delta\rangle$ such that Δ contains:

1. for each $i \in I$ the ND inference rules for propositional calculus and equality (see [Prawitz, 1965], and [Giunchiglia and Serafini, 1994])

2. for each $i \in I$ the rules for quantifiers:

$$\frac{i : \phi^y_x}{i : \forall x \phi}\, \forall\mathrm{I} \qquad\qquad \frac{i : \forall x \phi}{i : \phi^t_x}\, \forall\mathrm{E}$$

$$\frac{i : \phi^t_x}{i : \exists x \phi}\, \exists\mathrm{I} \qquad\qquad \frac{i : \exists x \phi \qquad \begin{matrix}[i : \phi^y_x] \\ i : \psi\end{matrix}}{i : \psi}\, \exists\mathrm{E}$$

3. for each interpretation constraint $i : \phi(x_1 \ldots, x_n) \to j : \psi(x_1, \ldots x_n) \in IC_{ij}$ the bridge rule C

$$\frac{i : \phi(x_1, \ldots, x_n)}{j : \psi(y_1, \ldots, y_n)}\, C$$

where if $n \neq 0$ (i.e., ϕ is not closed) then either, for each k, y_k is $x_k^{i\to}$ and $\mathsf{T} \in DC_{ij}$ or, for each k, x_k is $y_k^{\to j}$ and $\mathsf{S} \in DC_{ij}$;

4. the bridge rules for equality:

$$\frac{i : e_1 \vee e_2 \quad \begin{array}{c}[i : e_1]\\ j : \phi\end{array} \quad \begin{array}{c}[i : e_2]\\ j : \phi\end{array}}{j : \phi} \vee\mathrm{E}_{=} \qquad \frac{i : \exists x e \quad \begin{array}{c}[i : e^y_x]\\ j : \psi\end{array}}{j : \psi} \exists\mathrm{E}_{=}$$

where e, e_1, e_2 are pure equality formulae, i.e., formulae whose only predicate is the equality predicate $=$.

Restrictions C and bridge rules for equality are applicable only if the premise in i does not depend on assumptions in j. $\forall$I is applicable only if neither y nor $y^{\rightarrow i}$ occur in any assumptions which ϕ^y_x depends on. $\exists$E and $\exists\mathrm{E}_{=}$ are applicable only if neither y nor $y^{\rightarrow i}$ occur in $\exists x\phi$ and $\exists xe$ respectively, in ψ, and in any assumption which ψ depends on. Derivability in MS is denoted by $\vdash_{\mathrm{C}}$.

Rules for quantifiers defined above are similar to the ND rules for quantifiers introduced in first order logic. The only difference is that the applicability of $\forall$I and $\exists$E depends also on the occurrence of variables in other languages. This fact is formalized by restrictions on $\forall$I and $\exists$E. The bridge rule C expresses the combination of domain constraints DC_{ij} and interpretation constraints IC_{ij} from L_i to L_j. Notice that if $n = 0$ (i.e. the interpretation constraint above contains closed formulae), then C can be instantiated as follows:

$$\frac{i : \phi}{j : \psi} C$$

and it is independent of domain constraints in DC_{ij}. Otherwise if $n \neq 0$ the bridge rule C depends on the domain constraints in DC_{ij}. Domain constraints state how to connect the free variables contained in ϕ and in ψ. In particular if $\mathsf{T} \in DC_{ij}$ the bridge rule C can be instantiated as follows:

$$\frac{i : \phi(x_1, \ldots, x_n)}{j : \psi(x_1^{i\rightarrow}, \ldots, x_n^{i\rightarrow})} C$$

if $\mathsf{S} \in DC_{ij}$ the bridge rule C can be instantiated as follows:

$$\frac{i : \phi(y_1^{\rightarrow j}, \ldots, y_n^{\rightarrow j})}{j : \psi(y_1, \ldots, y_n)} C$$

$\vee\mathrm{E}_{=}$-rule and $\exists\mathrm{E}_{=}$-rule state that we can perform reasoning by case across different theories on equality. Reasoning by case allows one to infer a fact, say ϕ, from a disjunction, say $\psi \vee \theta$, by deriving ϕ from ψ and θ separately. The basic assumption of reasoning by case is that, if a model satisfies a disjunction, then it must satisfy at least one of the disjunct. In general this is not true in our logic as there are DFOL models S such that $S \models i : \psi \vee \theta$ and neither

$S \models i : \psi$ nor $S \models i : \theta$. Equality formulae, however, constitute an exception since the equality predicate $=$ has a unique interpretation on any domain. Therefore it can be proved that, if $S_i \models e_2 \vee e_2[a]$, then either $S_i \models e_1[a]$ or $S_i \models e_2[a]$. Analogous observations hold for the $\exists E_=$-rule.

As an explanatory example we give the deduction of property (2).

$$\dfrac{i : \exists x_1,\ldots x_n \bigwedge_{h\neq k} x_h \neq x_k \qquad \dfrac{\dfrac{\ldots \qquad \dfrac{\dfrac{[i : \bigwedge_{h\neq k} x_h \neq x_k]^{(1)}}{i : x_h \neq x_k}\,\wedge\mathrm{E}}{j : x_h^{i\to} \neq x_k^{i\to}}\,C \qquad \ldots}{j : \bigwedge_{h\neq k} x_h^{i\to} \neq x_k^{i\to}}\,\wedge\mathrm{I}}{j : \exists x_1,\ldots x_n \bigwedge_{h\neq k} x_h \neq x_k}\,\exists\mathrm{E}}{j : \exists x_1,\ldots x_n \bigwedge_{h\neq k} x_h \neq x_k}\,\exists\mathrm{I}_{=}(\text{discharging}^{(1)})$$

In the above deduction we start by taking n distinct individuals $x_1, \ldots x_n$ in the domain of i (assumption $^{(1)}$). Totality of the domain relation from i to j implies that each individual x_h in the domain of i is mapped in an individual $x_h^{i\to}$ in the domain of j. Interpretation constraint $i : x \neq y \to j : x \neq y$ implies that if x_h is distinct from x_k in i then $x_h^{i\to}$ is distinct form $x_k^{i\to}$ in j (application of C). We repeat the previous reasoning for each $h \neq k$ and apply $\wedge$I in j. We then introduce the existential quantifier in j and eliminate the existential quantifier in i by the $\exists E_=$ bridge rule.

Let us consider the mediator example described above. Suppose that the i-th database contains the statement *phone*(*John*, 12345), then we would like to infer *telno*(*John*, 39, 6, 12345) in M. This, however, is not possible only with the constraint (1), as this inference is based on the assumption that *John* and 12345 denote the same person and the same number respectively in i and M. This assumption should be made explicit as, i and M might adopt different encoding for person names and telephone numbers. What we are allowed to infer in M, however, is that there is a person from Rome who has a telephone number, i.e. $\exists xy.telno(x, 39, 6, y)$. The deduction looks as follows:

$$\dfrac{i : \exists xy, x = John \wedge y = 12345 \qquad \dfrac{\dfrac{\dfrac{i : phone(John, 12345) \quad [i : x = John \wedge y = 12345]^{(1)}}{i : phone(x, y)}\,=}{\mathrm{M} : telno(x^{i\to}, 39, 6, y^{i\to})}\,C}{\mathrm{M} : \exists xy.telno(x, 39, 6, y)}\,\exists\mathrm{I}}{\mathrm{M} : \exists xy.telno(x, 39, 6, y)}\,\exists\mathrm{I}_{=}(\text{discharging}^{(1)})$$

6 SOUNDNESS AND COMPLETENESS

The calculus defined in Section 5 can be proved sound and complete with regard to the class of models for $\{L_i\}$ defined in Section 3. For lack of space we report an outline of the formal proofs. For a complete version we refer the reader to [Ghidini, 1998].

Theorem 6.1 (Soundness) *If* $\Gamma \vdash_C i : \phi$, *then* $\Gamma \models_C i : \phi$.

Proof of Theorem 6.1. The proof of soundness is standard as far as the i rules are concerned. Soundness of bridge rule C immediately derives from the definition of satisfiability of compatibility constraints and of admissible assignment. Let's concentrate on the $\vee E_=$-rule. The equality predicate $=$ has a unique interpretation on any domain. Therefore if $S_i \models e_2 \vee e_2[a]$, then either $S_i \models e_1[a]$ or $S_i \models e_2[a]$. This ensures the soundness of the $\vee E_=$-rule. Let's consider the $\exists E_=$-rule. $S_i \models \exists x e[a]$ implies that there is an admissible assignment a', which agrees with a on all the variables but a new variable y not occurring in e, such that $S_i \models e^y_x[a']$. This ensures the soundness of the $\exists E_=$-rule. Q.E.D.

Theorem 6.2 (Completeness) *If* $\Gamma \models_C i : \phi$, *then* $\Gamma \vdash_C i : \phi$.

The proof of the theorem is a generalization of the completeness proof for first order logic [Chang and Keisler, 1973]. First we generalize the definition of inconsistency and witness.

Since in our logic there are many theories, there is not any notion of global (in)consistency. We introduce the notion of (in)consistency in the k-th theory.

Definition 6.1 (k-consistency) We say that a set of formulae Γ of $\{L_i\}$ is k-*consistent* with regard to the set of compatibility constraints C if $\Gamma \nvdash_C k : \bot$.

In the usual proof of completeness for first order logics (see for instance [Chang and Keisler, 1973]) witnesses are introduced for building the domain of the canonical model. In DFOL we consider a set of different domains and a family of domain relations. Therefore we need two classes of witnesses: the former used for building the domains and the latter for building the domain relations of the canonical model.

Definition 6.2 (Existential witnesses) Let Γ be a set of formulae and C a family of compatibility constraints on $\{L_i\}$. A set D of constants belonging to each language L_i is a set of witnesses for Γ in C if, for every formula $i : \phi(x)$ with only one free variable x, there is a constant $d \in D$ such that:

$$\Gamma \vdash_C i : \exists x \phi(x) \supset \phi(d)$$

Definition 6.3 (Constraint witnesses) Let Γ be a set of formulae and C a family of compatibility constraints on $\{L_i\}$. A set D of constants of $\{L_i\}$ is a set of constraint witnesses for Γ in C if for every closed term t of L_i and t' of L_j there are two constants $d, d' \in D$ such that for any interpretation constraint $\langle \phi(x), \psi(x) \rangle \in IC_{ij}$

1. if $\mathsf{T} \in DC_{ij}$ then, $\Gamma, i : \phi(t) \vdash_C j : \psi(d)$;

2. if $\mathsf{S} \in DC_{ij}$ then, $\Gamma, i : \phi(d') \vdash_C j : \psi(t')$;

For each language L_i existential witnesses are used to define the domains $\mathbf{dom}_i$ of the model of Γ (as in [Chang and Keisler, 1973]). Constraint witnesses instead are used to define the domain relations.

Similar to the proof of completeness for first order logic, the completeness statement can be rephrased as follows: if a set of sentences Γ is k-consistent, then it has a (canonical) model. Therefore the proof of completeness reduces to the construction of such a model, called the *distributed canonical model.* In the construction of the distributed canonical model for Γ, existential witnesses are used to define the domains $\mathbf{dom}_i$ of the local interpretations whereas constraint witnesses are used to define the domain relations. An outline of the construction of the distributed canonical model is as follows:

1. We show that any k-consistent set of formulae Γ can be extended to a k-consistent set Γ' with a set of existential witnesses.

2. We show that any k-consistent set of formulae Γ with regard to C can be extended to a set Γ' such that: (i) Γ' is k-consistent with regard to an extension C' of C and (ii) Γ' has existential and constraint witnesses in C'.

3. We show that a set of formulae Γ which is k-consistent with regard to C can be extended to a k-consistent set Γ' which is saturated with regard to equality (i.e. such that for any pure equality formula $i : e$, either $\Gamma' \vdash_C i : e$ or $\Gamma' \vdash_C i : \neg e$).

4. Starting from a k-consistent set of formulae Γ which is saturated with regard to equality and has existential and constraint witnesses in C, we build a distributed canonical model $\langle \{S_i\}, \{r_{ij}\} \rangle$ which satisfies C, Γ, and such that $S_k \neq \emptyset$.

7 RELATED WORK

We consider here four state of the art classes of formal systems for distributed knowledge and reasoning: Labelled Deductive Systems, Modal Logics, Annotated Logics, and Cooperative Information Systems.

LDS [Gabbay, 1994] are very general logical systems. We compare DFOL with LDSs for Quantified Modal Logics (LDSQML) [Basin *et al.*, 1998]. LDSQMLs deal with domains and relations between domains, as well as relations between formulae. LDSQMLs provide a formalization for modal logics with varying, increasing, decreasing, and constant domains. The main analogies between DFOL and LDSQML is that they both allow for distinct domains and basic relations between them. The main differences concern:

- *Reasoning on labels.* In LDSQML (and in LDS in general) it is possible to represent and reason about properties of labels (represented by *relwffs*). This is not possible in DFOL.

- *Reasoning on term existence.* In LDSQML the literal $w : t$ represents the fact that a term t denotes an individual in the domain of a world w. We cannot represent this fact in DFOL since we adopt first order semantics and each term of the language L_i denotes an object in $\mathbf{dom}_i$.

- *Reasoning about relation between individuals of different domains.* LDSQML allows for representing only few relations between domains: containment and equivalence. In DFOL the equality interpretation constraint $1 : x = t \rightarrow j : x = u$ represents the fact that a certain individual t in $\mathbf{dom}_i$ corresponds to another individual u in $\mathbf{dom}_j$. This fact allows DFOL to cope with more complex relations between domains than LDSQML.

- *Inconsistency propagation.* The $\bot E$ rule of LDSQML enables the propagation of false across worlds. This prevents LDSQML from representing local inconsistency.

A comparison of DFOL with quantified Modal Logics leads to the same observations given above.

Annotated Logic (AL) [Blair and Subrahmanian, 1987] is a formalism that has been applied to a variety of situations in knowledge representation, expert systems, quantitative reasoning, and hybrid databases. In annotated logics it is possible to integrate a set of logical theories in a unique amalgamated theory. The amalgamated theory is the disjoint union of the theories plus a set of clauses (called amalgamated clauses) which resolve conflicts due to inconsistent facts and compose uncertain information of different theories. One of the main similarities with our approach is the capability to cope with inconsistent knowledge bases. Annotated logics provide an explicit way to solve conflicts. The main difference between annotated logics and DFOL concerns the ability to represent different interpretation domains. Annotated logics has a unique logical language, and the same symbol in different knowledge bases is interpreted in the same object. This of course might be trivially solved by indexing the constant with the name of the knowledge base. In this case explicit relational symbols between objects of different knowledge bases should be introduced.

A Cooperative Information System (CIS) is a formal system based on description logics [Donini *et al.*, 1996]. A CIS copes with the semantics for heterogeneous information integration. A CIS is quite similar to a theory in DFOL. A CIS is composed of a set of description languages, a T-BOX for each language, and a set of so-called interschema assertions. Under the usual translation of description logics in first order logics and the translation of interschema assertion into DFOL compatibility constraints, it can be shown that a CIS can be embedded into DFOL. The main difference between CIS and DFOL concerns the semantics. A model for a CIS is defined over a "global" domain Δ which is the union of the domains Δ_i of the models of each description language. This implies that a constant c in different languages is interpreted in the same object c in the CIS. As a consequence various forms of relation between domains, e.g., abstraction, cannot be represented in CIS. A second difference is that CIS models complete databases and cannot express partiality. Totality affects directionality. Indeed in CIS every interschema assertion $L_1 \dot{\leq}_{ext} L_2$ ($L_1 \dot{\leq}_{int} L_2$) from a description language to another entails the converse interschema assertion $\neg L_2 \dot{\leq}_{ext} \neg L_1$ ($\neg L_2 \dot{\leq}_{int} \neg L_1$) in the opposite direction.

8 CONCLUSIONS

In this paper we have introduced a logic, called Distributed First Order Logic (DFOL), for the formalization of distributed knowledge representation systems. The major contributions of this paper are the following: First, we have defined a class of representational and structural requirements which must be satisfied by a logic formalizing distributed knowledge representation. Second, we have defined a model theoretic semantics for a family of languages based on the notions of possible model and of domain relation. Third, we have introduced the concepts of domain constraint and interpretation constraint. These concepts formalize relations between the semantics of different languages. Fourth, we have defined a notion of logical consequence among formulae of different languages, and we have defined a sound and complete calculus for DFOL based on ML systems. Finally, we have compared our formalism with other formalisms for the representation and integration of distributed knowledge and reasoning systems.

Acknowledgments. We thank all the people of the Mechanized Reasoning Group of IRST and DISA for useful discussions and feedback on the paper. We thank Massimo Benerecetti and Sergio Tessaris for carefully proofreading the paper. We also thank the anonymous referees for valuable comments. This work is part of the MRG project *Distributed Representations and Systems*; see `http://www.cs.unitn.it/~mrg/distributed-intelligence`.

REFERENCES

[Basin *et al.*, 1998] D. Basin, S. Matthews, and L. Viganò. Labelled Modal Logics: Quantifiers. *Journal of Logic, Language and Information*, 7(3):237–263, 1998.

[Blair and Subrahmanian, 1987] H.A. Blair and V.S. Subrahmanian. Paraconsistent Logic Programming. *Theoretical Computer Science*, 68:35–51, 1987.

[Borgida, 1996] A. Borgida. On the relative expressiveness of description logics and predicate logics. *Artificial Intelligence*, 82:353–367, 1996. Research Note.

[Catarci and Lenzerini, 1993] T. Catarci and M. Lenzerini. Representing and using interschema knowledge in cooperative information systems. *International Journal of Intelligent and Cooperative Information Systems*, 2(4):375–398, 1993.

[Chang and Keisler, 1973] C.C. Chang and J.M. Keisler. *Model Theory*. North Holland, 1973.

[D'Agostino, 1992] M. D'Agostino. Are Tableaux an Improvement on Truth-Tables? *Journal of Logic, Language and Information*, 1:235–252, 1992.

[Dinsmore, 1991] J. Dinsmore. *Partitioned Representations*. Kluwer Academic Publisher, 1991.

[Donini *et al.*, 1996] F. Donini, M. Lenzerini, D. Nardi, and A. Schaerf. Reasoning in description logics. In G. Brewka, editor, *Principles of Knowledge Representation and Reasoning*, Studies in Logic, Language and Information, pages 193–238. CLSI Publications, 1996.

[Fagin *et al.*, 1995] R. Fagin, J.Y. Halpern, Y. Moses, and M. Y. Vardi. *Reasoning about knowledge*. MIT Press, 1995.

[Gabbay, 1990] D. Gabbay. Labeled Deductive Systems. Technical Report CIS-Bericht-90-22, Universität München – Centrum für Informations und Sprachverarbeitung, 1990.

[Gabbay, 1994] D. Gabbay. Labeled Deductive Systems Volume 1 – Foundations. Technical Report MPI-I-94-223, Max-Planck-institut für Informatik, May 1994.

[Ghidini, 1998] C. Ghidini. *A semantics for contextual reasoning: theory and two relevant applications*. PhD thesis, Department of Computer Science, University of Rome "La Sapienza", March 1998.

[Giunchiglia and Ghidini, 1998] F. Giunchiglia and C. Ghidini. Local Models Semantics, or Contextual Reasoning = Locality + Compatibility. In *Proceedings of the Sixth International Conference on Principles of Knowledge Representation and Reasoning (KR'98)*, pages 282–289. Morgan Kaufmann, 1998.

[Giunchiglia and Serafini, 1994] F. Giunchiglia and L. Serafini. Multilanguage hierarchical logics (or: how we can do without modal logics). *Artificial Intelligence*, 65:29–70, 1994.

[Giunchiglia, 1993] F. Giunchiglia. Contextual reasoning. *Epistemologia, special issue on I Linguaggi e le Macchine*, XVI:345–364, 1993. Short version in Proceedings IJCAI'93 Workshop on Using Knowledge in its Context, Chambery, France, 1993, pp. 39–49.

[Masini, 1992] A. Masini. 2-Sequent calculus: a proof theory of modalities. *Annals of Pure and Applied Logic*, 58:229–246, 1992.

[Prawitz, 1965] D. Prawitz. *Natural Deduction - A proof theoretical study.* Almquist and Wiksell, Stockholm, 1965.

[Serafini and Ghidini, 1997] L. Serafini and C. Ghidini. Context Based Semantics for Federated Databases. In *Proceedings of the 1st International and Interdisciplinary Conference on Modeling and Using Context (CONTEXT-97)*, pages 33–45, Rio de Janeiro, Brazil, 1997.

[Subrahmanian, 1994] V.S. Subrahmanian. Amalgamating Knowledge Bases. *ACM Trans. Database Syst.*, 19(2):291–331, 1994.

[Wiederhold, 1992] G. Wiederhold. Mediators in the architecture of future information systems. *IEEE Computer*, 25(3):38–49, 1992.

Pushing the Frontiers of Combining Rewrite Systems Farther Outwards

Jürgen Giesl and Enno Ohlebusch

1 INTRODUCTION

Modularity is a well-known paradigm in computer science. Programs should be designed in a modular way, that is, as a combination of small programs. These so-called modules are implemented separately and are then integrated to form the whole program. Since TRSs have important applications in computer science, it is essential to know under which conditions a combined system inherits desirable properties from its constituent systems. For this reason modular aspects of term rewriting have been studied extensively. A property $\mathcal{P}$ of TRSs (like termination) is called *modular* if whenever $\mathcal{R}_1$ and $\mathcal{R}_2$ are TRSs both satisfying $\mathcal{P}$, then their combined system $\mathcal{R}_1 \cup \mathcal{R}_2$ also satisfies $\mathcal{P}$. The knowledge that (perhaps under certain conditions) a property $\mathcal{P}$ is modular facilitates software engineering because it allows an incremental development of programs. On the other hand, it provides a divide and conquer approach to establishing properties of TRSs. If one wants to know whether a large TRS has a certain modular property $\mathcal{P}$, then this system can be decomposed into small subsystems and one merely has to check whether each of these subsystems has property $\mathcal{P}$.

As all interesting properties are in general not modular, the starting-point of research was disjoint unions, i.e. combinations of TRSs without common function symbols. Toyama [1987b] proved that confluence is modular for disjoint systems, but termination and completeness lack a modular behavior [Toyama, 1987a]. So the question is what restrictions have to be imposed on the constituent TRSs so that their disjoint union is again terminating. The first results were obtained by investigating the distribution of collapsing rules and duplicating rules among the TRSs; see [Rusinowitch, 1987; Middeldorp, 1989]. In [Toyama *et al.*, 1995] it is shown that termination is modular for confluent and left-linear TRSs. Ever since, an abundance of modularity results for disjoint unions, constructor-sharing systems, composable systems, and hierarchical combinations has been published; see [Middeldorp, 1990; Ohlebusch, 1994a; Gramlich, 1996] for an overview. However, most of the modularity results are often not applicable in practice. For example, collapsing and duplicating rules occur naturally in most TRSs. In contrast to this, since most methods for *automated* termination proofs work with so-called *simplification* orderings [Dershowitz, 1987; Steinbach, 1995; Middeldorp and Zantema, 1997], Kurihara and Ohuchi's [1992] result for constructor-sharing systems is thus of practical relevance. They showed that the combination of finite simply terminating TRSs (systems whose termination can be verified by a simplification ordering) is again simply terminating. Their result was extended to composable systems [Ohlebusch, 1995] and to certain hierarchical combinations [Krishna Rao, 1994]. Moreover, all these results also hold for infinite TRSs; see [Middeldorp and Zantema, 1997].

However, there are numerous relevant TRSs where simplification orderings fail in proving termination. For that purpose, a new technique for automated termination proofs, viz. the so-called *dependency pair* approach, was developed by Arts and Giesl [1997a; 1997b; 1997c; 1998]. Given a TRS, this approach generates a set of constraints and the existence of a well-founded (quasi-)ordering satisfying these constraints is sufficient for termination. The advantage is that standard techniques can often generate such a well-founded ordering even if a direct termination proof with the same techniques fails. In this way, simplification orderings can now be used to prove termination of non-simply terminating TRSs. Several such systems from different areas of computer science (including many challenging problems from the literature) can for instance be found in [Arts and Giesl, 1997c].

Thus, the dependency pair approach pushed the frontier of TRSs whose termination is provable automatically a lot further. Now the TRSs where automated termination proofs are (potentially) feasible are no longer just the simply terminating systems, but the *DP-(quasi) simply terminating* systems, i.e. those systems whose termination can be verified by using simplification orderings in combination with dependency pairs. Hence, a natural question is whether the current frontier of modularity can be pushed further as well, by extending the modularity results from simple to DP-(quasi) simple termination. In this paper, we will show that this is indeed possible. Thus, the class of TRSs whose termination can be proved in a modular way is extended considerably.

The paper is organized as follows. First we briefly recall the basic notions of the combination of TRSs. Section 3 contains a short description of the dependency pair method. In Section 4 we introduce the concept of DP-(quasi) simple termination and show in Section 5 that DP-quasi simple termination is modular for disjoint unions. Section 6 contains similar results about constructor-sharing TRSs.

2 BASIC NOTIONS OF THE UNION OF TRSS

For an introduction to term rewriting see e.g. [Dershowitz and Jouannaud, 1990; Klop, 1992]. Let $\mathcal{R}$ be a TRS over the signature $\mathcal{F}$. A function symbol $f \in \mathcal{F}$ is called a *defined symbol* if there is a rewrite rule $l \to r \in \mathcal{R}$ such that $f = root(l)$. Function symbols from $\mathcal{F}$ which are not defined symbols are called *constructors*. Thus, if a TRS consists of the following two rules

$$\begin{array}{lrcl} (1) & f(0,1,x) & \to & f(s(x),x,x) \\ (2) & f(x,y,s(z)) & \to & s(f(0,1,z)), \end{array}$$

then f is the only defined symbol, whereas 0, 1, and s are constructors.

Let $\mathcal{R}_1$ and $\mathcal{R}_2$ be TRSs over the signatures $\mathcal{F}_1$ and $\mathcal{F}_2$, respectively. Their *combined system* is their union $\mathcal{R} = \mathcal{R}_1 \cup \mathcal{R}_2$ over the signature $\mathcal{F} = \mathcal{F}_1 \cup \mathcal{F}_2$. Its set of defined symbols is $\mathcal{D} = \mathcal{D}_1 \cup \mathcal{D}_2$ and its set of constructors is $\mathcal{C} = \mathcal{F} \setminus \mathcal{D}$, where $\mathcal{D}_i$ ($\mathcal{C}_i$) denotes the defined symbols (constructors) in $\mathcal{R}_i$.

(1) $\mathcal{R}_1$ and $\mathcal{R}_2$ are *disjoint* if $\mathcal{F}_1 \cap \mathcal{F}_2 = \emptyset$.

(2) $\mathcal{R}_1$ and $\mathcal{R}_2$ are *constructor-sharing* if $\mathcal{F}_1 \cap \mathcal{F}_2 = \mathcal{C}_1 \cap \mathcal{C}_2$ ($\subseteq \mathcal{C}$).

(3) $\mathcal{R}_1$ and $\mathcal{R}_2$ are *composable* if $\mathcal{C}_1 \cap \mathcal{D}_2 = \mathcal{D}_1 \cap \mathcal{C}_2 = \emptyset$ and both systems contain all rewrite rules that define a defined symbol whenever that symbol is shared: $\{l \to r \in \mathcal{R} \mid root(l) \in \mathcal{D}_1 \cap \mathcal{D}_2\} \subseteq \mathcal{R}_1 \cap \mathcal{R}_2$.

We next give a brief overview of the basic notions of disjoint unions. In the sequel let $t \in \mathcal{T}(\mathcal{F}_1 \cup \mathcal{F}_2, \mathcal{V})$. Let $\Box$ be a special constant $\notin \mathcal{F}_1 \cup \mathcal{F}_2$. A *context* C is a term in $\mathcal{T}(\mathcal{F}_1 \cup \mathcal{F}_2 \cup \{\Box\}, \mathcal{V})$ and $C[t_1, \ldots, t_n]$ is the result of replacing from left to right the $n \geq 0$ occurrences of $\Box$ with $t_1, \ldots, t_n$. We write $t = C[\![t_1, \ldots, t_n]\!]$ if $C \in \mathcal{T}(\mathcal{F}_i \cup \{\Box\}, \mathcal{V})$, $C \neq \Box$, and $root(t_1), \ldots root(t_n) \in \mathcal{F}_{3-i}$ for some $i \in \{1, 2\}$. In this case, the t_j are the *principal* subterms of t and C is the topmost $\mathcal{F}_i$-homogeneous part of t, denoted by $top_i(t)$ (whereas $top_{3-i}(t)$ is $\Box$). So for example, if $\mathcal{R}_1$ consists of the rules (1) and (2), and $\mathcal{R}_2$ contains the rules

$$\begin{array}{lrcl} (3) & g(x,y) & \to & x \\ (4) & g(x,y) & \to & y, \end{array}$$

then $\mathcal{R}_1$ and $\mathcal{R}_2$ are disjoint and a term like $f(g(0,0),x,g(y,y))$ can be written as $C[\![g(0,0), g(y,y)]\!]$, where C is $f(\Box, x, \Box)$. Thus $top_1(f(g(0,0),x,g(y,y))) = f(\Box, x, \Box)$ and $top_2(f(g(0,0),x,g(y,y))) = \Box$.

Moreover, for any term t its *rank* is the maximal number of alternating function symbols (from $\mathcal{F}_1$ and $\mathcal{F}_2$, resp.) in any path through the term, i.e.

$$rank(t) = 1 + \max\{rank(t_j) \mid 1 \leq j \leq n\} \text{ where } t = C[\![t_1, \ldots, t_n]\!]$$

and $\max \emptyset = 0$. So for example we have $rank(f(g(0,0),x,g(y,y))) = 3$. Our modularity results crucially depend on the fact that $s \to_{\mathcal{R}_1 \cup \mathcal{R}_2} t$ implies $rank(s) \geq rank(t)$ (the proof is straightforward by induction on $rank(s)$).

A rewrite step $s \to_{\mathcal{R}_1 \cup \mathcal{R}_2} t$ is *destructive at level 1* if $root(s) \in \mathcal{F}_i$ and $root(t) \in \mathcal{F}_{3-i}$ for some $i \in \{1,2\}$. A reduction step $s \to_{\mathcal{R}_1 \cup \mathcal{R}_2} t$ is *destructive at level* $m+1$ (for some $m \geq 1$) if $s = C[\![s_1, \ldots, s_j, \ldots, s_n]\!] \to_{\mathcal{R}_1 \cup \mathcal{R}_2} C[s_1, \ldots, t_j, \ldots, s_n] = t$ with $s_j \to_{\mathcal{R}_1 \cup \mathcal{R}_2} t_j$ destructive at level m. Obviously, if a rewrite step is destructive, then the rewrite rule applied is collapsing, i.e. the right-hand side of the rule is a variable. For example, the rewrite step $f(g(0,0),x,g(y,y)) \to f(0,x,g(y,y))$ is destructive at level 2.

Finally, we recall that for every signature $\mathcal{F}$ the TRS $\mathcal{E}mb(\mathcal{F})$ (which is important in the context of simple termination) is defined by

$$\mathcal{E}mb(\mathcal{F}) = \{f(x_1, \ldots, x_n) \to x_i \mid f \in \mathcal{F},\ f \text{ is } n\text{-ary and } 1 \leq i \leq n\}.$$

3 DEPENDENCY PAIRS

In the dependency pair approach of Arts and Giesl [1997a; 1997c; 1998] for showing termination, if $f(s_1, ..., s_n)$ rewrites to $C[g(t_1, ..., t_m)]$ (where g is a defined symbol), then one has to compare the argument tuples $s_1, ..., s_n$ and $t_1, ..., t_m$. To avoid the handling of *tuples*, a new *tuple symbol* $F \notin \mathcal{F}$ is introduced for every defined symbol f. Instead of comparing *tuples*, now the *terms* $F(s_1, ..., s_n)$ and $G(t_1, ..., t_m)$ are compared. Thus, to ease readability we assume that the signature $\mathcal{F}$ consists of lower case symbols only and that tuple symbols are denoted by the corresponding upper case symbols.

Definition 1 (Dependency Pair) If $f(s_1, ..., s_n) \to C[g(t_1, ..., t_m)]$ is a rule of a TRS $\mathcal{R}$ and g is a defined symbol, then $\langle F(s_1, ..., s_n), G(t_1, ...t_m)\rangle$ is a *dependency pair* of $\mathcal{R}$.

So for the TRS $\mathcal{R}_1 = \{(1), (2)\}$ we obtain the following dependency pairs

(5) $\langle F(0,1,x), F(s(x),x,x)\rangle$

(6) $\langle F(x,y,s(z)), F(0,1,z)\rangle$.

To trace those subterms which may start new reductions, we examine special sequences of dependency pairs, so-called *chains*. In the following, we consider substitutions whose domains may be infinite and assume that different (occurrences of) dependency pairs have disjoint sets of variables.

Definition 2 (Chain) A sequence of dependency pairs $\langle s_1, t_1\rangle\langle s_2, t_2\rangle$... is an $\mathcal{R}$*-chain* if there is a substitution σ such that $t_j\sigma \to^*_{\mathcal{R}} s_{j+1}\sigma$ holds for every two consecutive pairs $\langle s_j, t_j\rangle$ and $\langle s_{j+1}, t_{j+1}\rangle$ in the sequence.

For instance, in our example we have the chain

$$\langle F(0,1,x_1), F(s(x_1),x_1,x_1)\rangle\ \ \langle F(x_2,y_2,s(z_2)), F(0,1,z_2)\rangle\ \ \langle F(0,1,x_3), F(s(x_3),x_3,x_3)\rangle$$

because with $\sigma = \{x_1 \mapsto s(x_3), x_2 \mapsto s(s(x_3)), y_2 \mapsto s(x_3), z_2 \mapsto x_3\}$ we have $F(s(x_1), x_1, x_1)\sigma \to_{\mathcal{R}_1}^* F(x_2, y_2, s(z_2))\sigma$ and $F(0, 1, z_2)\sigma \to_{\mathcal{R}_1}^* F(0, 1, x_3)\sigma$. In fact, every finite alternating sequence of (5) and (6) is a chain. Arts and Giesl [1997a; 1997c] proved that the absence of *infinite* chains is a sufficient and necessary criterion for termination.

Theorem 3 (Termination Criterion) *A TRS $\mathcal{R}$ is terminating if and only if there exists no infinite $\mathcal{R}$-chain.*

Note that the first dependency pair (5) can never follow itself in a chain, because $F(s(x_1), x_1, x_1)\sigma \to_{\mathcal{R}_1}^* F(0, 1, x_2)\sigma$ does not hold for any substitution σ. To estimate which dependency pairs may occur consecutive in a chain, the *estimated dependency graph* has been introduced, cf. Arts and Giesl [1997a; 1997c; 1998]. We first recall the relevant notions. CAP(t) results from replacing all subterms of t that have a defined root symbol by different fresh variables and REN(t) results from replacing all variables in t by different fresh variables. Then, in order to determine whether $\langle u, v\rangle$ can follow $\langle s, t\rangle$ in a chain, we check whether REN(CAP(t)) unifies with u. The function REN is needed to rename multiple occurrences of the same variable x in t because when instantiated with σ, two occurrences of $x\sigma$ could reduce to different terms. So in our example, the estimated dependency graph contains an arc from (5) to (6) and arcs from (6) to (5) and to itself.

Definition 4 (Estimated Dependency Graph) The *estimated dependency graph* is the directed graph whose nodes are the dependency pairs and there is an arc from $\langle s, t\rangle$ to $\langle u, v\rangle$ if REN(CAP(t)) and u are unifiable.

A set $\mathcal{P}$ of dependency pairs is called a *cycle* if for any two dependency pairs $\langle s, t\rangle, \langle u, v\rangle \in \mathcal{P}$ there is a path from $\langle s, t\rangle$ to $\langle u, v\rangle$ and from $\langle u, v\rangle$ to $\langle s, t\rangle$ in the estimated dependency graph which traverses dependency pairs from $\mathcal{P}$ only. (In particular, there must also be a path from $\langle s, t\rangle$ to itself.) Thus, the only non-empty cycles in our example are $\{(6)\}$ and $\{(5), (6)\}$. In the remainder of the paper, we always restrict ourselves to finite TRSs (and to finite signatures). Then any infinite chain corresponds to a cycle, i.e. it suffices to prove that there is no infinite chain of dependency pairs *from any cycle*, cf. [Arts and Giesl, 1998].

For an automation of this criterion, we generate a set of inequalities such that the existence of a well-founded *quasi-ordering* satisfying these inequalities is sufficient for the absence of infinite chains. As usual, a *quasi-ordering* $\succsim$ is a reflexive and transitive relation. The corresponding *strict* relation $\succ^s$ is defined as $t \succ^s u$ iff $t \succsim u$ and $u \not\succsim t$. Moreover, we also define a corresponding *stable-strict* relation $\succ^{ss}$ as $t \succ^{ss} u$ iff $t\sigma \succ^s u\sigma$ holds for all ground substitutions σ, where a *ground* substitution is a substitution mapping all variables to ground terms. In other words, for all those substitutions σ we must have $t\sigma \succsim u\sigma$ and $u\sigma \not\succsim t\sigma$.

For instance, many useful quasi-orderings are constructed by using mappings $|.|$ from the set of ground terms to a well-founded set like the natural numbers $\mathbb{N}$, cf. e.g. [Lankford, 1979, "polynomial orderings"]. Then $\succsim$ is defined as $t \succsim u$ iff $|t\sigma| \geq_{\mathbb{N}} |u\sigma|$ holds for all ground substitutions σ. A natural way to define a corresponding irreflexive ordering $\succ$ is to let $t \succ u$ hold iff $|t\sigma| >_{\mathbb{N}} |u\sigma|$ for all ground substitutions σ. However, now $\succ$ is not the corresponding strict relation, but the corresponding stable-strict relation of $\succsim$. Thus, the irreflexive relation intuitively associated with a quasi-ordering is often the stable-strict one instead of the strict one. In particular, if the quasi-ordering $\succsim$ is stable under substitutions, then the corresponding stable-strict relation $\succ^{ss}$ is stable under substitutions too, whereas this is not necessarily true for the strict relation $\succ^{s}$.

For example, if $|a| = 0$, $|f_1(t)| = |t|$, and $|f_2(t)| = 2|t|$ for all ground terms t, then we have $f_2(x) \succsim f_1(x)$ and $f_1(x) \not\succsim f_2(x)$. Hence, this implies $f_2(x) \succ^{s} f_1(x)$. However, $\succ^{s}$ is not stable under substitutions because $f_2(a) \succ^{s} f_1(a)$ does not hold. This example also demonstrates that in general $\succ^{s} \subseteq \succ^{ss}$ is not true because for the stable-strict relation $\succ^{ss}$ we have $f_2(x) \not\succ^{ss} f_1(x)$.

Moreover, in general $\succ^{ss} \subseteq \succsim$ does not hold either (hence, $\succ^{ss} \subseteq \succ^{s}$ is false, too). If $\mathcal{R}$ is the TRS containing only the rule $h(a) \to a$ and $\succsim$ is defined as $\to_{\mathcal{R}}^{*}$, then we have $h(x) \succ^{ss} x$, but $h(x) \not\succsim x$.

The following lemma states some straightforward properties of stable-strict relations, where we always assume that our signature contains at least one constant (i.e. that there exist ground terms).

Lemma 5 (Properties of Stable-Strict Relations) *Let $\succsim$ be a quasi-ordering that is stable under substitutions. Then we have*

(i) $\succ^{ss}$ *is irreflexive*
(ii) $\succ^{ss}$ *is transitive*
(iii) $\succ^{ss}$ *is stable under substitutions*
(iv) if $\succ^{s}$ *is stable under substitutions, then* $\succ^{s} \subseteq \succ^{ss}$
(v) if $\succ^{s}$ *is well founded, then* $\succ^{ss}$ *is well founded, too*
(vi) $s \succsim t \succ^{ss} u$ *implies* $s \succ^{ss} u$
(vii) $s \succ^{ss} t \succsim u$ *implies* $s \succ^{ss} u$.

Proof. The conjectures (i) and (ii) follow from the reflexivity and the transitivity of $\succsim$. Conjectures (iii) and (iv) are direct consequences of the definition. For (v), every potential infinite descending sequence $t_0 \succ^{ss} t_1 \succ^{ss} \ldots$ would result in an infinite descending sequence $t_0\sigma \succ^{s} t_1\sigma \succ^{s} \ldots$ Conjectures (vi) and (vii) follow from the transitivity and stability of $\succsim$. □

In the following, instead of the corresponding strict relations we always consider the corresponding stable-strict relations of quasi-orderings $\succsim$. For the sake of brevity, we write $\succ$ instead of $\succ^{ss}$, i.e. in this paper $\succ$ always denotes the stable-strict relation corresponding to $\succsim$. Analogously, we will call a quasi-ordering *well-founded* if the corresponding stable-strict relation is well founded.

The following theorem is from [Arts and Giesl, 1998], where instead of the strict relation corresponding to the quasi-ordering we now use the stable-strict relation. Note that the present formulation of Theorem 6 with stable-strict relations is more powerful than the formulation with strict relations. To use the strict relation $\succ^s$ of a quasi-ordering in Theorem 6, $\succ^s$ would have to be stable under substitutions; cf. [Arts and Giesl, 1998, Theorem 6]. From Lemma 5 (iv) it follows that $s \succ^s t$ always implies $s \succ^{ss} t$. Hence, all constraints satisfied by $\succ^s$ are satisfied by the corresponding stable-strict relation $\succ^{ss}$ as well. Using Lemma 5 (vii), the proof for the if-part of this slightly modified theorem is identical to the corresponding one in [Arts and Giesl, 1998]. The proof for the only-if-part can be found in [Arts and Giesl, 1997c].

Theorem 6 (Dependency Pair Approach) *A TRS $\mathcal{R}$ is terminating iff for each cycle $\mathcal{P}$ in the estimated dependency graph there is a well-founded weakly monotonic quasi-ordering $\succsim$ stable under substitutions such that*

- $l \succsim r$ *for all rules* $l \to r$ *in* $\mathcal{R}$,
- $s \succsim t$ *for all dependency pairs* $\langle s,t\rangle$ *from* $\mathcal{P}$,
- $s \succ t$ *for at least one dependency pair* $\langle s,t\rangle$ *from* $\mathcal{P}$.

Thus, to prove the absence of infinite chains from the cycle $\{(6)\}$ we have to find a quasi-ordering satisfying

(7) $F(x,y,s(z)) \succ F(0,1,z)$

(8) $f(0,1,x) \succsim f(s(x),x,x)$

(9) $f(x,y,s(z)) \succsim s(f(0,1,z))$.

4 DP-(QUASI) SIMPLE TERMINATION

As mentioned, our aim is to use standard techniques to generate a suitable quasi-ordering satisfying the constraints of Theorem 6. However, most existing methods generate orderings which are *strongly* monotonic, whereas for the dependency pair approach we only need a *weakly* monotonic ordering. For that reason, before synthesizing a suitable ordering, some of the arguments of the function symbols can be eliminated, cf. Arts and Giesl [1997a; 1997c]. For instance, one may eliminate the first two arguments of the function symbol f. Then every term $f(t_1,t_2,t_3)$ in the inequalities is replaced by $f'(t_3)$, where f' is a new function symbol. So instead of (8) and (9) we would obtain the inequalities $f'(x) \succsim f'(x)$ and $f'(s(z)) \succsim s(f'(z))$. Now the resulting constraints are satisfied by the recursive path ordering (rpo) with the precedence $f' \rhd s \rhd 0 \rhd 1$. Similarly, (by eliminating the first two arguments of F) one can also prove the absence of infinite chains from the cycle $\{(5),(6)\}$. Hence, termination of the TRS consisting of the rules (1) and (2) is proved. Note that this TRS is not simply terminating. So in the dependency pair approach, simplification orderings like the rpo can be used to prove termination of TRSs where their direct application would fail.

Apart from eliminating arguments of function symbols, another possibility is to replace functions by one of their arguments. So instead of deleting the first two arguments of f, one could replace all terms $f(t_1, t_2, t_3)$ by f's third argument t_3. Then the resulting inequalities are again satisfied by the rpo. To perform this elimination of arguments and function symbols the following concept was introduced in [Arts and Giesl, 1997c].

Definition 7 (AFS) An *argument filtering system*[1] (AFS) over $\mathcal{F}$ is a TRS whose rewrite rules are of the form

$$f(x_1, \ldots, x_n) \to r$$

with $f \in \mathcal{F}$ and there is at most one such rule for every $f \in \mathcal{F}$. Here $x_1, \ldots, x_n$ are pairwise distinct variables and r is either one of these variables or it is a term $f'(y_1, \ldots, y_m)$, where $f' \notin \mathcal{F}$ is a fresh function symbol and $y_1, \ldots, y_m$ are pairwise distinct variables out of $x_1, \ldots, x_n$.

As proved in [Arts and Giesl, 1997c], in order to find a quasi-ordering satisfying a particular set of inequalities, one may first normalize the terms in the inequalities with respect to an AFS (where the AFS may also contain rules for the tuple symbols). Subsequently, one only has to find a quasi-ordering that satisfies these modified inequalities. Hence, by combining the synthesis of a suitable AFS with well-known techniques for the generation of (*strongly* monotonic) simplification orderings, now the search for a *weakly* monotonic ordering satisfying the constraints can be automated.

In this paper, we impose a (minor) restriction[2] on the AFSs used, viz. we restrict ourselves to AFSs $\mathcal{A}$ such that

- $\mathcal{V}ar(r\downarrow_{\mathcal{A}}) \subseteq \mathcal{V}ar(l\downarrow_{\mathcal{A}})$ and $l\downarrow_{\mathcal{A}} \notin \mathcal{V}$ or $l\downarrow_{\mathcal{A}} = r\downarrow_{\mathcal{A}}$ for all rules $l \to r$ in $\mathcal{R}$
- $\mathcal{V}ar(t\downarrow_{\mathcal{A}}) \subseteq \mathcal{V}ar(s\downarrow_{\mathcal{A}})$ and $s\downarrow_{\mathcal{A}} \notin \mathcal{V}$ or $s\downarrow_{\mathcal{A}} = t\downarrow_{\mathcal{A}}$ for all $\langle s, t\rangle$ in cycles

As already mentioned, most methods for the automated generation of well-founded orderings construct simplification orderings or quasi-simplification orderings [Dershowitz, 1987; Steinbach, 1995; Middeldorp and Zantema, 1997]. Here we use the following definition of [Middeldorp and Zantema, 1997]: A *simplification ordering* is an ordering (i.e. an irreflexive and transitive relation) that is monotonic (closed under contexts), closed under substitutions, and possesses the subterm property. It is a well-known consequence

[1] AFSs are a special form of recursive program schemes [Courcelle, 1990; Klop, 1992].

[2] This restriction is not very severe. If there exists a quasi-simplification ordering satisfying the constraints in Theorem 6 and if these constraints include at least one strict inequality with variables in its right-hand side, then $\mathcal{V}ar(r\downarrow_{\mathcal{A}}) \subseteq \mathcal{V}ar(l\downarrow_{\mathcal{A}})$ and $\mathcal{V}ar(t\downarrow_{\mathcal{A}}) \subseteq \mathcal{V}ar(s\downarrow_{\mathcal{A}})$ are always satisfied, because otherwise the constraints would imply $t \succ x$ for some term t with $x \notin \mathcal{V}ar(t)$.

of Kruskal's theorem that every simplification ordering on $\mathcal{T}(\mathcal{F}, \mathcal{V})$ is well founded provided that $\mathcal{F}$ is finite.[3]

Analogously, a *quasi-simplification ordering* (qso) is a quasi-ordering which is (weakly) monotonic, closed under substitutions, and has the (weak) subterm property. Since we restrict ourselves to finite signatures, every quasi-simplification ordering (more precisely, the corresponding stable-strict relation) is well founded, too.

Examples of simplification orderings and qso's include path orderings such as the rpo, the lexicographic path ordering (lpo), etc. [Dershowitz, 1987; Steinbach, 1995]. Polynomial orderings, however, are not qso's in general. For instance, if the constant 0 is associated with the number 0, $s(x)$ is associated with $x + 1$, and $f(x, y)$ is associated with the multiplication of x and y, then this polynomial ordering does not satisfy the subterm property (for example, $f(s(0), 0) \succsim s(0)$ does not hold). However, the following lemma shows that if the polynomial ordering respects some restrictions, then it is indeed a qso.

Lemma 8 (Polynomial Orderings as qso's) *Let $\succsim$ be a polynomial ordering where every function symbol is associated with a polynomial containing only non-negative coefficients.*

- *If every function symbol $f(x_1, \ldots, x_n)$ is associated with a polynomial containing all variables $x_1, \ldots, x_n$ and if every constant is associated with a number > 0, then $\succsim$ is a qso.*

- *If every function symbol $f(x_1, \ldots, x_n)$ is associated with a polynomial which contains a (non-mixed) monomial of the form $m\, x_i^k$ (with $m, k \geq 1$) for every $i = 1, \ldots, n$, then $\succsim$ is a qso.*

Proof. Straightforward. □

In fact, whenever polynomial orderings can be used in connection with the dependency pair approach, one can usually apply a polynomial ordering which satisfies one of the above conditions. By restricting ourselves to qso's, we obtain the following restricted notion of termination. Again, $\succ$ denotes the stable-strict relation corresponding to $\succsim$.

Definition 9 (DP-quasi simple termination) A TRS $\mathcal{R}$ is *DP-quasi simply terminating* iff for every non-empty cycle $\mathcal{P}$ in the estimated dependency graph there exists an AFS $\mathcal{A}$ and a qso $\succsim$ such that

(a) $l \downarrow_{\mathcal{A}} \succsim r \downarrow_{\mathcal{A}}$ for all rules $l \to r$ in $\mathcal{R}$,
(b) $s \downarrow_{\mathcal{A}} \succsim t \downarrow_{\mathcal{A}}$ for all dependency pairs $\langle s, t\rangle$ from $\mathcal{P}$,
(c) $s \downarrow_{\mathcal{A}} \succ t \downarrow_{\mathcal{A}}$ for at least one dependency pair $\langle s, t\rangle$ from $\mathcal{P}$.

[3]For details on infinite signatures see [Middeldorp and Zantema, 1997].

Definition 9 captures all TRSs where an automated termination proof using dependency pairs is potentially feasible. In fact, there are numerous DP-quasi simply terminating TRSs which are not simply terminating; cf. e.g. the collection in [Arts and Giesl, 1997c]. This observation motivated the development of the dependency pair approach and it also motivated the present work, as our aim is to extend well-known modularity results for simple termination to DP-quasi simple termination.

A straightforward way to generate a qso $\succsim$ from a simplification ordering $\succ$ is to define $t \succsim u$ iff $t \succ u$ or $t = u$, where $=$ is syntactic equality. In the following, we denote the reflexive closure of a relation by underlining, i.e. $\succeq$ denotes the reflexive closure of $\succ$. By restricting ourselves to this class of qso's, we obtain the notion of DP-simple termination.

Definition 10 (DP-simple termination) A TRS $\mathcal{R}$ is *DP-simply terminating* iff for every non-empty cycle $\mathcal{P}$ in the estimated dependency graph there is an AFS $\mathcal{A}$ and a simplification ordering $\succ$ such that

(a) $l\downarrow_{\mathcal{A}} \succeq r\downarrow_{\mathcal{A}}$ for all rules $l \to r$ in $\mathcal{R}$,
(b) $s\downarrow_{\mathcal{A}} \succeq t\downarrow_{\mathcal{A}}$ for all dependency pairs $\langle s, t\rangle$ from $\mathcal{P}$,
(c) $s\downarrow_{\mathcal{A}} \succ t\downarrow_{\mathcal{A}}$ for at least one dependency pair $\langle s, t\rangle$ from $\mathcal{P}$.

Note that (a) - (c) are equivalent to simple termination of the TRS

$$\begin{aligned} \mathcal{S}_{\mathcal{P}} \;=\; & \{l\downarrow_{\mathcal{A}} \to r\downarrow_{\mathcal{A}} \mid l \to r \in \mathcal{R} \text{ and } l\downarrow_{\mathcal{A}} \neq r\downarrow_{\mathcal{A}}\} \cup \\ & \{s\downarrow_{\mathcal{A}} \to t\downarrow_{\mathcal{A}} \mid \langle s,t\rangle \text{ is a dependency pair from } \mathcal{P} \text{ and } s\downarrow_{\mathcal{A}} \neq t\downarrow_{\mathcal{A}}\} \end{aligned}$$

provided that $s\downarrow_{\mathcal{A}} \neq t\downarrow_{\mathcal{A}}$ holds for at least one dependency pair $\langle s, t\rangle \in \mathcal{P}$.

It turns out that most of the examples in [Arts and Giesl, 1997c] are not only DP-quasi simply terminating but even DP-simply terminating. The following lemma illustrates the connections between the different notions.

Lemma 11 (Characterizing DP-(quasi) simple termination)
simple termination $\Rightarrow$ DP-simple termination $\Rightarrow$ DP-quasi simple termination $\Rightarrow$ termination

Proof. The second implication holds as $\succ$ is stable under substitutions and therefore $\succ$ is contained in the stable-strict relation of $\succeq$, cf. Lemma 5 (iv). The last implication follows from Theorem 6 by using the quasi-ordering $\succsim'$ where $u \succsim' v$ holds iff $u\downarrow_{\mathcal{A}} \succeq v\downarrow_{\mathcal{A}}$.

It remains to show the first implication. Let $\mathcal{R}$ be a simply terminating TRS over the signature $\mathcal{F} = \mathcal{C} \cup \mathcal{D}$ and let $\mathcal{T}up_{\mathcal{F}} = \{F \mid f \in \mathcal{D}\}$ be the set of tuple symbols. If $\mathcal{R}$ is simply terminating, then there exists a simplification ordering $\succ$ such that $l \succ r$ holds for all rules $l \to r$ of $\mathcal{R}$.

Let Ω be the function which in a term $s \in \mathcal{T}(\mathcal{F} \cup \mathcal{T}up_{\mathcal{F}}, \mathcal{V})$ replaces every tuple symbol F with its corresponding function symbol $f \in \mathcal{F}$. Then $\succ$ can be extended to a simplification ordering $\succ'$ on $\mathcal{T}(\mathcal{F} \cup \mathcal{T}up_{\mathcal{F}}, \mathcal{V})$ by defining

$t \succ' u$ iff $\Omega(t) \succ \Omega(u)$ holds. We claim that the simplification ordering $\succ'$ satisfies the constraints (a) - (c) of Definition 10 without applying an AFS.

Obviously, $l \succ' r$ holds for all rules $l \to r$ of $\mathcal{R}$. Thus $\succ'$ satisfies the constraint (a). Moreover, for every dependency pair $\langle s, t\rangle$ we have $s \succ' t$. The reason is that each dependency pair $\langle F(s_1, \ldots, s_n), G(t_1, \ldots, t_m)\rangle$ originates from a rule $f(s_1, \ldots, s_n) \to C[g(t_1, \ldots, t_m)]$ in $\mathcal{R}$. Thus, $f(\ldots) \succ C[g(\ldots)]$ implies $f(\ldots) \succ g(\ldots)$ which in turn implies $F(\ldots) \succ' G(\ldots)$. Hence, $\succ'$ also satisfies the constraints (b) and (c) of Definition 10. □

The following examples show that none of the converse implications of Lemma 11 holds.

Example 12 The system $\{f(f(x)) \to f(c(f(x)))\}$ is DP-simply terminating as the only dependency pair on a cycle is $\langle F(f(x)), F(x)\rangle$. Hence, the resulting constraints are satisfied by the rpo if one uses the AFS $c(x) \to x$. However, this TRS is not simply terminating. The TRS

$$\begin{array}{rclcrclcrcl} f(f(x)) & \to & f(c(f(x))) & \quad & g(c(x)) & \to & x & \quad & g(c(0)) & \to & g(d(1)) \\ f(f(x)) & \to & f(d(f(x))) & & g(d(x)) & \to & x & & g(c(1)) & \to & g(d(0)) \end{array}$$

is DP-quasi simply terminating as can be proved in a similar way using the AFS with the rules $c(x) \to x$ and $d(x) \to x$ and the rpo where 0 and 1 are equal in the precedence. However, it is not DP-simply terminating, because due to the first four rules, the AFS must reduce $c(x)$ and $d(x)$ to their arguments. But then $g(0) \geq g(1)$ and $g(1) \geq g(0)$ lead to a contradiction.

Finally, the system $\{f(0, 1, x) \to f(x, x, x)\}$ is terminating but not DP-quasi simply terminating. □

5 COMBINING DISJOINT SYSTEMS

In this section we show that DP-quasi simple termination is modular for disjoint TRSs. For the proof, we need the following lemma.

Lemma 13 (Transforming Reduction Sequences) *Let $\mathcal{R}_1$, $\mathcal{R}_2$ be two TRSs over disjoint signatures $\mathcal{F}_1$ and $\mathcal{F}_2$, respectively. Furthermore, let $\mathcal{R} = \mathcal{R}_1 \cup \mathcal{R}_2$ be their union. If u, v are terms over the signature $\mathcal{F}_1$ such that $u \to_{\mathcal{R}_1} v$ and $v\sigma \to^*_{\mathcal{R}} u\sigma$ hold for a ground substitution $\sigma : \mathcal{V}ar(u) \to \mathcal{T}(\mathcal{F}_1 \cup \mathcal{F}_2)$, then there is also a ground substitution $\tau : \mathcal{V}ar(u) \to \mathcal{T}(\mathcal{F}_1)$ such that $u\tau \to_{\mathcal{R}_1} v\tau \to^*_{\mathcal{R}_1 \cup \mathcal{E}mb(\mathcal{F}_1)} u\tau$.*

Proof. Clearly, all terms in the cyclic derivation

$$D: \quad u\sigma \to_{\mathcal{R}_1} v\sigma \to^*_{\mathcal{R}} u\sigma$$

have the same rank. Since the root symbol of u is in $\mathcal{F}_1$, the root symbol of every term in the reduction sequence D is also in $\mathcal{F}_1$ (reduction steps which are destructive at level 1 would decrease the rank).

Suppose first that every function symbol in $\mathcal{F}_1$ has arity ≤ 1. In this case, every reduction step in D which is destructive at level 2 strictly decreases the rank. Consequently, there is no reduction step of this kind in D. Hence

$$top_1(u\sigma) \to_{\mathcal{R}_1} top_1(v\sigma) \to^*_{\mathcal{R}_1} top_1(u\sigma)$$

is an $\mathcal{R}_1$-reduction sequence of ground terms over $\mathcal{F}_1$. Let $\mathcal{V}ar(u) = \{x_1, \ldots, x_n\}$ and recall $\mathcal{V}ar(v) \subseteq \mathcal{V}ar(u)$. In this case, we define the substitution τ by $\tau = \{x_i \mapsto top_1(x_i\sigma) \mid 1 \leq i \leq n\}$ and indeed

$$u\tau = top_1(u\sigma) \to_{\mathcal{R}_1} top_1(v\sigma) = v\tau \to^*_{\mathcal{R}_1} top_1(u\sigma) = u\tau$$

is the reduction sequence we are looking for.

Suppose otherwise that there is a function symbol f in $\mathcal{F}_1$ with arity $m > 1$. Let Cons be a binary function symbol which neither occurs in $\mathcal{F}_1$ nor in $\mathcal{F}_2$ and let $\mathcal{C}_\mathcal{E} = \{\mathrm{Cons}(x_1, x_2) \to x_1, \mathrm{Cons}(x_1, x_2) \to x_2\}$. By [Gramlich, 1994, Lemma 3.8] or [Ohlebusch, 1994b, Theorem 3.13], the reduction sequence D can be transformed by a transformation function[4] Φ into a reduction sequence

$$\Phi(u\sigma) \to_{\mathcal{R}_1} \Phi(v\sigma) \to^*_{\mathcal{R}_1 \cup \mathcal{C}_\mathcal{E}} \Phi(u\sigma)$$

of terms over $\mathcal{F}_1 \cup \{\mathrm{Cons}\}$. The transformation function Φ satisfies $\Phi(t) = C[\Phi(t_1), \ldots, \Phi(t_n)]$ for every term t with $root(t) \in \mathcal{F}_1$ and $t = C[\![t_1, \ldots, t_n]\!]$, cf. [Ohlebusch, 1994b]. In this case, we first define $\sigma' = \{x_i \mapsto \Phi(x_i\sigma) \mid 1 \leq i \leq n\}$ and obtain

$$u\sigma' = \Phi(u\sigma) \to_{\mathcal{R}_1} \Phi(v\sigma) = v\sigma' \to^*_{\mathcal{R}_1 \cup \mathcal{C}_\mathcal{E}} \Phi(u\sigma) = u\sigma' \ .$$

Let $u\sigma' = u_0, u_1, \ldots, u_k = u\sigma'$ be the sequence of terms occurring in the above reduction sequence. Now in each term u_i replace every $\mathrm{Cons}(t_1, t_2)$ with $f(t_1, t_2, z, \ldots, z)$, where z is a variable or constant, and denote the resulting term by $\Psi(u_i)$. The definition $\tau = \{x_i \mapsto \Psi(x_i\sigma') \mid 1 \leq i \leq n\}$ yields the desired reduction sequence

$$u\tau = \Psi(u\sigma') = \Psi(u_0) \to_{\mathcal{R}_1} \Psi(u_1) = \Psi(v\sigma') = v\tau \to^*_{\mathcal{R}_1 \cup \mathcal{E}mb(\mathcal{F}_1)} \Psi(u_k) = u\tau$$

in which $\Psi(u_i) \to_{\mathcal{R}_1 \cup \mathcal{E}mb(\mathcal{F}_1)} \Psi(u_{i+1})$ by the rule $f(x_1, \ldots, x_m) \to x_j$, $j \in \{1, 2\}$, if $u_i \to_{\mathcal{R}_1 \cup \mathcal{C}_\mathcal{E}} u_{i+1}$ by the rule $\mathrm{Cons}(x_1, x_2) \to x_j$. □

Now we are in a position to prove our first modularity theorem.

Theorem 14 (Modularity of DP-quasi simple termination) *Let $\mathcal{R}_1$ and $\mathcal{R}_2$ be two TRSs over disjoint signatures $\mathcal{F}_1$ and $\mathcal{F}_2$, respectively. Then their union $\mathcal{R} = \mathcal{R}_1 \cup \mathcal{R}_2$ is DP-quasi simply terminating iff both $\mathcal{R}_1$ and $\mathcal{R}_2$ are DP-quasi simply terminating.*

[4]More precisely, Φ is the transformation $\Phi_1^{u\sigma}$ defined in [Ohlebusch, 1994b, Definition 3.10].

Proof. The only-if direction is trivial. For the if direction, let $\mathcal{P}$ be a cycle in the estimated dependency graph of $\mathcal{R}$. Since $\mathcal{R}_1$ and $\mathcal{R}_2$ are disjoint, $\mathcal{P}$ is a cycle in the estimated dependency graph of $\mathcal{R}_1$ or of $\mathcal{R}_2$. Without loss of generality, let $\mathcal{P}$ be a cycle in the estimated dependency graph of $\mathcal{R}_1$.

As $\mathcal{R}_1$ is DP-quasi simply terminating, there is an AFS $\mathcal{A}$ such that inequalities (a) - (c) of Definition 9 are satisfied for $\mathcal{R}_1$, $\mathcal{P}$, and some qso $\succsim$. Let $\mathcal{F}_1'$ be the set of all function symbols occurring in the inequalities (a) - (c). Without loss of generality we may assume that $\mathcal{A}$ contains no rules with root symbols from $\mathcal{F}_2$. Now let[5]

$$\begin{aligned} \mathcal{S}_1 &= \{l\downarrow_{\mathcal{A}} \to r\downarrow_{\mathcal{A}} \mid l \to r \in \mathcal{R}_1 \text{ and } l\downarrow_{\mathcal{A}} \notin \mathcal{V}\} \cup \mathcal{E}mb(\mathcal{F}_1') \cup \\ &\quad\; \{s\downarrow_{\mathcal{A}} \to t\downarrow_{\mathcal{A}} \mid \langle s,t\rangle \in \mathcal{P} \text{ and } s\downarrow_{\mathcal{A}} \notin \mathcal{V}\} \\ \mathcal{S}_2 &= \mathcal{R}_2 \cup \mathcal{E}mb(\mathcal{F}_2). \end{aligned}$$

$\mathcal{S}_1$ is a TRS over the signature $\mathcal{F}_1'$. Hence $\mathcal{R}' = \mathcal{S}_1 \cup \mathcal{S}_2$ is a TRS over $\mathcal{F}_1' \cup \mathcal{F}_2$. Note that $\to_{\mathcal{R}'}^*$ is a qso.[6] Thus, (as the cycle $\mathcal{P}$ was chosen arbitrarily) to prove DP-quasi simple termination of $\mathcal{R}$, we only have to show

(a) $l\downarrow_{\mathcal{A}} \to_{\mathcal{R}'}^* r\downarrow_{\mathcal{A}}$ for all rules $l \to r$ in $\mathcal{R}$
(b) $s\downarrow_{\mathcal{A}} \to_{\mathcal{R}'}^* t\downarrow_{\mathcal{A}}$ for all dependency pairs $\langle s,t\rangle$ from $\mathcal{P}$
(c) there exists a dependency pair $\langle s,t\rangle$ from $\mathcal{P}$ such that
$t\downarrow_{\mathcal{A}}\sigma \not\to_{\mathcal{R}'}^* s\downarrow_{\mathcal{A}}\sigma$ holds for all ground substitutions σ.

Conditions (a) and (b) are obviously satisfied because for all $l \to r \in \mathcal{R}_2$ we have $l\downarrow_{\mathcal{A}} = l$ and $r\downarrow_{\mathcal{A}} = r$. Hence, we only have to show conjecture (c). Since $\succsim$ is the qso used for the DP-quasi simple termination proof of $\mathcal{R}_1$, we have $\to_{\mathcal{S}_1}^* \subseteq \succsim$. Let $\langle s,t\rangle$ be a dependency pair from $\mathcal{P}$ such that $s\downarrow_{\mathcal{A}} \succ t\downarrow_{\mathcal{A}}$. Suppose that there exists a ground substitution $\sigma : \mathcal{V}ar(s\downarrow_{\mathcal{A}}) \to \mathcal{T}(\mathcal{F}_1' \cup \mathcal{F}_2)$ such that $t\downarrow_{\mathcal{A}}\sigma \to_{\mathcal{R}'}^* s\downarrow_{\mathcal{A}}\sigma$. By Lemma 13, this implies the existence of a ground substitution $\tau : \mathcal{V}ar(s\downarrow_{\mathcal{A}}) \to \mathcal{T}(\mathcal{F}_1')$ such that $t\downarrow_{\mathcal{A}}\tau \to_{\mathcal{S}_1}^* s\downarrow_{\mathcal{A}}\tau$. This, however, would imply $t\downarrow_{\mathcal{A}}\tau \succsim s\downarrow_{\mathcal{A}}\tau$ and contradicts the fact that $s\downarrow_{\mathcal{A}} \succ t\downarrow_{\mathcal{A}}$. Thus, $t\downarrow_{\mathcal{A}}\sigma \not\to_{\mathcal{R}'}^* s\downarrow_{\mathcal{A}}\sigma$ holds for all ground substitutions σ. This proves conjecture (c). □

Thus, if $\mathcal{R}_1$ is the TRS consisting of the rules (1) and (2) and $\mathcal{R}_2$ contains the rules (3) and (4), then this theorem allows us to conclude termination of their combination because both systems are DP-quasi simply terminating. This example cannot be handled by any of the previous modularity results.

[5] $l\downarrow_{\mathcal{A}} = x \in \mathcal{V}$ implies $l\downarrow_{\mathcal{A}} = x = r\downarrow_{\mathcal{A}}$ and $x \succsim x$ is satisfied by every qso $\succsim$.

[6] If $\mathcal{R}$ is a TRS over the signature $\mathcal{F}$ then $\to_{\mathcal{R}\cup\mathcal{E}mb(\mathcal{F})}^*$ is the smallest qso containing $\to_{\mathcal{R}}$ (that is, if $\succsim$ is a qso with $\to_{\mathcal{R}} \subseteq \succsim$, then $\to_{\mathcal{R}\cup\mathcal{E}mb(\mathcal{F})}^* \subseteq \succsim$). Note however, that the *strict* part of a qso $\to_{\mathcal{R}\cup\mathcal{E}mb(\mathcal{F})}^*$ is not necessarily closed under substitutions. Hence, without the extension of the dependency pair approach in Theorem 6 and Definition 9 to *stable-strict* relations, such a qso cannot be used for termination proofs with dependency pairs. As this extension leads to a more powerful criterion, we did not investigate whether Theorem 14 would also hold for a formulation of Definition 9 with strict instead of stable-strict relations.

Note also that in this example, modularity of termination is far from being trivial because if $\mathcal{R}_1$'s rule $f(0,1,x) \to f(s(x),x,x)$ would be just slightly changed to $f(0,1,x) \to f(x,x,x)$, then $\mathcal{R}_1$ would still be terminating, but the union with $\mathcal{R}_2$ would not terminate any more, cf. [Toyama, 1987a].

DP-quasi simply terminating systems occur frequently in practice. For example, the two TRSs

$$
\begin{array}{rcl}
\mathcal{R}_1: \quad x - 0 & \to & x \\
s(x) - s(y) & \to & x - y \\
q(0, s(y)) & \to & 0 \\
q(s(x), s(y)) & \to & s(q(x - y, s(y)))
\end{array}
\qquad
\begin{array}{rcl}
\mathcal{R}_2: \quad app(nil, k) & \to & k \\
app(l, nil) & \to & l \\
app(x : l, k) & \to & x : app(l, k) \\
sum(x : nil) & \to & x : nil \\
sum(x : (y : l)) & \to & sum((x + y) : l) \\
sum(app(l, x : (y : k))) & \to & sum(app(l, sum(x : (y : k))))
\end{array}
$$

are not simply terminating, but they are both DP-quasi simply terminating, cf. Arts and Giesl [1997c; 1997a]. Hence, Theorem 14 now also allows one to conclude DP-quasi simple termination of their union.

6 COMBINING CONSTRUCTOR-SHARING SYSTEMS

It may be a bit surprising that Theorem 14 cannot be directly extended to constructor-sharing TRSs (even if we disallow the use of AFSs). In other words, there are constructor-sharing TRSs $\mathcal{R}_1$ and $\mathcal{R}_2$ which are both DP-quasi simply terminating, but their union $\mathcal{R} = \mathcal{R}_1 \cup \mathcal{R}_2$ is not DP-quasi simply terminating.

Example 15 Consider the following TRSs:

$$
\begin{array}{rcl}
\mathcal{R}_1: \quad f(c(x)) & \to & f(x) \\
f(b(x)) & \to & x
\end{array}
\qquad
\begin{array}{rcl}
\mathcal{R}_2: \quad g(d(x)) & \to & g(x) \\
g(c(x))) & \to & c(g(b(c(x))))
\end{array}
$$

The TRS $\mathcal{R}_1$ is simply terminating and $\mathcal{R}_2$ is DP-quasi simply terminating. The latter can be shown without any AFS by using a polynomial ordering which maps c, b, g, and G to the identity and which maps $d(x)$ to $x+1$. However, their union is not DP-quasi simply terminating. As $\langle F(c(x)), F(x) \rangle$ represents a cycle in the estimated dependency graph one would have to find a qso satisfying

$$
\begin{array}{lrcl}
(10) & F(c(x)) & \succ & F(x) \\
(11) & f(c(x)) & \succsim & f(x) \\
(12) & f(b(x)) & \succsim & x \\
(13) & g(d(x)) & \succsim & g(x) \\
(14) & g(c(x)) & \succsim & c(g(b(c(x)))).
\end{array}
$$

With empty AFS, no qso satisfies (10) - (14), since otherwise we would have

$$\begin{array}{lll} F(c(g(c(x)))) & \succ & F(g(c(x))) & \text{due to (10)} \\ & \succsim & F(c(g(b(c(x))))) & \text{due to (14)} \\ & \succsim & F(c(g(c(x)))) & \text{due to the subterm property.} \end{array}$$

By (10), the AFS cannot contain any c-rules. If the argument of b is eliminated, then (12) would be transformed into $f(b') \succsim x$. But as there exists the strict inequality (10) with a variable in its right-hand side, this results in the contradiction $F(c(f(b'))) \succ x$. Similarly, the argument of g cannot be eliminated either, since $g' \succsim c(g')$ would be a contradiction to (10).

Thus, the only possible rules in the AFS are f- and d-rules and rules that map b or g to its argument. But then we would again obtain $F(c(g(c(x)))) \succ F(c(g(c(x))))$ or $F(c(c(x))) \succ F(c(c(x)))$ as above. Hence, the TRS is not DP-quasi simply terminating. □

Thus, in the remainder of the section we will restrict ourselves to DP-simple termination instead of DP-quasi simple termination. Without applying AFSs, DP-simple termination of the TRS $\mathcal{R}_2$ cannot be proved. However, if one uses, for example, the AFS $b(x) \to b'$, then its resulting constraints are again satisfied by a simplification ordering. Thus, to achieve modularity results for constructor-sharing TRSs, we have to restrict ourselves to systems where the AFSs contain no rules for shared symbols like b.

But we need another requirement to ensure modularity. For example, let us eliminate the first rule $g(d(x)) \to g(x)$ from $\mathcal{R}_2$. Now there is no non-empty cycle in the estimated dependency graph of $\mathcal{R}_2$ any more and hence we obtain no constraints at all for $\mathcal{R}_2$. Thus, DP-simple termination of $\mathcal{R}_2$ can now even be proved without using AFSs, but the combined system $\mathcal{R}_1 \cup \mathcal{R}_2$ is still not DP-simply terminating. Here, the problem is due to the fact that TRSs without non-empty cycles are DP-simply terminating, even if there is no simplification ordering $\succ$ such that $l \succeq r$ holds for their rules. To exclude such TRSs we will demand that the constraint (a) of Definition 10 should also be satisfied for the *empty* cycle $\mathcal{P}$.[7]

Nonetheless, the following example shows that this restriction is not yet sufficient for obtaining a modularity result for DP-simple termination of constructor-sharing systems.

Example 16 Let $\mathcal{R}_1$ consist of the rules

$$\begin{array}{lllllll} g(s(x)) & \to & g(x) & \qquad & g(0) & \to & g(1) \\ g(s(x)) & \to & x & & f(0) & \to & g(f(s(0))) \end{array}$$

and let $\mathcal{R}_2$ consist of the rule $h(1) \to h(0)$. To prove DP-simple termination of $\mathcal{R}_1$ we have to use an AFS containing the rule $g(x) \to x$. This, however, would

[7]In fact, without this additional requirement DP-simple termination is not even modular for *disjoint* combinations. For reasons of space, a counterexample is omitted.

imply $0 \succ 1$ which is a contradiction to $h(1) \succeq h(0)$. Thus, the combination of both systems is not DP-simply terminating. □

So we also have to ensure that an application of the AFS to the resulting inequalities does not transform left-hand sides which had a non-shared root symbol like g into terms with a shared root symbol (like the former constructor 0).[8] The restrictions needed are captured by the notion of *C-restricted* DP-simple termination.

Definition 17 (C-restricted DP-simple termination) Let $\mathcal{R}$ be a TRS over $\mathcal{F}$ and let $C \subseteq \mathcal{F}$. $\mathcal{R}$ is *C-restricted* DP-simply terminating iff

(1) For every cycle $\mathcal{P}$ (including $\mathcal{P} = \emptyset$), there is an AFS $\mathcal{A}$ such that

$$\begin{aligned} \mathcal{S} \;=\; & \{l\downarrow_{\mathcal{A}} \to r\downarrow_{\mathcal{A}} \mid l \to r \in \mathcal{R} \text{ and } l\downarrow_{\mathcal{A}} \neq r\downarrow_{\mathcal{A}}\} \cup \\ & \{s\downarrow_{\mathcal{A}} \to t\downarrow_{\mathcal{A}} \mid \langle s,t\rangle \text{ is a dependency pair in } \mathcal{P} \text{ and } s\downarrow_{\mathcal{A}} \neq t\downarrow_{\mathcal{A}}\} \end{aligned}$$

is simply terminating and such that $s\downarrow_{\mathcal{A}} \neq t\downarrow_{\mathcal{A}}$ for some $\langle s,t\rangle \in \mathcal{P}$ if $\mathcal{P} \neq \emptyset$. (Some simplification ordering $\succ$ satisfies constraints (a), (b), (c) of Definition 10 for every cycle, where (a) is also required for $\mathcal{P} = \emptyset$.)

(2) For every rule $u\downarrow_{\mathcal{A}} \to v\downarrow_{\mathcal{A}}$ in $\mathcal{S}$: if $root(u) \notin C$, then $root(u\downarrow_{\mathcal{A}}) \notin C$.

(3) None of the AFSs contains a rule for a function symbol $f \in C$.

The following theorem shows that under this C-restriction, DP-simple termination is modular for constructor-sharing TRSs.

Theorem 18 (Modularity of C-restricted DP-simple termination) *Let $\mathcal{R}_1$ and $\mathcal{R}_2$ be constructor-sharing TRSs over the signatures $\mathcal{F}_1$ and $\mathcal{F}_2$, respectively. If $\mathcal{F}_1 \cap \mathcal{F}_2 \subseteq C$, then their combined system $\mathcal{R} = \mathcal{R}_1 \cup \mathcal{R}_2$ over the signature $\mathcal{F} = \mathcal{F}_1 \cup \mathcal{F}_2$ is C-restricted DP-simply terminating if and only if both $\mathcal{R}_1$ and $\mathcal{R}_2$ are C-restricted DP-simply terminating.*

Proof. The only-if direction is trivial. For the if direction, let $\mathcal{P}$ be a cycle in the estimated dependency graph of $\mathcal{R}$ (where $\mathcal{P}$ may also be empty). It is not difficult to prove that $\mathcal{P}$ is also a cycle in the estimated dependency graph of $\mathcal{R}_1$ or in the estimated dependency graph of $\mathcal{R}_2$ because $\mathcal{R}_1$ and $\mathcal{R}_2$ are constructor-sharing. Without loss of generality let $\mathcal{P}$ be a cycle in the estimated dependency graph of $\mathcal{R}_1$. We have to show that there is an AFS $\mathcal{A}$ such that the corresponding TRS $\mathcal{S}$ is simply terminating and moreover conditions (2) and (3) of Definition 17 are satisfied.

Since $\mathcal{R}_1$ is C-restricted DP-simply terminating and $\mathcal{P}$ is a cycle in the estimated dependency graph of $\mathcal{R}_1$, there is an AFS $\mathcal{A}_1$ such that the TRS

$$\begin{aligned} \mathcal{S}_1 \;=\; & \{l\downarrow_{\mathcal{A}_1} \to r\downarrow_{\mathcal{A}_1} \mid l \to r \in \mathcal{R}_1 \text{ and } l\downarrow_{\mathcal{A}_1} \neq r\downarrow_{\mathcal{A}_1}\} \cup \\ & \{s\downarrow_{\mathcal{A}_1} \to t\downarrow_{\mathcal{A}_1} \mid \langle s,t\rangle \text{ is a dependency pair in } \mathcal{P} \text{ and } s\downarrow_{\mathcal{A}_1} \neq t\downarrow_{\mathcal{A}_1}\} \end{aligned}$$

[8] If all AFSs used are non-collapsing, then this requirement is always fulfilled.

is simply terminating and such that $s\downarrow_{\mathcal{A}_1}\neq t\downarrow_{\mathcal{A}_1}$ holds for at least one dependency pair $\langle s,t\rangle$ from $\mathcal{P}$ if $\mathcal{P}\neq\emptyset$. On the other hand, there is an AFS $\mathcal{A}_2$ such that the TRS

$$\mathcal{S}_2=\{l\!\downarrow_{\mathcal{A}_2}\to r\!\downarrow_{\mathcal{A}_2}\mid l\to r\in\mathcal{R}_2 \text{ and } l\!\downarrow_{\mathcal{A}_2}\neq r\!\downarrow_{\mathcal{A}_2}\}$$

is simply terminating because $\mathcal{R}_2$ is C-restricted DP-simply terminating. (Here, $\mathcal{A}_2$ is the AFS corresponding to the empty cycle.)

Let $\mathcal{F}_i'$ be the set of all function symbols in $\mathcal{S}_i$ for $i\in\{1,2\}$ and let $\mathcal{A}=\mathcal{A}_1\cup\mathcal{A}_2$. Since $\mathcal{R}_1$ and $\mathcal{R}_2$ are C-restricted DP-simply terminating, $\mathcal{A}$ does not contain rules for elements of C. Consequently, it follows for every term $u\in\mathcal{T}(\mathcal{F}_i,\mathcal{V})$ that $u\downarrow_{\mathcal{A}}=u\downarrow_{\mathcal{A}_i}$. So in particular, $s\downarrow_{\mathcal{A}_1}\neq t\downarrow_{\mathcal{A}_1}$ implies $s\downarrow_{\mathcal{A}}\neq t\downarrow_{\mathcal{A}}$. Moreover, by requirements (2) and (3), the TRSs

$$\begin{aligned}\mathcal{S}_1 &= \{l\!\downarrow_{\mathcal{A}}\to r\!\downarrow_{\mathcal{A}}\mid l\to r\in\mathcal{R}_1 \text{ and } l\!\downarrow_{\mathcal{A}}\neq r\!\downarrow_{\mathcal{A}}\}\cup\\ &\quad\ \{s\!\downarrow_{\mathcal{A}}\to t\!\downarrow_{\mathcal{A}}\mid \langle s,t\rangle \text{ is a dependency pair in } \mathcal{P} \text{ and } s\!\downarrow_{\mathcal{A}}\neq t\!\downarrow_{\mathcal{A}}\},\\ \mathcal{S}_2 &= \{l\!\downarrow_{\mathcal{A}}\to r\!\downarrow_{\mathcal{A}}\mid l\to r\in\mathcal{R}_2 \text{ and } l\!\downarrow_{\mathcal{A}}\neq r\!\downarrow_{\mathcal{A}}\}\end{aligned}$$

are also constructor-sharing. This is because a former constructor can only be a defined symbol in $\mathcal{S}_1$ or $\mathcal{S}_2$ if it is not in C, i.e. if it is not shared. Thus by [Kurihara and Ohuchi, 1992, Theorem 3.10], their combined system $\mathcal{S}=\mathcal{S}_1\cup\mathcal{S}_2$ is also simply terminating.

Since the AFS $\mathcal{A}$ and the TRS $\mathcal{S}$ obviously satisfy the conditions (2) and (3) of Definition 17, $\mathcal{R}$ is C-restricted DP-simply terminating. □

For example, let $\mathcal{R}_2$ be the (simply terminating) TRS

$$\begin{aligned}x+0 &\to x\\ x+s(y) &\to s(x+y)\\ x+(y+z) &\to (x+y)+z.\end{aligned}$$

As the TRS $\mathcal{R}_1$ for subtraction and division from the end of Section 5 is DP-simply terminating, now Theorem 18 allows us to conclude DP-simple termination of the combined system $\mathcal{R}_1\cup\mathcal{R}_2$, since these two TRSs are constructor-sharing (they both contain the same constructors 0 and s).

There are even TRSs $\mathcal{R}_1\cup\mathcal{R}_2$ where DP-simple termination of both $\mathcal{R}_1$ and $\mathcal{R}_2$ can be proved with a standard technique, such as the lpo, whereas such standard orderings fail if one wants to prove DP-simple termination of their union directly. Hence, for such examples our result enables automatic termination proofs which were not possible before.

Example 19 Let $\mathcal{R}_1$ be the TRS

$$\begin{aligned}f(c(s(x),y)) &\to f(c(x,s(y)))\\ f(f(x)) &\to f(d(f(x)))\\ f(x) &\to x\end{aligned}$$

and let $\mathcal{R}_2$ consist of the rule $g(c(x,s(y))) \to g(c(s(x),y))$.

$\mathcal{R}_1$ is DP-simply terminating (using the AFS $d(x) \to x$ and the lpo comparing left-to-right), but it is not simply terminating. $\mathcal{R}_2$ is even simply terminating as can be shown with the lpo comparing right-to-left. Thus, DP-simple termination of both systems can be verified by the lpo.

By Theorem 18 their union is also DP-simply terminating. However, the constraints for the cycle $\{\langle G(c(x,s(y))), G(c(s(x),y))\rangle\}$ are not satisfied by any lpo (nor by any rpo nor by any polynomial ordering). Thus, there are indeed TRSs where termination of the subsystems can be shown with dependency pairs and the lpo, but (without our modularity result) termination of their union cannot be proved with dependency pairs and the lpo. □

We stress that Theorem 18 can be extended to composable systems. The proof is almost identical because simple termination is also modular for composable systems [Ohlebusch, 1995, Theorem 5.16]. Only the proof that $\mathcal{S}_1$ and $\mathcal{S}_2$ are indeed composable needs additional efforts. Therefore, C-restricted DP-simple termination is even modular for combinations of TRSs which have common defined symbols provided that both systems contain the same rules for shared defined symbols.

7 CONCLUSIONS

We have shown that the existing modularity results for simple termination of disjoint unions can be extended to DP-quasi simple termination. Under certain restrictions a similar modularity result also holds for constructor-sharing (and even composable) combinations. Future work will include an examination of whether such an extension is also possible for hierarchical combinations. In summary, the progress in automated termination proving which was made possible by the development of dependency pairs now also has a counterpart in the area of modularity. Dependency pairs enable automated termination proofs of non-simply terminating TRSs and now our results allow us to perform them in a modular way.

The present work completes the results in [Arts and Giesl, 1998], where the theory of dependency pairs was refined in a modular way (cf. Theorem 6) and where several new modularity criteria for *innermost* termination were presented. However, these criteria are only applicable for termination proofs of locally confluent overlay systems (and in particular, non-overlapping systems) [Gramlich, 1995]. But in practice there are many cases in which innermost termination is not sufficient for termination. Up to now, due to missing modularity results, the advantages of dependency pairs could not be fully exploited for these systems. So compared to previous work on modularity, the modularity criteria developed in the present paper represent a significant extension.

Acknowledgements. We thank Aart Middeldorp for many helpful remarks and hints. This work was partially supported by the DFG under grant Wa 652/7-2 as part of the focus program "Deduktion".

REFERENCES

[Arts and Giesl, 1997a] T. Arts and J. Giesl. Automatically Proving Termination Where Simplification Orderings Fail. In *Proc. TAPSOFT '97*, pages 261–272. LNCS 1214, 1997.

[Arts and Giesl, 1997b] T. Arts and J. Giesl. Proving Innermost Normalisation Automatically. In *Proc. RTA '97*, pages 157–171. LNCS 1232, 1997.

[Arts and Giesl, 1997c] T. Arts and J. Giesl. Termination of Term Rewriting Using Dependency Pairs. Technical Report IBN 97/46, TU Darmstadt, 1997. `http://www.inferenzsysteme.informatik.tu-darmstadt.de/~reports/notes/ibn-97-46.ps`. Revised version to appear in *Theoretical Computer Science.*

[Arts and Giesl, 1998] T. Arts and J. Giesl. Modularity of Termination Using Dependency Pairs. In *Proc. RTA '98*, pages 226–240. LNCS 1379, 1998.

[Courcelle, 1990] B. Courcelle. Recursive Applicative Program Schemes. In J. van Leeuwen, editor, *Handbook of Theoretical Computer Science*, volume B, pages 459–492. North-Holland, 1990.

[Dershowitz and Jouannaud, 1990] N. Dershowitz and J.-P. Jouannaud. Rewrite Systems. In J. van Leeuwen, editor, *Handbook of Theoretical Computer Science*, volume B, pages 243–320. North-Holland, 1990.

[Dershowitz, 1987] N. Dershowitz. Termination of Rewriting. *Journal of Symbolic Computation*, 3:69–115, 1987.

[Gramlich, 1994] B. Gramlich. Generalized Sufficient Conditions for Modular Termination of Rewriting. *Applicable Algebra in Engineering, Commutation and Computing*, 5:131–158, 1994.

[Gramlich, 1995] B. Gramlich. Abstract Relations Between Restricted Termination and Confluence Properties of Rewrite System. *Fund. Informaticae*, 24:3–23, 1995.

[Gramlich, 1996] B. Gramlich. *Termination and Confluence Properties of Structured Rewrite Systems.* PhD thesis, Universität Kaiserslautern, Germany, 1996.

[Klop, 1992] J. W. Klop. Term Rewriting Systems. In S. Abramsky, D. M. Gabbay, and T. S. E. Maibaum, editors, *Handbook of Logic in Computer Science*, volume 2, pages 1–116. Oxford University Press, New York, 1992.

[Krishna Rao, 1994] M. R. K. Krishna Rao. Simple Termination of Hierarchical Combinations of Term Rewriting Systems. In *Proc. STACS '94*, pages 203–223. LNCS 789, 1994.

[Kurihara and Ohuchi, 1992] M. Kurihara and A. Ohuchi. Modularity of Simple Termination of Term Rewriting Systems with Shared Constructors. *Theoretical Computer Science*, 103:273–282, 1992.

[Lankford, 1979] D. S. Lankford. On Proving Term Rewriting Systems are Noetherian. Technical Report MTP-3, Louisiana Tech. Univ., Ruston, LA, 1979.

[Middeldorp and Zantema, 1997] A. Middeldorp and H. Zantema. Simple Termination of Rewrite Systems. *Theoretical Computer Science*, 175:127–158, 1997.

[Middeldorp, 1989] A. Middeldorp. A Sufficient Condition for the Termination of the Direct Sum of Term Rewriting Systems. In *Proc. LICS '89,* p. 396-401, 1989.

[Middeldorp, 1990] A. Middeldorp. *Modular Properties of Term Rewriting Systems.* PhD thesis, Vrije Universiteit te Amsterdam, 1990.

[Ohlebusch, 1994a] E. Ohlebusch. *Modular Properties of Composable Term Rewriting Systems.* PhD thesis, Universität Bielefeld, 1994.

[Ohlebusch, 1994b] E. Ohlebusch. On the Modularity of Termination of Term Rewriting Systems. *Theoretical Computer Science*, 136:333–360, 1994.

[Ohlebusch, 1995] E. Ohlebusch. Modular Properties of Composable Term Rewriting Systems. *Journal of Symbolic Computation*, 20:1–41, 1995.

[Rusinowitch, 1987] M. Rusinowitch. On Termination of the Direct Sum of Term Rewriting Systems. *Information Processing Letters*, 26:65–70, 1987.

[Steinbach, 1995] J. Steinbach. Simplification Orderings: History of Results. *Fundamenta Informaticae*, 24:47–87, 1995.

[Toyama, 1987a] Y. Toyama. Counterexamples to Termination for the Direct Sum of Term Rewriting Systems. *Information Processing Letters*, 25:141–143, 1987.

[Toyama, 1987b] Y. Toyama. On the Church-Rosser Property for the Direct Sum of Term Rewriting Systems. *Journal of the ACM*, 34:128–143, 1987.

[Toyama *et al.*, 1995] Y. Toyama, J. W. Klop, and H. P. Barendregt. Termination for the Direct Sum of Left-Linear Term Rewriting Systems. *Journal of the ACM*, 42:1275–1304, 1995.

Toward Sharing Libraries of Mathematics between Theorem Provers

Douglas J. Howe

1 INTRODUCTION

We have designed and implemented a link [Howe, 1996a] between the interactive theorem-provers Nuprl [Constable, et al., 1986] and HOL [Gordon and Melham, 1993] that allows us to use HOL mathematics in Nuprl proofs. We have used this link in two substantial applications of Nuprl [Felty and Howe, 1997; Felty *et al.*, 1998]. Our experience has shown that it can be practical to share basic formal mathematics between interactive theorem provers.

There is another lesson to be drawn from our experience with the Nuprl-HOL link. Even though the soundness of the connection is justified externally by a semantic argument, importing mathematics from HOL to Nuprl requires substantial amounts of theorem-proving. In fact, most of our effort in implementing the connection went into developing automation for this. This need for theorem-proving comes from two sources. First, when a new library of mathematics is imported, there are proof obligations arising from soundness concerns. Typically, this is because HOL constants are *interpreted* in Nuprl. For example, arithmetic operators in HOL become interpreted as the corresponding Nuprl operators. For soundness, we require Nuprl proofs which demonstrate that the interpretations have the necessary properties. Second, the imported mathematics is usually not of a form that can be directly applied in Nuprl proofs. Theorem-proving is required to transform the imported theorems into a more usable representation.

We do not advocate implementing a link like ours for every pair of theorem provers. Clearly it is desirable to avoid such a duplication of effort, and instead to mediate sharing using a single system into which the other logics can be embedded. One possibility for such a system is a logical framework, for example [Harper *et al.*, 1993; Paulson, 1994; Matthews *et al.*, 1993; Martí-Oliet and Meseguer, 1996]. Logical frameworks typically embed other formal systems by representing their deductive apparatus.

Another possibility is to use a theorem prover that supports so-called *shallow* embeddings of other logics. In a shallow embedding, one represents the *semantics* of the formal system to be embedded. This is the most common method used in the HOL community. For example, one represents temporal logic by defining a type of trace, and then defining the temporal operators as particular functions over these traces. The rules of temporal logic become derived rules about these defined operators. The soundness of this embedding can be justified by appealing the semantics of HOL and temporal logic. One advantage of shallow embedding is that it allows the deductive apparatus of the embedded system to be ignored. This is valuable in the case of a logic like PVS [Owre *et al.*, 1996], whose set-theoretic semantics is much simpler than its proof system. Another advantage is that often machinery and mathematics libraries of the ambient system can be applied to the embedded logic. Examples of the use of Nuprl for shallow embedding include [Constable and Howe, 1990] and [Felty *et al.*, 1998].

We believe that our classical variant of Nuprl [Howe, 1996b; Howe, 1997] is a plausible medium for sharing mathematics via shallow embedding. Nuprl has a rich type system, but typing is semantic, in the sense that no structural typing discipline is imposed on the syntax. This gives the type system a great deal of flexibility, and lets it directly model a wide range of other type systems, including, for example, the type theories HOL, PVS, and Coq [Werner, 1997], as well as various record calculi. It also combines an executable sublanguage with a straightforward set-theoretic semantics.

In Section 2 we present ν_0, a simple "core" version of the classical variant of Nuprl, in order to illustrate the above features. We give a simple example of shallow embedding in ν_0, and sketch a semantic proof that the embedding is sound. The design of ν_0 and the related semantic simplifications are the only new technical content in this paper.

In Section 3, we review the HOL/Nuprl effort. In Section 4 we extrapolate a bit from our experience and discuss, among other things, connecting Nuprl and PVS.

2 THE CLASSICAL TYPE THEORY ν_0

2.1 The Semantics of ν_0

We first give the semantics of our core logic ν_0. We build the semantics in two stages. First, we construct a set-theoretic "universe" V that contains the objects that will serve as the denotations of the terms and types of ν_0. Essentially, V is built by iterating power set and generalized cartesian product from a particular base set. The types of ν_0 are modeled directly in terms of these sets: function types are modeled as generalized cartesian products, and power sets let us model subtype formation.

Our semantics differs from the usual set-theoretic semantics of classical higher-order type theory (as in, say, HOL[Gordon and Melham, 1993]) in two important ways. The first difference is due to the general form of subtyping in ν_0. In particular, ν_0 supports contravariant subtyping, so if A is a subtype of A', and B is a subtype of B', then $A' \to B$ is a subtype of $A \to B'$. Contravariant subtyping is useful in modeling subtyping of records. Partly to model this kind of subtyping in the semantic domain V, we introduce an ordering $\preceq$ on V. Because of the requirements of the second stage of the semantics, this ordering is not the expected one, *i.e.* the subtype relation on objects denoting types, but, instead, an ordering on all objects, and we will use the upward closure of γ under $\preceq$ to model the set of objects that will be considered to be members, via subtyping, of a type-denotation $\gamma \in V$.

The second important difference is that ν_0 contains a programming language. To deal with this, in the second stage of the semantics we layer an *operational* semantics on top of V. In particular, we inject the members of V into the syntax of ν_0, and then give an operational-style inductive definition of an evaluation relation over the augmented set of terms. This will simultaneously give us both the computational meaning of terms, as well as their set-theoretic semantics. Although we do not make much use of the computational aspects of ν_0 here, they are critical for modeling Nuprl because of its constructive character and its particular approach to reasoning about computation.

We give no proofs here. Full proofs of analogous results for Nuprl can be found in [Howe, 1997]. The paper [Howe, 1996b] contains a preliminary version of [Howe, 1997].

We now form V by iteratively applying certain set building operations, and then collecting all the sets together with their members. Choose distinct sets c_{true}, c_{false} and c_i for $i \in \omega$. Let B be the set of these sets. These are the "base" objects in the semantics. For X a set, define $F(X)$ as the union of

1. $\{B\}$,
2. $\{X\}$,
3. $Pow(Z)$ for all $Z \in X$ (where $Pow(Z)$ is the power set of Z), and
4. $\Pi x{:}\, Z \,.\, f(x)$ for all $Z \in X$ and all $f \in Z \to X$.

In clause 4 above, $\Pi x{:}\, Z \,.\, f(x)$ is the *generalized Cartesian product*, defined to be the set of all $g \in Z \to \cup_{x \in Z} f(x)$ such that for all $x \in Z$, $g(x) \in f(x)$.

Let σ_0 be the limit of a countable sequence $\tau_1 < \tau_2 \ldots$ of inaccessible cardinals. Define sets V_σ, indexed by ordinals σ, by $V_{\sigma+1} = V_\sigma \cup F(V_\sigma)$, and $V_\tau = \bigcup_{\sigma<\tau} V_\sigma$ if τ is a limit ordinal. Define

$$V = V_{\sigma_0} \cup \bigcup_{X \in V_{\sigma_0}} X.$$

Also, define $\gamma_N = \{c_1, c_2, \ldots\}$, $\gamma_B = \{c_{true}, c_{false}\}$, and $\gamma_i = V_{\tau_i}$ $(1 \leq i)$. The γ_i will be used to model universes of types.

In what follows, the letter γ will range over the sets introduced by clauses 1-4 in the definition of F. We will use ϕ to range over the members of all sets introduced by clause 4, and c will range over the members of B. The letters α and β will be used for arbitrary members of V. For technical reasons, we need to assume that the set-theoretic encoding of Cartesian product and the choice of B are such that the sets $\{\gamma \mid \gamma \in V\}$, $\{\phi \mid \phi \in V\}$ and $\{c \mid c \in V\}$ are disjoint. (We can accomplish this by adding "tags" to the standard set theoretic encodings.)

If B' is a proper subset of B, then although any function in $B \to B$ can be restricted to one in $B' \to B$, we do not have $B \to B \subseteq B' \to B$. Partly because we desire this kind of *contravariant subtyping*, we introduce an ordering on objects that is intended to capture the idea of one object "extending" another, as in the case of a function on B extending its restriction to B'. This ordering is slightly complicated by the fact that we need to lift it to higher types. More precisely, define $\alpha \preceq \beta$, for $\alpha, \beta \in V$, by induction on the rank of α as follows.

1. $\gamma \preceq \gamma$.
2. $c \preceq c$
3. $\phi \preceq \phi'$ if for all $(\alpha, \beta) \in \phi$ there exists $(\alpha', \beta') \in \phi'$ such that $\alpha' \preceq \alpha$ and $\beta \preceq \beta'$.

If $\alpha \preceq \beta$ then we say that β *subsumes*, or *extends* α. It is straightforward to show that $\preceq$ is a partial order.

When β subsumes α, then we will consider β to be a member of any set α is a member of. The notion of function application can be extended to this generalized membership: to apply a ϕ to an α, look for a pair $(\alpha', \beta) \in \phi$ such that α subsumes α', and return β. This is well-defined because of the following lemma.

Lemma 1 *For all $\gamma \in V$, all $\alpha, \alpha' \in \gamma$ and all $\beta \in V$, if $\alpha \preceq \beta$ and $\alpha' \preceq \beta$ then $\alpha = \alpha'$.*

Now that we have defined V, we proceed to the second stage of the semantics. We first define an extended syntax for ν_0. The set T of terms of the extended language of ν_0 are inductively defined as follows.

1. N, B, the booleans *true* and *false*, the numerals $0, 1, \ldots$, and the *universes* $U_1, U_2, \ldots$ are terms.
2. If x is a variable and e, e', e'' are terms then $\lambda x.\ e$, ee', $\epsilon x{:}e\,.\,e'$, $e =_{e''} e'$, $x : e \to e'$ and $\{x{:}e \mid e'\}$ are terms.
3. For all $\phi \in V$ and all $\gamma \in V$, κ_ϕ and κ_γ are terms.

If we close under the first two clauses above only, we get the syntax of ν_0 itself. In addition, there are terms for various programming language constructs, such as the basic arithmetic and boolean operations. From the perspective of the semantics, there is nothing interesting about these additional terms, and it is easy to include them. The usual kind of variable binding applies here. Let T_0 be the set of closed members of T.

There is an obvious injection of the members of V into T_0. Define $i[c_{true}] = \mathit{true}$, $i[c_{false}] = \mathit{false}$, $i[c_k] = k$, $i[\phi] = \kappa_\phi$, and $i[\gamma] = \kappa_\gamma$. We will usually write $\hat{\alpha}$ for $i[\alpha]$.

The "operational" semantics simultaneously inductively defines two binary relations. The first is the evaluation relation $\Downarrow$ over T_0. When $e \Downarrow v$ we say that e *evaluates* to v. The second relation, $\preceq \subseteq V \times T_0$, is a generalization of $\preceq \subseteq V \times V$ to terms. As before, when $\alpha \preceq e$, we say that e *extends*, or *subsumes*, α. When $e \Downarrow \hat{\gamma}$, we will say that e is a *type*. In accordance with the discussion above, the members of the type e will be taken to be the upward closure of γ under subsumption, so the members will be the e' such that $\alpha \preceq e'$ for some $\alpha \in \gamma$.

We first give the rules for subsumption.

$$\frac{e \Downarrow \hat{\gamma}}{\gamma \preceq e} \qquad \frac{e \Downarrow \hat{c}}{c \preceq e} \qquad \frac{e \Downarrow \hat{\phi} \quad \phi' \preceq \phi}{\phi' \preceq e}$$

$$\frac{e \Downarrow \lambda x.\, e' \quad \forall (\alpha, \beta){:}\phi.\ \beta \preceq e'[\hat{\alpha}/x]}{\phi \preceq e}$$

In the last rule, a λ-abstraction extends a set-theoretic function ϕ if its values subsume the values of ϕ. Although only arguments in the domain of ϕ are considered in this rule, it will turn out that this is enough, i.e. subsumption of values will hold for all arguments e' such that $\hat{\phi}(e')$ evaluates to something.

For evaluation, we have the usual rules for the lazy λ-calculus. (We can also handle the call-by-value calculus.)

$$\frac{e \Downarrow \lambda x.\, e' \quad e'[a/x] \Downarrow v}{ea \Downarrow v} \qquad \frac{}{\lambda x.\, e \Downarrow \lambda x.\, e} \qquad \frac{}{\hat{\alpha} \Downarrow \hat{\alpha}}$$

As mentioned before, we omit the rules for computing applications of boolean and arithmetic operators.

For applying a set-theoretic function, we have the following rule.

$$\frac{e \Downarrow \phi \quad (\alpha, \beta) \in \phi \quad \alpha \preceq a}{ea \Downarrow \hat{\beta}}$$

Thus, to evaluate $\hat{\phi}a$, find α in the domain of ϕ such that a extends α, and return the corresponding value.

Rules for the remaining constructs are given in Figure 1. We use the abbreviation $\mathit{pred}(\gamma, x, e, \phi)$ to mean that $\phi \in \gamma \to \{c_{true}, c_{false}\}$ and for all

$$\frac{e \Downarrow \gamma \quad \gamma \neq \emptyset \quad \mathit{pred}(\gamma, x, e', \phi)}{\epsilon\, x{:}e\,.\,e' \Downarrow \widehat{\gamma^{\phi}}}$$

$$\frac{e'' \Downarrow \gamma \quad \alpha \in \gamma \quad \alpha \preceq e \quad \alpha \preceq e'}{e =_{e''} e' \ \Downarrow\ \mathit{true}} \qquad \frac{e'' \Downarrow \gamma \quad \alpha \neq \beta \in \gamma \quad \alpha \preceq e \quad \beta \preceq e'}{e =_{e''} e' \ \Downarrow\ \mathit{false}}$$

$$\frac{}{U_i \Downarrow \gamma_i} \qquad \frac{}{N \Downarrow \gamma_N} \qquad \frac{}{B \Downarrow \gamma_B}$$

$$\frac{e \Downarrow \hat{\gamma} \quad \forall\,\alpha{:}\gamma.\ e'[\hat{\alpha}] \Downarrow \widehat{\gamma_\alpha}}{x : e \to e' \ \Downarrow\ i[\Pi\alpha{:}\gamma\,.\,\gamma_\alpha]} \qquad \frac{e \Downarrow \hat{\gamma} \quad \mathit{pred}(\gamma, x, e', \phi)}{\{\, x{:}e \mid e' \,\} \ \Downarrow\ i[\{\, \alpha{:}\gamma \mid \phi(\alpha) = c_{true} \,\}]}$$

Figure 1: Evaluation rules.

$\alpha \in \gamma$, $e[\hat{\alpha}/x] \Downarrow \widehat{\phi(\alpha)}$. Also, if, in addition, $\gamma \in V$ is non-empty, then let γ^{ϕ} be some $\alpha \in \gamma$ such that $\phi(\alpha) = c_{true}$, or, if no such α exists, then let γ^{ϕ} be an arbitrary member of γ. The rules in Figure 1 are straightforward if one keeps in mind the definitions of typehood and type membership given earlier. The operator $\epsilon\, x{:}e\,.\,e'$ is a *choice* operator. Note that equality can be used to define universal quantification:

$$\forall\, x{:}e.\ e' \quad \equiv \quad \lambda x.\ e' =_{e \to B} \lambda x.\ \mathit{true}.$$

Some of the properties of the semantics we have just given are summarized in the following:

Theorem 1 *The following hold.*

1. *If $\gamma \preceq e$ and $\gamma' \preceq e$, then $\gamma = \gamma'$.*
2. *If $\alpha, \alpha' \in \gamma$, $\alpha \preceq e$ and $\alpha' \preceq e$, then $\alpha = \alpha'$.*
3. *If $e \Downarrow v$ and $e \Downarrow v'$, then $v = v'$.*
4. *If $\alpha \preceq e[\beta/x]$ and $\beta \preceq e'$ then $\alpha \preceq e[e'/x]$.*

Because of parts 1 and 2 of Theorem 1, we can make the following definitions of the (partial) meaning functions that assign expressions of ν_0 their set-theoretic meaning.

Definition 1 If $e \Downarrow \hat{\gamma}$, define $M_*[e] = \gamma$. If $M_*[e] = \gamma$, and if there exists $\alpha \in \gamma$ such that $\alpha \preceq e'$, define $M_e[e'] = \alpha$.

Thus M_* assigns meaning to type expressions, and M_e assigns the meaning of expressions considered as members of a given type e. Because of part 4 of Theorem 1, we get analogs of the usual substitutivity property of a denotational semantics. In particular, we have the following.

Corollary 1 *The following equations hold whenever the right-hand sides are defined.*

$$\begin{aligned} M_*[e[e'/x]] &= M_*[e[\widehat{M_*[e']}/x]] \\ M_a[e[e'/x]] &= M_a[e[\widehat{M_{a'}[e']}/x]] \end{aligned}$$

2.2 Inference Rules of ν_0

Since we justify embeddings of other logics via the semantics, for our purposes here there is no need to give a set of inference rules for ν_0. However, this could easily be done by making minor adaptations to the rules of Nuprl. In this section we briefly discuss such a proof system in order to bring out some of the ideas implicit in the semantic account, and to illustrate the flexibility of the logic.

The inference rules relate *sequents*, which have the form

$$\psi_1, \ldots, \psi_n \vdash \psi.$$

Such a sequent is defined to be true if: for all substitutions σ of closed terms in T_0 for the free variables of $\psi_1, \ldots, \psi_n, \psi$, if $M_B[\sigma(\psi_i)] = c_{true}$ $(1 \leq i \leq n)$, then $M_B[\sigma(\psi)] = c_{true}$. Note that there is no separate notion of "well-formed" for sequents — this is built into the definition of sequent truth. Consider, for example, the sequent $\vdash e =_N e$. By the evaluation and subsumption rules, this is true if and only if $e \Downarrow n$ for some numeral n.

Abbreviate $e =_{e'} e$ by $e \in e'$. There are several ways of proving sequents of the form $\Gamma \vdash e \in e'$. One is to use rules for the operators in e and e'. For example, we have a rule

$$\frac{\Gamma \vdash e \in U_i \qquad \Gamma, x \in e \vdash B \in U_i}{\Gamma \vdash x : e \to e' \in U_i}$$

Note that we use membership in a universe to express typehood. Another is by deriving a contradiction from the hypotheses Γ. For example, we can prove

$$\lambda x.\, e \;\in\; \{x{:}N \,|\, 0 =_N 1\} \to N$$

for any e whatsoever since in the proof of $e \in N$ we can assume $0 = 1$. Finally, one can use *computational reasoning.* In particular, we can define a computational equivalence $\sim$ from the operational semantics which validates a rule that infers ψ' from ψ whenever $\psi \sim \psi'$. The significance of this is that the equivalence includes the redex-contractum relation of the operational semantics, so that, for example, $(\lambda x.\ e)e' \sim e'[e/x]$.

We can get contravariant subtyping for functions via an *extensionality* rule, which, in simplified form, is as follows.

$$\frac{\vdash f \in x : e_1' \to e_2' \quad x \in e_1 \vdash fx \in e_2 \quad \vdash e_1 \in U_i}{\vdash f \in x : e_1 \to e_2}$$

The key point is that the function type in the first premise is unrelated to the one in the conclusion. Thus, to show an object is in a function type, it suffices to show that it is a function, and that for inputs of the right type it produces outputs of the right type.

2.3 An Embedding Example

As a simple example of embedding another logic in ν_0, consider the simply typed λ-calculus theory Λ with a set of base types Ty, a set of constants C with corresponding types τ_c $(c \in C)$, a set of equational axioms $\mathcal{E}$, and a set of equational theorems $\mathcal{T}$ that follow from the axioms.

Because we can quantify over types, we could capture the set-theoretic semantics of Λ exactly, by formalizing the statement that for all (non-empty) sets assigned to base types $\iota \in Ty$, and for all values for the constants, if the axioms are true then the theorems are true. Another possibility is to directly interpret the types and constants in Nuprl. Since this second approach is closer to what we have done for embedding HOL in Nuprl, we focus on this one; the first approach can be dealt with similarly.

Thus, we pick closed terms $E[\iota]$ of ν_0 for each $\iota \in Ty$, and terms $E[c]$ for each $c \in C$. We extend E to all types of Λ by defining $E[\tau \to \tau']$ to be $E[\tau] \to E[\tau']$, where the latter $\to$ is the non-dependent special case of ν_0's function type. E can be extended to terms m of Λ in the obvious way, mapping application to applications, and mapping terms $\lambda x\!:\!\tau.\, m$ to $\lambda x.\ E[m]$. If ξ is an equation $m = m'$ in Λ over type τ, then define $E[\xi]$ to be $E[m] =_{E[\tau]} E[m']$.

Theorem 2 *Suppose the following sequents are true in ν_0:*

1. $\vdash\ E[\iota] \in U_1$, *for each* $\iota \in Ty$.

2. $\vdash\ \neg\forall x\!:\!E[\iota].\ \mathit{false}$, *for each* $\iota \in Ty$.

3. $\vdash\ E[c] \in E[\tau_c]$, *for each* $c \in C$.

4. $\vdash\ E[\xi]$ *for all* $\xi \in \mathcal{E}$.

Then for all $\xi \in \mathcal{T}$, $\vdash\ E[\xi]$.

The proof of this theorem is straightforward and we only sketch it here. The only reason we discuss it at all is to illustrate that despite the unconvential semantics, the proof proceeds the way one might expect. Because of conditions 1 and 2 above, for each $\iota \in Ty$ we get $M_*[E[\iota]] = \gamma_\iota$ for some non-empty γ. Let $M[\tau]$ be the meaning function on types of Λ induced by $\iota \mapsto \gamma$. Because of the semantics of the function type in ν_0, for each $c \in C$ we get, from 3, an $\alpha_c \in M[\tau_c]$ such that $M_{E[\tau]}[E[c]] = \alpha_c$. Let $M[m]$ be the meaning function on terms m of Λ induced by $c \mapsto \alpha_c$. To avoid having to deal with environments mapping free variables to their meanings, assume M is defined on abstractions via substitution of argument values.

All we need to show to complete the proof is that for all terms m of type τ of Λ, $M[m] = M_{E[\tau]}[E[m]]$. This can be proved by induction on the size of m, using Corollary 1.

Note that we get a computability property here for free. For example, if $E[\iota] = N$, then for any m of type ι in Λ, $E[m] \Downarrow n$ for some n. So, if all the terms $E[c]$, $c \in C$, use only computationally sensible operators (*i.e.*, exclude operator such as equality and choice), then $E[m]$ will compute to a number.

The semantics of Λ given by a particular interpretation in ν_0 is just one possible semantics. If we prove $E[\xi]$ in ν_0 for other equations ξ, we do not in general know that ξ follows from the equational axioms of Λ, but only that ξ is true in a particular model of Λ. In the case of general purpose logics, this does not seem to be a significant practical drawback to passing results back from ν_0 (or Nuprl) to the embedded logic L, since it turns out in most cases of interest that the model of L induced by the embedding is an acceptable "standard" model of L. We will return to this point when we discuss the Nuprl-HOL connection.

2.4 Extensions to ν_0

It is a simple matter to make certain extensions to ν_0. For example, we can easily add new base types, sigma types, and inductive types. These can all be integrated smoothly into the semantic account described above. With this extension, we can embed PVS [Owre *et al.*, 1996]. Another extension we can easily make is to a sequent system like Nuprl's where priority is given to judgments about type membership, and logic is encoded in the type system via propositions-as-types. The point of this encoding of logic is that it permits programs to be synthesized from proofs. In [Howe, 1996a] we sketched the design of a mechanism that puts a "firewall" between constructive and classical mathematics, so that programs can be extracted from constructive proofs, but classical mathematics can be used in computationally irrelevant parts of the proofs (which is almost everywhere). This mechanism has now been implemented.

The one feature of Nuprl that substantially complicates the semantics is the quotient type. The semantics we give in [Howe, 1997] gives an account of quotienting in which quotients can be interpreted classically as sets of equivalence classes, but also permit computable operations to be defined over them. Quotienting is only significant if one wishes to give constructive interpretations of certain theories. For example, to embed an abstract theory of the rationals in Nuprl, one would like to give a computationally useful representation, *e.g.* as sets of pairs, but the equality on such a type must be quotiented with the usual notion of the equality of rationals in order to match the equality expected by the abstract theory. For classical interpretations, representing equivalence classes by their characteristic functions is sufficient.

3 NUPRL/HOL

From a semantic perspective, embedding HOL's logic in ν_0 or Nuprl is very similar to the example embedding in Section 2.3. The main differences between HOL and the simply-typed case are as follows:

- Constants may be polymorphic, *i.e.* have type parameters.
- Axioms and theorems, instead of equations, are expressions of boolean type and may contain type variables.
- Constants for base types and new type constructors may be introduced.

An HOL theory is simply a collection of such constants and axioms, along with theorems that follow from the axioms. To embed HOL theories in ν_0 or Nuprl, we interpret the constants as before, but condition 3 in Theorem 2 will now quantify over the subtype S of U_1 consisting of all non-empty members of U_1. Type constants are interpreted, and we add a condition requiring the interpretations to have the right types (in terms of S). For condition 4 of Theorem 2, translations of axioms have their type variables universally closed. HOL also requires a particular semantics for several constants, such as choice and equality. The semantics for these constants agrees with the ones in ν_0 (and is definable in Nuprl) so we can simply interpret them directly.

The HOL system supports a discipline for introducing new axioms that guarantees consistency. Essentially, the discipline restricts new axioms to a few definition schemes. For any theory built using this discipline, an interpretation in ν_0 can always be constructed, and in fact is easy to compute. However, as will be discussed further below, this particular interpretation is often not desirable in practice. In particular, we will usually want to interpret with objects that are in Nuprl but not in the fragment corresponding to ν_0.

The connection we have implemented between HOL and Nuprl supports incremental interpretation of HOL theories in Nuprl. It guarantees soundness, essentially by checking that the conditions corresponding to 1-4 of Theorem 2 are satisfied, before any translations of consequent theorems are accepted as Nuprl theorems. The implementation has two other main tasks. One is to aid users in constructing instantiations and proving the correctness conditions. The other is to rewrite the imported theorems into a form more usable in Nuprl proofs, since the imported theorems are themselves almost never directly usable.

In the remainder of this section, we (1) briefly describe the support for and some of the practical issues related to constructing interpretations, (2) explain why rewriting of imported theorems is needed and describe what kind of rewriting is done, and (3) summarize some of the practical applications made of the Nuprl/HOL connection.

3.1 Constructing Interpretations

By convention, whenever a new type is introduced in HOL, the only axiom that may be assumed about it is that it is isomorphic to a subset of an existing type. This isomorphism, and its use in defining primitive operations over the type, amount to a particular implementation of the type in terms of the simple types formed over two base types (essentially the booleans and the natural numbers). This is fine for HOL purposes, but these implementations are not computationally reasonable (*e.g.* products are implemented as functions, and trees are implemented as integers). The desired interpretations in Nuprl for HOL type constants are almost never the implementations given in HOL. Supplying a different interpretation gives rise to a proof obligation, sometimes highly non-trivial, to establish the isomorphism axiom, *i.e.* to prove that the two implementations are isomorphic. This is most of the work in building an interpretation.

Fortunately, proving this isomorphism is difficult only for a few types, in particular for some of the most basic types, like products, integers and trees. The basic types aside, it is typical for a new type to be introduced via a *recursive datatype package*, which takes an ML-like specification of a datatype and extends the current HOL theory with appropriate new constants and axioms that implement the datatype (in terms of labelled trees). Our implementation recognizes such collections of constants and axioms, and automatically determines the appropriate interpretations in terms of Nuprl's type of trees, and proves the correctness conditions for the interpretation. A similar treatment is given to HOL's inductive predicate package.

Our implementation also has support for user-chosen interpretations of constants. Usually the correctness conditions are proven automatically, and the theorem-rewriter (described below) is automatically updated with information about the constants.

3.2 Rewriting Imported Theorems

We illustrate why rewriting is needed by considering an example. In particular, consider the HOL theorem

$$\vdash \forall l.\ \neg(null\ l) \Rightarrow (cons\ (hd\ l)\ (tl\ l) = l).$$

The formula gets interpreted in Nuprl as

$$\forall A{:}S.\ \forall l{:}A\ list.\ \neg(null'\ A\ l) \Rightarrow (cons'\ A\ (hd'\ A\ l)\ (tl'\ A\ l) =_A l),$$

where we have given the same names (using Nuprl's definition facility) to the interpretations of HOL's logical connectives. This form is not useful in typical Nuprl proofs about properties of lists for the following reasons.

1. The interpretations *cons'*, *hd'*, *tl'* and *null'* all take type arguments, whereas the preferred Nuprl analogs do not (*i.e.* they are implicitly polymorphic).

2. The formula uses boolean logic, while Nuprl inference machinery is tuned to the propositions-as-types encoding of logic.

3. The correctness condition for the interpretation of *hd'* requires that it be a *total* function on lists, so that it must be given the non-constructive definition that uses the choice operator for the case where the list is empty.

Fortunately, all differences between the direct interpretation and the desired form can be eliminated by term rewriting based on an extensible set of rewrite rules. Typically the set of rules is automatically updated by the support for instantiation construction. In the case of partial functions like *hd'*, users must explicitly supply an appropriate rewrite rule, which will usually be conditional. For example, in the case of *hd'*, we have the theorem

$$\forall A{:}S.\ \forall l{:}A\ list.\ \neg(null\ l) \Rightarrow hd'\ A\ l = hd\ l$$

where *hd* is Nuprl's standard head function (which is undefined on empty lists). The set of rewrite rules has a (very) small number of ad hoc entries as well.

3.3 Applications

We have applied our Nuprl/HOL connection in two substantial applications. The applications demonstrate that imported mathematics can be effectively employed, and that with the infrastructure we have built, importing a body of mathematics from HOL is far less work than reimplementing it in Nuprl.

In [Felty and Howe, 1997], we imported classical theories of combinatory logic and minimal logic from HOL, and used them in Nuprl, in combination with Nuprl's constructive mathematics, to prove a constructive normalization property. Because the new types of these theories were introduced using the recursive datatype package, interpreting them in Nuprl required little effort.

In [Felty *et al.*, 1998], we imported HOL's extensive library of list theory, and used it heavily in our verification of the SCI cache coherence protocol.

4 DISCUSSION

In this section we briefly discuss some further elaborations of our approach to sharing mathematics that we feel are within our reach (and which are the subject of ongoing or planned work).

Reversing the Nuprl/HOL link. In principle this is straightforward. If we add new constants, but no axioms to an HOL theory, extend the interpretation E to these constants, and check that the analogs of 1-3 in Theorem 2 hold, then there is a "standard model" of the extended theory which validates all formulas ϕ such that $E(\phi)$ is proven in Nuprl. Thus pulling back a suitable set of theorems from Nuprl to HOL is essentially just a matter of introducing some new constants.

Of course, there are some practical issues. The desired theorems will not usually be of the required form $E(\phi)$. Rewriting may be required to put them in this form. Since Nuprl is a more expressive logic, therc will be examples where the simple kinds of rewriting done with the existing Nuprl/HOL link are not sufficient. If mathematics built in Nuprl is intended to be exported to HOL, then it may be desirable to follow guidelines, perhaps with assistance from the implementation, that ensure the mathematics can be rewritten.

PVS/Nuprl. Including PVS in a sharing scheme is valuable since it is one of the most widely used systems. Since PVS is also easily embedded in Nuprl, the issues in linking Nuprl with PVS are not very different from linking with HOL. In some ways it should be easier, since PVS and Nuprl are much closer as type theories than are HOL and Nuprl. One difference is that, unlike HOL, PVS has a theory mechanism. HOL has objects it calls "theories", but this is just an organizational device and there are no operations on them other than union. Theories in PVS can take type or individual parameters. This parameterization is easily modeled in Nuprl: the parameters are added as new arguments to all defined constants, and as quantified variables in all theorems and axioms.

PVS/Nuprl/HOL. Having embeddings of PVS and HOL opens the possibility of sharing mathematics between all three systems. The main open question here is how much mathematics can be usefully shared. Going from PVS (or Nuprl) to HOL requires eliminating dependent types and subtyping. There are translations that can eliminate many of these types (see, for example [Jacobs and Melham, 1993]), but it is not clear how often, in a practical sense, they can be applied, or how useful the results would be. Experience suggests that there is a great deal of lower-level, basic mathematics that does not stress the expressive power of type systems and hence is sharable.

REFERENCES

[Constable and Howe, 1990] Robert Constable and Douglas Howe. Nuprl as a general logic. In P. Odifreddi, editor, *Logic in Computer Science*, pages 77–90. Academic Press, 1990.

[Constable, et al., 1986] Robert L. Constable, et al. *Implementing Mathematics with the Nuprl Proof Development System.* Prentice-Hall, Englewood Cliffs, New Jersey, 1986.

[Felty and Howe, 1997] Amy P. Felty and Douglas J. Howe. Hybrid interactive theorem proving using Nuprl and HOL. In *Fourteenth International*

Conference on Automated Deduction, volume 1249 of *Lecture Notes in Computer Science*, pages 351–365, Berlin, 1997. Springer-Verlag.

[Felty *et al.*, 1998] Amy Felty, Douglas Howe, and Frank Stomp. Protocol verification in nuprl. In *CAV'98*, Lecture Notes in Computer Science. Springer-Verlag, 1998.

[Gordon and Melham, 1993] M. J. C. Gordon and T. F. Melham. *Introduction to HOL: A Theorem Proving Environment for Higher Order Logic.* Cambridge University Press, Cambridge, UK, 1993.

[Harper *et al.*, 1993] Robert Harper, Furio Honsell, and Gordon Plotkin. A framework for defining logics. *Journal of the ACM*, 40(1), January 1993.

[Howe, 1996a] Douglas J. Howe. Importing mathematics from HOL into Nuprl. In J. von Wright, J. Grundy, and J. Harrison, editors, *Theorem Proving in Higher Order Logics*, volume 1125 of *Lecture Notes in Computer Science*, pages 267–281, Berlin, 1996. Springer-Verlag.

[Howe, 1996b] Douglas J. Howe. Semantics foundations for embedding HOL in Nuprl. In Martin Wirsing and Maurice Nivat, editors, *Algebraic Methodology and Software Technology*, volume 1101 of *Lecture Notes in Computer Science*, pages 85–101, Berlin, 1996. Springer-Verlag.

[Howe, 1997] Douglas J. Howe. A classical set-theoretic model of polymorphic extensional type theory. (Submitted to *Information and Computation*), 1997.

[Jacobs and Melham, 1993] B. Jacobs and T. Melham. Translating dependent type theory into higher order logic. In *Proceedings of the Second International Conference on Typed Lambda Calculi and Applications*, volume 664 of *Lecture Notes in Computer Science*, pages 209–229. Springer, 1993.

[Martí-Oliet and Meseguer, 1996] N. Martí-Oliet and J. Meseguer. Rewriting logic as a logical and semantic framework. *Electronic Notes in Theoretical Computer Science*, 4, 1996.

[Matthews *et al.*, 1993] S. Matthews, A. Smaill., and D. Basin. *Experience with FS_0 as a Framework Theory*, pages 61–82. Cambridge University Press, 1993.

[Owre *et al.*, 1996] S. Owre, S. Rajan, J.M. Rushby, N. Shankar, and M. Srivas. PVS: Combining specification, proof checking, and model checking. In *Proceedings of CAV'96*, Lecture Notes in Computer Science. Springer Verlag, 1996.

[Paulson, 1994] Lawrence C Paulson. *Isabelle: A Generic Theorem Prover*, volume 828 of *Lecture Notes in Computer Science.* Springer-Verlag, 1994.

[Werner, 1997] Benjamin Werner. Sets in types, types in sets. In *International Symposium on Theoretical Aspects of Computer Software*, Lecture Notes in Computer Science. Springer-Verlag, 1997.

Negation in Combining Constraint Systems

Stephan Kepser

1 INTRODUCTION

One idea behind constraint solving is to use specialised formalisms and inference mechanisms to solve domain-specific tasks. In many applications, however, one is faced with a complex combination of different problems, which means that a system tailored to solving a single problem can only be applied if it is possible to combine it with other specialised systems. In a recent paper [Baader and Schulz, 1998], Baader and Schulz present a general method for the combination of constraint systems over disjoint signatures. Their method is applicable to a large class of structures, the so called quasi-free structures. Quasi-free structures comprise many important infinite non-numerical domains such as term algebras and quotient term algebras, rational tree algebras, vector spaces, hereditarily finite wellfounded and non-wellfounded lists, sets and multi sets as well as features structures of the Smolka & Treinen variety [Smolka and Treinen, 1994]. The combined solution domain they present, the so called free amalgamated product, is characterised by the property of being the most general combined solution domain in the sense that every other combined solution domain contains a homomorphic image of the free amalgamated product. The authors also give an algorithm to reduce solvability of "mixed" constraints, i.e., constraints over the joint signature, in the free amalgamated product to pure constraints in the components. With the help of this algorithm, they show that the positive theory of the free amalgamated product is decidable provided the positive theory of both components is decidable. Negation is not addressed in their paper. Indeed, the proof methods used are such that they cannot be extended to include negation.

Negation, however, plays an important role in constraint solving. For example, it is required for implication and constraint entailment, both of which are used heavily in actual implementations of constraint solvers when reducing sets of constraints (see, e.g., [Henz *et al.*, 1993]). Its usefullness for practical applications of constraint techniques is obvious. Hence it is the aim of this paper to investigate to what extent constraint systems with negation can be combined. We show how solvability of positive and negative mixed constraints in the free amalgamated product can be reduced to solvability of pure constraints in the components using an extension of the algorithm by Baader and Schulz. The existential theory of the free amalgamation is decidable provided solvability of conjunctions of literals with so called linear constant restrictions is decidable in the components. When solving negative constraints, one is often interested in the existence of ground solutions. Therefore we present sufficient conditions for solutions in the components to give a ground solution in the free amalgamation. The existence of a ground solution of mixed constraints in the free amalgamated product is decidable, if the existence of so-called restrictive solutions of pure constraints with linear constant restrictions is decidable in the components.

In the second half of this paper, we deal with the independence property of negative constraints. A constraint system is said to have the independence property, if the solvability of a conjunction of positive constraints and a conjunction of negative constraints can be reduced to solving each single negative conjunct together with the conjunction of positive constraints. In other words, if for some conjunction of negative constraints, each of them – plus the positive constraints – is solvable separately, then their conjunction is solvable. This property plays an important role in real world constraint solvers that cope with negative constraints (see, e.g., [Henz *et al.*, 1993]).

The aim of this paper is to give a modularity result for the independence property, that is, to state under which conditions the free amalgamated product has the independence property provided the component structures do so. Finally, we will show that the free amalgamation of two unitary regular and non-collapsing quasi-free structures has the independence property. We look at a particular subclass of quasi-free structures, namely free algebras defined by equational theories, since there is not yet any discussion of this topic in the literature for these prototypical quasi-free structures. We find that unitary equational theories have the independence property, while finitary equational theories do not. And the combination of unitary regular and collapse-free equational theories is again unitary and hence has the independence property. We lift these results to quasi-free structures in general. Unitary quasi-free structures have the independence property. And the free amalgamation of two unitary regular and non-collapsing quasi-free structures is unitary and hence has the independence property.

Some of the results we present are inspired by [Baader and Schulz, 1995], which describes combination techniques for equational disunification algorithms. For example, our decomposition algorithm (cf. 4.4) is a generalisation of the one given in [Baader and Schulz, 1995]. But it should be noted that the proof methods of that paper – heavy use of rewriting techniques – are *not* extendable to the general, algebraic view that we take here.

For lack of space, we had to omit most proofs. They can be found in

Chapter 6 of [Kepser, 1998].

2 PRELIMINARIES

A signature Σ consists of a set Σ_F of function symbols and a disjoint set Σ_P of predicate symbols (not containing "="), each of fixed arity. Expressions $\mathfrak{A}^\Sigma$ denote Σ-structures over the carrier set A, and $f_\mathfrak{A}$ ($p_\mathfrak{A}$) stands for the interpretation of $f \in \Sigma_F$ ($p \in \Sigma_P$) in $\mathfrak{A}^\Sigma$. Σ-terms $(t, t_1, \ldots)$ and atomic Σ-formulas (of the form $t_1 = t_2$, or of the form $p(t_1, \ldots, t_n)$) are built as usual. A Σ-formula φ is written in the form $\varphi(v_1, \ldots, v_n)$ in order to indicate that the set $Var(\varphi)$ of free variables of φ is a subset of $\{v_1, \ldots, v_n\}$. An *assignment* a for a formula φ is a mapping $a : Var(\varphi) \to A$; an assignment a is a *solution* for φ, if $\mathfrak{A}^\Sigma \models \varphi[v_1/a(v_1), \ldots, v_n/a(v_n)]$.

Let $\mathfrak{A}^\Sigma$ and $\mathfrak{B}^\Sigma$ be two Σ-structures. A mapping $h : A \to B$ is a Σ-homomorphism, iff for all n-ary functions $f \in \Sigma_F$, all m-ary predicates $p \in \Sigma_P$ and all $a_1, \ldots, a_n, a_m \in A$: $h(f(a_1, \ldots, a_n)) = f(h(a_1, \ldots, h(a_n))$ and $p(a_1, \ldots, a_m) \Longrightarrow p(h(a_1), \ldots, h(a_m))$. A homomorphic mapping $h : A \to A$ is called a Σ-endomorphism. A bijection $h : A \to B$ is a Σ-isomorphism, if both h and its inverse h^{-1} are Σ-homomorphisms. A Σ-automorphism is an endomorphic Σ-isomorphism. For more information, see e.g., [Mal'cev, 1971]. With $End_\mathfrak{A}^\Sigma$ we denote the monoid of all endomorphisms of $\mathfrak{A}^\Sigma$, with composition as operation. If $g : A \to B$ and $h : B \to C$ are mappings, then $h \circ g : A \to C$ denotes their composition. If M is a set, then $|M|$ denotes its cardinality.

The constraint problems we consider in this paper are conjunctions of literals. Clearly, every such constraint problem Γ can be written as $\Gamma^+ \wedge \bigwedge_{i=1}^k C_i^-$ where Γ^+ is the conjunction of all positive constraints (atoms) of Γ and $\bigwedge_{i=1}^k C_i^-$ is the conjunction of all negative constraints (negated atoms) of Γ. If we do not need to refer to specific negative constraints, we abbreviate $\bigwedge_{i=1}^k C_i^-$ by Γ^-.

Definition 2.1 A structure $\mathfrak{A}$ has the *independence of negative constraints* property, iff for every constraint problem Γ we have:

$\Gamma = \Gamma^+ \wedge \bigwedge_{i=1}^k C_i^-$ is solvable in $\mathfrak{A}$, iff
for every $i = 1, \ldots, k$ the conjunction $\Gamma^+ \wedge C_i^-$ is solvable in $\mathfrak{A}$.

3 QUASI-FREE STRUCTURES AND FREE AMALGAMATED PRODUCTS

In this section we introduce the class of structures we use as solution domains for constraint solving. First we recall the definition of quasi-free structures given in [Baader and Schulz, 1998]. We consider a fixed Σ-structure $\mathfrak{A}^\Sigma$.

Definition 3.1 Let A_0, A_1 be subsets of A, the carrier of $\mathfrak{A}^\Sigma$. Then A_0 *stabilises* A_1 iff all elements m_1 and m_2 of $End^\Sigma_\mathfrak{A}$ that coincide on A_0 also coincide on A_1. For $A_0 \subseteq A$ the *stable hull* of A_0 is the set

$$\mathrm{SH}^\mathfrak{A}(A_0) := \{a \in A \mid A_0 \text{ stabilises } \{a\}\}.$$

$\mathrm{SH}^\mathfrak{A}(A_0)$ is always a Σ-substructure of $\mathfrak{A}^\Sigma$, and $A_0 \subseteq \mathrm{SH}^\mathfrak{A}(A_0)$. The stable hull of A_0 can be larger than the Σ-subalgebra generated by A_0.

Definition 3.2 The set $X \subseteq A$ is an *atom set* for $\mathfrak{A}^\Sigma$ if every mapping $X \to A$ can be extended to an endomorphism.

Definition 3.3 A countably infinite Σ-structure $\mathfrak{A}^\Sigma$ is a *quasi-free structure* iff $\mathfrak{A}^\Sigma$ has an infinite atom set X where every $a \in A$ is stabilised by a finite subset of X. We denote this quasi-free structure by $(\mathfrak{A}^\Sigma, X)$.

Examples 3.4 The class of quasi-free structures contains many important non-numerical infinite solution domains. For example, all free structures (see, e.g., [Mal'cev, 1971]), rational tree algebras ([Colmerauer, 1984; Maher, 1988]), feature structures with arity ([Smolka and Treinen, 1994; Backofen and Treinen, 1994]), domains with nested, finite or rational lists (rational lists are used in Prolog III, see [Colmerauer, 1990]), and domains with nested, finite or rational sets (as introduced in [Aczel, 1988] and used in [Rounds, 1988]) are quasi-free structures. For details we refer to [Baader and Schulz, 1998].

A fundamental property of quasi-free structures is the following (see [Baader and Schulz, 1998], Lemma 3.19): for each $a \in A$ there exists a *unique minimal* finite set $Y \subseteq X$ such that $a \in \mathrm{SH}^\mathfrak{A}(Y)$.

Definition 3.5 The *stabiliser* of $a \in A$, $\mathrm{Stab}^\mathfrak{A}(a)$, is the unique minimal finite subset Y of X such that $a \in \mathrm{SH}^\mathfrak{A}(Y)$. The stabiliser of $A' \subseteq A$ is the set $\mathrm{Stab}^\mathfrak{A}(A') := \bigcup_{a \in A'} \mathrm{Stab}^\mathfrak{A}(a)$.

For example in a term algebra, the stabiliser of a term is the set of variables occurring in that term.

Let $(\mathfrak{A}^\Sigma, X)$ be a quasi-free structure. $(\mathfrak{A}^\Sigma, X)$ is called *quasi-free for a structure* $\mathfrak{D}^\Sigma$, iff every mapping $X \to D$ can be extended to a homomorphism $\mathfrak{A}^\Sigma \to \mathfrak{D}^\Sigma$.

Definition 3.6 Let $(\mathfrak{A}^\Sigma, X)$ and $(\mathfrak{B}^\Delta, X)$ be two quasi-free structures. The structure $\mathfrak{D}^{\Sigma \cup \Delta}$ is an *amalgamation* of $\mathfrak{A}^\Sigma$ and $\mathfrak{B}^\Delta$, iff there exist homomorphisms $d_A : \mathfrak{A}^\Sigma \to \mathfrak{D}$ and $d_B : \mathfrak{B}^\Delta \to \mathfrak{D}$ such that $d_A(x) = d_B(x)$ for all $x \in X$.

The following definition introduces the combined solution domain for our constraints.

Definition 3.7 Let $(\mathfrak{A}^\Sigma, X)$ and $(\mathfrak{B}^\Delta, X)$ be two quasi-free structures. The amalgamation $(\mathfrak{A}^\Sigma \otimes \mathfrak{B}^\Delta, e_A, e_B)$ is the *free amalgamated product*, iff for every amalgamation $(\mathfrak{D}^{\Sigma\cup\Delta}, d_A, d_B)$ such that $(\mathfrak{A}^\Sigma, X)$ is quasi-free for $\mathfrak{D}^\Sigma$ and $(\mathfrak{B}^\Delta, X)$ is quasi-free for $\mathfrak{D}^\Delta$ there exists a unique homomorphism $h! : \mathfrak{A}^\Sigma \otimes \mathfrak{B}^\Delta \to \mathfrak{D}^{\Sigma\cup\Delta}$ such that $h! \circ e_A = d_A$ and $h! \circ e_B = d_B$.

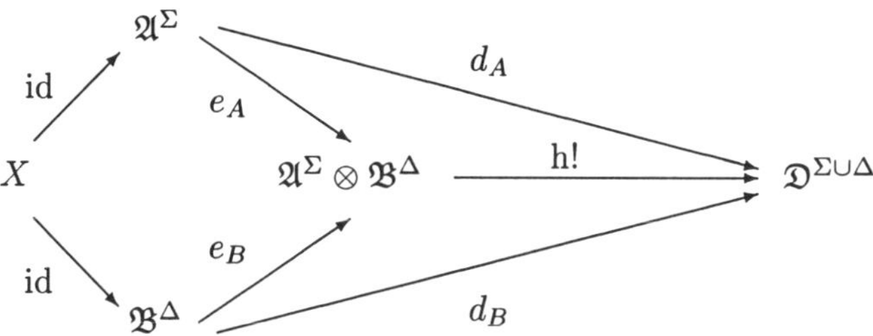

In [1998], Baader and Schulz provide a construction method to obtain the free amalgamated product of two arbitrary quasi-free structures which we will sketch in the next section.

4 FREE AMALGAMATION OF NEGATIVE CONSTRAINTS: THE GENERAL CASE

In this section, we show that solving mixed constraints in the free amalgamation can be reduced to solving pure constraints in the components, even if the constraints contain negations. The type of problems we have to solve in the components are constraint problems with generalised linear constant restrictions. Problems with linear constant restrictions are defined in [Baader and Schulz, 1995].

Definition 4.1 Let Γ be a constraint problem of signature Σ and Δ be some signature disjoint from Σ. A *generalised linear constant restriction* $L = (\Pi, Lab, <_L)$ consists of a partition Π of $Var(\Gamma)$, a labelling function $Lab : Var(\Gamma) \to \{\Sigma, \Delta\}$ and a linear order $<_L$ on $Var(\Gamma)$ such that for all $x, x', y, y' \in Var(\Gamma)$: if $x \equiv_\Pi x'$[1], $y \equiv_\Pi y'$ then $Lab(x) = Lab(x')$ and $x <_L y$ implies $x' <_L y'$.

An assignment σ solves (Γ, L) in $(\mathfrak{A}^\Sigma, X)$, iff it solves Γ and
for each $x, y \in Var(\Gamma) : \sigma(x) = \sigma(y)$ iff $x \equiv_\Pi y$,
for each $x \in Var(\Gamma)$ with $Lab(x) = \Delta : \sigma(x) \in X$, and
for each $x, y \in Var(\Gamma)$ with $Lab(x) = \Delta, Lab(y) = \Sigma, y <_L x$: $\sigma(x) \notin \mathrm{Stab}^{\mathfrak{A}}(\sigma(y))$.

Now we can present the main theorems of the first part of this paper.

[1] $x \equiv_\Pi x'$ means x and x' belong to the same equivalence class of Π.

Theorem 4.2 *Let $(\mathfrak{A}^\Sigma, X)$ and $(\mathfrak{B}^\Delta, X)$ be two quasi-free structures over disjoint signatures. The solvability of mixed $\Sigma \cup \Delta$ constraint problems, i.e., conjunctions of literals, in the free amalgamated product $\mathfrak{A}^\Sigma \otimes \mathfrak{B}^\Delta$ is decidable, if solvability of constraint problems with generalised linear constant restrictions is decidable in both components $(\mathfrak{A}^\Sigma, X)$ and $(\mathfrak{B}^\Delta, X)$.*

Since existential quantifiers distribute over disjunctions, we can strengthen our result.

Theorem 4.3 *Let $(\mathfrak{A}^\Sigma, X)$ and $(\mathfrak{B}^\Delta, X)$ be two quasi-free structures over disjoint signatures. The existential theory of the free amalgamated product $\mathfrak{A}^\Sigma \otimes \mathfrak{B}^\Delta$ is decidable, if solvability of constraint problems with generalised linear constant restrictions is decidable in both components $(\mathfrak{A}^\Sigma, X)$ and $(\mathfrak{B}^\Delta, X)$.*

Decomposition Algorithm

In order to show the above theorems, we present a decomposition algorithm that reduces solvability of mixed constraints in the free amalgamated product to solvability of pure constraints with generalised linear constant restrictions in the components. The algorithm is a simple generalisation of the one given in [Baader and Schulz, 1995]. We assume that the input problem Γ is in *decomposed form*, i.e., $\Gamma = \Gamma_\Sigma \wedge \Gamma_\Delta \wedge \Gamma_{\neq}$ where Γ_Σ (respectively Γ_Δ) contains only pure Σ- (respectively Δ-) literals and no disequations; and $\Gamma_{\neq}$ contains only disequations between variables. Every input problem can be transformed into an equivalent one in decomposed form by means of a polynomial algorithm which is now standard in the field (see, e.g., [Baader and Schulz, 1995]).

Algorithm 4.4 The input Γ is a constraint problem over signature $\Sigma \cup \Delta$ in decomposed form.

Step 1: variable identification

Non-deterministically choose a partition Π of the set of all variables of the constraint problem such that whenever the constraint problem contains a disequation $x \dot{\neq} y$ then x and y belong to different classes of the partition.

Step 2: variable labelling and ordering

For a given system, choose a mapping *Lab* from the set of variables into the set of theory labels $\{\Sigma, \Delta\}$ and a strict linear order $<_L$ on the variables such that both respect the partition Π. This triple $L = (\Pi, Lab, <_L)$ gives rise to a generalised linear constant restriction.

Step 3: split systems

A given system Γ is split into two systems $\Gamma_{3,\Sigma} \wedge \Gamma_{3,\Delta}$ where $\Gamma_{3,\Sigma} = \Gamma_\Sigma \wedge \Gamma_{\neq}$ and $\Gamma_{3,\Delta} = \Gamma_\Delta \wedge \Gamma_{\neq}$. The subsystems can now be considered as constraint problems with generalised linear constant restriction.

Proposition 4.5 *The input problem Γ is solvable, if and only if there exists an output pair $(\Gamma_{3,\Sigma}, \Gamma_{3,\Delta})$ and a generalised linear constant restriction L such that $(\Gamma_{3,\Sigma}, L)$ is solvable in $\mathfrak{A}^\Sigma$ and $(\Gamma_{3,\Delta}, L)$ is solvable in $\mathfrak{B}^\Delta$.*

The difficult part of this proposition is to show that the existence of solutions for $(\Gamma_{3,\Sigma}, L)$ and $(\Gamma_{3,\Delta}, L)$ implies the existence of a solution for Γ. We will sketch the proof here. To do so, we have to present how the free amalgamated product of two quasi-free structures $(\mathfrak{A}^{\Sigma}, X)$ and $(\mathfrak{B}^{\Delta}, X)$ is constructed. The full details of the construction can be found in [Baader and Schulz, 1998]. We embed both into isomorphic super structures $(\mathfrak{A}^{\Sigma}_{\infty}, Y_1)$ and $(\mathfrak{B}^{\Delta}_{\infty}, Y_2)$ such that $X \subset Y_i$ and $Y_i \setminus X$ is infinite for $i = 1, 2$. These structures are fibred in the following way. A *fibre* is either of the form $\{x\}$ with $x \in X$ or $\{b, y\}$ where $y \in Y_1 \setminus X$ and $b \in B_{\infty} \setminus Y_2$ or $y \in Y_2 \setminus X$ and $b \in A_{\infty} \setminus Y_1$. Let $d_1, d_2, d_3, d_4, \ldots$ be some enumeration of $A_{\infty} \cup B_{\infty}$. We define an ascending tower of fibre sets $\mathcal{F}_0 \subseteq \mathcal{F}_1 \subseteq \mathcal{F}_2 \subseteq \ldots$ where each $\mathcal{F}_i$ is a set of mutually disjoint fibres. We start with $\mathcal{F}_0 := \{\{x\} \mid x \in X\}$. Suppose $\mathcal{F}_k$ is already defined. To define $\mathcal{F}_{k+1}$ we distinguish two cases.

Case 1: There exists an element $d \in A_{\infty} \cup B_{\infty}$, say in A_{∞}, such that each element of the stabiliser $\mathrm{Stab}_{\Sigma}^{\mathfrak{A}_{\infty}}(d)$ belongs to a fibre in $\mathcal{F}_k$, but d itself does not. Then we proceed as follows: Let d_{min} be the smallest such element (in the enumeration) and, say, $d_{min} \in A_{\infty}$. We select an atom $y \in Y_2 \setminus X$ which does not belong to any fibre in $\mathcal{F}_k$ and define $\mathcal{F}_{k+1} := \mathcal{F}_k \cup \{\{d_{min}, y\}\}$.

Case 2: Otherwise define $\mathcal{F}_{k+1} := \mathcal{F}$.

Let $\mathcal{F} := \bigcup_{k \geq 0} \mathcal{F}_k$. Then $D := \bigcup_{F \in \mathcal{F}} F$ is a subset of $A_{\infty} \cup B_{\infty}$. Let $A_{\diamond} := D \cap A_{\infty}$, $B_{\diamond} := D \cap B_{\infty}$ and $Z_i = D \cap Y_i$. $(\mathfrak{A}^{\Sigma}_{\diamond}, Z_1)$ (respectively $(\mathfrak{B}^{\Delta}_{\diamond}, Z_2)$) is an isomorphic substructure of $(\mathfrak{A}^{\Sigma}_{\infty}, Y_1)$ (respectively $(\mathfrak{B}^{\Delta}_{\infty}, Y_2)$). Since each element $d \in D$ belongs to a unique fibre F_d of $\mathcal{F}$, the fibring gives rise to a bijection $h : A_{\diamond} \to B_{\diamond}$ that is used to superimpose the Σ-structure of $\mathfrak{A}^{\Sigma}_{\diamond}$ onto $B_{\diamond}$ and the Δ-structure of $\mathfrak{B}^{\Delta}_{\diamond}$ onto $A_{\diamond}$. Thus h can be seen as a $\Sigma \cup \Delta$-isomorphism $\mathfrak{A}^{\Sigma \cup \Delta}_{\diamond} \to \mathfrak{B}^{\Sigma \cup \Delta}_{\diamond}$ and $\mathfrak{A}^{\Sigma} \otimes \mathfrak{B}^{\Delta} := \mathfrak{A}^{\Sigma \cup \Delta}_{\diamond}$.

The *shadow* of an element $a \in D$ is recursively defined: Let $a \in A_{\diamond}$, then $\mathrm{Sd}(a) := \{a\} \cup \mathrm{Stab}_{\Sigma}^{\mathfrak{A}_{\diamond}}(a) \cup \bigcup \{\mathrm{Sd}(h(x)) \mid x \in \mathrm{Stab}_{\Sigma}^{\mathfrak{A}_{\diamond}}(a)\}$, so it contains the element a, its stabiliser, the fibre images of the stabiliser elements, their stabiliser, the fibre images thereof and so on, everything pending below a via stabilisers and fibre images. For $a \in B_{\diamond}$, the definition is analogous. Due to the inductive definition of the fibring construction, the shadow of each element is finite.

Soundness proof: suppose there is a solution σ_{Σ} of $(\Gamma_{3,\Sigma}, L)$ in $\mathfrak{A}^{\Sigma}$ and a solution σ_{Δ} of $(\Gamma_{3,\Delta}, L)$ in $\mathfrak{B}^{\Delta}$. Then there is a solution σ'_{Σ} of $(\Gamma_{3,\Sigma}, L)$ in $\mathfrak{A}^{\Sigma}_{\diamond}$ and a solution σ'_{Δ} of $(\Gamma_{3,\Delta}, L)$ in $\mathfrak{B}^{\Delta}_{\diamond}$. If we had $h(\sigma'_{\Sigma}(v)) = \sigma'_{\Delta}(v)$ for all $v \in \mathit{Var}(\Gamma)$, then our proof would be complete, because then, obviously, σ'_{Σ} would be a solution of $\Gamma_{3,\Delta}$ in $\mathfrak{A}^{\Sigma \cup \Delta}_{\diamond}$. Since the solutions σ_{Σ} and σ_{Δ} are found independently of each other, we cannot expect this condition to hold a priory. And we do not want to restrict the type of admitted solutions in the components. So the task is to show that the given solutions σ'_{Σ} and σ'_{Δ} can be transformed by means of automorphisms in such a way that finally the value of a variable $v \in \mathit{Var}(\Gamma)$ under σ'_{Σ} is the fibre image of the value under σ'_{Δ}. We call this the fibring condition for v. The use of automorphisms is required to handle negative constraints. While endomorphisms preserve only the validity of positive formulae, automorphisms preserve validity of arbitrary formulae.

The generalised linear constant restriction L contains a linear order $<_L$ of the variables in Γ. Let $v_1, v_2, v_3, \ldots, v_n$ be the enumeration of the variables alongside the order. By induction on this enumeration we prove that there exist solutions $l_{i,\Sigma}$ of $(\Gamma_{3,\Sigma}, L)$ and $l_{i,\Delta}$ of $(\Gamma_{3,\Delta}, L)$ such that for all variables v_j with $j \leq i$ we have $h \circ l_{i,\Sigma}(v_j) = l_{i,\Delta}(v_j)$, that is the first i variables fulfil the fibring condition. But this simple statement alone is weak. In an induction step, we may need to apply an automorphism exchanging two atoms in order to establish the fibring condition for the current variable. The difficult part consists in showing that this automorphism does not dissolve the fibring conditions for variables already handled. This is where the shadows of elements come in. We assume that every variable occurring in the shadows of the first i variables already fulfils the fibring condition. And we show that the automorphisms are the identity on the shadows of the first i variables and hence do not dissolve an already established fibring condition.

We first look at a simple case. Let $V := \mathit{Var}(\Gamma)$ and $n := |V|$. We partition V as follows. Define $V_Z := \{v \in V \mid \sigma'_\Sigma(v) \in Z_1 \text{ and } \sigma'_\Delta(v) \in Z_2\}$ as the set of all variables that are mapped to atoms in both solutions, and let $V_{\bar{Z}} := V \setminus V_Z$ be its complement. Let X be partitioned as $X_0 \uplus X_1 \uplus X_2$ such that X_1 and X_2 are infinite and X_0 has at least n elements. For a variable $v \in V_Z$, we pick some atom $z \in X_0$, the common atom set, and apply the automorphism that transposes z and $\sigma'_\Sigma(v)$ and the automorphism that transposes z and $\sigma'_\Delta(v)$. Iterating this process with other variables if necessary, we can ensure that if both solutions map a variable to an atom, this atom is one and the same in the common atom set X_0, and thus the fibring condition is fulfilled. These are our new solutions σ''_Σ and σ''_Δ.

Now, we prove by induction the following properties for each $i \leq n$:

There are solutions $l_{i,\Sigma}$ of $(\Gamma_{3,\Sigma}, L)$ in $\mathfrak{A}^\Sigma_\diamond$ and $l_{i,\Delta}$ of $(\Gamma_{3,\Delta}, L)$ in $\mathfrak{B}^\Delta_\diamond$ such that

1. For each variable $v \in V_Z$ holds $h \circ l_{i,\Sigma}(v) = l_{i,\Delta}(v) \in X_0$. We took care of these variables before. Now we have to ensure that this property is not lost.

2. Let $S_i := \bigcup_{j \leq i} \mathrm{Sd}(l_{i,\Sigma}(v_j)) \cup \bigcup_{j \leq i} \mathrm{Sd}(l_{i,\Delta}(v_j))$ be the union of all shadows of all solutions for variables up to v_i in the enumeration. For each variable v such that $l_{i,\Sigma}(v) \in S_i$ or $l_{i,\Delta}(v) \in S_i$ it is the case that $h \circ l_{i,\Sigma}(v) = l_{i,\Delta}(v)$. So, we do not just demand that for all variables in the enumeration up to v_i the fibring condition holds, we also demand this for all elements in their shadows that are solutions for some variables (maybe occurring only later in the enumeration).

3. $\mathrm{Stab}(l_{i,\Sigma}(V_{\bar{Z}})) \subseteq S_i \cup X_1, \mathrm{Stab}(l_{i,\Delta}(V_{\bar{Z}})) \subseteq S_i \cup X_2$. This is a technical condition on the stabilisers of solution elements for variables where one of the two solutions is a non-atom. It gives a strong control over the stabilisers

of non-atom solutions. Either they are in S_i, which means we have already taken care of them. Or they are in X_1 (respectively X_2), which means they are harmless.

Proving these properties for the base case poses no difficulties. The induction step is done by a complicated and technical argument by cases. We have two simple cases: the next variables $v_{i+1} \in V_Z$ (the set of all variables) were mapped to atoms in both solutions, or $v_{i+1} \in V_{\bar{Z}}$ and $h \circ l_{i,\Sigma}(v_{i+1}) = l_{i,\Delta}(v_{i+1})$. In both cases we define $l_{i+1,\Sigma} := l_{i,\Sigma}$ and $l_{i+1,\Delta} := l_{i,\Delta}$ and can check without great effort that Properties 1 to 3 hold for $i + 1$.

The important and interesting case is where $v_{i+1} \in V_{\bar{Z}}$ and $h \circ l_{i,\Sigma}(v_{i+1}) \neq l_{i,\Delta}(v_{i+1})$. We assume that $Lab(v_{i+1}) = \Sigma$, which implies that $y_{i+1} := l_{i,\Delta}(v_{i+1}) \in Z_2$, and $z_{i+1} := h \circ l_{i,\Sigma}(v_{i+1}) \in Z_2$ by definition of the fibres. The transposition that exchanges y_{i+1} and z_{i+1} can be extended to a unique automorphism τ_{i+1} of $\mathfrak{B}^{\Delta}_{\diamond}$. Thus we define $l_{i+1,\Delta} := \tau_{i+1} \circ l_{i,\Delta}$ and set $l_{i+1,\Sigma} := l_{i,\Sigma}$. Then, clearly, $l_{i+1,\Sigma}$ is a solution of $(\Gamma_{3,\Sigma}, L)$ and $l_{i+1,\Delta}$ one of $(\Gamma_{3,\Delta}, L)$. And $h \circ l_{i+1,\Sigma}(v_{i+1}) = l_{i+1,\Delta}(v_{i+1})$ by definition of $l_{i+1,\Delta}$. Showing that the Properties 1 to 3 still hold is technical. It is based on the following observations.

(a) $z_{i+1} \in Z_2 \setminus X$ and especially $z_{i+1} \notin X_2$ or X_1. This is a consequence of the fibring constructions.

(b) $z_{i+1} \notin S_i$. If z_{i+1} were in S_i, then $z_{i+1} = y_{i+1}$ by induction hypothesis.

(c) $z_{i+1} \notin \mathrm{Stab}(l_{i,\Sigma}(V_{\bar{Z}}))$ and $z_{i+1} \notin \mathrm{Stab}(l_{i,\Delta}(V_{\bar{Z}}))$, because, clearly, $\mathrm{Stab}(l_{i,\Sigma}(V_{\bar{Z}})) \subset S_i \cup X_1$ and $\mathrm{Stab}(l_{i,\Delta}(V_{\bar{Z}})) \subset S_i \cup X_2$.

(d) $y_{i+1} \in X_2, y_{i+1} \notin S_i$. $y_{i+1} \in S_i \cup X_2$ by induction hypothesis and $y_{i+1} \in S_i$ would imply $y_{i+1} = z_{i+1}$.

It is important to see what τ_{i+1} does. By (a) and (d), τ_{i+1} is the identity on X_0. Therefore for all $v \in V_Z : l_{i+1,\Delta}(v) = \tau_{i+1} \circ l_{i,\Delta}(v) = l_{i,\Delta}(v) = h \circ l_{i+1,\Sigma}(v) \in X_0$. This is Property 1.

By (c) and (d) and the above, τ_{i+1} is the identity on $\mathrm{Stab}(l_{i+1,\Sigma}(V))$ and therefore also on $l_{i+1,\Sigma}(V)$. That means τ_{i+1} does not "touch" the solution $l_{i+1,\Sigma}$. By (b) and (d), τ_{i+1} is also the identity on S_i. This is very important, since it states that all the 'nice' properties of S_i are preserved. This insight together with the fact that $S_{i+1} = S_i \cup \mathrm{Sd}(l_{i+1,\Delta}(v_{i+1}))$ can then be used to show Properties 2 and 3 the technicalities of which are not included in this paper.

For $i = n$ this induction shows there exists a solution $l_{n,\Sigma}$ of $\Gamma_{3,\Sigma}$ in $\mathfrak{A}^{\Sigma}_{\diamond}$ and a solution $l_{n,\Delta}$ of $\Gamma_{3,\Delta}$ in $\mathfrak{B}^{\Delta}_{\diamond}$. Furthermore $l_{n,\Delta} = h \circ l_{n,\Sigma}$, hence $l_{n,\Sigma}$ is also a solution of $\Gamma_{3,\Delta}$ in $\mathfrak{A}^{\Sigma\cup\Delta}_{\diamond} = \mathfrak{A}^{\Sigma} \otimes \mathfrak{B}^{\Delta}$. Therefore there exists a solution of Γ.

A Stronger Combination Result

We showed that the *existential* fragment of the free amalgamated product is decidable provided conjunctions of literals with linear constant restrictions are decidable in the components. A natural question to ask is: can we decide a larger quantifier prefix fragment than just the existential one? This issue appears even more interesting, since Baader and Schulz [1998] showed that for the case of purely positive constraints, the full positive theory, i.e., arbitrary quantifier prefixes, can be decided in the free amalgamated product. The answer we find is a negative one. We will provide counterexamples from the field of equational unification.

Solvability of disunification problems with linear constant restrictions for the free theory and the theory of an AC-function symbol is shown to be decidable by Baader and Schulz [1995]. Therefore the existential theory of the free algebra of one AC-symbol and any finite number of free function symbols is decidable. Even better, the full first order theory for both the free theory [Maher, 1988] and the theory of an AC-function symbol and any number of constants [Skolem, 1970] is decidable.

On the other hand, the Σ_3-fragment of the first order theory of an AC-function symbol, one free function symbol and one constant over the free algebra is shown to be undecidable by Treinen [1992].

Proposition 4.6 *There exists an instance of combination, namely the combination of the theory of an AC-symbol and the free theory, the Σ_3-fragment of which is undecidable, though the full first order theories of both components are decidable.*

5 FREE AMALGAMATION AND NEGATIVE CONSTRAINTS: GROUND SOLVABILITY

To discuss ground solvability, we first have to extend the notion of a ground solution known from equational unification to the more general case of quasi-free structures. In an equational theory, a solution is ground, if no solution term contains a variable. For quasi-free structures, this means that the stabiliser of every solution element has to be empty. We also define the notion of a restrictive solution for quasi-free structures. A corresponding notion was introduced by Baader & Schulz [1995] for equational disunification problems; they state that a solution is restrictive, if whenever a variable is assigned a complex term in the solution, then this term is not equivalent (modulo the equational theory) to a variable.

Definition 5.1 Let Γ be a constraint problem. A solution is called *ground*, iff every solution element has an empty stabiliser.

A solution σ is called *restrictive*, iff for every variable $v \in Var(\Gamma)$ such that $Lab(v) = \Sigma$: $\sigma(v)$ is not an atom.

Obviously, every ground solution is restrictive. On the basis of this definition we can now present the main result of this section.

Theorem 5.2 *Let $(\mathfrak{A}^{\Sigma}, X)$ and $(\mathfrak{B}^{\Delta}, X)$ be two quasi-free structures over disjoint signatures with infinitely many ground elements. The existence of a* *ground solution of mixed $\Sigma \cup \Delta$-constraint problems in the free amalgamated product $\mathfrak{A}^\Sigma \otimes \mathfrak{B}^\Delta$ is decidable, if the existence of restrictive solutions of pure constraint problems with generalised linear constant restrictions is decidable in both components $(\mathfrak{A}^\Sigma, X)$ and $(\mathfrak{B}^\Delta, X)$.*

It would of course be desirable to reduce ground solvability of mixed constraints in the free amalgam to ground solvability in the components. But the decomposition algorithm given in the last section does not permit this action. The algorithm is correct, i.e., the existence of ground solutions in the components implies the existence of a ground solution in the free amalgamated product. But unfortunately, the existence of a ground solution in the free amalgamated product does not imply the existence of ground solutions in the components. In [Baader and Schulz, 1995], Baader and Schulz give an example for equational disunification (4.2 on page 243) that demonstrates this fact. Therefore we have to demand the existence of restrictive solutions in the components in order to get ground solvability in the combined domain.

Proposition 5.3 *Assume that both $\mathfrak{A}^\Sigma$ and $\mathfrak{B}^\Delta$ contain an infinite number of ground elements. The input problem Γ has a ground solution, if and only if there exists a generalised linear constant restriction L such that both $(\Gamma_{3,\Sigma}, L)$ has a restrictive solution in $\mathfrak{A}^\Sigma$ and $(\Gamma_{3,\Delta}, L)$ has a restrictive solution in $\mathfrak{B}^\Delta$.*

6 INDEPENDENCE PROPERTIES OF EQUATIONAL THEORIES

In equational theories, the only negative constraints are of course disequations. For a general introduction to disunification, we refer the reader to [Comon, 1991].

Unification Type and the Independence Property

Theorem 6.1 *Let E be a unitary equational theory. Then E has the independence of negative constraints property.*

The reason for this theorem is that every solution σ_i of $\Gamma^+ \wedge C_i^-$ is an instance of μ, the most general unifier of Γ^+, and μ is also a solution of $\Gamma^+ \wedge C_i^-$. Hence μ is a unifier of $\Gamma^+ \wedge \Gamma^-$. It is noteworthy that there are non-unitary equational theories which have the independence property. An example is the theory *AMh* of Abelian Monoids with a homomorphism (see [Kepser, 1998]).

Next, we present the following statement: If an equational theory is such that one of its unification problems has a minimal complete set of unifiers of finite cardinality, but no single most general unifier, then this theory does not have the independence property. The incomparability of the solutions is used to construct disequations that can be solved individually, but not collectively.

Lemma 6.2 *Let E be a non-unitary equational theory. Let Γ^+ be a unification problem with variables X and $\{\mu_1, \ldots, \mu_n\}$ (where $n > 1$) be a minimal complete set of solutions of Γ^+ with domain X. Then there exists a set of disequations $\{d_1, \ldots, d_n\}$ such that for each $i = 1, \ldots, n$ the problem $\Gamma^+ \wedge d_i$ is solvable, but $\Gamma^+ \wedge \{d_1, \ldots, d_n\}$ is not.*

Instead of giving the proof, we explain the lemma by means of an example. Consider the infinitary equational theory A with $\circ$ being the associative function symbol and a a constant. The A-unification problem $\Gamma^+ := x \circ y \doteq a \circ a \circ a$ has exactly two A-unifiers: $\mu_1 := \{x \mapsto a, y \mapsto a \circ a\}$ and $\mu_2 := \{x \mapsto a \circ a, y \mapsto a\}$. If we define $d_1 := x \dot{\neq} a$, then $\Gamma^+ \wedge d_1$ clearly has a solution: μ_2. Also, if we set $d_2 := y \dot{\neq} a$, then $\Gamma^+ \wedge d_2$ has μ_1 as solution. But $\Gamma^+ \wedge d_1 \wedge d_2$ is unsolvable.

The next theorem can easily be proven with the above lemma.

Theorem 6.3 *Let E be an equational theory that has a unification problem where the cardinality of the minimal complete set of unifiers is finite and larger than 1. Then E does* not *have the independence property.*

Corollary 6.4 *Let E be a finitary equational theory. Then E does* not *have the independence property. The theory A of an associative function symbol and constants does* not *have the independence property.*

The theory A is an example of an infinitary theory without independence property.

Combining Equational Theories and the Independence Property

In this section, we present sufficient conditions under which the independence property of equational theories is preserved under combination. This result seems interesting in itself, since it shows that even for the specific case of equational theories the conditions are quite restrictive, as the following theorem shows.

Theorem 6.5 *Let E and F be two unitary regular and collapse-free theories over disjoint signatures. Then the combined theory $E \cup F$ is again unitary and hence has the independence property.*

It seems the requirement that E and F be regular and collapse-free[2] can hardly be weakened as the following example shows. Consider the theory of Boolean rings, which is neither regular nor collapse-free. It is known (see [Baader and Siekmann, 1994]) that in this theory, unification with constants is unitary. But general unification is finitary. And in any theory,

[2] An equational theory is *regular*, if in each axiom the left and the right side contain the same set of variables, and *collapse-free*, if it does not contain an axiom $x = t$ where x is a variable and t a non-variable term.

general unification can be regarded as an instance of combining unification with constants in that theory with syntactic unification.

The above demanded requirements are rather restrictive. There are only two "natural" theories which come to mind that are unitary and regular and collapse-free. One is of course syntactic unification. The other is single-sided distributivity, such as distributivity to the left ($D_L = \{f(g(x,y),z) = g(f(x,z), f(y,z))\}$) and distributivity to the right.

7 INDEPENDENCE PROPERTIES OF QUASI-FREE STRUCTURES

Before we can start to extend the results of the previous sections to quasi-free structures, it is necessary to generalise some basic notions of unification theory to quasi-free structures. Remember that our constraint problems are existentially quantified conjunctions of literals.

Definition 7.1 Let Γ be a positive constraint problem of the quasi-free structure $(\mathfrak{A}^\Sigma, X)$. Let σ and τ be two solutions of Γ. We say σ is *more general* than τ, or τ is an instance of σ, with respect to the variables in Γ (and write $\sigma \leq_\Gamma \tau$), iff there exists an endomorphism $m \in End_{\mathfrak{A}}^{\Sigma}$ such that for all $x \in Var(\Gamma)$ we have $m \circ \sigma(x) = \tau(x)$.

A quasi-free structure $(\mathfrak{A}^\Sigma, X)$ is called *unitary*, iff for every solvable positive constraint problem Γ there exists a solution μ such that every solution of Γ is an instance of μ.

Examples of unitary quasi-free structures are free algebras defined by unitary equational theories, since all instantiating substitutions are indeed endomorphisms. Rational tree algebras are also unitary.

Theorem 7.2 *Let $(\mathfrak{A}^\Sigma, X)$ be a unitary quasi-free structure. Then $(\mathfrak{A}^\Sigma, X)$ has the independence property.*

It is worth mentioning that Smolka and Treinen [1994] give an example of a quasi-free structure which is not an equational theory, is not unitary, but has the independence property, namely the algebra of feature trees with arity. This structure is interesting, because it has unique syntactic solved forms, although it is not unitary. This is a consequence of the sort system chosen by the authors. With a different sort system (e.g., a flat hierarchy) the structure would be unitary.

To lift the modularity result, we also have to generalise the notions "collapse-free" and "regular" to quasi-free structures.

Definition 7.3 A quasi-free structure $(\mathfrak{A}^\Sigma, X)$ is called *non-collapsing*, iff every endomorphism maps non-atoms to non-atoms, i.e., $m(a) \in A \setminus X$ for all $m \in End_{\mathfrak{A}}^{\Sigma}$ and all $a \in A \setminus X$.

In order to define the notion *regular* for quasi-free structures, we observe the following.

Lemma 7.4 *Let* $(\mathfrak{A}^{\Sigma}, X)$ *be a quasi-free structure,* $m \in End_{\mathfrak{A}}^{\Sigma}$ *an endomorphism, and* $a \in A$ *some element. Suppose* $\mathrm{Stab}^{\mathfrak{A}}(a) = \{x_1, \ldots x_k\}$. *Then* $\mathrm{Stab}^{\mathfrak{A}}(m(a)) \subseteq \bigcup_{i=1}^{k} \mathrm{Stab}^{\mathfrak{A}}(m(x_i)) = \mathrm{Stab}^{\mathfrak{A}}(m(\mathrm{Stab}^{\mathfrak{A}}(a)))$.

Definition 7.5 A quasi-free structure $(\mathfrak{A}^{\Sigma}, X)$ is *regular*, iff for all $m \in End_{\mathfrak{A}}^{\Sigma}$ and all $a \in A : \mathrm{Stab}^{\mathfrak{A}}(m(a)) = \mathrm{Stab}^{\mathfrak{A}}(m(\mathrm{Stab}^{\mathfrak{A}}(a)))$.

Now, we can give sufficient conditions for the component structures in order for the free amalgamated product to have the independence property.

Theorem 7.6 *Let* $(\mathfrak{A}^{\Sigma}, X)$ *and* $(\mathfrak{B}^{\Delta}, X)$ *be two unitary regular non-collapsing quasi-free structures over disjoint signatures. Then the free amalgamated product* $\mathfrak{A}^{\Sigma} \otimes \mathfrak{B}^{\Delta}$ *is again unitary and hence has the independence property.*

The theorem is based on the fact that there exists a *deterministic* combination algorithm for unitary regular and non-collapsing quasi-free structures, which was developed for the more specific case of the combination of equational theories by Schulz in [Schulz, 1996]. This algorithm computes most general solutions (see [Kepser, 1998]).

8 CONCLUSIONS

We analysed the role of negation in the combination of quasi-free structures. The contribution of this paper is twofold. In the first half, we showed that the existential fragment of the free amalgamated product $\mathfrak{A}^{\Sigma} \otimes \mathfrak{B}^{\Delta}$ is decidable provided that the solvability of conjunctions of literals with linear constant restrictions is decidable in both components $\mathfrak{A}^{\Sigma}$ and $\mathfrak{B}^{\Delta}$. The existence of a ground solution for a mixed $\Sigma \cup \Delta$ constraint problem in the free amalgamated product is decidable, if the existence of restrictive solutions of conjunctions of literals with linear constant restrictions is decidable in the components. This shows that the free amalgamated product is a suitable combined solution domain for combining constraint systems which have negation in their languages. The reader familiar with Rational Amalgamation, another general combination method introduced in [Kepser and Schulz, 1996], may be interested to hear that Rational Amalgamation is not very well suited for combining constraint systems with negation (see [Kepser, 1998]).

We also investigate the independence of negative constraints property in quasi-free structures and the free amalgamated product. Looking at the particular subclass of equational theories, we found that unitary theories have the independence property while finitary theories do not. We saw, too, that the union of two signature-disjoint, unitary, regular and collapse-free equational theories is again unitary and hence has the independence property. For the general case of quasi-free structures, we showed that unitary quasi-free structures have the independence property. We also obtained a modularity result which states that the free amalgamated product of two signature-disjoint, unitary, regular and non-collapsing quasi-free structures is again unitary, and thus has the independence property.

Acknowledgement

The author would like to thank in particular Klaus U. Schulz for suggestions and comments on drafts of this paper and Franz Baader for discussing certain aspects of it. We are also grateful to two anonymous referees whose comments and suggestions helped improve the paper.

This work was supported by a DFG grant (SPP "Deduktion") and by the Esprit working group 22457 – CCL II of the EU.

REFERENCES

[Aczel, 1988] Peter Aczel. *Non-wellfounded Sets.* Number 14 in CSLI Lecture Notes. CSLI, Stanford University, USA, 1988.

[Baader and Schulz, 1995] Franz Baader and Klaus U. Schulz. Combination techniques and decision problems for disunification. *Theoretical Computer Science*, 142:229–255, 1995.

[Baader and Schulz, 1998] Franz Baader and Klaus U. Schulz. Combination of constraint solvers for free and quasi-free structures. *Theoretical Computer Science*, 192:107–161, 1998.

[Baader and Siekmann, 1994] Franz Baader and Jörg H. Siekmann. Unification theory. In Dov M. Gabbay, Christopher J. Hogger, and John Alan Robinson, editors, *Handbook of Logic in Artificial Intelligence and Logic Programming*, volume 2, pages 41–125. Oxford University Press, 1994.

[Backofen and Treinen, 1994] Rolf Backofen and Ralf Treinen. How to win a game with features. In Jean-Pierre Jouannaud, editor, *1st International Conference on Constraints in Computational Logics*, volume 845 of *LNCS*, pages 320–335. Springer-Verlag, September 1994.

[Colmerauer, 1984] Alain Colmerauer. Equations and inequations on finite and infinite trees. In Institute for New Generation Computer Technology, editor, *Proceedings of the 2nd International Conference on Fifth Generation Computing Systems*, pages 85–99, Tokyo, 1984. Ohmsha et al.

[Colmerauer, 1990] Alain Colmerauer. An introduction to PROLOG III. *Communications of the ACM*, 33:69–90, 1990.

[Comon, 1991] Hubert Comon. Disunification: A survey. In Jean-Louis Lassez and Gordon Plotkin, editors, *Computational Logic*, pages 322–359. MIT Press, 1991.

[Henz *et al.*, 1993] Martin Henz, Gert Smolka, and Jörg Würtz. Oz-a programming language for multi-agent systems. In Ruzena Bajcsy, editor, *13th International Joint Conference on Artificial Intelligence*, volume 1, pages 404–409. Morgan Kaufmann Publishers, 1993.

[Kepser and Schulz, 1996] Stephan Kepser and Klaus U. Schulz. Combination of constraint systems II: Rational amalgamation. In *Principles and Practice of Constraint Programming, Proceedings CP96*, volume 1118 of *LNCS*. Springer–Verlag, 1996.

[Kepser, 1998] Stephan Kepser. *Combination of Constraint Systems.* PhD thesis, CIS, Universität München, 1998. Available at `ftp://ftp.cis.uni-muenchen.de/pub/kepser/ccl/diss.ps.gz`.

[Maher, 1988] Michael J. Maher. Complete axiomatizations of the algebras of finite, rational and infinite trees. In *3rd Logic in Computer Science Conference*. IEEE, 1988.

[Mal'cev, 1971] Anatolij Ivanovič Mal'cev. *The Metamathematics of Algebraic Systems*, volume 66 of *Studies in Logic*. North-Holland Publishing Company, 1971.

[Rounds, 1988] William C. Rounds. Set values for unification based grammar formalisms and logic programming. Technical Report CSLI-88-129, CSLI, Stanford University, 1988.

[Schulz, 1996] Klaus U. Schulz. Combining unification and disunification algorithms—tractable and intractable instances. Technical Report CIS-Bericht-96-99, CIS, Universität München, 1996.

[Skolem, 1970] Thoralf Skolem. *Selected Works in Logic.* Universitetsforlaget, Oslo, 1970.

[Smolka and Treinen, 1994] Gert Smolka and Ralf Treinen. Records for logic programming. *Journal for Logic Programming*, 18(3):229–258, 1994.

[Treinen, 1992] Ralf Treinen. A new method for undecidability proofs of first order theories. *Journal of Symbolic Computation*, 14:437–457, 1992.

Optimisation Techniques for Combining Constraint Solvers

Stephan Kepser and Jörn Richts

1 INTRODUCTION

One idea behind constraint solving is to use specialised formalisms and inference mechanisms to solve domain-specific tasks. In many applications, however, one is faced with a complex combination of different problems. Therefore constraint solvers tailored to solving a single problem can only be applied, if it is possible to combine them with others. Concrete examples of the combination of constraint solvers can be found, e.g., in [Dovier *et al.*, 1996; Colmerauer, 1990]. In a recent paper [Baader and Schulz, 1998], Baader and Schulz present a general method for the combination of constraint systems. Their method is applicable to a large class of structures, the so-called quasi-free structures. Quasi-free structures comprise many important infinite non-numerical solution domains such as (quotient) term algebras [Malćev, 1971], rational tree algebras [Colmerauer, 1984], vector spaces, hereditarily finite wellfounded and non-wellfounded lists, sets [Aczel, 1988] and multi sets as well as certain types of feature structures [Smolka and Treinen, 1994]. The combined solution domain the authors present in [Baader and Schulz, 1998], the so called free amalgamated product, has the characterising property of being the most general combination in the sense that every other combined domain contains a homomorphic image of it.

The question of how to combine specialised methods was first discussed in the field of unification theory (see [Baader and Siekmann, 1994] for an overview). Equational unification algorithms, which can be seen as an instance of constraint solvers, were built into resolution-based theorem provers [Plotkin, 1972] and rewriting engines [Jouannaud and Kirchner, 1986] to improve their handling of equality. Since the unification problems occurring in these applications usually contain function symbols from various equational theories, the question of how to combine equational unification algorithms became important. For algorithms that compute complete sets of unifiers for equational theories over disjoint signatures, this problem was solved by Schmidt-Schauß [1989] and Boudet [1993]. With the development of constraint-based approaches to theorem proving [Bürckert, 1990; Nieuwenhuis and Rubio, 1994] and rewriting [Kirchner and Kirchner, 1989], the interest in combining unification algorithms extended towards combinations of decision procedures, for which Baader and Schulz [1996] finally presented a general algorithm.

As a generalisation of the one given in [Baader and Schulz, 1996], the algorithm for combining constraint solvers in [Baader and Schulz, 1998] inherits the old weakness of being so highly non-deterministic that it is of very limited practical use. The aim of this paper is to provide optimisation techniques for the combination algorithm by Baader and Schulz that make the combination of constraint solvers practically usable and are still general enough to be applicable to a large class of constraint solvers. The method we propose is the so called deductive method which is based on the insight that many decisions of the combination algorithm do not really need to be made non-deterministically, but can rather be deduced on the base of the constraint domains involved, the input problem and other decisions made earlier. In our deductive combination method the component solvers are consulted to gain information on what further steps can be made deterministically. This obviously requires component solvers capable of providing this information. The strength of this combination method lies in the interchange of information between the component algorithms. The impact of this interchange is highlighted by the fact that, although developed for the more general case, our combination algorithm turns out to be an implementation of the *PTIME* combination algorithm given in [Schulz, 1996; Kepser, 1998] for a special subclass of constraint solvers. We also present a selection strategy for choosing the next non-deterministic decision. In order to detect unsolvability of a single component faster, we first make all non-deterministic decisions relative to one component before we proceed to the next one. The run time tests we present in this paper show the enormous effect of our optimisation methods, making us confident that combination of constraint solvers is feasible in practice.

In this paper, we present our combination method as an algorithm for combining constraint solvers, but our optimisation techniques are nevertheless useful for the special case of equational unification. Moreover our method can be directly extended to compute complete sets of unifiers.

2 PRELIMINARIES

Quasi-free Structures and the Free Amalgamated Product

A signature Σ consists of a set Σ_F of function symbols and a disjoint set Σ_P of predicate symbols (not containing "="), each of fixed arity. Σ-structures over the carrier set A are denoted by $\mathfrak{A}^\Sigma$. Σ-terms $(t, t_1, \ldots)$ and atomic Σ-formulae (of the form $t_1 = t_2$, or of the form $p(t_1, \ldots, t_n)$) are built as usual from Σ and a countable set of variables $\mathcal{V}$. A Σ-formula φ is written in the form $\varphi(v_1, \ldots, v_n)$ in order to indicate that the set $Var(\varphi)$ of free variables of φ is a subset of $\{v_1, \ldots, v_n\}$. A mapping $\sigma : \mathcal{V} \to A$ from the set of variables to the carrier set of $\mathfrak{A}^\Sigma$ is called an *assignment.* A *constraint problem over signature* Σ is a set of atomic Σ-formulae. An assignment σ is a *solution* for a constraint problem Γ in $\mathfrak{A}^\Sigma$ iff $\varphi(\sigma(v_1), \ldots, \sigma(v_n))$ becomes true in $\mathfrak{A}^\Sigma$ for all formulae $\varphi(v_1, \ldots, v_n) \in \Gamma$.

Σ-homomorphisms and Σ-endomorphisms are defined as usual, see e.g., [Malćev, 1971]. With $End_{\mathfrak{A}^\Sigma}$ we denote the monoid of all endomorphisms of $\mathfrak{A}^\Sigma$, with composition as operation.

We will now introduce the solution domains for constraint solving we consider here, namely quasi-free structures. Quasi-free structures, a generalisation of free structures, were introduced by Baader and Schulz [1998]. We consider a fixed Σ-structure $\mathfrak{A}^\Sigma$.

Let A_0, A_1 be subsets of $\mathfrak{A}^\Sigma$. Then A_0 *stabilises* A_1 iff all elements m_1 and m_2 of $End_{\mathfrak{A}^\Sigma}$ that coincide on A_0 also coincide on A_1. For $A_0 \subseteq A$ the *stable hull* of A_0 is the set $\mathrm{SH}^{\mathfrak{A}}(A_0) := \{a \in A \mid A_0 \text{ stabilises } \{a\}\}$.

$\mathrm{SH}^{\mathfrak{A}}(A_0)$ is always a Σ-substructure of $\mathfrak{A}^\Sigma$, and $A_0 \subseteq \mathrm{SH}^{\mathfrak{A}}(A_0)$. The stable hull of A_0 can be larger than the Σ-subalgebra generated by A_0.

The set $X \subseteq A$ is an *atom set* for $\mathfrak{A}^\Sigma$ if every mapping $X \to A$ can be extended to an endomorphism of $\mathfrak{A}^\Sigma$.

Definition 2.1 A countably infinite Σ-structure $\mathfrak{A}^\Sigma$ is a *quasi-free structure* iff $\mathfrak{A}^\Sigma$ has an infinite atom set X where every $a \in A$ is stabilised by a finite subset of X. We denote this quasi-free structure by $(\mathfrak{A}^\Sigma, X)$.

The class of quasi-free structures contains many important non-numerical infinite solution domains. For example, all free structures (see, e.g., [Malćev, 1971]), rational tree algebras ([Colmerauer, 1984]), feature structures with arity ([Smolka and Treinen, 1994]), domains with nested, finite or rational lists (rational lists are used in Prolog III, see [Colmerauer, 1990]), and domains with nested, finite or rational sets ([Aczel, 1988]) are quasi-free structures. For details we refer to [Baader and Schulz, 1998].

A fundamental property of quasi-free structures is the following: for each $a \in A$ there exists a *unique minimal* finite set $Y \subseteq X$ such that $a \in \mathrm{SH}^{\mathfrak{A}}(Y)$. The *stabiliser* of $a \in A$, $\mathrm{Stab}^{\mathfrak{A}}(a)$, is the unique minimal finite subset Y of X such that $a \in \mathrm{SH}^{\mathfrak{A}}(Y)$. The stabiliser of $A' \subseteq A$ is the set $\mathrm{Stab}^{\mathfrak{A}}(A') := \bigcup_{a \in A'} \mathrm{Stab}^{\mathfrak{A}}(a)$.

We extend the notions *regular* and *collapse-free*, known from equational unification, to quasi-free structures.

Definition 2.2 A quasi-free structure $(\mathfrak{A}^\Sigma, X)$ is called *collapse-free*, iff every endomorphism maps non-atoms to non-atoms, i.e., $m(a) \in A \setminus X$ for all $m \in End_{\mathfrak{A}^\Sigma}$ and all $a \in A \setminus X$. The quasi-free structure $(\mathfrak{A}^\Sigma, X)$ is *regular*, iff for all $m \in End_{\mathfrak{A}^\Sigma}$ and all $a \in A : \mathrm{Stab}^{\mathfrak{A}}(m(a)) = \mathrm{Stab}^{\mathfrak{A}}(m(\mathrm{Stab}^{\mathfrak{A}}(a)))$.

Note that $m(\mathrm{Stab}^{\mathfrak{A}}(a))$, the image of $\mathrm{Stab}^{\mathfrak{A}}(a)$ under m, can contain non-atoms; therefore we have to apply $\mathrm{Stab}^{\mathfrak{A}}$ again.

Baader and Schulz [1998] present a combined solution domain of two or more quasi-free structures, the so-called free amalgamated product, which is characterised amongst all combined solution domains as being the most general in the sense that every domain contains a homomorphic image of it. The authors also provide a construction method to obtain the free amalgamated product of arbitrary quasi-free structures. If $(\mathfrak{A}_1^{\Sigma_1}, X), \ldots, (\mathfrak{A}_n^{\Sigma_n}, X)$ are n quasi-free structures over paiwise disjoint signatures, we write $\mathfrak{A}_1^{\Sigma_1} \otimes \ldots \otimes \mathfrak{A}_n^{\Sigma_n}$ for their free amalgamated product. If the quasi-free structures one combines are free algebras defined by equational theories over disjoint signatures, then their free amalgamated product is the free algebra defined by the theory over the union of the axiom sets.

In this paper, we investigate "mixed" constraint problems. For $i = 1, \ldots, n$ $(n \geq 2)$, let Σ_i be pairwise disjoint signatures and let $(\mathfrak{A}_i^{\Sigma_i}, X)$ be a quasi-free structure over signature Σ_i. A "mixed" constraint problem is a conjunction of atomic formulae over the joined signature $\Sigma_1 \cup \ldots \cup \Sigma_n$. A constraint problem Γ is in *decomposed form*, if Γ has the form $\bigcup_{i=1}^n \Gamma_i$ where each Γ_i is a pure constraint problem over the signature Σ_i. Any constraint problem Γ can be transformed into a constraint problem in decomposed form that is solvable, iff the original problem is solvable, by a simple deterministic preprocessing step (variable abstraction, see [Baader and Schulz, 1996]). In the following, we will therefore always assume that a constraint problem is in decomposed form $\bigcup_{i=1}^n \Gamma_i$.

Only variables occurring in more than one component system Γ_i have to be considered by the combination algorithm. Hence we define the set of *shared variables* $\mathcal{U} := \{x \mid \exists i, j : i \neq j, x \in \mathit{Var}(\Gamma_i) \cap \mathit{Var}(\Gamma_j)\}$. The combination algorithm presented in the next section imposes some restrictions on the shared variables in order to prevent conflicts between the solutions of the component structures (like a variable being assigned to different elements by solutions of different structures). The solutions of the component problems Γ_i have to obey these so-called *linear constant restrictions.*

Definition 2.3 A *linear constant restriction* $L = (\Pi, \mathit{Lab}, <_L)$ for variables $\mathcal{U}$ consists of a partition[1] Π of $\mathcal{U}$, a labelling function $\mathit{Lab} : \mathcal{U}/_\Pi \rightarrow \{\Sigma_1, \ldots, \Sigma_n\}$ and a linear order $<_L$ on $\mathcal{U}/_\Pi$. We use $\mathit{Lab}(x)$ and $x <_L y$ instead of $\mathit{Lab}([x]_\Pi)$ and $[x]_\Pi <_L [y]_\Pi$.

An assignment σ of $\mathcal{U}$ into $\mathfrak{A}_i^{\Sigma_i}$ is a *solution* for the *constraint problem with linear constant restrictions* (Γ_i, L) in $(\mathfrak{A}_i^{\Sigma_i}, X)$, iff it is a solution for Γ_i and for each $x, y \in \mathcal{U}$:

- $\sigma(x) = \sigma(y)$ if $x \equiv_\Pi y$,
- $\sigma(x) \in X$ if $\mathit{Lab}(x) \neq \Sigma_i$, and

[1] The equivalence relation induced by Π is denoted by $\equiv_\Pi$, $[x]_\Pi$ is the equivalence class of a variable x, and $\mathcal{U}/_\Pi$ is the set of all equivalence classes of variables in $\mathcal{U}$.

◇ $\sigma(x) \notin \mathrm{Stab}^{\mathfrak{A}}(\sigma(y))$ if $Lab(x) \neq \Sigma_i, Lab(y) = \Sigma_i, y <_L x$.

Intuitively speaking, item two guarantees that all variables receiving a label different from Σ_i are treated as constants by σ. By item three, the use of these constants in σ is further restricted in order to prevent cycles. Two linear constant restrictions L_1 and L_2 over $\mathcal{U}$ are called *equivalent*, if they have identical partitions and labelling functions and their orders differ at most in the ordering of variables with an identical label. This definition induces an equivalence relation on all linear constant restrictions for a given set of variables $\mathcal{U}$. If L_1 and L_2 are equivalent and an assignment σ solves (Γ, L_1), then σ also solves (Γ, L_2).

The Original Combination Algorithm

In the following we describe the combination algorithm given by Baader and Schulz in [1998]. Here we give a straightforward generalisation of this algorithm to the case where more than two structures are combined. Additionally, we include basic optimisations similar to those described in [Baader and Schulz, 1996].

Let Γ be a constraint problem in decomposed form. We assume the constraints in Γ are connected by shared variables, i.e., there is no partition $\Gamma = \Gamma' \cup \Gamma''$ where Γ' and Γ'' do not have variables in common. Otherwise Γ' and Γ'' can be solved separately. The algorithm consists of three non-deterministic steps which result in a linear constant restriction for the constraint problem.

Step 1: Variable identification. Non-deterministically choose a partitioning Π of $\mathcal{U}$.

Step 2: Labelling. Non-deterministically choose a labelling function $Lab : \mathcal{U}/_\Pi \to \{\Sigma_1, \ldots, \Sigma_n\}$.

Step 3: Ordering. Non-deterministically choose a linear order $<_L$ on $\mathcal{U}/_\Pi$. $L = (\Pi, Lab, <_L)$ constitutes a linear constant restriction. Note that for each equivalence class of linear constant restrictions, it suffices to choose just one member. The output tuple determined by these three steps is $((\Gamma_1, L), \ldots, (\Gamma_n, L))$.

Theorem 2.4 *The input problem Γ has a solution in the free amalgamated product $\mathfrak{A}_1^{\Sigma_1} \otimes \ldots \otimes \mathfrak{A}_n^{\Sigma_n}$, if and only if there exists an output tuple $((\Gamma_1, L), \ldots, (\Gamma_n, L))$ such that for each $i = 1, \ldots, n$, the constraint problem with linear constant restriction (Γ_i, L) has a solution in $\mathfrak{A}_i^{\Sigma_i}$.*

Decision Sets

The original algorithm makes all non-deterministic decisions first, and only thereafter does it call the component algorithms to determine whether the input problem with the thus chosen constant restriction is solvable. Our optimisations interleave these two parts. Hence we have to deal with linear constant restrictions which are only partially specified, i.e., restrictions representing the choices already made but making no statements about the decisions still open. In order to describe these partial constant restrictions and to have a framework for describing our optimisations on a formal level we introduce the notion of decision sets. A decision describes a single non-deterministic choice. There exist five different types of decisions.

Definition 2.5 Let $\mathcal{U}$ be the set of variables. A *decision* is an expression of the form $x \doteq y$, $x \not\doteq y$, $x \dot{\leq} y$, $x \dot{\mapsto} \Sigma_i$, or $x \dot{\not\mapsto} \Sigma_i$, where $x, y \in \mathcal{U}$ and $1 \leq i \leq n$. The decision $x \dot{<} y$ is used as an abbreviation for $x \dot{\leq} y, x \not\doteq y$.

We speak about sets of decisions (for a set of variables $\mathcal{U}$) which are—as usual—read conjunctively. In order to represent the two options when making a non-deterministic choice, we define the negation of a decision.

Definition 2.6 Let d be a decision. Its *negation* $\neg d$ is defined as follows:

$$\begin{aligned} \neg x \doteq y &:= x \not\doteq y, & \neg x \not\doteq y &:= x \doteq y, \\ \neg x \dot{\mapsto} \Sigma_j &:= x \dot{\not\mapsto} \Sigma_j, & \neg x \dot{\not\mapsto} \Sigma_j &:= x \dot{\mapsto} \Sigma_j, \\ \neg x \dot{\leq} y &:= y \dot{<} x. \end{aligned}$$

These rules of negation reflect the three non-deterministic steps of the algorithm: Two variables have to be identified or treated as different variables; each variable has to be treated as a variable or a constant in a particular component system; and two variables with distinct labels have to be ordered in one way or the other. In the following we formally define this correspondence between sets of decisions and linear constant restrictions.

Definition 2.7 Let $\mathcal{U}$ be a set of variables. A linear constant restriction $L = (\Pi, Lab, <_L)$ over $\mathcal{U}$ *satisfies* a decision set D, if the following holds:

$$\begin{aligned} x \equiv_\Pi y \;\text{ if }\; & x \doteq y \in D, & x \not\equiv_\Pi y \;\text{ if }\; & x \not\doteq y \in D, \\ Lab(x) = \Sigma_i \;\text{ if }\; & x \dot{\mapsto} \Sigma_i \in D, & Lab(x) \neq \Sigma_i \;\text{ if }\; & x \dot{\not\mapsto} \Sigma_i \in D, \\ x <_L y \text{ or } x \equiv_\Pi y \;\text{ if }\; & x \dot{\leq} y \in D. \end{aligned}$$

The set of linear constant restrictions satisfying D is denoted by $\mathcal{L}(D)$. A set D is called *inconsistent* if $\mathcal{L}(D) = \emptyset$.

So, the decisions are interpreted by a linear constant restriction in a straightforward way. We can now use decision sets to represent constraint problems with partially specified linear constant restrictions.

Definition 2.8 A *constraint problem with decision set* (Γ, D) consists of a constraint problem Γ together with a set of decisions D. An assignment σ is a *solution* of (Γ, D) if σ is a solution of (Γ, L) for some $L \in \mathcal{L}(D)$.

Since decision sets represent linear constant restrictions, they inherit some properties like $\dot{<}$ representing an ordering. This is reflected by the following definition.

Definition 2.9 A decision set D is called *closed* if $D = \{d \mid \text{every } L \in \mathcal{L}(D) \text{ satisfies } \{d\}\}$.

This definition implies that for each decision set D there is exactly one closed set which is equivalent to D; this set is called the *closure* of D. This closure can be computed efficiently; one has to consider that $\doteq$ denotes a congruence, $\dot{<}$ stands for an ordering, and $x \dot{\mapsto} \Sigma_i$ represents a functional relation. For example, a closure always contains $x \doteq x$ for all variables $x \in \mathcal{U}$, the two decisions $x \doteq y \in D$ and $y \dot{<} z \in D$ imply that $x \dot{<} z$ is in the closure of D, and the closure of $\{x \dot{\mapsto} \Sigma_i\}$ contains $x \dot{\not\mapsto} \Sigma_j$ for all $i \neq j$. In the following we will always assume that sets of decisions are closed, i.e., when adding decisions to a set we assume that the closure is formed immediately.

We need a criterion to tell when a set of decisions already represents one linear constant restriction, i.e., when no more decisions have to be made.

Definition 2.10 A set of decisions D is *complete*, if all linear constant restriction in $\mathcal{L}(D)$ are equivalent.

From this definition and the one above it follows that there is a one-to-one correspondence between the equivalence classes of linear constant restrictions over $\mathcal{U}$ and closed and complete sets of decisions for $\mathcal{U}$. In order to test inconsistency and completeness of decision sets by an algorithm, we need a syntactic formulation of these properties. This is provided by the following lemma.

Lemma 2.11

1. *A closed set of decisions D is inconsistent iff $d \in D$ and $\neg d \in D$ for some decision d.*
2. *A closed and consistent set of decisions D (for variables $\mathcal{U}$) is complete iff for all $x, y \in \mathcal{U}$*
   either $x \doteq y \in D$ *or* $x \dot{\neq} y \in D$, *and*
   either $x \dot{<} y \in D$ *or* $y \dot{<} x \in D$ *if* $x \dot{\mapsto} \Sigma_i, y \dot{\not\mapsto} \Sigma_i \in D$, *and*
   $x \dot{\mapsto} \Sigma_i \in D$ *for one* Σ_i.

3 DEDUCTIVE METHOD

In this section we will show how information deduced from the component systems and their individual structures can be used to prune the search space. The power of the method lies in the interchange of this information between the components.

Interchanging Decisions

A severe disadvantage of the original combination algorithm is that all non-deterministic decisions are made blindfoldedly without respecting the requirements that the component structures may impose. For example, if a component structure $\mathfrak{A}_i$ is collapse-free and the problem contains an equation

$x = f(\ldots y \ldots)$ where $f \in \Sigma_i$, then x must receive label Σ_i. If $\mathfrak{A}_i$ is also regular then the problem is unsolvable if $y \,\dot{\not\mapsto}\, \Sigma_i \in D$ and $x \dot{<} y \in D$. Hence the algorithm can choose $x \dot{\mapsto} \Sigma_i \in D$ deterministically and take into account that $y \,\dot{\not\mapsto}\, \Sigma_i \in D$ implies $y \dot{<} x \in D$.

As the example shows, some decisions that have been deduced earlier in one component can be used to deduce new decisions in another component. This possible interplay between different structures suggests that one should use a method where component algorithms computing new decisions are called alternately in the beginning of the combination algorithm and whenever a non-deterministic choice has been made: Starting with some initial decisions, each component algorithm computes new decisions; these new decisions are added to the current set of decisions, which is used when calling the other component algorithms. When this process comes to an end because no new decisions can be deduced, the next non-deterministic choice has to be made by the combination algorithm. After this choice the process of computing new consequences can be started again. At any step of computing the consequences, a component algorithm may return that its subproblem has become unsolvable with the current set of decisions. Thereby, unsolvable branches of the search tree can be detected earlier.

Obviously, this method requires new component algorithms that are capable of computing consequences implied by the component structures, the problem, and the decisions computed so far. A structure for which such an algorithm does not exist can still be used in this method, but it cannot contribute to the deductive process. It is clearly the quality of the deductive component algorithms that determines the amount of optimisation achieved. The optimisations of our component algorithms go quite beyond using only syntactic properties of structures as in the example above. The goal is to deduce as much information as is possible with reasonable effort.

The Algorithm

First we define the task of the new deductive component algorithms. Their input is a pure constraint problem and a set of decisions which need not be complete. The result is a set of decisions that follows from the constraint problem and the input decisions. If the input is unsolvable, the result may also be an inconsistent set of decisions.

Definition 3.1 Let (Γ, D) be a constraint problem with decision set C. The decision set is a *consequence* of (Γ, D), iff it is contained in every complete decision set $D' \supseteq D$ such that (Γ, D') is solvable, that is, iff

$$C \subseteq \bigcap \{D' \mid D \subseteq D', D' \text{ is complete, and } (\Gamma, D') \text{ is solvable}\}.^2$$

[2] $\bigcap\{\}$ is the (inconsistent) set of all decisions over $\mathcal{U}$.

Note that $C = \emptyset$ is always a consequence and that the consequence need not be inconsistent if (Γ, D') is unsolvable for all complete extensions D' of D. Therefore, the standard algorithms for constraint solving with linear constant restrictions must be called in the end when a complete set of decisions is reached. See Section 4 for a discussion on how deductive component algorithms co-operate with standard ones.

```
D := ∅
loop: Repeat
    Deduce consequences:
    Repeat
        For each system i
            call the component algorithm of system i to calculate
            new consequences C of (Γ_i, D),
            set the new current set of decisions D := D ∪ C.
            If D is inconsistent
                break loop.        /* exit from outer loop */
    Until no component algorithm computes new decisions.
    If D is not complete
        Select next choice:
        Select a decision d ∉ D such that D ∪ {d} is consistent.
        Non-deterministically choose either
        D := D ∪ {d} or
        D := D ∪ {¬d}.
Until D is complete.
Return D.
```

Figure 1: The deductive combination algorithm

Figure 1 shows the combination algorithm. As before, we present the method as a non-deterministic algorithm, i.e., the algorithm contains non-deterministic steps for which both alternatives have to be regarded. In the algorithm, D denotes the current set of decisions. The termination condition in case of success is that the set of decisions is complete, as given in Lemma 2.11.

Proposition 3.2 *The input problem Γ is solvable, iff the algorithm computes a consistent set D such that for each $i = 1, \ldots, n$ the constraint problem with decision set (Γ_i, D) is solvable.*

Again, testing (Γ_i, D) for solvability can be performed by the component algorithms used in the original combination algorithm. Since a consequence is a decision that is contained in every solvable complete decision set, it is clear that we prune those branches of the search space that are unsolvable. Hence correctness of the algorithm is an immediate consequence of the correctness of the original combination algorithm in Theorem 2.4.

The deductive method additionally allows us to reduce certain redundancies in the search space. We can prune some solvable branches that would only lead to redundant solutions. For example, let $\Gamma_1 = \{x = a, y = a\}$ and $\Gamma_2 = \{z = x + y\}$ where $+$ is associative and commutative. Clearly, $Lab(x) = Lab(y) = \Sigma_1$ and $Lab(z) = \Sigma_2$. And the order must be such that x and y are below z. But there are two different partitions that lead to a solution: We can identify x and y or leave them different. The resulting solution looks the same in both cases. Hence we compute only one partition. Other more complicated examples occur in ordering decisions. It turns out that sometimes it is useful to order variables of the same label to avoid the computation of superfluous orders that only lead to redundant solutions. A longer discussion of this side issue would be beyond the scope of this paper.

Deterministic Combination

It is interesting to observe that there exists a class of constraint systems for which the deductive combination algorithm has *PTIME* complexity, which entails that all steps can be made deterministically. In [Schulz, 1996], Schulz gives a general description of a *PTIME* combination algorithm for certain equational theories. This algorithm is extended to the combination of quasi-free structures in [Kepser, 1998]. The class of structures that are deterministically combinable is quite restricted. Currently, only unitary regular collapse-free structures are known to belong to it.

Although our deductive component algorithm is designed for the general case, it turns out to be an implementation of the deterministic algorithm when applied to component algorithms satisfying the conditions imposed in [Schulz, 1996] and [Kepser, 1998]. Our component algorithms for unification in the empty theory, for rational tree algebras, and for feature structures meet these conditions. Thus, when applied to these structures, our combination algorithm runs deterministically. This deterministic behaviour shows the great impact of interchanging decisions between component algorithms.

4 COMPONENT ALGORITHMS

In order to prune the search space significantly, new component algorithms are needed for the deductive method. When designing these algorithms one should take into account the special way in which they are called. Constraint solvers are usually designed to work incrementally (e.g., [Colmerauer, 1990]). But standard unification algorithms are "one shot" algorithms: they are started only once with all necessary information given and they compute the final results. Deductive component algorithms must be able to cope with partial information and deliver a meaningful but not necessarily the final result. More importantly, when receiving new information the algorithms should not restart computation from scratch but rather continue on the base of their prior internal states. Otherwise, the search space would be partially shifted from the combination algorithm to the deductive component algorithms. The same holds for the standard component algorithms for problems with linear constant restrictions that perform a complete test at the end of the combination algorithm: they should take into account the information already computed by the corresponding deductive component algorithms.

Note that there is no need for completeness in the deductive component algorithm: the algorithm need not compute all decisions implied by the input and it need not return an inconsistent set if the problem is unsolvable. Thus an algorithm always returning the empty set would be correct, though it would not contribute to the deductive process. This, however, enables us to use every structure that is suitable for the original algorithm. In the other extreme it might not be advisable to compute new decisions at any cost; there should be a careful consideration between optimisations of the combination algorithm resulting from new decisions and a higher complexity of the deductive component algorithm.

We have developed deductive component algorithms for the free theory, A, AC, and ACI and for rational trees and feature structures. A detailed description of these algorithms would be beyond the scope of this paper. In the following, we outline the ideas underlying the algorithms for the free theory, a theory in which one can deduce many decisions, and for ACI as a more complicated example.

Syntactic Unification

The deductive algorithm for the free theory is based on the quasi-linear algorithm described in [Baader and Siekmann, 1994] where terms and unifiers are represented as directed acyclic graphs. We assume that the reader is familiar with this representation. When the deductive component algorithm is called for the first time, the dag is built, which is then used again for all further calls of this component algorithm. Decisions of the form $x \dot{\neq} y$, $x \dot{\mapsto} \Sigma_i$, $x \dot{\not\mapsto} \Sigma_i$, or $x \dot{\leq} y$ do not initiate any computation. Only identification decisions $x \doteq y$ cause a call of the corresponding unification procedure, which updates the existing dag. The decision set to be returned by the component algorithm can be computed from the dag: $x \doteq y$ is returned if x and y are identified in the dag; $x \dot{\mapsto} \Sigma_{Free}$ is returned if x is connected to a non-variable term; $x \dot{<} y$ is returned if x can be reached from y. Additionally $x \dot{\neq} y$ is returned if x and y are *certainly not unifiable*. The algorithm does not test real unifiability of x and y since it would be too costly to do this for all pairs of variables; instead it tests if the variables are connected to non-variable terms with different top symbol. The dag is also used by the decision procedure for problems with linear constant restrictions. This algorithm works in exactly the same manner as the deductive component algorithm, except that it does not compute a decision set but returns solvable or unsolvable.

The deductive algorithm for rational trees works similarly to this algorithm. It does not perform an occur-check and it returns $x \dot{<} y$ only if x can be reached from y and y has been labelled by another structure.

The Theory *ACI*

In the theory of Abelian monoids, *ACI*, the binary function symbol + is associative, commutative and idempotent. In [Kapur and Narendran, 1992], an algorithm was given that decides solvability of *ACI*-unification with constants. The main idea is to set up Horn clauses which describe the solvability of the equations. The Horn clauses are built from propositional variables $P_{x,a}$ which are true iff the constant a does not occur in a solution for the variable x. A clause $P_{x,a} \wedge P_{y,a} \Rightarrow \mathsf{False}$ means that the problem is unsolvable if a appears neither in x nor in y, or equivalently: if we can deduce that a does not occur in x, then it must appear in y.

We extend the algorithm given in [Kapur and Narendran, 1992] for our situation where the set of variables and constants is not fixed in the beginning. In this way we prevent new Horn clauses from having to be set up when a new labelling decision is made. Let $\mathcal{V}_{ACI}$ be the set of variables in Γ_{ACI}; note that there are no constants in Γ_{ACI}. We introduce a new constant $\bar{x}$ for each variable $x \in \mathcal{V}_{ACI}$ and construct two types of Horn clauses:

- $\bigwedge_{y \in \mathcal{V}_{ACI}} P_{x,\bar{y}} \Rightarrow \mathsf{False}$ for each variable $x \in \mathcal{V}_{ACI}$,
- $P_{x_1,\bar{y}} \wedge \ldots \wedge P_{x_k,\bar{y}} \Leftrightarrow P_{u_1,\bar{y}} \wedge \ldots \wedge P_{u_l,\bar{y}}$
  for each $y \in \mathcal{V}_{ACI}$ and each equation
  $x_1 + \ldots + x_k = u_1 + \ldots + u_l \in \Gamma_{ACI}$.

The first type of clauses guarantees that the solution for each variable contains at least one constant. The second type represents the equations of Γ_{ACI}: if a constant does not appear on the left hand side, it must not appear on the right hand side, and vice versa. A decision $x \not\mapsto \Sigma_{ACI}$ introduces the Horn clauses $P_{x,\bar{x}} \Rightarrow \mathsf{False}$ and $\Rightarrow P_{x,\bar{y}}$ for each $y \in \mathcal{V}_{ACI}$ with $y \neq x$, i.e., the propositional variables are set to False and True, respectively. The effect of these clauses is that $\bar{x}$ is the only constant that appears in x, i.e., x is identified with $\bar{x}$ and is treated like a constant by the algorithm. A decision $x \dot{<} y$ causes the atom $P_{x,\bar{y}}$ to be set to True.

The constraint problem with linear constant restrictions is solvable iff the set of Horn clauses is solvable. This can be tested efficiently by an algorithm which constructs a graph from the Horn clauses and propagates True and False through this graph (see [Kapur and Narendran, 1992]). The set of Horn clauses (and the corresponding constraint problem) is unsolvable if True meets False during this propagation. New decisions can be deduced from the atoms mapped to True or False: $x \dot{\mapsto} \Sigma_{ACI}$ is returned if $P_{x,\bar{y}}$ is set to False and

$x \dot{\neq} y$ has been already deduced or if $P_{x,\bar{y}}$ and $P_{x,\bar{z}}$ have been set to False for three different variables x, y, and z. The decision $x \dot{<} y$ is returned if $P_{x,\bar{y}}$ has been set to False with $x \neq y$.

Like the dag for syntactic unification, the Horn clauses and the state of the propositional variables are stored and used again for each further call of the component algorithm; only when a new identification decision $x \doteq y$ is deduced by another component do the clauses have to be set up anew.

5 A SELECTION STRATEGY FOR NON-DETERMINISTIC DECISIONS

The deductive method is a method of calculating the consequences of a non-deterministic decision, once it is made. It does not state how to select the next non-deterministic decision. In this section, we will describe a strategy for selecting the next decision called the *iterative strategy*. It is based on the insight that trying to find a set of decisions for the whole problem is best done by looking at one component at a time. We assume that the component systems are linearly ordered by some heuristics. One such heuristic is to place the more deterministic systems in front. The first non-deterministic choices are made for the first system only. And one proceeds to the next system after all choices for the first system are made and a set of decisions is found with which the first component problem is solvable. Suppose all decisions for the first k systems are already made. The next non-deterministic choice is made for system $k + 1$, if there is one left. These choices are made locally, which means that the decisions are made only on variables of system $k + 1$, the labelling solely determines if the variable will be assigned to the current system or not; and an order is determined only between two variables, if one belongs to the current system while the other does not. Implicitly, we have already given a priority order of the decisions: first identification or discrimination decisions, then labelling decisions and finally ordering decisions for the current system. It should be clear that after each non-deterministic decision the deductive process is started to deduce its consequences.

The main effect of this selection strategy is that we proceed to non-deterministic choices for the next system only after we have found a set of choices with which all previous systems are solvable. This leads to earlier detection of failure, when one component problem is unsolvable. As a side effect, the search space is reduced in that certain superfluous non-deterministic choices such as the ordering of two variables that do not occur together in at least one system are never made due to localisation of choices.

6 TESTS

The combination method and component algorithms for the free theory, A, AC, and ACI as well as for rational tree algebras and feature structures are implemented[3] in Common Lisp using the Keim toolkit [Huang *et al.*, 1994]. In the following we show some results of our optimisations. As already stated, the constraint solvers for rational tree algebras and feature structures are such that one can even combine them deterministically. Hence we do not present any test data for them. In order to test our algorithms with examples that occur in practice we used the REVEAL theorem prover [Chen and Anantharaman, 1995]. For some example theorems, we collected all unification problems that are generated and solved by REVEAL while proving this theorem. These theorems (and the corresponding set of unification problems) contain free function symbols and constants and one or two AC-symbols.

Table 1 gives an overview of the run time for some sets of unification problems. The first six lines contain all unification problems that have to be solved by REVEAL during the proof search or completion of the respective example. All examples except the first one contain two AC-symbols and several free symbols. The last three examples, containing several AC- and ACI-symbols, are added to demonstrate the potential of the iterative selection strategy. In order to see the effect of the iterative selection strategy on its own, we integrated it into the original algorithm (column 'it'). An empty cell in the columns indicates that the algorithm was aborted after one hour.

		Time in seconds						Bktrk	
Example	Size	i+d	ded	i+d-	ded-	it	orig	i+d	ded
Abelian group	29	3.7	3.7	5.0	5.0	11.6	17.2	4	4
Boolean ring	51	3.2	3.2	4.8	4.8	3.5	3.3	0	0
Boolean algebra	122	15.8	15.7	20.5	24.5			12	12
`exboolston`	87	12	12	948	997			17	14
`exgrobner`	1002	154	155	1442	1488			65	66
`exuqsl2`	404	109	108					74	74
AC^*–ACI^* 1	1	16	101	74	385	15		16	103
AC^*–ACI^* 2	1	31	407	393		841		13	205
AC^*–ACI^* 3	1	67	557			248		22	192

Legend

Size	Number of unification problems
Bktrk	Number of backtracking steps
i+d	Iterative selection strategy in deductive method
ded	Deductive method
i+d-, ded-	Same as i+d/ded, but AC-component replaced by one that uses only collapse-freeness and regularity
it	Iterative selection strategy in original algorithm
orig	Original unoptimised algorithm

Table 1: Run time of some example sets

[3]The implementation can be found at `http://www-lti.informatik.rwth-aachen.de/Forschung/unimok.html`.

We want to emphasise the differences between column 'ded' and 'ded-'. Column 'ded-' shows the run time of the algorithm when using only syntactic properties as described in [Baader and Schulz, 1996]; a comparison with column 'ded' demonstrates the power of the deductive method and the deductive component algorithms. The run time decreases dramatically for most examples and some examples cannot even be solved in suitable time when using only syntactic properties.

The use of the iterative selection strategy does not lead to a performance increase in the deductive algorithm in the first six example sets, because these examples are too simple: They contain too few component theories. The last three examples show that the use of the iterative selection strategy can lead to a speed-up by more than one order of magnitude. The equations in these examples contain several AC and ACI-function symbols besides free function symbols. It is a general observation that the iterative selection strategy is advantageous, if the number of systems is large or the deductive component algorithms do not deduce many decisions.

7 RELATED WORK AND CONCLUSIONS

The work that is most closely related to ours is Boudet [1993]. He presents an optimised algorithm for the combination of finitary equational theories. Our method is considerably more general, we are neither restricted to equational theories nor to structures for which minimal complete sets of solutions must be finite. But since combining unification algorithms is such an important instance of our methods, we want to compare the two approaches in more detail. Boudet's algorithm computes a complete set of unifiers for each theory, subsequently treats arisen conflicts between the theories (like one variable getting assigned to different terms in different systems), and repeats these two steps until all conflicts have been solved. Thus there is an important difference in the way the non-determinism inherent in most constraint problems is handled. Our algorithm prophylactically makes a choice for all possible conflict situations before solving the component systems. We showed that many of these choices can be made deterministically, but some have to be made non-deterministically. Boudet follows another approach: his algorithm only makes a non-deterministic choice if a conflict actually arises. But as a drawback his approach introduces another source of non-determinism: in order to detect actual conflicts, the algorithm has to compute complete sets of unifiers for the component systems and it has to choose one of the unifiers non-deterministically if the computed set contains more than one solution. The set of unifiers can by very large, e.g., doubly-exponential in the number of variables of the input problem for the theory AC.

Both algorithms have to perform several rounds of computation for the component systems, i.e., consequences (in our algorithm) or complete sets of unifiers (in Boudet's algorithm) have to be computed more than once for each component system. In our algorithm the constraint problem to be solved by a component has the same size in each round. In Boudet's algorithm the computation of a complete set of unifiers is based on the unifier found in the previous round. This means that the unification problem to be solved by a component theory can grow in each round, e.g., the number of variables in an AC-unifier can be exponential in the number of variables of the input problem. This can result in a higher worst-case complexity of Boudet's algorithm: it may well be non-elementary, even though the inherent complexity of combination is in NP. Our algorithm on the other hand has singly exponential complexity. Despite its high worst-case complexity, Boudet's algorithm performs quite well in many practical examples. It seems to be a promising line of research to try to integrate some of our optimisation ideas into Boudet's algorithm.

We presented an optimised algorithm for combining constraint solvers. Our empirical analysis indicates that the combined constraint solvers obtained this way can indeed be used in practice. It should be noted, however, that some of the non-determinism is inherent in the combination problem, which means that even the best optimisation methods cannot avoid this complexity, unless the structures to be combined are severely restricted, as pointed out in the subsection on deterministic combination.

Acknowledgments. We would like to thank Franz Baader and Klaus Schulz for helpful discussions and comments on drafts of this paper. We are also grateful to two anonymous referees whose comments helped improve the paper.

This work was funded by the "Schwerpunkt Deduktion" of the *Deutsche Forschungsgemeinschaft* (DFG) and was supported by the Esprit working group 22457 – CCL II of the European Union.

REFERENCES

[Aczel, 1988] Peter Aczel. *Non-wellfounded Sets.* Number 14 in CSLI Lecture Notes. CSLI, Stanford University, USA, 1988.

[Baader and Schulz, 1996] Franz Baader and Klaus U. Schulz. Unification in the union of disjoint equational theories: Combining decision procedures. *Journal of Symbolic Computation*, 21:211–243, 1996.

[Baader and Schulz, 1998] Franz Baader and Klaus U. Schulz. Combination of constraint solvers for free and quasi-free structures. *Theoretical Computer Science*, 192:107–161, 1998.

[Baader and Siekmann, 1994] Franz Baader and Jörg H. Siekmann. Unification theory. In Dov M. Gabbay, C. J. Hogger, and J. A. Robinson, editors,

Handbook of Logic in Artificial Intelligence and Logic Programming, volume 2, Deduction Methodologies. Clarendon Press, Oxford, 1994.

[Boudet, 1993] Alexandre Boudet. Combining unification algorithms. *Journal of Symbolic Computation*, 16(6):597–626, 1993.

[Bundy, 1994] Alan Bundy, editor. *Automated Deduction — CADE-12*, Proceedings of the 12th Conference on Automated Deduction, Nancy, France, 1994. Springer-Verlag LNAI 814, Berlin, Germany.

[Bürckert, 1990] Hans-Jürgen Bürckert. A resolution principle for clauses with constraints. In Mark E. Stickel, editor, *10th Conference on Automated Deduction*, Proceedings, pages 178–192, Kaiserslautern, Germany, 1990. Springer-Verlag LNAI 449, Berlin, Germany.

[Chen and Anantharaman, 1995] Ta Chen and Siva Anantharaman. Storm: A many-to-one associative-commutative matcher. In Jieh Hsiang, editor, *Rewriting Techniques and Applications*, Proceedings of the 6th International Conference, RTA-95, pages 414–419, Kaiserslautern, Germany, 1995. Springer-Verlag LNCS 914, Berlin, Germany.

[Colmerauer, 1984] Alain Colmerauer. Equations and inequations on finite and infinite trees. In *Proceedings of the International Conference on Fifth Generation Computer Systems*, pages 85–99, ICOT, Tokyo, Japan, 1984. North Holland.

[Colmerauer, 1990] Alain Colmerauer. An introduction to PROLOG III. *Communications of the ACM*, 33:69–99, 1990.

[Dovier *et al.*, 1996] Agostino Dovier, Alberto Policriti, and Gianfranco Rossi. Integrating lists, multisets, and sets in a logic programming framework. In Franz Baader and Klaus U. Schulz, editors, *Frontiers of Combining Systems*, Proceedings of the 1st Int. Workshop, FroCoS'96. Kluwer Academic Publishers, 1996.

[Huang *et al.*, 1994] Xiaorong Huang, Manfred Kerber, Michael Kohlhase, Erica Melis, Dan Nesmith, Jörn Richts, and Jörg Siekmann. KEIM: A Toolkit for Automated Deduction. In Bundy [1994], pages 807–810.

[Jouannaud and Kirchner, 1986] Jean-Pierre Jouannaud and Hélène Kirchner. Completion of a set of rules modulo a set of equations. *SIAM J. Computing*, 15:1155–1195, 1986.

[Kapur and Narendran, 1992] Deepak Kapur and Paliath Narendran. Complexity of unification problems with associative-commutative operators. *Journal of Automated Reasoning*, 9:261–288, 1992.

[Kepser, 1998] Stephan Kepser. *Combination of Constraint Systems.* PhD thesis, CIS, Universität München, 1998. Available at `ftp://ftp.cis.uni-muenchen.de/pub/kepser/ccl/diss.ps.gz`.

[Kirchner and Kirchner, 1989] Claude Kirchner and Hélène Kirchner. Constrained equational reasoning. In Gaston H. Gonnet, editor, *Proceedings of the ACM-SIGSAM 1989 International Symposium on Symbolic and Algebraic Computation: ISSAC '89*, pages 382–389, Portland, Oregon, 1989. ACM Press, New York, USA.

[Malćev, 1971] Anatolij Ivanovič Malćev. *The Metamathematics of Algebraic Systems*, volume 66 of *Studies in Logic*. North-Holland Publishing Company, 1971.

[Nieuwenhuis and Rubio, 1994] Robert Nieuwenhuis and Albert Rubio. *AC*-superposition with constraints: no *AC*-unifier needed. In Bundy [1994], pages 545–559.

[Plotkin, 1972] Gordon D. Plotkin. Building in equational thories. In Bernard Meltzer and Donald Michie, editors, *Machine Intelligence*, volume 7, pages 73–90. University Press, Edinburgh, United Kingdom, 1972.

[Schmidt-Schauß, 1989] Manfred Schmidt-Schauß. Unification in a combination of arbitrary disjoint equational theories. *Journal of Symbolic Computation*, 8(1,2):51–99, 1989.

[Schulz, 1996] Klaus U. Schulz. Combining unification- and disunification algorithms—tractable and intractable instances. Research Report CIS-Rep-96-99, CIS, LMU Munich, Germany, 1996.

[Smolka and Treinen, 1994] Gert Smolka and Ralf Treinen. Records for logic programming. *Journal for Logic Programming*, 18(3):229–258, 1994.

The Constraint Solver Collaboration Language of BALI

Eric Monfroy

1 INTRODUCTION

The need for *solver collaboration*, i.e., solver combination and cooperation, has by now been well recognized: several solvers collaborate to process constraints that cannot be solved efficiently by a single solver. Informally, combination [Nelson and Oppen, 1979; Kirchner and Ringeissen, 1994; Ringeissen, 1996] focuses on building a solver for the union of theories, whereas cooperation [Marti and Rueher, 1995; Monfroy *et al.*, 1996; Granvilliers, 1998] concerns data exchange between solvers devoted to a single domain. Distributed cooperative problem solving has also been studied in the field of artificial intelligence for multi-agent systems [Yokoo *et al.*, 1992], in advanced knowledge processing systems [Aiba *et al.*, 1995], and for cooperation of symbolic computation and automated deduction [Homann and Calmet, 1995; Dalmas *et al.*, 1996].

In this paper, we are concerned with the need for a general scheme of integration, and re-usability of heterogeneous solvers, together with some strategies and a language for realizing their collaborations. Some systems (such as [Marti and Rueher, 1995; Monfroy *et al.*, 1996]) significantly ease integration and cooperation, but they require some modifications of the system itself to add/replace solvers. Moreover, these systems provide only one "fixed" cooperation mechanism for a unique constraint system. The ILOG "library" approach enables cooperation of specific problem-solving strategies [Beringer and De Backer, 1995]. However, the solvers must be built into ILOG and no high-level language is provided for their cooperation. Concurrent constraint programming [Saraswat, 1993] provides a framework where several homogeneous solvers cooperate. Furthermore, the solving process and the programming process are mixed in the same language. Similar comments can be made about Oz [Mehl *et al.*, 1995]. In [Benhamou and Older, 1997], a Constraint Logic Programming (CLP) [Jaffar and Lassez, 1987] language, CLP(BNR), is proposed for expressing constraints on reals, integers, and Booleans in a unified framework. Solvers are expressed as narrowing algorithms and are, thus, restricted to interval methods. A formal framework for heterogeneous constraint solving is presented in [Benhamou, 1996]. However, no language of strategy is provided for improving the efficiency of the cooperating solvers.

BALI is an environment for solver collaboration that separates constraint programming (the *host language*) from constraint solving (the *solver collaboration language*). The host language is a constraint programming language and possibly a constraint logic programming language [Jaffar and Maher, 1994] which enables us to execute solver collaborations with respect to three *solving strategies*. The first one consists in determining the satisfiability of the constraint store each time a new constraint occurs ("incremental use of a solver"). The alternative to this method is to solve the constraint store when reaching a final state (e.g. the end of resolution for logic programming). The last strategy allows the user to trigger the solvers when he needs. Furthermore, BALI allows several solver collaborations, associated to different solving strategies, to coexist in a single system.

This paper focuses on constraint solving techniques, i.e., the constraint solver collaboration language of BALI. (The constraint programming part of BALI is described in [Monfroy, 1996].) This domain independent language has been designed for realizing a solving mechanism in terms of solver collaborations with solving strategies. The basic objects handled by the language are heterogeneous solvers. They are used inside *collaboration primitives* that integrate several paradigms (such as sequentiality, parallelism and concurrency) commonly used in solver combination or cooperation. In order to write finer strategies, we have also introduced some *control primitives* (such as iterator, fixed-point and conditional) in the collaboration language.

A prototype showed the feasibility of our approach and its interest in practice as well. We have determined the primitives required in the collaboration language of BALI after some discussions with potential users about their needs, and after some studies of the cooperation (as well as combination) mechanisms of some working systems. Thus, although most of the current realizations are related to non-linear constraints, the field of potential applications of BALI is wide.

BALI allows us to describe solver collaborations with a high-level language, and automatically implement new prototypes of cooperative solvers. The ease with which collaborations are realized modifies the programming methodology: with a classic system, the program is transformed when the solver is not efficient enough, with BALI, the program is unchanged, but the collaboration is modified at a high level.

The paper is organized as follows. In Section 2, the syntax of the collaboration language of BALI is given. Then (Section 3), we enlarge and improve the framework. In Section 4, we describe an operational semantics of the language. The implementation of BALI is described in Section 5 and some applications illustrate its use in Section 6. Finally, we conclude in Section 7.

2 THE SOLVER COLLABORATION LANGUAGE OF BALI

The expressions written in this language describe *collaboration of solvers*, i.e., complex solvers built from several component solvers together with some strategies for their applications. The solver collaboration language of BALI is composed of: *collaboration primitives* that correspond to phases known as computation phases, and *control primitives* that are strategies for composing computation phases.

First, we define the objects manipulated by the solver collaboration language of BALI, i.e., solvers, and constraints.

Definition 2.1 (Constraint system) A constraint system is a quadruple $(\Sigma, \mathcal{D}, V, \mathcal{L})$ where:

- Σ is a many-sorted first-order signature given by a set of sort symbols $\mathcal{S}_\Sigma$, a set of function symbols $\mathcal{F}_\Sigma$, and a set of predicate symbols $\mathcal{P}_\Sigma$,
- $\mathcal{D}$ is a Σ-structure whose domain is the union of the pairwise disjoint domains $|\mathcal{D}_{s_i}|$ for each sort s_i of $\mathcal{S}_\Sigma$ $(|\mathcal{D}| = \bigcup_{s_i \in \mathcal{S}_\Sigma} |\mathcal{D}_{s_i}|)$,
- V is the union of pairwise disjoint nonempty sets V_{s_i} of $\mathcal{S}_i$-sorted variables, $V = \bigcup_{s_i \in \mathcal{S}_\Sigma} V_{s_i}$
- $\mathcal{L}$ is a set of constraints: it is a non-empty set of (Σ, V)-formulas closed under conjunction and disjunction. An atomic (Σ, V)-formula of $\mathcal{L}$ is called an *atomic constraint.* The unsatisfiable constraint is denoted by $\perp$ and the truth constraint is denoted by $\top$.

The many-sorted nature of a constraint system will be exploited in Section 3.1 for extending the use of solvers. A sort is a set of elements. V_{s_i} denotes the subset of V containing the variables of sort s_i, and $|\mathcal{D}_{s_i}|$ denotes the domain associated to the sort s_i, i.e., the subset of $|\mathcal{D}|$ on which variables of sorts s_i can take their values.

Intuitively, a *component solver* is an algorithm which transforms a constraint C into C', a constraint which is "simpler" than C, but equivalent to C in $\mathcal{D}$[1], i.e., a solver preserves the solutions. This concept of solver is intentionally quite general: component solvers can be "conventional" solvers (such as Gröbner bases, Simplex, unification), or transformation/simplification functions (such as normal form, expansion of polynomials, transformation or rewrite rules,), or domain reduction functions (for solving constraint satisfaction problems based on propagation). Hence, we can make collaborate at the same level numerous mechanisms that are generally considered to be of different natures.

[1]We will use $\mathcal{D} \models C \Leftrightarrow S(C)$ to denote that the application of the solver S to C does not change the set of solutions of C.

Definition 2.2 (Component solver) A *component solver* (or *solver*) on a constraint system $(\Sigma, \mathcal{D}, \mathcal{V}, \mathcal{L})$ is a computable function $S : \mathcal{L} \to \mathcal{L}$ such that:

$$\forall C \in \mathcal{L},\ \mathcal{D} \models S(C) \Leftrightarrow C.$$

2.1 The Language

Consider now several component solvers, each associated with a fragment of $\mathcal{L}$. (In Section 3.1 we give conditions that are needed to use a solver on different constraint systems.) We are interested in expressing their collaboration formally. The syntax of the language is given in Figure 1.

$\mathcal{E}$ represents the set of solver collaboration expressions (derived from the non-terminal E) that can be written with the language. $\mathcal{S}$ is the set of *component solvers*. We call the *constraint store* the current constraint of $\mathcal{L}$ that must be treated by a collaboration. $\mathcal{S}tores$ is the set of all the stores. The identifiers $\mathcal{I}$ are names of solver expressions. Their major uses are (1) several collaborations associated to different solving strategies can coexist in the host language, and can be called by their name [Monfroy, 1996], (2) solver expressions can be collected in a library and used as part of an expression. Col allows us to name collaborations. We assume that a component solver collaboration is a component solver (see [Monfroy, 1996] for proofs). Thus, the notation $E(C)$ represents the application of the collaboration E of $\mathcal{E}$ to the constraint C of $\mathcal{S}tores$. $\Diamond$ denotes the *identity solver*: $\forall C \in \mathcal{L},\ \Diamond(C) = C$.

$Id \in \mathcal{I}$ (identifiers)

$S \in \mathcal{S}$ (solvers)

$\psi \in \Psi$ (concurrency functions)

$n \in \mathbb{N}$ (positive integers)

$OA \in \mathcal{OA}$ (arithmetic observation functions)

$OB \in \mathcal{OB}$ (boolean observation functions)

$Col ::= Id = E$

$E ::= \Diamond \mid Id \mid S \mid E; E \mid \text{psi\_c}(\psi, SE) \mid \text{par}(SE) \mid \text{f\_p}(E) \mid \text{rep}(Ar, E) \mid \text{if}(B, E, E)$

$SE ::= E \mid E, SE$

$Ar ::= n \mid Ar + Ar \mid Ar - Ar \mid Ar * Ar \mid OA$

$B ::= true \mid false \mid Ar < Ar \mid Ar \leq Ar \mid Ar = Ar \mid B \wedge B \mid \mid B \vee B \mid \neg B \mid OB$

Figure 1: Grammar of the solver collaboration language of BALI

2.2 Collaboration Primitives

Informally, combination [Nelson and Oppen, 1979; Kirchner and Ringeissen, 1994; Ringeissen, 1996] focuses on building a solver for the union of theories with solvers already defined on each of these theories. On the other hand, cooperation [Marti and Rueher, 1995; Monfroy *et al.*, 1996; Granvilliers, 1998] concerns communication problems between solvers devoted to a single domain. In this case the cooperating solvers share the constraints according to their languages and specificities. A solver collaboration is either a solver combination, or a solver cooperation (see [Monfroy, 1998a] for formal definitions of these notions).

Collaboration primitives define concepts for applying solvers on parts of constraints (see Section 3.1), and re-composing constraints that are equivalent but simpler than the input ones. **BALI** provides three collaboration primitives, each one presenting different mechanisms for combination and cooperation of solvers.

Sequentiality. *Sequentiality* (denoted by ;) is similar to composition of functions. Let $E_1; \ldots; E_n$ be a sequential composition of solvers. Then

$$\forall C \in \mathcal{L},\ E_1; \ldots; E_n(C) \ = \ E_1 \circ \ldots \circ E_n(C)$$

Parallelism. This primitive (denoted by $\mathrm{par}(E_1, \ldots, E_n)$) aims at executing several solvers in parallel. Its result is the conjunction of the solutions of each solver. This primitive is interesting for several applications, i.e., simultaneously treating equalities and inequations with different solvers, applying several specialized solvers (such as quadratic solvers) on the store, simultaneously solving constraints over disjoint sorts with the adequate solvers. Let $\mathrm{par}(E_1, \ldots, E_n)$ be a parallel collaboration of solvers. Then,

$$\forall C \in \mathcal{L},\ \mathrm{par}(E_1, \ldots, E_n)(C) = \bigwedge_{i=1}^{n} E_i(C)$$

In the current state, this primitive is close to the "AND-parallelism" concept of logic programming. In Section 3.2, we improve the solution recomposition in order to remove most of the redundancies introduced by the parallel primitive; thereby, the primitive deviates from the notion of AND-parallelism.

ψ_concurrency. This primitive (denoted by $\mathrm{psi\_c}(\psi, E_1, \ldots, E_n)$) provides a non-deterministic choice upon which we can act by introducing methods (i.e., *concurrency functions* ψ of Ψ) to select one store among the results of all the solvers. This primitive is similar to the don't care indeterminism but also provides control for choosing the new store. This primitive is interesting when considering several similar solvers (especially when the solvers are very expensive in time) without having any clue about their efficiency for solving a given problem.

During execution, the ψ functions determine the constraint returned by a solver that satisfies some imposed conditions (such as the quickest solver, properties on the form of the solutions, etc.). They are defined as follows:

$$\psi : \mathcal{E}^n \times \mathcal{L} \to \mathcal{L}$$
$$\psi(\{E_1, \ldots, E_n\}, C) \in \{E_1(C), \ldots, E_n(C), C\}$$

The constraint returned by ψ is the solution of one of the solvers. The choice depends on the execution time of the E_j and on the form of the $E_j(C)$. When no pair (execution time, $E_j(C)$) validates the imposed conditions, then $\psi(\{E_1, \ldots, E_n\}, C) = C$. ψ_{basic} is a standard function of Ψ that returns the result of the first solver that finishes executing. Some more complex ψ functions can be considered, such as ψ_{solved_form} which selects the result of the first solver whose solution is in solved form on the computation domain.

Let psi_c($\psi, E_1, \ldots, E_n$) be a ψ-concurrent primitive. Then,

$$\forall C \in \mathcal{L},\ \text{psi\_c}(\psi, E_1, \ldots, E_n)(C) = \psi(\{E_1, \ldots, E_n\}, C)$$

2.3 Control Primitives

Conditionals. "if(γ, E_1, E_2)" means that if the application to the store of the function γ is true, then E_1 is executed, otherwise E_2 is executed. γ is a Boolean expression that is evaluated at run-time as it makes use of *observation functions* of the constraint store. These functions may be either arithmetic (set $\mathcal{OA}$) or Boolean (set $\mathcal{OB}$). Arithmetic observation functions have the profile: $\mathcal{S}tores \to \mathbb{N}$. The following are 3 such functions: (1) *card_var* computes the number of distinct variables in the constraint store. This is interesting for solvers that are sensitive to the number of variables. (2) *card_c* returns the number of atomic constraints that compose the store. This is really important for solvers whose complexity is a function of the number of constraints (such as solvers based on propagation). (3) *card_uni_var* returns the number of univariate atomic constraints. This is essential for solvers whose efficiency is improved with univariate constraint, such as interval propagation.

Boolean observation functions have the profile: $\mathcal{S}tores \to Boolean$. The following are 3 Boolean observation functions: (1) *linear* tests whether all the atomic constraints do not contain more than two occurrences of the same variable. This tests the interest of applying a linear solver. (2) *uni_var* tests whether there is at least one univariate equality in the store. This information is important since, for example, univariate constraints are generally the starting point of interval propagation. (3) *tri* tests whether the store is in triangular form (i.e., there are some equality constraints over a variable X, some over variables X and Y, some over X, Y and Z, etc.). This is interesting for eliminating variables, or determining an ordering for Gröbner bases.

Some other observation functions (such as *card_eq*, or even constraint domain specific functions such as *max_degree*) can be added to complete $\mathcal{OA}$, and $\mathcal{OB}$.

- if$(\gamma, E_1, E_2)(C)$ is equivalent to $E_1(C)$ if $\gamma(C) = true$ and to $E_2(C)$ if $\gamma(C) = false$.

Iterators. The strategy "rep(δ, E)" allows one to apply n times the collaboration E on the store. n is the result of the application of the arithmetic function δ (that uses *arithmetic observation functions* of $\mathcal{OA}$) to the store. Since this primitive takes into account the constraint and its form at run-time, it improves the dynamic aspect of the collaboration language. When treating rep(δ, E), if the evaluation of δ returns a negative value, rep(δ, E) is replaced by the identity solver $\diamondsuit$. rep$(\delta, E)(C)$ is equivalent to $E^{\delta(C)}(C)$.

Fixed-point. "f_p(E)" iteratively applies the collaboration E to the store until a fixed-point is reached. f_p$(E)(C)$ is equivalent to $E^n(C)$, where n is defined by:

$$\exists n,\ E^{n+1}(C) = E^n(C)\ \wedge\ \forall m,\ E^{m+1}(C) = E^m(C) \Rightarrow n \leq m$$

This primitive allows one to create an idempotent solver from a non-idempotent solver. For example, if the solver T is a transformation rule (similar to a rewrite rule), or a domain reduction function, f_p(T) applies T until the store is in canonical form with regard to T. As a second example: if S is a solver that eliminates one linear variable, f_p(S) eliminates all the linear variables.

3 IMPROVEMENTS AND ENRICHMENTS

We now discuss several aspects that enlarge the class of solvers that can be used in **BALI**, and improve the efficiency.

3.1 Solver Enrichment

Consider a solver S defined on a constraint system CS. Now, we want to use S on a more "complex" constraint system CS^+ without modifying the solutions over CS^+. This enlarges the class of solvers we can use in **BALI**, and allows for stores to be composed of constraints on different domains (Boolean, rational, integers, etc.).

Definition 3.1 (Constraint system enrichment) Let $CS = (\Sigma, \mathcal{D}, \mathcal{V}, \mathcal{L})$ and $CS^+ = (\Sigma^+, \mathcal{D}^+, \mathcal{V}^+, \mathcal{L}^+)$ be two constraint systems. Then, CS^+ is an enrichment of CS if:

- $\mathcal{S}_\Sigma \subseteq \mathcal{S}_{\Sigma^+}$, $\mathcal{F}_\Sigma \subseteq \mathcal{F}_{\Sigma^+}$, and $\mathcal{P}_\Sigma \subseteq \mathcal{P}_{\Sigma^+}$
- $\forall s_i \in \mathcal{S}_\Sigma$, $|\mathcal{D}_{s_i}| = |\mathcal{D}^+_{s_i}|$ and $\forall r \in \mathcal{F}_\Sigma \cup \mathcal{P}_\Sigma$, $r_{\mathcal{D}^+} = r_{\mathcal{D}}$ [2]
- $\mathcal{V} \subseteq \mathcal{V}^+$, and $\mathcal{L} \subseteq \mathcal{L}^+$.

[2] $r_{\mathcal{D}}$ (respectively $r_{\mathcal{D}^+}$) represents the interpretation of r on the Σ-structure $\mathcal{D}$ (respectively $\mathcal{D}^+$).

Obviously, all the constraints of CS^+ cannot be treated by S. Thus, we use the concept of "admissible constraints" [3], i.e., parts of constraints over CS^+ that S is effectively able to handle.

Definition 3.2 (S_admissible constraint) Let S be a solver on $CS = (\Sigma, \mathcal{D}, V, \mathcal{L})$, $CS^+ = (\Sigma^+, \mathcal{D}^+, V^+, \mathcal{L}^+)$ be an enrichment of CS, and C be a constraint of $\mathcal{L}^+$. Then, the S_admissible constraint of C over CS (denoted $\phi_S(C)$) is defined as follows: $\phi_S(c \wedge C') = c \wedge \phi_S(C')$ if $c \in \mathcal{L}$, otherwise $\phi_S(C')$, and $\phi_S(\top) = \top$.

Similarly, we define $\bar{\phi}_S(C)$, the S_non-admissible constraint, of C over CS: $\bar{\phi}_S(c \wedge C') = c \wedge \bar{\phi}_S(C')$ if $c \notin \mathcal{L}$, otherwise $\bar{\phi}_S(C')$, and $\phi_S(\top) = \top$.

Property 3.3 (Solver enrichment) *Let S be a component solver on CS, $CS^+ = (\Sigma^+, \mathcal{D}^+, V^+, \mathcal{L}^+)$ be an enrichment of CS, and $S^\star$, the enrichment of S on CS^+, be defined as follows: $\forall C \in \mathcal{L}^+,\ S^\star(C) = S(\phi_S(C)) \wedge \bar{\phi}_S(C)$. Then, $S^\star$ is a component solver on CS^+.*

This allows us to use solvers on enrichments of their constraint system, without modifying the set of solutions. Although the notion of admissible constraint could seem "rough", it is well suited if we let BALI manage the disjunctions.

3.2 Managing Backtracking in BALI

Many solvers do not allow disjunctions of constraints as input. Thus, we now assume that BALI manages disjunctions and backtracking itself. We consider another class of solvers. $\mathcal{L}^\wedge$ denotes the set of conjunctions of atomic constraints ($\mathcal{L}^\wedge \subseteq \mathcal{L}$).

Definition 3.4 (Component $\wedge$_solver) A *component $\wedge$_solver* on a constraint system $(\Sigma, \mathcal{D}, \mathcal{V}, \mathcal{L})$ is a computable function $S : \mathcal{L}^\wedge \rightarrow \mathcal{L}$ such that $\forall C \in \mathcal{L}^\wedge,\ \mathcal{D} \models S(C) \Leftrightarrow C$.

The mechanism of solver enrichment also holds for component $\wedge$_solvers. This new class of solvers does not restrict, but enlarges the set of solvers BALI can use. The only difference is that component solvers do not handle disjunctions themselves. Let C be a constraint in $\mathcal{L}$, C' be C in DNF, and S be a $\wedge$_solver. BALI sends one by one each disjunct of C' to S (see Section 4). Recompositions of constraints (especially in the case of parallel collaboration) can be handled more efficiently by removing some of the redundancies. Similarly, the treatment of solver enrichments is refined: admissible and non-admissible parts of constraints can be taken into account during the recomposition process.

[3] The notion of admissible/non_admissible part of a constraint is close to the notion of active/passive constraint of the CLP scheme [Jaffar and Maher, 1994].

Moreover, the solving process is homogenized, depth-first or breadth-first,[4] and it becomes a parameter of the semantics.

3.3 Normal Form

In order to simplify constraints and propagate information in between constraints and in between results of several solvers, we now consider a normal form computation, nf (described as a set of rules in Figure 2), which is applied to the store after each application of a solver/collaboration (see Section 4). Rules of Figure 2 can be used with component ($\wedge$_)solvers, and solver enrichments (even **Rules for disjunctions** although they are of no use with $\wedge$_solvers). When considering only component solvers, nf is more useful after a parallel primitive: it removes some of the redundancies, and propagates information in between results of different solvers. With solver enrichments, and $\wedge$_solvers, the use of nf is more significant: information deduced in the admissible part of a constraint can be propagated to the non-admissible part, and some more redundancies can vanish.

Constraint simplification rules:	Idempotence	$C \wedge C$	$\rightarrow$	C
	Bottom	$C \wedge \bot$	$\rightarrow$	$\bot$
	Top	$C \wedge \top$	$\rightarrow$	C
	DNF	C	$\rightarrow$	$DNF(C)$
Propagation rule:	Propagation	$X \in V, d \in (\mathcal{D}, \mathcal{L}),\ X = d \wedge C$	$\rightarrow$	$C[d/X]$
Rules for disjunctions:	$\vee$ – Idempotence	$C \vee C$	$\rightarrow$	C
	$\vee$ – Top	$C \vee \top$	$\rightarrow$	$\top$
	$\vee$ – Bottom	$C \vee \bot$	$\rightarrow$	C
Domain specific rules:	simplify ineq	$A \geq B \wedge B \geq A$	$\rightarrow$	$A = B$
	$\cap$–domains	$x \in D_x \wedge x \in D'_x$	$\rightarrow$	$x \in (D_x \cap D'_x)$
	domain fail	$x \in \emptyset$	$\rightarrow$	$\bot$

Figure 2: Normal Form

Some transformations of nf are domain-independent, whereas others are simplifiers on special domains. The rule DNF puts a constraint into disjunctive normal form. This form is important when BALI manages the disjunctions itself: for example, after a parallel collaboration the store is no longer in DNF and the rule DNF must be applied (see Figure 4 and 5). This operation is purely syntactic and does not perform any simplification. *Simplification rules*

[4] With component solvers, the global solving process (in-between solvers) is breadth-first. But, the local solving processes (inside solvers) depend on the solvers and, thus, can be different.

are applied modulo the commutativity and associativity of $\wedge$. The Propagation rule is similar to a variable elimination: it propagates the value of a variable inside a constraint of $\mathcal{L}^{\wedge}$. Most of the solvers perform this operation, but their propagation is limited to their own admissible constraints. Thus, Propagation extends this mechanism. The *rules for disjunctions* can be added to nf in order to limit the number of disjunctions and obtain a mechanism similar to collaboration of decision procedures. The rule $\vee$-Idempotence removes multiple solutions and the rule $\vee$-Top deduces the satisfiability of the constraint, when at least one disjoint is $\top$. $\vee$-Bottom removes unsatisfiable disjuncts from the constraint. We give examples of rules to complete nf for specific domains of computation. For interval propagation, we have generally to deal with the special membership constraint $\in$. The rule $\cap$-domains reduces two membership constraints[5] into only one. domain fail transforms $x \in \emptyset$ into $\bot$ which then can be used by other rules of nf. simplify ineq transforms two inequations into one equality. This last rule is valid for constraint languages that include inequations and equalities.

Finally, nf can be considered as a solver which is automatically applied after each solver computation. Since they perform only syntactical transformations, simplification and propagation rules can be seen as a generic solver that can be used for all constraint systems. nf is not unique and must at least contain the rule DNF when using component $\wedge$_solvers. Hence, **BALI** can manage the backtracking itself (Section 4).

4 OPERATIONAL SEMANTICS

The operational semantics of the collaboration language takes into account the enrichments presented in Section 3. Thus, the backtracking mechanism is integrated.

A *configuration*, denoted $\langle E\,,\,C\rangle$ represents a constraint $C \in \mathcal{L}^{\wedge}$, waiting to be treated by a solver expression E of $\mathcal{E}$. A transition between two configurations:

$$\langle E\,,\,C\rangle \longrightarrow \langle E'\,,\,C'\rangle$$

means "the first sub-collaboration of E has been applied on the constraint C and transformed it into C'; we still have to apply E' on C'."[6]

The *backtrack mechanism* manages the disjunctions (due to the non-determinism of the solvers, or multiplicities of solutions) inside and between collaborations. To get constraints in DNF, we use the normal form nf. A *state* P is a multi-set of configurations: it stores for each disjunct of a constraint the disjunct itself and the collaboration that must be applied to it. P represents all the configurations that have to be treated. At the beginning of the

[5]They can be the results of two solvers (in this case, domain reduction functions) in parallel.

[6]$\langle E\,,\,C\rangle \longrightarrow \langle\Diamond\,,\,C'\rangle$ means that the complete collaboration E has been applied to C, and $C' = E(C)$.

solving process, P is composed of one configuration: a designed solver collaboration E together with an input store of constraints to be solved using E. Then, P stores all the configurations that appear during the solving process: sub-collaborations associated with constraint stores, and choice points (i.e., configurations) created by disjunctions (results of solvers/collaborations). At the end, P is empty. For a given configuration $\langle E_1; E_2\,,\, C\rangle$, if E_1 returns a disjunction of constraints $C' = C_1 \vee \ldots \vee C_n$, then $\langle E_2\,,\, C_1\rangle$, $\ldots\langle E_2\,,\, C_n\rangle$ (denoted $\lceil E_2, C'\rceil$) must be added to P (denoted $\biguplus(\lceil E_2, C'\rceil, P)$). We call $\biguplus$ the adding operator.

We consider a *computation rule*: at each transition step, this rule selects an element of P as the configuration to be treated. We denote by $\langle E\,,\, C\rangle \uplus P'$ the choice of the configuration $\langle E\,,\, C\rangle$ in the multi-set $P = \{\langle E\,,\, C\rangle\} \cup P'$.

The adding operator (together with Auxiliary rules of Figure 3) and the computation rule describe the backtracking mechanism and the resolution strategy. To specialize this semantics in order to get a depth-first resolution, a stack can be used for representing states. Then, the computation rule is a "pop" and the adding operator $\biguplus$ is a sequence of "push."

Transitions between configurations are extended to states:

$$\langle E; E'\,,\, C\rangle \uplus P \longrightarrow \biguplus(\lceil E', C''\rceil, P)$$

To describe the behavior of the system, the operational semantics associates to each possible state a new state obtained by applying the first solver/collaboration of the selected configuration to its associated store. Constraint solving becomes a sequence of transitions called *derivations*:

$$\langle C_1\,,\, E_1\rangle \uplus \emptyset \longrightarrow \langle C_2\,,\, E_2\rangle \uplus P_2 \longrightarrow \ldots \longrightarrow \langle C_n\,,\, E_n\rangle \uplus P_n \longrightarrow \ldots$$

A *successful* state is a state of the form $\langle \Diamond\,,\, C\rangle \uplus P$ with $C \neq \bot$. A derivation is *successful* if at least one of its states is successful. The *constraint solution* (with regard to the solver collaboration) is the disjunction of all the stores of successful states of a derivation. A derivation is *failed* if it is finite and not successful.

The operational semantics (Figures 3, 4 and 5) is given in SOS form. The transition relation is inductively defined, i.e., an action is performed if the relevant conditions are satisfied. Due to space limitation, we do not "duplicate" rules of Figures 4 and 5 to take into account the last primitive of a collaboration. However, if we consider that a collaboration is always a sequential collaboration that terminates with the identity solver $\Diamond$ (this does not modify the result of the collaboration), the semantics is complete: the last primitive is always $\Diamond$, and is treated by the auxiliary rules of Figure 3.

The Auxiliary rules (Figure 3) do not modify the constraint store. Either they return the current constraint store (printing of the solutions for the rules success and next), or they perform the backtrack mechanism (backtrack and next). If one of the rules fail or success is applied, then the derivation is finite. The rules fail and backtrack mean that this resolution branch does not give any solution. With fail the resolution is finished, whereas with backtrack there

fail	$$\frac{}{\langle E\,,\,\bot\rangle \uplus \emptyset \longrightarrow \text{fail}}$$
backtrack	$$\frac{P \neq \emptyset}{\langle E\,,\,\bot\rangle \uplus P \longrightarrow P}$$
next	$$\frac{C \neq \bot \quad \wedge \quad P \neq \emptyset}{\langle \Diamond\,,\,C\rangle \uplus P \longrightarrow P}$$
success	$$\frac{C \neq \bot}{\langle \Diamond\,,\,C\rangle \uplus \emptyset \longrightarrow \text{success}}$$

Figure 3: Auxiliary rules

are still some configurations to treat. The rules next and success are similar, but in the case of a branch that gives a solution.

sequential elementary solve	$$\frac{\langle S\,,\,C\rangle \longrightarrow \langle \Diamond\,,\,S^\star(C)\rangle \;\wedge\; S \in \mathcal{S}}{\langle S; E\,,\,C\rangle \uplus P \longrightarrow \biguplus(\lceil E, nf(S^\star(C))\rceil, P)}$$
parallel solve	$$\frac{\langle E_1\,,\,C\rangle \longrightarrow \langle \Diamond\,,\,C_1\rangle \;\ldots\; \langle E_n\,,\,C\rangle \longrightarrow \langle \Diamond\,,\,C_n\rangle \;\wedge\; C' = \bigwedge_{i=1}^{n} C_i}{\langle \text{par}(E_1,\ldots,E_n); E\,,\,C\rangle \uplus P \longrightarrow \biguplus(\lceil E, nf(C')\rceil, P)}$$
ψ_concurrent solve	$$\frac{\langle E_i\,,\,C\rangle \longrightarrow \langle \Diamond\,,\,C'\rangle \quad \psi(\{E_1\ldots,E_n\},C) = C'}{\langle \text{psi\_c}(\psi, E_1,\ldots,E_n); E\,,\,C\rangle \uplus P \longrightarrow \biguplus(\lceil E, nf(C')\rceil, P)}$$

Figure 4: Transition rules for collaboration primitives

Figure 4 groups together the rules for managing the collaboration primitives. The recomposition of the constraint store after each kind of collaboration primitive is to be noticed. For sequential elementary solve and concurrent solve the reconstruction is immediate since $\uplus$ takes into account disjunctions by creating and adding corresponding configurations to the state P. On the other hand, parallel solve first rebuilds the store with regard to solutions of each solver before the adding operator $\uplus$ treats the disjunctions. sequential elementary solve is special as it is the only rule which applies a component solver.

Notice that normal form computation (nf) is applied in each rule where a solver/collaboration is effectively applied, i.e., each time a disjunction can be created (rules of Figure 4 and repeat fixed point in Figure 5). Thus, $\biguplus$ always adds configurations using constraints in DNF.

Figure 5 treats control primitives. The rule repeat fixed point performs no progress in the solver expressions. However, the constraint is modified and becomes "smaller" with regard to the solver. The fixed-point is reached (fixed-point) when E_0 no longer modifies the constraint. At that time, the

repeat fixed-point
$$\frac{\langle E_0\,,\,C\rangle \longrightarrow \langle \Diamond\,,\,C'\rangle \quad \wedge \quad C' \neq C}{\langle \mathrm{f\_p}(E_0); E\,,\,C\rangle \uplus P \longrightarrow \biguplus(\lceil \mathrm{f\_p}(E_0); E, nf(C')\rceil, P)}$$

fixed-point
$$\frac{\langle E_0\,,\,C\rangle \longrightarrow \langle \Diamond\,,\,C\rangle}{\langle \mathrm{f\_p}(E_0); E\,,\,C\rangle \uplus P \longrightarrow \biguplus(\lceil E, C\rceil, P)}$$

δ_repeat
$$\frac{n = \delta(C)}{\langle \mathrm{rep}(\delta, E_0); E\,,\,C\rangle \uplus P \longrightarrow \biguplus(\lceil \underbrace{E_0; \ldots; E_0}_{n \text{ times}}; E, C\rceil, P)}$$

γ then
$$\frac{\gamma(C)}{\langle \mathrm{if}(\gamma, E_1, E_2); E\,,\,C\rangle \uplus P \longrightarrow \biguplus(\lceil E_1; E, C\rceil, P)}$$

γ else
$$\frac{\neg\gamma(C)}{\langle \mathrm{if}(\gamma, E_1, E_2); E\,,\,C\rangle \uplus P \longrightarrow \biguplus(\lceil E_2; E, C\rceil, P)}$$

Figure 5: Transition rules for control primitives

treatment of f_p(E_0) is finished (for the current constraint), and f_p(E_0) is removed from the solver expression of the current configuration.

The condition of δ_**repeat** is not the application of a solver, but the evaluation of the observation function δ applied to C. The iteration primitive is replaced by a sequential primitive that takes into account the number of iterations to be performed ($\delta(C)$). If $\delta(C) \leq 0$, then $\mathrm{rep}(\delta, E_0)$ has no effect.

The last set of rules concerns the conditional collaborations. The rule γ_**then** replaces the conditional primitive by its "then" part (solver collaboration E_1), if the evaluation of the boolean observation function γ is true. Otherwise, γ_**else** replaces the conditional primitive with E_2.

These rules can be optimized in order to manage recomposition of the store better and avoid some redundancies (see [Monfroy, 1996]).

5 IMPLEMENTATION

The implementation of **BALI** automatically transforms an expression written with the solver collaboration language into a distributed system that behaves and executes like a classic solver collaboration: inputs are constraints, and outputs are solved forms (with regard to the collaborations).

From the operational semantics we have derived an organizational model for the implementation of **BALI**. This model describes (1) a distributed environment for integrating heterogeneous solvers, (2) communications between solvers in spite of their differences[7], and (3) coordination of their executions.

[7]Each solver has its own data representation, is written in a different language, and

Solvers and collaborations are encapsulated into ECL^{i}PS^e [Meier and Schimpf, 1993] processes to create homogeneous agents. ECL^{i}PS^e launches the solvers and re-connects the input and output through channels. The encapsulation also provides some data-structure converters, a constraint store, an admissibility function (for solver enrichment) and a re-composition function (which is more complex in encapsulation of collaboration). The agents execute independently in a distributed environment composed of a network of machines. Some ECL^{i}PS^e processes coordinate the whole system. This implementation turns solver collaborations into *servers* to which *clients*, called *host languages*, can connect. Clients can be constraint programming languages (such as CLP languages, imperative languages with constraints), or all kinds of processes that may require a solver (such as spreadsheets or theorem provers).

To validate our approach that consists in separating constraint programming from constraint solving, we have also implemented a host language. It is a CLP(X) system that integrates three solving strategies associated to solver collaborations: (1) the *incremental strategy* applies the associated collaboration each time the store is modified, (2) the *final strategy* solves the store with the "final collaboration" at the end of the logic resolution, and (3) the *user strategy* which enables one to use self-triggering collaborations.[8] Our CLP system is based on ECL^{i}PS^e [Meier and Schimpf, 1993], its meta-terms, and the CHRs [Frühwirth, 1995] for managing constraints and the store, and implementing the primitives to associate strategies to collaborations of **BALI**. More details concerning the implementation can be found in [Monfroy, 1996; Arbab and Monfroy, 1998].

6 APPLICATIONS

Simulation of CoSAc. **CoSAc** [Monfroy *et al.*, 1996] is a CLP system over the domain of non-linear constraints (equalities and inequations of polynomials). As the constraint programming language of **CoSAc** is ECL^{i}PS^e, we use the host language described in Section 5 for the simulation. In **CoSAc** five heterogeneous solvers cooperate through a client/server architecture: *chr_lin* [Frühwirth, 1995] for solving linear constraints, *gb [Faugere, 1994]* for computing Gröbner bases, *maple_uni* for computing roots of univariate polynomials, *maple_exp* for simplifying and transforming constraints, and *ecl* for testing closed inequalities. These solvers cooperate in three collaborations: S_{inc}, S_{fin} and S'_{fin}. S_{inc} is triggered each time the store is modified: *maple_exp* transforms polynomials so *eq_lin* can propagate information and simplify the linear equations. This gives in **BALI**:

$$S_{inc} = maple\_exp \ ; \ eq\_lin$$

executes on a different architecture and operating system.

[8]In this case, *solve(a)* means: apply collaboration *a* on the current constraint store.

After Prolog resolution, S_{fin} is applied to the remaining constraints. They are simplified (*maple_exp*) before computing their Gröbner base. Then, variables are eliminated from univariate polynomials (*maple_uni*), solutions are propagated, and linearized equations are solved (*eq_lin*). This process terminates when all the variables have been eliminated or when there are no more univariate polynomials. In BALI, this is written:

$$S_{fin} = maple\_exp \ ; \ gb \ ; \ \text{f\_p}(maple\_uni; eq\_lin)$$

S'_{fin} is an alternative to S_{fin} and is more efficient when eliminations of non-linear variables do not linearize any other constraints.

$$S'_{fin} = maple\_exp \ ; \ gb \ ; \ \text{f\_p}(maple\_uni); ecl$$

As soon as solvers are integrated (i.e., data converters) in BALI only two lines of code are required to create the collaborations. This is negligible compared to the thousands of lines of CoSAc, and the simulation of CoSAc in BALI is nearly as efficient as CoSAc itself. We improved CoSAc using the ψ_concurrent primitive. The first solvers of S_{fin} and S'_{fin} are "factorized", and we obtain:

$$S''_{fin} = maple\_exp; gb; \text{psi\_c}(\psi_{basic}, \text{f\_p}(maple\_uni; eq\_lin), \text{f\_p}(maple\_uni); ecl)$$

S''_{fin} is always as efficient as the most efficient of the two collaborations S_{fin} and S'_{fin}. Thus, the simulation of CoSAc with BALI becomes more efficient than CoSAc itself.

Interval arithmetic and the dependency problem. We now consider *Rint* [Monfroy, 1998b] a solver for non-linear numerical constraints over the real numbers, based on interval propagation methods (and more precisely on box-consistency [Van Hentenryck *et al.*, 1997]). Solvers of this type are strongly dependent on the way constraints are expressed. For example, let us consider the polynomial $2*x-3*x+x$ with $x \in [0,10]$. Its evaluation should give 0 although its interval evaluation is $[-30,30]$. In fact, the occurrences of x are considered to be different variables: this problem is known as the *dependency problem.* In [Moore, 1966], it is suggested that the number of occurrences of variables be reduced before interval evaluation. Thus, we create the following collaboration:

$$a = maple\_exp \ ; \ Rint$$

maple_exp reduces the dependency problem and, consequently, the work of *Rint* is simplified: since constraint evaluation is more accurate, *Rint* performs less iterations and local splittings to compute solutions. To sum up, the collaboration a gives the same solutions as *Rint*, but in a shorter time.[9]

Symbolic pre-processing for numerical solvers. Propagation based solvers are often the iteration of two steps: pruning (reduction of the constraint

[9]When *maple_exp* does not simplify any constraint, the running time of a is a little bit longer than the running time of *Rint*. However, the difference is negligible.

satisfaction problem (CSP)), and splitting (creation of disjoint CSPs) to compute isolated solutions. Consider the following problem: $x^3 - x * y + 2 = 0, x^2 - y + 2 = 0$. *Rint* requires splitting to isolate the solution. However, with the following collaboration,

$$b = gb;\ Rint$$

Rint finds the solutions without splitting. *gb* acts as a symbolic pre-processing for the numerical solver (see [Benhamou and Granvilliers, 1996; Granvilliers, 1998] for details about combining symbolic rewriting and interval methods): *gb* reduces the dependency problem more than *maple_exp*, and also transforms the problem in a more suitable form for *Rint*. However, the speed-up of *Rint* does not always compensate for the additional cost of *gb*. Thus, to overcome this problem, we create the following collaboration:

$$c = \text{psi\_c}(\psi_{basic}, gb; Rint\ ,\ maple\_exp; Rint)$$

Now, let's consider the following collaboration:

$$d = \text{psi\_c}(\psi_{basic}, gb; Rint\ ,\ maple\_exp; Rint\ ,\ Rint)$$

d should "theoretically" be the most efficient collaboration. However, *d* requires more resources than *c*, and in practice, *d* is generally less efficient than *c*.

Rational constraint solving. We now consider two new solvers: *Simplex* to solve linear inequalities, and *Gauss* which is a Gaussian elimination to solve equalities. Solvers for rational constraint are generally based on these two methods (e.g., see [Lim, 1994]). Since information deduced in *Simplex* can be used by *Gauss* and vice-versa, we have to iterate the application of both solvers till a fixed-point is reached. Furthermore, these two solvers can be applied in parallel without adding redundancies. Thus, we create the following collaboration for rational constraint solving:

$$e = \text{f\_p}(\text{par}(Simplex, Gauss))$$

Dynamic adaptation of solvers. With non-linear constraints, one can never be sure that the store is indeed non-linear[10]: it can be exclusively composed of linear constraints. Thus, we tried another collaboration which avoids using "heavy" non-linear solvers when the store is purely linear:

$$f = \text{if}(linear, e, c)$$

We re-used the previously defined collaborations *e* and *c*, but we could also have used S_{inc} as the linear collaboration, and S_{fin} as the non-linear collaboration. *f* is basically as efficient as *c* (a small additional cost is given by the test and the communications) when the store is non-linear, but is much more efficient when the store is purely linear.

[10]This is due to the fact that constraints are generally not given as an "input set", but are stated and collected during executions of the host language (constraint programming language).

7 CONCLUSIONS

The solver collaboration language of **BALI** provides collaboration primitives and control primitives to build complex solvers from component solvers. The motivations for realizing such an environment were issued from the need of a general scheme for integrating, composing and re-using heterogeneous solvers in order to efficiently solve constraints and to tackle new problems. **BALI** fulfills the requirements: integrating solvers is facilitated, solver collaborations are realized quite declaratively and their implementation is then automated. As soon as solvers are integrated, creating a new collaboration becomes an easy and quick task — one line of code and a few minutes for launching the collaboration — compared to the thousands of lines of code for implementing a cooperation and the considerable development time. This also modifies the programming methodology: with classic constraint systems, the users modify their programs; with **BALI** they modify the solving process itself. Generally, solving a problem is done as follows. The users specify their problem for a given solver. When the solver is not powerful enough, and not able to solve the problem, the users re-design their application: they try to simplify the constraints by hand, they re-organize the constraints, or they even have to simplify the problem. In fact, they modify their program until the system can solve it. With **BALI**, the approach is different. First the users specify their problem, and then they choose a collaboration to solve it. If the collaboration cannot solve the given constraints, then the users design more powerful collaborations but do not modify the program. Hence, **BALI** saves the bother of designing specialized solvers and re-modeling problems.

BALI has been successfully used to solve problems on non-linear constraints. The primitives of the language were designed with respect to users' needs, and cooperation/combination techniques of existing systems. Thus, we are confident in the use of **BALI** for other constraint domains. Although they have not already been tested, some applications on other domains (such as constraint propagation) have been formalized with **BALI**.

In the future, we plan to develop a new implementation based on a coordination language [Arbab and Monfroy, 1998]. Not only can the system be improved in terms of robustness, stability and required resources, but also the constraint solving activity itself can be improved through the resulting clarity of search, efficient handling of the disjunctions, and modularity. Furthermore, the encapsulation mechanism of the current implementation is rather heavy and it is not realistic to use light weight solvers (such as domain reduction functions for constraint propagation). With this new implementation, light weight and heavy solvers will be mixed inside collaborations without jeopardizing the global efficiency.

We plan to enrich the collaboration language with more control on the part of the store to which solvers are applied. Some filters could be used to select specific parts of the store. Thus, some highly complex solvers (such as quantifier elimination [Collins, 1975]) that are generally too slow to be applied to the store, could be successfully applied to a part of it. This could also lead to splitting primitives to simultaneously execute one solver on several specific parts of the store. Thus, techniques such as in [Benhamou and Granvilliers, 1996] (where Gröbner bases are computed on sub-parts of the store to simplify the work of the interval solver) could also be expressed in **BALI**. We also plan to add a primitive that controls (at run-time) the application order of solvers because this primitive cannot always be simulated with the current primitives of **BALI**.

Acknowledgements. We are grateful to K.R. Apt who helped us to better articulate and clarify this paper.

REFERENCES

[Aiba *et al.*, 1995] A. Aiba, K. Yokota, and H. Tsuda. Heterogeneous Distributed Cooperative Problem Solving System Helios and its Cooperation Mechanisms. *International Journal of Cooperative Information Systems*, 4(4):369–385, 1995.

[Arbab and Monfroy, 1998] F. Arbab and E. Monfroy. Using Coordination for Cooperative Constraint Solving. In *Proc. of ACM SAC'98, Atlanta (USA)*, pages 139–148. ACM Press, 1998.

[Benhamou and Granvilliers, 1996] F. Benhamou and L. Granvilliers. Combining local consistency, symbolic rewriting, and interval methods. In *Proc. of AISMC3*, Steyr (Austria), 1996.

[Benhamou and Older, 1997] F. Benhamou and W. Older. Applying Interval Arithmetic to Real, Integer and Boolean Constraints. *Journal of Logic Programming*, 32(1):1–24, July 1997.

[Benhamou, 1996] F. Benhamou. Heterogeneous Constraint Solving. In *Proc. of ALP'96*, volume 1139 of *LNCS*, pages 62–76, Aachen, Germany, 1996. Springer-Verlag.

[Beringer and De Backer, 1995] H. Beringer and B. De Backer. Combinatorial problem solving in constraint logic programming with cooperative solvers. In *Logic programming: formal methods and practical applications*. Elsevier Science Publisher B.V., 1995.

[Collins, 1975] G. E. Collins. Quantifier Elimination for Real Closed Fields by Cylindrical Algebraic Decomposition. In *Proc. of the Second GI Conference on Automata Theory and Formal Languages*, volume 33 of *LNCS*. Springer Verlag, 1975.

[Dalmas *et al.*, 1996] S. Dalmas, M. Gaëtano, and C. Huchet. A database for mathematical formulas. In *Proc. of DISCO'96*, 1996.

[Faugere, 1994] J-C. Faugere. *Résolution des systèmes d'équations algébriques.* PhD thesis, Université Paris 6, 1994.

[Frühwirth, 1995] T. Frühwirth. Constraint handling rules. In A. Podelski, editor, *Constraint Programming: Basics and Trends*, volume 910 of *LNCS*. Springer Verlag, 1995.

[Granvilliers, 1998] L. Granvilliers. A symbolic-numerical branch and prune algorithm for solving non-linear polynomial systems. *Journal of Universal Computer Science, Springer*, 4:125–146, 1998.

[Homann and Calmet, 1995] K. Homann and J. Calmet. Combining Theorem Proving and Symbolic Mathematical Computing. In J. Calmet and J.A. Campbell, editors, *Proc. of AISMC-2*, volume 958 of *LNCS*, pages 18–29. Springer Verlag, 1995.

[Jaffar and Lassez, 1987] Joxan Jaffar and Jean-Louis Lassez. Constraint logic programming. In *Proc. of POPL'87*, pages 111–119. ACM Press, Jan. 1987.

[Jaffar and Maher, 1994] J. Jaffar and M. Maher. Constraint Logic Programming: a Survey. *Journal of Logic Programming*, 19,20:503–581, 1994.

[Kirchner and Ringeissen, 1994] H. Kirchner and C. Ringeissen. Combining symbolic constraint solvers on algebraic domains. *Journal of Symbolic Computation*, 18(2):113–155, 1994.

[Lim, 1994] P Lim. Implementation of the ECLiPSe rational Constraint Solver. Report ECRC-94-23, ECRC, Munich, Germany, 1994.

[Marti and Rueher, 1995] P. Marti and M. Rueher. A Distributed Cooperating Constraints Solving System. *International Journal on AI Tools*, 4(1&2):93–113, 1995.

[Mehl *et al.*, 1995] M. Mehl, T. Müller, K. Popov, and R. Scheidhauer. *DFKI Oz User's manual.* DFKI, Sarrebrucken (Germany), May 1995.

[Meier and Schimpf, 1993] M. Meier and J. Schimpf. ECLiPSe User Manual. Report ECRC-93-6, ECRC, Munich, Germany, 1993.

[Monfroy *et al.*, 1996] E. Monfroy, M. Rusinowitch, and R. Schott. Implementing Non-Linear Constraints with Cooperative Solvers. In *Proc. of ACM SAC'96, Philadelphia, USA*, pages 63–72, Feb. 1996.

[Monfroy, 1996] E. Monfroy. *Collaboration de solveurs pour la programmation logique à contraintes.* PhD Thesis, Université Henri Poincaré-Nancy I, Nov. 1996. (Also available in english.)

[Monfroy, 1998a] E. Monfroy. An environment for designing/executing constraint solver collaborations. In *Proc. of COTIC'98, Nice, France*, ENTCS. Elsevier Science Publishers, 16(1), 1998.

[Monfroy, 1998b] E. Monfroy. Using "weaker" functions for constraint propagation over real numbers. In *Proceedings of The 14th ACM Symposium on Applied Computing* (SAC'99), Scientific Computing Track, pages 553–559, San Antonio, Texas, USA, March 1999.

[Moore, 1966] R. E. Moore. *Interval Analysis.* Series in Automatic Computation. Prentice Hall, Englewood Cliffs, N. J., 1966.

[Nelson and Oppen, 1979] C. G. Nelson and D. C. Oppen. Simplifications by cooperating decision procedures. *ACM Transactions on Programming Languages and Systems*, 1(2), 1979.

[Ringeissen, 1996] C. Ringeissen. Cooperation of decision procedures for the satisfiability problem. In *Frontiers of Combining Systems*, Applied Logic. Kluwer Academic Publishers, 1996.

[Saraswat, 1993] V. Saraswat. *Concurrent Constraint Programming.* MIT Press, Cambridge, London, 1993.

[Van Hentenryck *et al.*, 1997] P. Van Hentenryck, D. McAllester, and D. Kapur. Solving polynomial systems using a branch and prune approach. *SIAM Journal on Numerical Analysis*, 34(2), 1997.

[Yokoo *et al.*, 1992] M. Yokoo, H. Durfee, T. Ishida, and K. Kuwabara. Distributed Constraint Satisfaction for Formalizing Distributed Problem Solving. In *Proc. of the IEEE ICDCS'92*, pages 614–621, June 1992.

A Hybrid Language for the Analysis of Aspectual and Temporal Phenomena in Natural Language

Ralf Naumann

1 DATA AND EVIDENCE

1.1 Aspectual Variety

It is well-known that the properties of noun phrases (NPs) can have an influence on the aspectual behaviour of expressions they combine with; witness the examples (1a,b) with the (accomplishment) verb 'eat':

(1) a. John ate an apple in ten minutes/*for ten minutes.
    b. John ate apples *in ten minutes/for ten minutes.
    c. John pushed a cart *in ten minutes/for ten minutes.
    d. John pushed carts *in ten minutes/for ten minutes.

If both arguments are singular and non-mass, (1a), modification with an *in*- but not with a *for*- adverbial is possible. If the internal argument is a bare plural, (1b), the admissibility is reversed: only modification with a *for*-adverbial is possible. An activity verb like 'push', on the other hand, behaves differently: only modification with a *for*-adverbial is possible, independent of whether the internal argument is non-mass and singular or a bare plural. This difference in aspectual behaviour shows that 'eat' and 'push' must be assigned different semantic properties. This distinction is paralleled by a distinction between accomplishment- and activity-expressions according to four criteria, Vendler [1967].

(2)

	in-adv.	*for*-adv.	*at*-adv.	progressive
accomplishment	+	-	-	+
activity	-	+	-	+

A central question of a theory of aspect concerns the problem of how this distinction can be formally defined. One kind of data is supplied by the perfect of result (PR).

1.2 The Perfect of Result

The perfect of result (PR) is defined by Comrie informally as follows: "In the perfect of result, a present state is referred to as being the result of some past situation." (Comrie [1976], p. 56). In contrast to the experiential perfect, this type of perfect imposes aspectual restrictions. Only expressions of type accomplishment (3a) or achievement (3b) but not of type activity (3c), state (3d) or point (3e) admit a PR.

(3) a. John has crossed the street/eaten an apple.
    b. John has arrived/reached the top.
    c. ?John has run in the park/has worked in the garden.
    d. *John has been dead/?The room has been empty.
    e. ?John has hit Mary/?The star has twinkled.

According to Comrie's informal definition, this difference must be explained by the fact that the expressions in (3a,b) do semantically define a result whereas this is not the case for the expressions in (3c-e). Furthermore, the notion of result must be related to the aspectual properties that define the aspectual classes of accomplishment and achievement.

The discussion so far may suggest that expressions belonging to the same aspectual class behave alike with respect to the PR. They either all admit a PR (accomplishments and achievements) or they all do not admit a PR (activities, states and points). This assumption is not borne out by the data.

(4) a. John ate apples ⇒ PR is possible
    b. Students arrived at the university ⇒ PR is possible
    c. Mary has played the sonata. ⇒ PR is not possible

According to the aspectual criteria in (2), (4a,b) are of type activity, yet they both allow a PR ('John has eaten apples', 'Students have arrived at the university'). (4c) shows that expressions of type accomplishment do not always admit the PR. If someone plays a sonata, there is no result which is brought about by the playing and which can continue to hold up to speech time. What is important therefore in order that a result is semantically defined are the aspectual properties of the verb and not those of the whole expression. For instance, a distinction must be made between elements of the class of activities in which the verb itself is of this type and those elements for which this does not hold. This leads to the distinction between lexical and non-lexical activities. An element of the class of activity-expressions is lexical just in case its underlying verb is of type activity ('(Bill) push the cart'). An example of a non-lexical activity-expression is '(John) eat apples'. The former expressions never define a result whereas the latter do define a result if the underlying verb is of type accomplishment or achievement. Note that not all lexical accomplishments semantically define a result, witness (4c). What a nominal (subcategorized) argument can add at the semantic level is a property that is sufficient for the whole expression to be of type (non-lexical) accomplishment. What it cannot contribute is a property that is sufficient to semantically define a result.

1.3 Manner-Adverbs

The notion of result equally plays an important role in the interpretation of a particular class of manner-adverbs which admit of two different interpretations.

(5) a. Hans put the chairs properly/neatly.

b. Peter dressed Mary elegantly.

On the first interpretation of (5a), the procedural reading, 'properly' specifies the way the agent, i.e. Hans, executed the event. He proceeded 'properly'. On the second interpretation, the result reading, it is not the way Hans proceeded but rather the resulting state that is modified by the adverb. The chairs were standing 'properly'. Both interpretations are independent of each other. Hans can act in a proper manner without it being the case that the chairs are standing properly as a result of his action. Conversely, even if Hans does not act properly, the chairs can nevertheless be standing properly. A similar argument holds for (5b).

In order that, in addition to a procedural interpretation, a result interpretation is possible, first, the property expressed by the adverb must be defined for the object that undergoes the change. This excludes a result interpretation of 'John crossed the street quickly' because the property expressed by 'quickly' is not defined for the location of an object, which is the property that gets changed by an event of type 'cross'. Second, the expression must semantically define a result. This excludes 'John pushed five carts inconspiciuously'. Although a cart can be inconspiciuous, no result modification is possible because the pushing of a cart does not have a result that can endure beyond the pushing.

2 THE NOTION OF RESULT IN A DYNAMIC EVENT STRUCTURE

2.1 Changes as Events and State-Transformers

In Naumann [1995, 1996, 1997, 1998a, b] and Naumann and Mori [1998] a theory of aspectual phenomena has been developed which is based on the intuition that non-stative verbs like 'eat' express changes. One way of making this intuition precise is to interpret a change as a transformation of state: some input-state s is transformed into an output-state s' such that some condition ϕ which did not hold in s comes to hold in s'. This perspective is captured in Dynamic Logic (DL) where changes are programs that are interpreted as binary relations on the domain S of states. The disadvantage of this perspective is that changes are not (basic) objects but are reduced to input-output relations defined in terms of S. There are no objects in the model which are interpreted as bringing about the change (transformation) from the input-state to the output-state. This disadvantage can be overcome by assuming a second domain E of events (transitions, actions) besides the domain S of states. Each event e is assigned an execution-sequence σ, i.e. a finite sequence of states. Furthermore, each event e is of a particular type P. Events of type P bring about changes with respect to some property Q. Events of type P bring about Q in some particular way that corresponds to some program from DL, for instance a (variant of the) **while**-loop or a form of iteration. This relationship between the result Q and a binary relation can be expressed by so-called dynamic modes. In (6) two examples are given ($<$ is an ordering on S, $s < s'$).

(6) a. $R_{Con\text{-}BEC}$: $\lambda Q \lambda ss'[\neg Q(s) \wedge Q(s') \wedge \forall s''[s < s'' < s' \rightarrow Q(s'')]]$

b. $R_{Min\text{-}BEC}$: $\lambda Q \lambda ss'[\neg Q(s) \wedge Q(s') \wedge \forall s''[s < s'' < s' \rightarrow \neg Q(s'')]]$

The Min-BEC mode (partly) characterizes the class of accomplishments, and the Con-BEC mode (partly) characterizes the class of activities. Other aspectual classes are characterized by other modes (see below section (2.4)). The dynamic mode assigned to an event-type P determines the way in which a change effected by an event of this type is basically brought about. The principle idea therefore is to reify the programs from DL. Each event-type P corresponds to a binary relation R on S. Each event $e \in P$ brings about the transformation of some state s into some state s' such that $(s, s') \in R$.

The picture sketched in the previous paragraph only accounts for expressions like 'eat an apple' or 'push a cart' where the arguments are singular and non-mass. For instance, the former expression can be analyzed as denoting a set of (completed) events of type eating, the execution-sequences of which are characterized by the Min-BEC mode. For expressions with plural objects, like '(John) eat five apples' or 'Three boys pushed five carts', the situation is different. Whereas '(John) eat an apple' denotes a set of events (or a set of singleton sets of events), '(John) eat five apples' denotes a set of sets of events. In this case the postcondition must be brought about (by John) for each of the five apples. At the procedural level where changes are transformations of states this means that instead of a relation between states one gets a relation between sets of states. Instead of a sequential program from DL a parallel program having five (sequential) programs as parts must be executed. At the level of changes as objects (events) an analogous distinction must be made: sequential events, i.e. single events ('eat an apple'), are distinguished from parallel events, i.e. sets of events ('eat five apples'), Naumann [1998b].

2.2 A Dynamic Event Structure

A dynamic event structure **ES** is a tuple

$$<\mathbf{E}, \mathbf{S}, \mathbf{O}, \alpha, \beta, \gamma, \delta, \kappa, \rho, \rho^*, \{AD_{ADV}\}_{ADV \in ADVERB}, \{\theta_{TR}\}_{TR \in THR}>$$

such that

- $\mathbf{E} = \langle E, \sqsubseteq_E, \{P_v\}_{v \in VERB} \rangle$ is an eventuality structure where

  E is a (non-empty) set of events

  $\sqsubseteq_E \subseteq E \times E$ is a reflexive, antisymmetric and transitive relation (the part-of relation on E)

  The P_v are unary relations on E (called event-types);

  $VERB = \{eat, drink, run, push, hit, \ldots\}$ corresponds to (a subset of) the (non-stative) verbs in English

- $\mathbf{S} = \langle S, <, S^*, \{R_{OP(p)}\}_{OP \in DM, p \in VAR}, \{Q_p\}_{p \in VAR} \rangle$ is a transition structure where

  S is a (non-empty) set of states

  $<$ is a linear and discrete ordering on S

  S^* is the set of finite sequences based on S that respect the ordering $<$ on S (i.e., S^* is the set of finite convex subsets of S). Elements from S^* will be written σ (possibly primed) or as (s, s') where s and s' are the beginning- and end-point of the sequence, respectively

  Each $R_{OP(p)}$ is a binary relation on S;

  $DM = \{Min\text{-}BEC, Con\text{-}BEC, CHANGE, HOLD, Min\text{-}BEC_{suf}\}$;

  S is *standard* if the $R_{OP(p)}$ are interpreted as follows:

  $R_{Min\text{-}BEC(p)} =$
  $\{(s, s') \in S^* \mid s < s' \wedge \neg Q_p(s) \wedge Q_p(s') \wedge \forall s''[s < s'' < s' \rightarrow \neg Q_p(s'')]\}$

  $R_{Con\text{-}BEC(p)} =$
  $\{(s, s') \in S^* \mid s < s' \wedge \neg Q_p(s) \wedge Q_p(s') \wedge \forall s''[s < s'' < s' \rightarrow Q_p(s'')]\}$

  $R_{CHANGE(p)} =$
  $\{(s, s') \in S^* \mid s < s' \wedge \neg Q_p(s) \wedge Q_p(s') \wedge \forall s''[s \leq s'' \leq s' \rightarrow s'' = s \vee s'' = s']\}$

  $R_{HOLD(p)} =$
  $\{(s, s') \in S^* \mid s \leq s' \wedge \forall s''[s \leq s'' \leq s' \rightarrow Q_p(s'')]\}$

  $R_{Min\text{-}BEC_{suf}(p)} =$
  $\{(s, s') \in S^* \mid s = s' \wedge Q_p(s') \wedge \exists s''[s'' < s \wedge \forall s^*[s'' \leq s^* < s' \rightarrow \neg Q_p(s^*)]]\}$

  The $R_{OP(p)}$ are used to characterize aspectual classes

  Each Q_p is a unary relation on S (a property of states)

- $\mathbf{O} = <O, \sqsubseteq_O>$ is an object structure where

  O is a (non-empty) set of objects

  $\sqsubseteq_O \subseteq O \times O$ is a reflexive, antisymmetric and transitive relation (the part-of relation on O)

- α and β are functions $E \to S$ which assign to each event its source-state $\alpha(e)$ and target-state $\beta(e)$, respectively. The product-mapping $\langle \alpha, \beta \rangle$ ($= \tau) : E \to S \times S$ assigns to each event e its execution-sequence $\tau(e)$

- γ is a function that assigns to each P_v a dynamic mode, i.e. a function that maps properties of states to binary relations on S (or, equivalently, to a subset of S^*); the assignment is done in accordance with the aspectual class to which P_v belongs; for instance, $\gamma(P_{eat}) = R_{Min\text{-}BEC}$ and $\gamma(P_{push}) = R_{Con\text{-}BEC}$, where $R_{Min\text{-}BEC}$ and $R_{Con\text{-}BEC}$ are defined as given in (6) above

- δ is a function that assigns to each P_v and each $d \in O$ some Q_p, i.e. an element from $\{Q_p \mid p \in VAR\}$; if $d = [\rho(P_v)](e)$ for some completed event $e \in P_v$, $\delta(P_v)(d)$ is the postcondition brought about by e

- κ is a function that assigns to each P_v an n-tuple of thematic roles θ_{TR} for which elements of P_v are defined

- ρ is a function that assigns to each P_v the thematic relation with respect to the value of which a change is effected by an event of type P_v; there is the following condition on ρ. If an event $e \in E$ is an element of two different event-types P_v and $P_{v'}$, i.e. $P_v \neq P_{v'}$, then $\rho(P_v)(e) = \rho(P_{v'})(e)$. Thus, the object with respect to which an event e brings about a change does not depend on the particular event-types to which it belongs. It is therefore possible to define a function ρ^* from E to O that assigns to each e the object $\rho^*(e)$ that undergoes the change brought about by e

- each AD_{ADV} is a partial function from $E \cup O$ to $\wp(S)$; they are used in the interpretation of the translations of manner-adverbs

- each θ_{TR} is a partial function from E to O that corresponds to a thematic relation that is used to link the 'ordinary' arguments to the event argument of a verb in event-semantics.

The main aim of the present paper is to construct a language L that is interpreted with respect to the dynamic event structure **ES** and in which it is possible to succinctly express aspectual distinctions between different expressions (of category sentence in particular). This will make it necessary to abstract from all those semantic aspects of an expression which do not contribute to the aspectual properties of expressions they are a constituent of. The result will be a 'shallow' analysis of a sentence that reveals only sufficient semantic structure to account for aspectual differences. The procedure is best explained by means of some examples.

The verb plays an important role in the determination of the aspectual properties of a sentence; witness the difference in aspectual behaviour between 'eat an apple' and 'push a cart'. It is therefore necessary to express its contribution in L. From what has been said in Section 2.1 above it follows that this contribution consists in determining a binary relation. Thus, it must be possible in L to distinguish between different types of binary relations, corresponding to the different aspectual classes. On the other hand, if the arguments are non-mass and singular, the information that is contributed by the argument is aspectually not relevant (if no PR or manner-adverb is present). For instance, "John eat an apple" has the same aspectual properties as the underlying verb 'eat'. What is added by the arguments is specific information about the objects that are related to the event by the thematic relations Patient ('an apple') and Agent ('John'). Even if an argument influences the aspectual properties, e.g. a bare plural in 'eat apples', the head noun of the NP is aspectually not relevant: 'eat apples' and 'eat sandwiches' show no difference in aspectual behaviour despite the different head nouns. On the other hand, an argument-NP can introduce an object (discourse-referent) on which the determiner inside the NP imposes a cardinality condition that can influence the aspectual properties of the resulting expression. It is therefore not possible to completely abstract from the semantic properties of an argument-NP.

What is needed is a mechanism that allows us to refer to the objects that are introduced by NPs and that are related to the event denoted by the verb by some thematic relation. This need for being able to refer to individuals from different domains of **ES** is further confirmed by the data concerning the PR and manner-adverbs. The resultative reading of the present perfect makes reference to the speech-time, i.e. the time of utterance, because the result brought about by e is required to still hold at that time. Although the time of utterance can be taken to be an element of S, it is independent of the relation between e and its execution-sequence. Manner-adverbs can be interpreted either with respect to the event (procedural interpretation) or with respect to the object that undergoes the change (result interpretation). In these cases too it must be possible to refer to specific individuals.

Thus, besides having the means to express different types of binary relations on S, the language is required to have in addition a mechanism that admits reference to elements from the different domains of **ES**. This can be achieved by extending L to a hybrid language L^h which is no longer purely propositional but in which there are variables (of different sorts) that can be bound by different types of binders. The advantage of this way of semantically analyzing sentences is that many aspectual distinctions can be expressed on the basis of a propositional modal language that is augmented by some binding mechanism for variables.

2.3 The Notion of a Nucleus Structure

In this section the notions of result and consequent-state of an event will be defined in **ES**. From what has been said so far it follows that it is not possible to simply define these notions in terms of the mere input/output behaviour of the binary relation that corresponds to an event-type P_v. The execution-sequence $\tau(e)$ of each event $e \in P_v$ is characterized by the fact that the property brought about by e does hold in e's target-state $\beta(e)$ and fails to hold in its source-state $\alpha(e)$ (except for events that belong to some type P_v of type achievement). From this it immediately follows that the notion of result cannot be identified with the postcondition $\delta(P_v)([\rho(P_v)](e))$ because the latter is defined for each P_v. Each execution-sequence (s, s') can be split into three parts: IS = input-state (= s), OS = output-state (= s') and INT-SEQ = $\{s'' \in S \mid s < s'' < s'\}$ (i.e. the sequence of states in between s and s'). For INT-SEQ there are the following possibilities. First, there are no intermediate states, i.e., the execution-sequence consists of exactly two states. In this case INT-SEQ is empty. This holds for the execution-sequences of programs characterizing a point-verb like 'hit'. Second, there are intermediate states. Then three cases can be distinguished: (i) the postcondition is required to be constantly false on INT-SEQ; (ii) the postcondition is required to be constantly true on INT-SEQ or (iii) there is no condition on how the postcondition is evaluated on INT-SEQ.

From what has been said in Section 1, together with the definitions given in Section 2.2, it follows that only those event-types P_v define a result for which first the postcondition determined by δ does not hold at all states $s \in INT - SEQ$ and for which second it is possible that there are execution-sequences such that INT-SEQ is non-empty. Thus, the definition of the notion of result must refer not to the mere input/output behaviour but rather to properties which the execution-sequences of type P_v have, due to the way the postcondition is evaluated at intermediate states, i.e. at those states that belong to INT-SEQ. This idea is made precise in terms of the notion of a nucleus structure, Moens and Steedman [1988]. A nucleus structure consists maximally of four parts: inception-point (IP), development-portion (DP), culmination-point (CP) and consequent-state (CS).

| - - - - - - - - - - - | - - - - - - - - - - |

IP $\quad DP \quad$ CP $\quad CS$

What elements of the nucleus structure are defined by an event-type P_v depends on the properties the type of execution-sequence has that is assigned to it by γ. The notions of IP, DP and CP are defined as properties of event-types. The notions of result and consequent-state are defined in terms of those of DP and CP. Result is a partial function that maps an event to a property of states, whereas CS is a partial function that maps an event to a sequence of states. The definitions are given in (7).

(7) a. $P_v \in IP$, for all P_v

b. $P_v \in DP$ iff (i) $\gamma(P_v) = R_{Min\text{-}BEC}$ or (ii) $\forall e \in P_v \exists e'[e \sqsubset_E e' \wedge \tau(e') \in R_{Min\text{-}BEC}(\delta(P_v)(\rho(P_v)(e)))]$

c. $P_v \in CP$ iff (i) $P_v \in DP$ and (ii) $\neg Event - related(\delta(P_v))$ where $Event - related(\delta(P_v))$ iff
$\forall d \forall s[s \in \delta(P_v)(d) \rightarrow \exists e[s \in \tau(e) \wedge [\rho(P_v)](e) = d]]$

d. $Result : E \rightarrow \wp(S)$ is a partial function that is defined for an $e \in P_v$ only if $P_v \in CP$ and $\tau(e) \in \gamma(P_v)(\delta(P_v)([\rho(P_v)](e)))$. In this case $Result(e) = \delta(P_v)([\rho(P_v)](e))$

c. $CS : E \rightarrow S^*$ is a partial function that is defined for an $e \in P_v$ only if $Result(e)$ is defined. In this case
$CS(e) =$

$$\iota\sigma \left[\begin{array}{c} \beta(e) = first(\sigma) \wedge \forall s[s \in \sigma \rightarrow s \in Result(e)] \wedge \\ \forall\sigma' \left[\begin{array}{c} \beta(e) = first(\sigma') \wedge \\ \forall s[s \in \sigma' \rightarrow s \in Result(e)] \rightarrow prefix(\sigma', \sigma) \end{array} \right] \end{array} \right]$$

where $first(\sigma) = s$ iff $s \in \sigma \wedge \forall s'[s' \in \sigma \rightarrow s \leq s']$

Comments. First, each event-type P_v defines an IP. Second, an event-type P_v defines a DP if it is possible that there are intermediate states at which the postcondition Q is required to be false (first disjunct: $\gamma(P_v) = R_{Min\text{-}BEC}$). Consequently, event-types that define a DP impose a condition on the post-condition for elements of INT-SEQ. An event-type P_v equally defines a DP if each event $e \in P_v$ is a proper part of an event e' that brings about the postcondition of e according to the way determined by $R_{Min\text{-}BEC}$. This second possibility accounts for verbs belonging to the class of achievements that admit a PR although the nucleus structure determined by the corresponding event-types does not admit of intermediate states. According to (7b), no event-type P_v belonging to the class of activities, states or points defines a DP. Third, the notion of a CP presupposes that of a DP. An event-type P_v defines a CP only if it also defines a DP. Yet, the existence of a DP is only a necessary condition for a CP, and not a sufficient condition. In addition, the postcondition determined by δ must not be event-related. The latter notion intuitively means that the postcondition brought about by an event e of this type only holds for the object that undergoes the change during the execution of the event. This condition excludes accomplishment-verbs like 'play' ('play the sonata') which do neither admit a PR nor a result reading of manner-adverbs. For 'play the sonata' the postcondition specifies that the musical score must have been 'gone through' completely. Thus, what has to be done in order to play a sonata is to 'traverse' a path which consists of the playing of all the notes (in the appropriate order) of the musical score. This path is completely traversed only if all notes have been played, i.e. if the last note is reached. In this respect there is no difference between the execution-sequences of events of type 'play the sonata' and, say, 'eat an apple' or 'run to the station': the postcondition only holds in the output-state of the execution. The difference consists of the fact that for 'play' the postcondition can only be defined in terms of the playing itself. For 'eat' and 'run to the station', on the other hand, the postcondition can be specified in a way that does not rely on events of any type because an object can have the property independently of being involved in an event ('be at the station', e.g.). Fourth, an event $e \in P_v$ has a result just in case P_v defines a CP (and, consequently, a DP) and it brings about $\delta(P_v)([\rho(P_v)](e))$. If P_v defines a CP, the result of (completed) events of this type is identified with the postcondition determined by δ for P_v and e (7d). From what has been said it follows that the existence of a result depends both on the type of execution-sequence determined by γ for the event-type P_v to which e belongs and on the property determined by δ for P_v. Fifth, an event $e \in P_v$ has a consequent-state only if it has a result. As an event e has a result only if the event-type to which it belongs defines both a DP and a CP, the notion of a consequent-state presupposes both that of a DP and that of a CP. If e has a consequent-state, it is defined as the longest sequence starting with $\beta(e)$ on which the result brought about by e holds, (7e).

2.4 The Languages L and $\mathbf{L}^h$

According to the theory presented in the previous sections, an event e can be said to have three different sorts of components. First, there is the dynamic component of e. This component is its execution-sequence $\tau(e)$. The dynamic component links e to an element from S^* and therefore the eventuality structure **E** to the transition structure **S**. Second, there is the static component of e. Each event e is related to an n-tuple of objects from O. These objects are the values for e of the thematic relations for which the event-type P_v, to which e belongs, is defined. They correspond to the objects that are denoted by the subcategorized arguments of the verb v. In **ES** this relationship is captured by the function κ that maps an event-type P_v to an n-tuple of thematic relations. Substituting the values of these thematic relations for e for the thematic relations yields the n-tuple of objects to which e is related by subcategorized arguments. The static component relates the eventuality structure **E** to the object structure **O**. Third, there is the external component of e. The interpretation of a sentence not only depends on an event e and the relation both to its dynamic and its static component but also on other individuals. One example is given by the relation between e and speech-time, i.e. the time of utterance. Although the latter can be taken to be an element of S, it is not determined by the event e or its execution-sequence but is independent. A second example is the consequent-state of e. If CS is defined for e, it is not part of e's execution-sequence (except for its first state).

Aspectual distinctions can be made with respect to each of the three components. The distinctions between verbs concern the dynamic component of e. They are not directly explained at the level of the eventuality structure **E** and therefore at the level of changes as objects. Rather they are defined at the level of changes as transformations of states, that is at the level of the corresponding transition structure **S**. Verbs belonging to the class of accomplishments are distinguished from those belonging to the class of activities by the way they bring about the change (transformation of state) and therefore by the basic procedure that is assigned to the corresponding event-type. For instance, both 'John eat an apple' and 'Bill push a cart' can aspectually be interpreted as denoting sets of sequences that are brought about by events of type P_{eat} and P_{push}, respectively. It is therefore possible to translate these sentences uniformely as expressions, of sort procedure, in some language L that are interpreted with respect to the dynamic event structure **ES** and a sequence σ $(= (s, s')) \in S^*$ (the reference to **ES** will be dropped).

(8) $(s, s') \models \phi$

In (8), ϕ is a σ-procedure-expression (or σ-procedure, for short). The interpretation of ϕ corresponds to the application of a dynamic mode to a property of states. Let $OP \in \{Min\text{-}BEC, Con\text{-}BEC\}$. The σ-procedure ϕ will then be either of the form $Min\text{-}BEC(p)$ or $Con\text{-}BEC(p)$ for $p \in VAR$. The disadvantage of using (8) is that there is no reference to an event e that brought about the change denoted by ϕ. This shortcoming can be remedied by extending L to a two-sorted language in which besides σ-procedures there are also e-formulas, the base clause of which can be defined as:

(9) $e \models SEQ(\phi)$ iff $\tau(e) \models \phi$

According to (9), the base clause for an e-formula combines the perspective of a change as an object with that of a change as a transformation of state.

What has so far been said in this section can be further illustrated by comparing our account to the way verbs are interpreted in event-semantics. What is here called the static component of an event e is captured in event-semantics by interpreting an n-place verb v as an $n+1$-ary relation, (10).

(10) $\lambda d_1 \ldots \lambda d_n \lambda e[P_v(e) \wedge \theta_{TR_1}(e) = d_1 \wedge \ldots \wedge \theta_{TR_n}(e) = d_n]$

In **ES**, (10) can be defined by means of the function κ. What is missing in event-semantics is the second aspect of the notion of change, namely that it is a transformation of state, bringing about a particular condition that holds in e's target-state $\beta(e)$ and that is brought about in a particular way. This aspect corresponds to the dynamic structure of an event and can be expressed by a function that maps each P_v to the binary relation in (11).

(11) $\lambda s \lambda s' \lambda e[P_v(e) \wedge \alpha(e) = s \wedge \beta(e) = s' \wedge \gamma(P_v)(\delta(P_v)(\rho(P_v)(e)))(s)(s')]$

Note that in general not for each $e \in P_v$ will there be a pair (s, s') that satisfies (11) for P_v. This is the case because not each event e brings about the postcondition determined by δ. Thus,

$$\lambda e[P_v(e) \wedge \gamma(P_v)(\delta(P_v)(\rho(P_v)(e)))(\alpha(e))(\beta(e))]$$

need not be equal to P_v (identifying a set with its characteristic function). Each event $e \in P_v$ that satisfies

$$\lambda e[P_v(e) \wedge \gamma(P_v)(\delta(P_v)(\rho(P_v)(e)))(\alpha(e))(\beta(e))]$$

is a completed, non-aborting event of type v. Combining the two perspectives yields (12), where $(1 \leq i \leq n)$.

(12) $\lambda d_1 \ldots \lambda d_n \lambda e[P_v(e) \wedge \theta_{TR_i}(e) = d_i \wedge \gamma(P_v)(\delta(P_v)(\rho(P_v)(e)))(\alpha(e))(\beta(e))]$

(12) can be interpreted as a decompositional elaboration of (10). What is added is the dynamic structure: firstly, an event has an execution-sequence, correlating e to a particular type of binary relation and, secondly, e brings about a particular postcondition.

What cannot be expressed in the language sketched so far are the contributions made by other constituents of a sentence. (9) only captures the relationship between e and its dynamic component. The other two components are completely left out. The static component of e is primarily related to the introduction of objects by argument-NPs. As was already explained above, these objects are aspectually relevant because, for instance, they can play a role in the interpretation of manner-adverbs that admit of two interpretations and that are distinguished by the individual they refer to: either the event e or the object that undergoes a change. It must therefore be possible to refer to individuals other than e.

The interpretation of the PR and manner-adverbs primarily concerns the external component of e. First, the aspectual restriction on both types of expression is formulated in terms of the notions of result and consequent-state. As was shown in Section 2.3, they can be formulated in terms of the functions δ and γ. Secondly, if this presupposition is satisfied, the interpretation of the PR relates the speech-time (or, more generally, the time of orientation, see Section 2.7 and Naumann [1998a]) to the consequent-state of e. The interpretation of the PR therefore determines the relation between elements of e's external structure. The interpretation of a manner-adverb, on the other hand, imposes a further condition on either the execution-sequence (procedural interpretation) or the consequent-state of e (result interpretation).

What is determined during the interpretation of a sentence are two things: first, the relationship between $\tau(e)$ and speech time and second the relationship between e and the elements of its static component. These two relationships must be determined for each sentence because otherwise the sentence cannot be assigned a truth value.

2.4.1 The Basic Language L

The basic language L captures the relationship between e and its dynamic component and therefore aspectual distinctions that can be expressed in terms of the properties of execution-sequences of events. Recall from Section 2.3 that the execution-sequences of events can be split into three different parts: IS, INT-SEQ and OS. The execution-sequences of (completed) events of type P_v are elements of the binary relation $R = \gamma(P_v)(Q)$ for some Q such that Q is evaluated on each part in a particular way that is determined by $\gamma(P_v)$. It must therefore be possible to express in L properties of each of the three parts of an execution-sequence. A logic in which this can be done is the propositional modal logic of time intervals developed by Halpern and Shoham [1986], a fragment of which will be used as the basic language L.

Syntax of L. The language L is two-sorted. There are σ-procedures that are evaluated on sequences of states and e-formulas that are evaluated with respect to elements from E.

(a) The set of (well-formed) σ-procedures is inductively defined as follows: (i) each propositional letter $p \in VAR$ is a σ-procedure, (ii) if ϕ,ψ are σ-procedures, then $\neg\phi$, $\phi \wedge \psi$, $<B>\phi$, $<B^*>\phi$, $<E>\phi$ and $<E'>\phi$ are σ-procedures. The dual operators $[B]$, $[B^*]$, $[E]$ and $[E']$ are defined as usual: $[X] \equiv \neg\langle X\rangle\neg$ for $X \in \{B, B^*, E, E'\}$.

(b) The set of (well-formed) e-formulas is defined as follows: if ϕ is a σ-procedure of the form $OP(p)$ for $OP \in DM - \{HOLD\}$ and $p \in VAR$, then $SEQ(\phi)$ is an e-formula (see (19) below for a definition of $OP(p)$ in terms of the σ-procedures defined in (a)).

According to (ii), there is only one type of e-formula in L. This will be different for the extended language L^h defined in the next section. The restriction to σ-procedures of the form $OP(p)$ has to do with the fact that the relevant σ-procedures are those that express possible execution-sequences of events.

Semantics of L. The satisfaction relation $\models$ for σ-procedures is defined as follows.

(13)
a. $(s, s') \models p$ iff $s' \in Q_p$ for $p \in VAR$
b. $(s, s') \models \neg\phi$ iff not $(s, s') \models \phi$
c. $(s, s') \models \phi \wedge \varphi$ iff $(s, s') \models \phi$ and $(s, s') \models \varphi$
d. $(s, s') \models \top$ iff $s' \in S$, where $\top$ is some tautology
e. $(s, s') \models \langle B\rangle\phi$ iff $\exists s''$ such that $s \leq s'' \wedge s'' < s' \wedge (s, s'') \models \phi$
f. $(s, s') \models \langle E\rangle\phi$ iff $\exists s''$ such that $s < s'' \wedge s'' \leq s' \wedge (s'', s') \models \phi$
g. $(s, s') \models \langle E'\rangle\phi$ iff $\exists s''$ such that $s'' < s \wedge (s'', s) \models \phi$
h. $(s, s') \models \langle B^*\rangle\phi$ iff $\exists s''$ such that $s < s'' \wedge s'' < s' \wedge (s, s'') \models \phi$

To the clauses for the modal-operators correspond the binary relations in (14).

(14) a. $R_{\langle B\rangle\phi} = \{(s,s') \in S^* \mid \exists s''[s \leq s'' \wedge s'' < s' \wedge (s,s'') \models \phi]\}$

b. $R_{\langle E\rangle\phi} = \{(s,s') \in S^* \mid \exists s''[s < s'' \wedge s'' \leq s' \wedge (s'',s') \models \phi]\}$

c. $R_{\langle E'\rangle\phi} = \{(s,s') \in S^* \mid \exists s''[s'' < s \wedge (s'',s) \models \phi]\}$

d. $R_{\langle B^*\rangle\phi} = \{(s,s') \in S^* \mid \exists s''[s < s'' \wedge s'' < s' \wedge (s,s'') \models \phi]\}$

Intuitively, $\langle B\rangle$ and $\langle E'\rangle$ mean 'at some sequence during the current one, beginning when the current one begins' and 'at some sequence of which the current one is an end'. $\langle B\rangle\phi$ is true on a sequence σ just in case ϕ is true of a proper prefix σ' of σ, i.e. of a sequence (s,s'') such that $s'' < s'$ and $\sigma = (s,s')$. The B^*-variant excludes the (strict) minimal prefix (s,s). Similarly, $\langle E'\rangle\phi$ is true on a sequence (s,s') just in case ϕ is true of some sequence (s'',s') of which (s,s') is an end. As both the B-operator and the two E-operators refer to strict subsequences, it follows that $[B]\bot$ is true only on minimal sequences of the form (s,s). It is therefore possible to define a beginning-point modal operator and an end-point modal operator.

(15) a. $[[BP]]\phi \equiv ((\phi \wedge [B]\bot) \vee \langle B\rangle(\phi \wedge [B]\bot))$

b. $[[EP]]\phi \equiv ((\phi \wedge [B]\bot) \vee \langle E\rangle(\phi \wedge [B]\bot))$

The binary relations corresponding to these two modal-operators are given in (16).

(16) a. $R_{[[BP]]\phi} = \{(s,s') \in S^* \mid (s,s) \models \phi\}$

b. $R_{[[BP]]\phi} = \{(s,s') \in S^* \mid (s',s') \models \phi\}$

In terms of $[B]\bot$, σ-procedures can be defined expressing the length of a sequence.

(17) a. $length(0) \equiv [B]\bot$

b. $length(1) \equiv \langle B\rangle\top \wedge [B][B]\bot$

The execution-sequences of completed events of an event-type P_v belonging to the aspectual class

$$AC \in \{Accomplishment, Activity, State, Point, Achievement\}$$

must satisfy the corresponding procedure in (18).

(18) a. Accomplishment: $[[BP]]\neg\phi \wedge [B^*]\neg\phi \wedge [[EP]]\phi$

b. Activity: $[[BP]]\neg\phi \wedge [B^*]\phi \wedge [[EP]]\phi$

c. State: $[[BP]]\phi \wedge [B^*]\phi \wedge [[EP]]\phi$

d. Point: $[[BP]]\neg\phi \wedge [[EP]]\phi \wedge length(1)$

e. Achievement: $\langle E'\rangle([[BP]]\neg\phi \wedge [B^*]\neg\phi \wedge [[EP]]\phi) \wedge length(0)$

For a given verb-type v, ϕ in (18) expresses the postcondition brought about by an event e of type P_v (i.e. $Q_\phi = \delta(P_v)(\rho(P_v)(e)))$ such that $\phi \in VAR$. Thus, the procedures expressed in (18) correspond to dynamic modes. The σ-procedures of the form $OP(\phi)$ for $OP \in DM$ can now be defined as given in (19).

(19) a. Accomplishment: $Min\text{-}BEC(\phi) \equiv [[BP]]\neg\phi \wedge [B^*]\neg\phi \wedge [[EP]]\phi$
b. Activity: $Con\text{-}BEC(\phi) \equiv [[BP]]\neg\phi \wedge [B^*]\phi \wedge [[EP]]\phi$
c. State: $HOLD(\phi) \equiv [[BP]]\phi \wedge [B^*]\phi \wedge [[EP]]\phi$
d. Point: $CHANGE(\phi) \equiv [[BP]]\neg\phi \wedge [[EP]]\phi \wedge length(1)$
e. Achievement: $Min\text{-}BEC_{suf}(\phi) \equiv \langle E' \rangle([[BP]]\neg\phi \wedge [B^*]\neg\phi \wedge [[EP]]\phi) \wedge length(0)$

In the sequel, $OP(\phi)$, $OP \in DM$ will be used to denote one of the five possibilities. The elements from $\{R_{OP(p)} \mid OP \in DM,\ p \in VAR\}$ can now be defined in terms of the relations corresponding to the different modal-operators. For $R_{Min\text{-}BEC(p)}$ and $R_{Con\text{-}BEC(p)}$ one gets the definitions in (20).

(20) a. $R_{Min\text{-}BEC(p)} = R_{[[BP]](\neg p)} \cap R_{[B^*](\neg p)} \cap R_{[[EP]](p)}$
b. $R_{Con\text{-}BEC(p)} = R_{[[BP]](\neg p)} \cap R_{[B^*](p)} \cap R_{[[EP]](p)}$

Semantics of *e*-formulas.

(21) $e \models SEQ(\phi)$ iff $\tau(e) \models \phi$

According to the formation rule for *e*-formulas, ϕ in (21) is always of the form $OP(p)$ for $OP \in \{Min\text{-}BEC, Con\text{-}BEC, CHANGE, Min\text{-}BEC_{suf}\}$ and $p \in VAR$. This reflects the fact that *e*-formulas are supposed to always relate an event e with its execution-sequence $\tau(e)$. Therefore, of interest are only those procedures that express a binary relation, the execution-sequence of an event can possibly be an element of. Consequently, the translation of a verb is (22).

(22) $v \rightsquigarrow OP(p)$ where $R_{OP} = \gamma(P_v)$ and $p \in VAR$

Note that according to (22), there is no single translation of a verb v into L. Rather for each $p \in VAR$ one gets a procedure as translation. This is a consequence of the assumption that changes as state-transformers are interpreted as a particular way of bringing about some condition that is not restricted to particular properties of states but applies in principle to all properties of states.

It may be argued that (21) does not capture the intended interpretation, namely that e brought about Q_p according to the way denoted by OP. (21) can be true although $R_{OP(p)} \neq \gamma(P_v)(\delta(P_v)(\rho(P_v)(e)))$. For instance, it is possible that (21) is true because some event e' with $\tau(e') = \tau(e)$ brought about Q_p. This shortcoming can be overcome by strengthening (21) to (23).

(23) $e \models SEQ(OP(p))$ iff $R_{OP} = \gamma(P_v), Q_p = \delta(P_v)(\rho(P_v)(e))$ and $\tau(e) \models OP(p)$

2.4.2 The Extended Language $\mathbf{L}^h$

The language L defined in the previous section is not expressive enough to account for all the data discussed in Section 1. What is needed is an extension of L that allows us to refer to particular elements of **ES**. From a linguistic perspective this means that more (semantic) structure of a sentence must be represented in the object-language. The structure that becomes 'visible' concerns objects that are introduced by various constituents. For instance, argument-NPs introduce objects and the tenses are always interpreted with respect to speech-time. The extension of L consists essentially in admitting variable binding operators (or binders for short). Syntactically, they apply to e-formulas and yield other e-formulas. This means that one goes from a purely propositional system to a hybrid system in which there are variables that can be bound and that are interpreted with respect to the different domains of **ES**. The resulting language will be called L^h where the index h is an acronym for hybrid.

Syntax of L^h.

(i) Variables. Variables in L^h are sorted: V_O is a set of variables of sort object; elements from V_O will be x (possibly primed). V_S is a set of variables of sort sequence; elements from V_S will be v (possiby primed) as well as t_O and t_E. V_E is a set of variables of sort event; elements from V_E are ε (possibly primed). Finally, there is a set V of variables of sort event or object; they are written α (possibly primed). The variable assignment g assigns to each variable of a given sort an element of the corresponding domain (in particular, variables from V are assigned by g either an element from E or an element from O).

(ii) Procedures and Formulas. (a) The definition of σ-procedures in L^h is identical to that in L

(b) e-formulas: (i) if ϕ is a σ-procedure of the form $OP(p)$, then $SEQ(\phi)$ is an e-formula; (ii) if π is an e-formula and x, v, α are variables of sort object, sequence and object/event, respectively, then $A_v\pi$, $RES_v\pi$, $G_v\pi$, $ADV_\alpha\pi$ ($ADV \in ADVERB$), $AR_x\pi$ and $\mathbf{P}\pi$ are e-formulas

Semantics of L^h. In contrast to L, e-formulas in L^h are not interpreted with respect to a model and an event e but rather with respect to a model, a variable assignment g, an event e, a sequence ϵ of objects and a sequence σ such that the evaluation-context is a triple $< e, \epsilon, \sigma >$. The second element ϵ corresponds to the function of an argument-NP described above, namely that of introducing an object. Each occurrence of the binder AR, that is used in the translation of NPs, changes ϵ by adding an object to it (see Section 2.5). Furthermore, ϵ makes it possible to define a notion of scope. If a binder syntactically follows an occurrence of AR, the value of its bound variable can be required to be bound to the first element of ϵ and therefore to the same value to which the variable of AR is bound. This mechanism can be used to account for the ambiguity of manner-adverbs that can be interpreted either with respect to e or with respect to the object that undergoes the change. At the beginning of the evaluation-process, ϵ is initialized to the empty sequence. The third element of the evaluation-context, the sequence σ, is used to model the temporal (or dynamic) dimension that is independent of the event e. It therefore concerns the interpretation of the tenses as well as the perfect (see Section 2.7). The relationship between the first and third element of the evaluation-context is the following. Both are determined contextually. At the beginning of the evaluation process, σ is always the speech time S which is taken to be a sequence of length 0: (s, s). The semantic function of the translations of the different tenses as well as that of the perfect is to determine the relation between S and the execution-sequence $\tau(e)$ of e. It is therefore necessary that the third element of the evaluation-context can be changed by (the interpretation of the translations of) the tenses and the perfect (in effect, it is only (the interpretation of the translation of) the perfect which changes the value of the third element). In contrast to the second and the third element of the evaluation-context, the first element is not changed during the process of evaluation but remains constant. At the end of the evaluation process the value of the second element is required to be identical to e's static component, that is to the n-tuple of elements from O to which e is related by the subcategorized arguments of the verb. The third element of the evaluation-context is required to be the execution-sequence $\tau(e)$ of e. These requirements are expressed in the base-clause for e-formulas.

(24) $e, \epsilon, \sigma \models SEQ(OP(p))\ [g]$ iff $\tau(e) = \sigma, \theta^*(e) = \epsilon$ and $\tau(e) \models OP(p)$

θ^* is a function that assigns to each $e \in P_v$ an n-tuple of objects such that $\theta^*(e) = <d_1, \ldots, d_n>$ iff $\kappa(P_v) = <\theta_1, \ldots, \theta_n>$ and $\theta_i(e) = d_i$ for $1 \leq i \leq n$.

Recall that syntactically binders apply to e-formulas and yield e-formulas. In L^h two different types of binders can be distinguished. The first type of binder neither changes the evaluation-context nor does it reset the variable assignment g. Its function consists in testing whether the value assigned to the variable bound by the binder satisfies a particular condition. To this type belong the binders A, RES and ADV for $ADV \in ADVERB$. The former two are used in the translations of the tenses and the PR whereas the latter are used in the translation of manner-adverbs. The difference consists in what is tested. The binder A requires the value of the variable that is bound by it to be identical to the third element of the current evaluation-context. The binder RES, on the other hand, does not require the value of the variable bound by it to be identical to an element of the current evaluation-context. Rather it requires its value to satisfy a particular condition with respect to the first element of the evaluation-context (see Section 2.7 for details). The second type of binder resets the assignment but does not change the evaluation-context. To this type belong the binders G and AR. G binds the value of its variable to the third element of the current evaluation-context. It is used in the translation of the PR. The binder AR, on the other hand, does not bind the value of its variable to an element of the current evaluation-context. Rather, it adds an object to ϵ. Together with the base clause for e-formulas this object must be identical to the value that is assigned to e by some thematic relation (see Section 2.5 below). In addition to these binders there is the modal operator **P** which changes the evaluation-context by changing its third element but which does not reset the assignment.

2.5 The Translation and Interpretation of NPs

According to the analysis of NPs developed by Van der Does and Verkuyl [1991], they are syntactically of the form [[SPEC [DET]] N]. For instance, 'the three carts' is parsed as 'the$_{SPEC}$ three$_{DET}$ carts$_N$'. The head noun ('cart') only contributes sortal information that is aspectually not relevant. The specifier SPEC has two functions. First, it introduces a set W of objects (e.g., a set of five carts). The second function of SPEC is to spawn a set of processes which are running in parallel, Naumann [1995, 1998b]. This function is closely related to that of DET which imposes a cardinality condition on the set W which at the same time restricts the number of processes. For example, in the case of 'push five carts' each of the five carts must be pushed separately. As each process is the execution-sequence of an event e, SPEC in effect spawns a set E of events that are running in parallel. In the present context where only singular, non-mass arguments are considered, which can be taken to denote elements from O, SPEC will semantically be interpreted as introducing an object $d \in O$ such that the contribution of an argument NP reduces to that of introducing some object d which is related to e by some thematic relation θ_{TR}: $\theta_{TR}(e) = d$. NPs are translated by the binder AR that binds a variable of sort object.

(25) a. $NP \rightsquigarrow AR_x$

b. $e, \epsilon, \sigma \models AR_x\pi\ [g]$ iff there is a g' such that $g' \stackrel{x}{=} g$ and $e, \langle g'(x)|\epsilon\rangle, \sigma \models \pi\ [g']$

($g' \stackrel{x}{=} g$ means that g' differs from g at most in the value that is assigned to x.) The semantic function of the AR operator is first to reset the assignment and secondly to add the value assigned to the variable bound by AR to ϵ. Together with the condition on the relation between e and ϵ in the base clause (24), (25b) requires that AR binds the value of its bound variable to that object which stands in the thematic relation that corresponds to the argument position in which AR occurs, i.e., $g'(x)$ must be bound to the value of that thematic relation that corresponds to the position of ϵ in which AR adds the value of its variable.

2.6 The Translation and Interpretation of Manner-Adverbs

Recall from Section 1.3 that a sentence with a manner-adverb can have two different readings, the procedural and the result reading. This ambiguity (or underspecification) must be accounted for by the interpretation of this type of adverb. A result reading is possible only if the underlying event-type P_v defines a CP such that the function *Result* is defined for (completed) events of this type. On the result reading (if defined), (the translation of) a manner-adverb is interpreted as expressing a property of the object with respect to which a change is effected. For instance, if John put the chairs properly, the chairs are said to have the property of standing in a particular way that is specified by the adverb. (Alternatively, the adverb can be interpreted as modifying the result itself, i.e. the property of the chairs that they have when they are standing (and not lying). Yet this interpretation will make the definition of the functions AD_{ADV} more complex. Therefore the first interpretation given in the text is chosen for sake of simplicity). This property must hold for some initial prefix of e's consequent-state. A property of individuals can be defined as a partial function from O to $\wp(S)$ that assigns to an object $d \in O$, for which it is defined, the set of states such that d has the property expressed by the function at s. On the procedural interpretation, (the translation of) a manner-adverb is interpreted as expressing a property of events. In contrast to properties of individuals, a property of events is defined as a partial function that maps an event e for which it is defined to a set of sequences such that each sequence is of length ≥ 1. This condition on the length of the sequence reflects the fact that events are changes which can occur only on non-minimal sequences, i.e. sequences of non-zero length. For the sake of simplicity it is assumed that an event e is either mapped to its execution-sequence $\tau(e)$ (if the adverb is defined for e and applies to e) or to the empty sequence $()$, which will be identified with the empty set (if the adverb is defined for e and does not apply to e). It is therefore possible to define a property of events as a partial function from E to S^*. As $S^* \subseteq \wp(S)$, each AD_{ADV} can be defined as a partial function from $E \cup O$ to $\wp(S)$ (see Section 2.2 above).

The adverb *Adv* is translated by the operator $ADV \in ADVERB$ that binds a variable α of sort event or object, the interpretation of which involves the partial function AD_{ADV}. One gets a result interpretation if α is interpreted with respect to $[\rho(P_v)](e)$ (if defined at all) and a procedural interpretation if it is interpreted with respect to e.

(26) a. $Adv \leadsto ADV_\alpha$

b. $e, \epsilon, \sigma \models ADV_\alpha \pi \ [g]$ is defined only if

$CS(e)$ and $AD_{ADV}(e)$ or $AD_{ADV}(\rho(P_v)(e))$ is defined; if all three are defined, one gets:

$e, \langle d|\epsilon\rangle, \sigma \models ADV_\alpha\ \pi\ [g]$ iff

(i) either $g(\alpha) = d, d = \rho(P_v)(e)$ and there is a $\sigma' \in S^*$ such that

(a) $prefix(\sigma', CS(e))$ and

(b) $\forall s \in \sigma' : s \in AD_{ADV}(\rho(P_v)(e))$

or $g(\alpha) = e$ and $\tau(e) = AD_{ADV}(e)$ and

(ii) $e, \langle d|\epsilon\rangle, \sigma \models \pi\ [g]$

According to (26), ADV_α functions like a test. The value of the variable must either be bound to e, i.e. the first element of the evaluation-context, or to the first element of the second element of the evaluation-context. In the latter case it is in addition required that the value is the object that undergoes the change. In contrast to the binder A it is furthermore tested whether this object satisfies a paticular condition ($\tau(e) = AD_{ADV}(e)$ or $\forall s' \in \sigma' : s \in AD_{ADV}(\rho(P_v)(e))$). (26bia) and (26bib) together imply that AD_{ADV} must hold of $g(\alpha)$ ($= \rho(P_v)(e)$) only on some initial prefix of $CS(e)$.

2.7 The Interpretation of the Simple Tenses and the PR

The simple tenses and the PR are interpreted as denoting a relation between two temporal parameters: orientation time (OT) and evaluation time (ET). OT is the time from where one counts. For the simple tenses this time is always the speech time. ET is the time for which a claim is made by the utterance of a sentence and which is temporally related to the first time in a particular way (determined by the tense- or the perfect-operator). In order to account for the referential or context-sensitive character of the simple tenses, they are translated by the operator A which requires the value of its variable to be bound to the current sequence (Richards et al. [1989]).

(27) $e, \epsilon, \sigma \models A_v\pi\ [g]$ iff $g(v) = \sigma$ and $e, \epsilon, \sigma \models \pi\ [g]$

The $PAST_{v,v'}$-operator is defined as a combination of A and the usual past-modality $\mathbf{P}$. In (38) t_O and t_E are two fixed (and distinct) variables (of sort sequence) corresponding to OT and ET, respectively.

(28) a. $PAST_{t_O,t_E} \equiv A_{t_O}\mathbf{P}A_{t_E}$

b. $e, \epsilon, \sigma \models PAST_{t_O,t_E}\pi\ [g]$ iff there is a σ' such that $e, \epsilon, \sigma' \models \pi[g]$, $g(t_E) = \sigma'$, $\sigma' < \sigma$ and $g(t_O) = \sigma$.

In (28) σ is the speech time, which can be taken to be a minimal sequence of the form (s, s). According to definition (28), it is first tested whether the value of t_O is identical to σ (and thus to speech time). Next the index is shifted indefinitely into the past to a sequence σ' by the $\mathbf{P}$-operator. Finally it is tested whether the value assigned to t_E by g is identical to the sequence σ'. Thus, **P** functions like a shift-component and A as a referential device. The tense-operators do not change the values of the variables t_O and t_E but only test whether they satisfy a particular condition. In contrast to the tenses, the perfect-operator changes the values of both variables. This is done by the G-operator that binds its variable to the current sequence.

(29) $e, \epsilon, \sigma \models G_v \pi \; [g]$ iff there is a g' such that $g' \overset{v}{=} g, g'(v) = \sigma$ and $e, \epsilon, \sigma \models \pi \; [g']$

The G-operator was already used by Richards et al. [1989]. An analogous operator is defined in Blackburn and Seligman [1995]. The PERF-operator is defined in (30).

(30) a. $PERF_{t_O, t_E} \equiv G_{t_O} \mathbf{P} G_{t_E}$

b. $e, \epsilon, \sigma \models PERF_{t_O, t_E} \pi \; [g]$ iff $e, \epsilon, \sigma' \models \pi \; [g']$ for some σ' and g' such that $\sigma' < \sigma$, $g' \overset{t_O, t_E}{=} g, g'(t_O) = \sigma$ and $g'(t_E) = \sigma'$

So far the interpretation in (30) does not account for the aspectual restriction on the PR, that it is admissible only for those verb-types v which define a result, or for the condition that this result must still hold at the current t_O (which is the speech time in the case of the present perfect). The latter condition will be accounted for by means of the operator RES_v defined in (31) (where $\sqsubseteq_{S^*}$is the part-of relation on S^*: $\sigma \sqsubseteq_{S^*} \sigma'$ iff $\forall s[s \in \sigma \rightarrow s \in \sigma']$).

(31) $e, \epsilon, \sigma \models RES_v \pi \; [g]$ is defined only if $CS(e)$ is defined; if $CS(e)$ is defined, one gets

$e, \epsilon, \sigma \models RES_v \pi \; [g]$ iff $g(v) \sqsubseteq_{S^*} CS(e)$ and $e, \epsilon, \sigma \models \pi \; [g]$

RES_v imposes a condition on the current value $g(v)$ of the variable v: the result brought about by e is required to still hold at $g(v)$. If $v = t_O$, the result must still hold at the current orientation-time which is the speech-time for the present perfect. The evaluation fails if this test fails. Note that this requirement induces a definite interpretation on the current value of t_E although the indexical operator A is not used in its interpretation. If the current value of t_O is required to be a part of $CS(e)$, then the result must have been brought about by e because otherwise there would have been another event e' which had undone the result such that t_O could not be a part of e's CS. The PR is defined in (32).

(32) $PR_{t_O, t_E} \equiv G_{t_O} \mathbf{P} G_{t_E} RES_{t_O}$

(See Naumann [1998a] for an analysis of the tenses and the PR that uses Reinhard Muskens' 'Logic of Change'; in this analysis no events are used.)

REFERENCES

[Blackburn and Seligman, 1995] P. Blackburn and J. Seligman Hybrid Languages. *Journal of Logic, Language and Information*, 4:251–272, 1995.

[Comrie, 1976] B. Comrie. *Aspect.* Cambridge University Press, 1976.

[Halpern and Shoham, 1986] J. Halpern and Y. Shoham. A propositional Logic of Time Intervals. In: *Proc. IEEE Symposium on Logic in Computer Science.* Cambridge, 272–292, 1986.

[Krifka, 1992] M. Krifka. Nominal Reference, Temporal Constitution and Thematic Relations. In: I. Sag et al., editors, *Lexical Matters.* CSLI Publications, Stanford, 29–53, 1992.

[Moens and Steedman, 1988] M. Moens and M. Steedman. Temporal Ontology and Temporal Reference. *Journal of Computational Linguistics*, 14, 1988.

[Naumann, 1995] R. Naumann. *Aspectual Composition and Dynamic Logic.* PhD thesis, University of Düsseldorf, 1995.

[Naumann, 1996] R. Naumann. Aspectual Composition in Dynamic Logic. In: P. Dekker et al., editors, *Proc. 10th Amsterdam Colloquium*, ILLC, University of Amsterdam, 1996.

[Naumann, 1998] R. Naumann. A Dynamic Temporal Logic for Aspectual Phenomena in Natural Language. To appear in: D. Gabbay and H. Barringer, editors, *Proc. 2nd Int. Conference on Temporal Logic*, Kluwer, 1998.

[Naumann, 1998a] R. Naumann. A Dynamic Approach to the Present Perfect in English. *Theoretical Linguistics*, 24(2/3), 1998.

[Naumann, 1998b] R. Naumann. A Dynamic Logic of Events and States for the Analysis of the Interaction between Plural Quantification and Verb Aspect in Natural Language. In: *Proc. 5th WoLLIC*, Sao Paulo, 1998.

[Naumann and Mori, 1998] R. Naumann and Y. Mori. An Analysis of Definite and Indefinite Japanese Aspectualities in Dynamic Decompositional Event Semantics. In: *Proceedings 6th Symposium on Logic and Language*, Budapest, 1998.

[Richards et al., 1989] B. Richards et al. *Temporal Representation and Inference.* London, Academic Press, 1989.

[Van Benthem and Bergstra, 1995] J. van Benthem and J. Bergstra. Logic of Transition Systems. *Journal of Logic, Language and Information*, 3:247–283, 1995.

[Van der Does and Verkuyl, 1991] J. van der Does and H. Verkuyl. The Semantics of Plural Noun Phrases. In: J. Van der Does et al., editors, *Generalized Quantifier Theory and Application.* ILLC, University of Amsterdam, 1991.

[Vendler, 1967] Z. Vendler. *Linguistics in Philosophy.* Ithaca, Cornell University Press, 1967.

Combining Semantical and Syntactical Theory Reasoning

Uwe Petermann

1 INTRODUCTION

This paper presents a framework for constructing theory reasoning calculi which are based on hybrid theories. A hybrid theory consists of several subtheories. Some of them may be given by sets of axioms, others by classes of models. In other words, hybrid theories may be given in part syntactically and in part semantically.

1.1 Hybrid Reasoning

In applications we often have to answer the question whether a formula is valid *in a theory* or not. Experimental results (e.g. [Moser and Steinbach, 97]) show that specialized approaches to solving the question whether a formula is $\mathcal{T}$-valid are advantageous over applying a general reasoner for answering the question if $T \to F$ is valid.

Theories are not always homogeneous. They may consist of different subtheories, which, moreover, may be given in different form. Following the paradigm of *hybrid reasoning* we will apply a general approach for coupling a *foreground reasoner*, which takes care of the general logical structure of a formula to be proved, with a *background reasoner*, which is consulted whenever the meaning of special built-ins has to be considered. The dedicated reasoner may use meta-knowledge about the theory $\mathcal{T}$ in order to improve the reasoning process[1]. We say that the theory is a built-in theory.

1.2 Hybrid Theories Combining Semantical and Syntactical Information

Nevertheless, the concept of a homogeneous built-in theory does not cover all applications. In [Petermann, 1996; Petermann, 1997] the notion of a hy-

[1]The STE-transformation presented in [Moser and Steinbach, 97] uses meta-knowledge (only) in a preprocessing step relativizing away the theory of equality.

$\mathcal{T}_1$:

$\neg p(X) \vee \neg np(X)$

$\Re_1 = \{\mathcal{N}\}$ where

$l(X,Y)$: X less than Y

$r(X,Y,Z)$: Z is the remainder of X divided by Y

M_1 :

(1) $$np(X) \vee \neg l(1,Y) \vee \neg l(Y,X) \vee \neg r(X,Y,0) \wedge p(100)$$

Figure 1: A simple hybrid theory and a formula unsatisfiable in that theory

brid theory has been introduced. A hybrid theory is combined from different sub-theories which are related to syntactically different fragments of queries.[2] The assumption that different sub-theories refer to different fragments of formulas raises the hope that a dedicated reasoner for a hybrid theory may be constructed as a rather simple combination of the reasoning procedures for the sub-theories.

While [Petermann, 1996; Petermann, 1997] only treat hybrid theories which are given syntactically, i.e. by sets of axioms, in the present paper we also consider the case that one of the constituents of a hybrid theory is given by a class of models. This representation suggests that reasoning in this sub-theory will be done by computation in the models rather than by deduction relying on an axiomatization. This is a generalization of Bürckert's [Bürckert, 1994] approach to constraint reasoning. That approach allows us to combine a theory, which is given by a class of models of a certain signature Δ, with a theory which, admittedly, enriches signature Δ, but which is not allowed to introduce new axioms. Our approach gets rid of this restriction.

Let us illustrate our generalization by the example in Figure 1. Here we have a hybrid theory constructed from $\mathcal{T}_1$ given by one first-order axiom and $\Re_1$ given by the interpretation of the notions "X is less than Y" and "Z is the remainder of X divided by Y" in the set of natural numbers (upper case letters denote variables). The signature Δ_1 of $\Re_1$ consists of the predicate symbols l and r and the natural numbers as constants. The signature Σ_1 of $\mathcal{T}_1$ consists of the predicate symbols p and np. Models of the hybrid theory $\mathcal{T}_1$, $\Re_1$ are the Σ_1, Δ_1-structures which are extensions of $\mathcal{N}$ and valid in $\mathcal{T}_1$. Formula M_1 displayed in the same figure is unsatisfiable in that hybrid theory $\mathcal{T}_1$, $\Re_1$. In Figure 2 we display formula M_1 in matrix form together with the substitution $\sigma = \{X \mapsto 100\}$. In order to prove the $\mathcal{T}_1$, $\Re_1$-unsatisfiability of Formula M_1 the following argument may be given. There exists a substitution σ' such

[2]By a query we mean a formula to be proved.

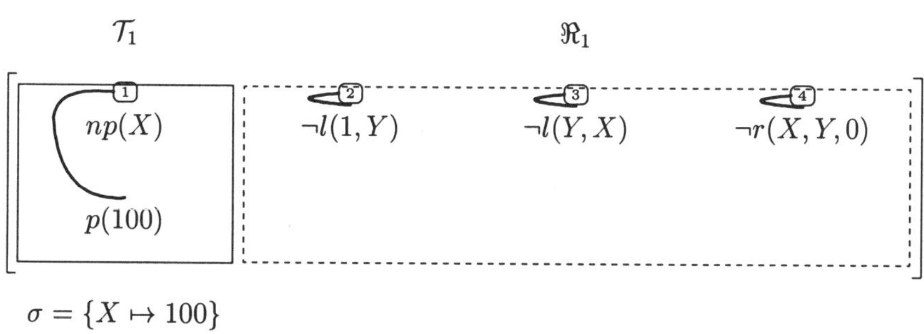

Figure 2: An instantiated matrix with a spanning mating

that every disjunct of the disjunctive normal form of the instance $M_1\sigma$ of M_1 **either** is $\mathcal{T}_1$-unsatisfiable, because it contains a certain $\mathcal{T}_1$-unsatisfiable sub-formula $u\sigma$, **or** becomes $\mathfrak{R}_1$-unsatisfiable after the application of σ', because it contains a certain sub-formula u' such that $u'\sigma'$ is $\mathfrak{R}_1$-unsatisfiable. In our example $\sigma' = \{Y \mapsto 2\}$ would be appropriate. The disjuncts of the normal form will be called paths, the mentioned sub-paths $\mathcal{T}_1$- or $\mathfrak{R}_1$-*connections*. The mentioned substitution σ will be called a $\mathcal{T}_1$-*unifier* for u. Analogously σ' is a theory unifier for u'. In Figure 2 the elements of $\mathcal{T}_1$- and $\mathfrak{R}_1$-connections are connected by arcs. With each of them has been associated one of the labels 1, 2, 3, and 4. Note that, unlike reasoning under the empty theory, theory connections may consist of one or more literals.

It is the task of the background reasoner either to decide the existence of a theory unifier for a multi-set of literals u or to construct a representation of all theory unifiers for u. However, in the present case the background reasoner consists of two sub-devices. One of them is responsible for reasoning under the syntactically given theory $\mathcal{T}_1$, the other for computing under the semantically given theory $\mathfrak{R}_1$. Because $\mathfrak{R}_1$ is given by a class of models we may not assume that there is syntactical characterization of all $\mathfrak{R}_1$ -connections. Rather for each model $\mathfrak{A} \in \mathfrak{R}_1$ the elements of the carrier of $\mathfrak{A}$ will be considered as new constants. We assume that for each literal L in the language extended by the new constants it may be known whether it has a $\mathfrak{A}$-unsatisfiable ground instance or not. Following [Bürckert, 1994], those literals may be seen as constraints, and the existence problem of an $\mathfrak{A}$-unsatisfiable ground instance as a constraint solving problem. The argument given for the $\mathcal{T}_1$, $\mathfrak{R}_1$-unsatisfiability of formula M_1 will be repeated for each model of $\mathfrak{R}_1$ if there is more than one such model. This will be illustrated by an example in Section 3. Consequently also the $\mathcal{T}_1$-reasoner has to be parameterized by the models of $\mathfrak{R}_1$. Again the elements of the carrier of a model $\mathfrak{A} \in \mathfrak{R}_1$

will be treated as additional constants. Here it will be important that the unification procedure for $\mathcal{T}_1$-connections may be combined with free constant symbols. By the treatment of the individuals of $\mathfrak{K}_1$-models as constants it will be possible to use the technique which has been developed for building in theories, which are given syntactically, also to those given semantically (cf. Section 4).

The foreground reasoner should be able to detect multi-sets of literals as candidates for theory connections. It has to keep track that for every path through a formula has been found a connection, and that there exists a common theory unifier for all those connections. In the terminology of the matrix or connection method we say that every path is spanned by a connection. The set of connections is then called a spanning mating. Its existence makes sure that the instance in Figure 2 and consequently Formula (1) are $\mathcal{T}_1$, $\mathfrak{K}_1$-unsatisfiable.

1.3 Building In Hybrid Theories Using a General Framework

In order to combine theories given semantically or syntactically we use a general approach for constructing complete theory reasoning calculi. For constructing a complete total theory reasoning calculus we need first of all a sufficiently rich set of theory connections and secondly a unification procedure for those theory connections.

It is not always necessary that a theory reasoning calculus is able to prove all theorems of the considered theory. Restrictions on the set of formulas (called *queries* further-on) treated as potential theorems may simplify the set of theory connections to be considered. Because of this observation the notion of a set of theory connections which is complete with respect to a given *query language* (see Subsection 5.1) has been introduced. The ability to construct all necessary theory unifiers for theory connections has been formalized by the notion of a solvable unification problem within a given set of theory connections (see Subsection 5.2). Having at hand a complete set of theory connections with a solvable unification problem a complete total theory reasoning calculus may be constructed (see Subsection 5.3).

Finally, the apparatus, which has been developed in Section 5, will be applied in Section 6. Given a class $\mathfrak{K}$ of Δ-models and a theory $\mathcal{T}$ expressed by axioms over a signature Σ, for each $\mathfrak{A} \in \mathfrak{K}$ will be constructed complete sets of theory connections for the theory $th(\mathfrak{A})$[3] and for the theory $\mathcal{T} \cup E(\mathfrak{A})$, where $E(\mathfrak{A})$ is the set of equations valid in $\mathfrak{A}$.

1.4 Related Work

[Bibel, 1982] proposed the use of special purpose connections, exemplifying this approach by building in equality via the use of eq-connections, which are

[3]The set of all sentences expressed in Δ and valid in $\mathfrak{A}$.

instances of theory connections. [Stickel, 1985] viewed many refinements of resolution as instances of theory resolution. [Murray and Rosenthal, 1987] enhanced their path resolution method, which actually is a matrix method, by theory links. [Bürckert, 1994] introduced resolution based reasoning in theories given by classes of models, which provide interpretations of constraints. In the substitutional framework of [Frisch, 1991] solving constraints becomes part of the unification algorithm. For implementations see, for example, [Baumgartner and Stolzenburg, 1995] or [Rigó, 1995]. Along this line [Baumgartner, 1992] and [Petermann, 1990] started the development of a general framework for theory reasoning that is not restricted to syntactically given theories, provides complete calculi for open first-order theories and led to efficient implementations. A translation of Bürckert's approach to theory model elimination may be found in [Baumgartner and Stolzenburg, 1995].

2 PRELIMINARIES

For syntax and semantics of first-order logic we refer the reader to Chapter I.1.1 and for the notion of a *clause* to Chapter I.1.2 of [Bibel and Schmitt, 1998].

A clause is represented as the multi-set of its literals. Multi-sets will be denoted as sequences of their elements. Clauses will be abbreviated also by Γ, C, D etc. Γ_1, Γ_2 denotes the union $\Gamma_1 \cup \Gamma_2$, whereas Γ, L denotes $\Gamma \cup \{L\}$ etc. A clause with at most (exactly) one positive literal will be called a *Horn (definite) clause.* A definite clause consisting only of equational literals will be called a *conditional equation.* Using a terminology dual to [Bibel, 1982], a conjunction of clauses $C_1, \ldots, C_n$ will be called a *matrix.* A copy of a clause is a clause obtained by variable renaming. A set of copies of clauses of a matrix M will be called an *amplification* of M (cf. [Miller, 1983]). For example, the formula displayed in Figure 2 is obtained by instantiating an amplification consisting of exactly one copy of each clause of formula M_1 in Figure 1.

A *(partial) path (in) through* a matrix M is a multi-set containing (at most) exactly one literal from each clause of M. Paths will be abbreviated by p or q. A set of partial paths in a matrix M is called a *mating* in M. A partial path u in a matrix M is *spanning* a path p through M if $u \subseteq p$. A mating U in a matrix M is *spanning* M if for every path through M there exists an element of U spanning it. The mating consisting of the four $\mathcal{T}$-connections labeled by 1, 2, 3 and 4 in Figure 2 is spanning Formula (1).

If matrix M consists of the multi-set of the clauses $C_1, \ldots, C_n$ then $M' = C'_1, \ldots, C'_k$ is called a *sub-matrix* of M if and only if there is a sequence of pairwise different indices $i_1, \ldots, i_k$ such that C'_l is a sub-multi-set of C_{i_l} for each l with $1 \leq l \leq k$. A set of matrices which is closed with respect to the application of substitutions, forming amplifications and sub-matrices, will be called a *query language.* For a path $p = L_1, \ldots, L_n$ and a query language $\mathcal{Q}$ we write $p \in \mathcal{Q}$ as an abbreviation for $\{\{L_1\}, \ldots, \{L_n\}\} \in \mathcal{Q}$.

$\mathcal{T}_2$:

$$\neg s(X,Y,Z) \lor ly(X,Y) \lor ly(X,Z)$$

$$\neg s(X,Y,Z) \lor \neg ly(X,Y) \lor \neg ly(X,Z)$$

$\Re_2 = \{\mathfrak{A}_x\}_{x \in \{mo,tu,we,th,fr,sa,su\}}$ with

mo, tu, we, th, fr, sa, su: constants — the days of the week

to : "today" — a constant, equal to day x in model $\mathfrak{A}_x$

$X \succ Y$: "X is the day after Y"

$X \in Y$: "X is element of set Y"

M_2 :

$$\begin{array}{ll} & ly(X,Y) \lor \neg X = li \lor \neg Y \in \{mo,tu,we\} \\ \land & ly(X,Y) \lor \neg X = un \lor \neg Y \in \{th,fr,sa\} \\ \land & \neg ly(X,Y) \lor \neg X = li \lor \neg Y \in \{th,fr,sa,su\} \\ \land & \neg ly(X,Y) \lor \neg X = un \lor \neg Y \in \{mo,tu,we,su\} \\ \land & s(X,Y,Z) \lor \neg X = li \lor \neg Y = to \lor \neg Y \succ Z \\ \land & s(X,Y,Z) \lor \neg X = un \lor \neg Y = to \lor \neg Y \succ Z \\ \land & \neg to = th \end{array}$$

Figure 3: The Lion and Unicorn Puzzle

For a language $\mathcal{L}$, entailment operator $\models$ and set of formulas $\mathcal{A}$ the triple $\mathcal{T} = (\mathcal{L}, \models, \mathcal{A})$ will be called a *theory*. The elements of set $\mathcal{A}$ are called the *axioms of* $\mathcal{T}$. The set of predicate (function) symbols occurring in the axioms of a theory $\mathcal{T}$ will be denoted by $\mathcal{P}(\mathcal{T})$ ($\mathcal{F}(\mathcal{T})$ respectively).

Alternatively, a theory may be given as a triple $\mathcal{T} = (\mathcal{L}, \models, \mathcal{M})$ where $\mathcal{M}$ is a class of interpretations of a given signature Σ. The notions of free and bound variable, term, atom, literal, (immediate) sub-formula, substitution, and sentence (a formula not containing free variables), satisfiability, unsatisfiability and validity in a theory are defined as usual. If in doubt, the reader is referred to (Fitting, 1996) for the exact definitions. Unbound variables in axioms of a theory or within a formula to be proved are treated as universally quantified. By $\mathcal{E}$ we denote the theory of equality, i.e. the theory with a set of axioms consisting of clauses expressing reflexivity, symmetry, transitivity and functional and predicative substitutivity of the equality sign $=$.

3 A MORE DETAILED EXAMPLE

The Lion and Unicorn Puzzle tells us about the lion and unicorn, who both are sometimes lying and sometimes telling truth depending on the day of the week. For the original formulation of the Lion and Unicorn Puzzle see [Smullyan, 1978], for a later discussion [Ohlbach and Schmidt-Schauß, 1985].

Puzzle 3.1 *1. The lion lies on Monday, Tuesday and Wednesday and tells the truth on the other days. 2. The unicorn lies on Thursday, Friday and Saturday and tells the truth on the other days. 3. Both say they were lying yesterday. Question: Is it true that today is Thursday?*

In Figure 3 we give a formalization of this puzzle. The formulation of the puzzle has been translated in a conjunction of clauses M_2. $ly(X,Y)$ means that X is lying on day Y. $s(X,Y,Z)$ means that X says on Y that he is lying on day Z. The words lion, unicorn and today have been abbreviated to li, un and to. For the days of the week their usual abbreviations are used. Common knowledge about "lying" has been formulated by two axioms in the theory $\mathcal{T}_2$. The knowledge about the days of the week is given by the models of the class $\mathfrak{R}_2$. Each of the models has a universe consisting of the days of the week, and a successor relation $\succ$ modeling the notion "yesterday". Each model represents one day of the week, in particular, the model $\mathfrak{A}_x$ represents the day x. Formally, the equation $to = x$ is true in this model, and for all days y different from x the $to = y$ is not satisfied.

In Figure 4 we present a solution to the Lion and Unicorn Puzzle. Instances of the 5 last disjunctions of formula M_2 form a matrix in that figure. For the third and fourth clause occur two instances. The matrix is divided into two parts, one surrounded by a box the other by a dashed line. The first part is related to the syntactically given theory $\mathcal{T}_2$. The latter part is related to theory $\mathfrak{R}_2$, which is given as a class of models. For each of the models a refutation of the matrix has to be found. For the models $\mathfrak{A}_{mo}$, $\mathfrak{A}_{tu}$ and $\mathfrak{A}_{we}$ the $\mathcal{T}_2$-connection 1 denotes the common first derivation step. This theory connection consists of three literals. It is $\mathcal{T}_2$-unified by substitution σ_1. After this step also for the remaining goals (all of them can be considered as constraints) exist solutions, however, they are different for each of the mentioned models. For model $\mathfrak{A}_x$ with $x \in \{mo, tu, we\}$ they have the form

$$\sigma_{1,x} = \{X''', X^{iv}, X^v \mapsto li, Y''', Y^v \mapsto x, Y^{iv}, Z^v \mapsto x'\} \tag{2}$$

where x is the day after x'. Let us observe that $\sigma_{1,x}$ plays a role like σ' in Figure 2. Similarly, $\mathcal{T}_2$-connection 2 denotes the common first inference step for the models $\mathfrak{A}_{fr}$, $\mathfrak{A}_{sa}$ and $\mathfrak{A}_{su}$. It is $\mathcal{T}_2$-unified by σ_2. Finally, the singleton $\mathfrak{R}_2$-connection 3 covers the model $\mathfrak{A}_{th}$.

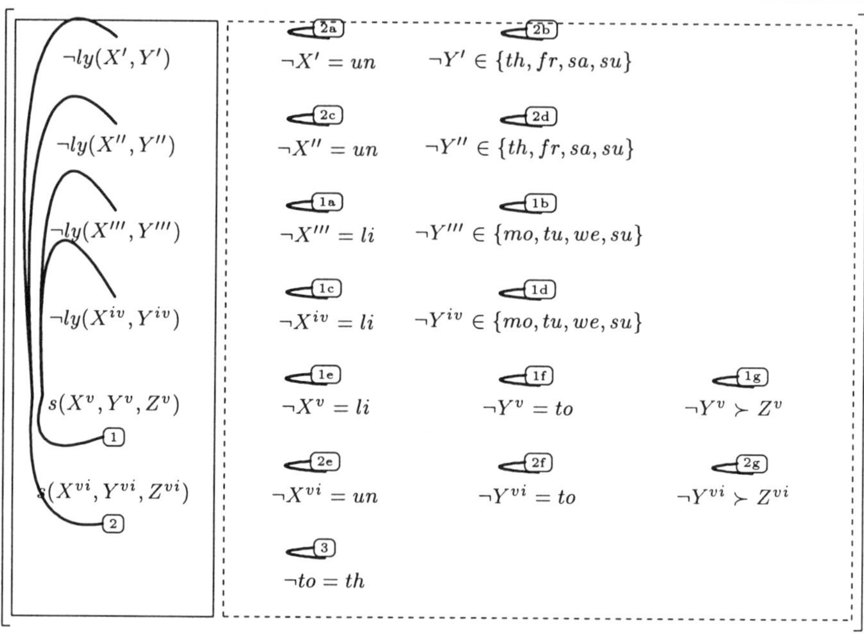

$$\sigma_1 = \{X''', X^{iv} \mapsto X^v, Y''' \mapsto Y^v, Y^{iv} \mapsto Z^v\}$$

$$\sigma_2 = \{X', X'' \mapsto X^{vi}, Y' \mapsto Y^{vi}, Y'' \mapsto Z^{vi}\}$$

$$\sigma_3 = \emptyset$$

Figure 4: A solution to the Lion and Unicorn Puzzle

4 COMBINING SEMANTICAL AND SYNTACTICAL INFORMATION

In this section we outline how to relate semantical and syntactical information. We assume that there are given two theories, $\mathcal{T}$, given by a set of axioms, which are expressed in a signature Σ, and $\mathfrak{R}$, given by a class of structures, which are interpretations of a signature Δ. Both signatures are disjoint. In order to combine the information contained in $\mathcal{T}$ and $\mathfrak{R}$ for each model $\mathfrak{A} \in \mathfrak{R}$, the signatures Δ and Σ may be enriched by treating the elements of the carrier of $\mathfrak{A}$ as new constants. The set of all formulas over the enriched signature which are valid in $\mathfrak{A}$ may be considered. This way we transform that fragment of semantic information given in $\mathfrak{A}$, which is first-order expressible, into syntactical information. We will obtain a calculus for reasoning under theory $\mathcal{T}$, which is parameterized by the Δ-structure $\mathfrak{A} \in \mathfrak{R}$. This is a generalization of Bürckert's resolution principle for constraint logics [Bürckert, 1994]. In Bürckert's approach theory $\mathcal{T}$ may provide only new symbols but no axioms, i.e. Bürckert's approach may be obtained if $\mathcal{T}$ is the empty theory. Having this generalization in mind we have to generalize the Herbrand theorem 5.5, similar to Bürckert's generalization of the Herbrand theorem for the case without built-in theory.

5 A GENERIC APPROACH TO THEORY REASONING

In the present section we introduce a formal framework for constructing complete total theory reasoning calculi for open, i.e. quantifier free, theories. A complete theory reasoning calculus for an open theory needs the following key capabilities: (1) finding theory connections, (2) computing unifiers for theory connections, and (3) managing amplifications and representations of sets of paths which are not spanned by a currently found theory mating. The ingredients for constructing a complete theory reasoning calculus — a complete set of theory connections with a solvable unification problem and a calculus managing amplifications of matrices and keeping track of unsolved goals — will be introduced in subsections 5.1, 5.2 and 5.3 respectively. In Section 6 the results of this section and of Section 4 will be combined. Implementation issues are discussed in short in Section 7.

5.1 Complete Sets of Theory Connections

In order to formulate sufficient conditions for the completeness of the calculus we introduce the notion of a set of theory connections which is complete with respect to a given query language. Having this notion, a Herbrand theorem (5.5) may be proved.

Let $\mathcal{T}$ be an open first-order theory. Without lost of generality we may assume that it is given as a set of clauses. First of all (see Definition 5.3) we formalize what it means to have "enough" theory connections in order to refute all theory unsatisfiable matrices, which belong to a given query language.

Definition 5.1 ((Minimally) $\mathcal{T}$-complementary, $\mathcal{T}$-unifier) A path u is called *$\mathcal{T}$-complementary* if and only if the existential closure of the conjunction of the elements of u, $\bar{\exists}\left(\bigwedge_{L\in u} L\right)$, is $\mathcal{T}$-unsatisfiable. If, moreover, no proper subset of u is $\mathcal{T}$-complementary then u is called *minimally $\mathcal{T}$-complementary.* A substitution σ is a *$\mathcal{T}$-unifier* of u if and only if $u\sigma$ is $\mathcal{T}$-complementary. If $u\sigma$ is minimally $\mathcal{T}$-complementary then σ is called a *minimal $\mathcal{T}$-unifier.*

Obviously, a path u is $\mathcal{T}$-complementary if and only if the universal closure of the disjunction of the negations of the elements of u, $\bar{\forall}\,(\bigvee_{L\in u}\bar{L})$, is $\mathcal{T}$-valid. The following definition introduces the central notion of a $\mathcal{T}$-connection for a given theory $\mathcal{T}$. We will speak simply about theory connections if it is clear from the context which theory is meant.

Definition 5.2 ($\mathcal{T}$-Connection) Let $\mathcal{T}$ be a theory and M a matrix. A partial path u in M will be called a *$\mathcal{T}$-connection* in M if for u exists a $\mathcal{T}$-unifier.

Often we do not need all theory connections in order to prove all theorems out of a considered query language. Rather it may be sufficient (and even desired) to deal only with a certain subset $\mathcal{U}$ of the set of all theory connections. Treating theories given by sets of axioms we used to require $\mathcal{U}$ to be decidable. This requirement has to be relaxed if theories are given by model classes. For each model $\mathfrak{A}$ a calculus has to be constructed where the elements of the carrier of $\mathfrak{A}$ will be treated as constants. Clearly, $\mathfrak{A}$ may be of arbitrary cardinality.

Definition 5.3 (Complete set of theory connections) Let $\mathcal{T}$ be a theory, M a matrix, $\mathcal{U}$ a set of $\mathcal{T}$-connections and $\mathcal{Q}$ a query language. (1) Any set of $\mathcal{T}$-connections in M which are elements of $\mathcal{U}$ is called a *$\mathcal{U}$-mating* in M. (2) If $\mathcal{U}$ is closed with respect to the application of substitutions then it will be called *$\mathcal{T}$-complete with oracle with respect to $\mathcal{Q}$* if (2.1) for each $\mathcal{T}$-complementary ground path $p \in \mathcal{Q}$ there exists $u \in \mathcal{U}$ such that $u \subseteq p$ and (2.2) for each $\mathcal{T}$-complementary ground path of the form $u\sigma \in \mathcal{U}$ with $u \in \mathcal{Q}$ holds $u \in \mathcal{U}$. (3) If, additionally to (2), $\mathcal{U}$ is decidable then it will be called *$\mathcal{T}$-complete with regard to $\mathcal{Q}$*.

Example 5.4 (1) For theory $\mathcal{T}_1$ from Figure 1 the query language is the set of all first-order clauses constructed from np and p as unary predicate symbols. A set of $\mathcal{T}_1$-connections which is complete with respect to this query language consists of the set of all literal sets $\{n(s), \neg n(t)\}$, $\{np(s), \neg np(t)\}$, and $\{n(s), np(t)\}$ where s and t are syntactically unifiable terms. (2) If A is the set of real numbers, then the set $\mathcal{U}_A$ consisting of all $\mathcal{T}_1$-connections of the $u\sigma$, where σ assigns an element of A to each of its elements is complete with oracle with respect to $\mathcal{Q}_A$ obtained analogously to $\mathcal{U}_A$.

A complete set of theory connections also contains all minimal theory connections (cf. [Petermann, 1998] or Chapter I.2.6 in [Bibel and Schmitt, 1998]). This is important because the fewer literals a connection consists of the more paths it may span. Every extra literal may cause additional sub-goals to be solved. The Herbrand theorem follows from [Petermann, 1993] and the closedness of complete sets of theory connections with regard to minimal theory connections.

Theorem 5.5 (Herbrand's theorem) *Let $\mathcal{T}$ be an open theory, $\mathcal{Q}$ a query language, $\mathcal{U}$ a set of $\mathcal{T}$-connections complete (with or without oracle) with respect to $\mathcal{Q}$. Then for every $\mathcal{T}$-unsatisfiable matrix $M \in \mathcal{Q}$ there exists an amplification M' of M, a $\mathcal{U}$-mating U spanning M' and a ground substitution σ such that $u\sigma$ is minimally $\mathcal{T}$-complementary for each $u \in U$.*

5.2 Unification of Theory Connections

The Herbrand theorem (5.5) gives neither a hint of how to find for a given amplification M' of a matrix M the substitution σ such that $M'\sigma$ is theory complementary nor how to decide the existence of such a substitution. In the case of the empty theory, the substitution σ may be computed as the most general unifier for literals $p(t_1, \ldots, t_n)$ and $p(s_1, \ldots, s_n)$. By application of the unifier we obtain a complementary connection. In the case of theory unification, there need not be a most general unifier. Complete sets of unifiers have to be considered instead. For constructing a proof calculus for a given complete set of $\mathcal{T}$-connections $\mathcal{U}$ we need to be able to compute or to represent for every $u \in \mathcal{U}$ all substitutions σ such that $u\sigma$ is $\mathcal{T}$-complementary. This leads to the notion of a solvable unification problem. Like in the notion of a complete set of theory connections we have to take care that theories may be given by model classes. Therefore we also need the notion of a unification problem solvable by oracle.

Definition 5.6 ($\mathcal{T}$-unifier, solvable $\mathcal{T}$-unification problem in $\mathcal{U}$) Let $\mathcal{U}$ be a set of multi-sets of literals. (1) Let ϱ and σ be $\mathcal{T}$-unifiers of a path $u \in \mathcal{U}$ such that $D(\varrho), D(\sigma) \subseteq Var(u)$. Then ϱ is called *more general than* σ if there exists η such that $\varrho\eta =_{Var(u)} \sigma$ (notation: $\varrho \leq \sigma$). (2) A set S of $\mathcal{T}$-unifiers of a multi-set $u \in \mathcal{U}$ will be called *complete* if for each $\mathcal{T}$-unifier σ of u there exists a substitution $\varrho \in S$ such that $\varrho \leq \sigma$. (3) We say that the *$\mathcal{T}$-unification problem in $\mathcal{U}$ is solvable by an oracle* if (3a) for every $u \in \mathcal{U}$ there exists a complete set S_u of $\mathcal{T}$-unifiers for u. If additionally to (3a) S_u is always enumerable and (3b) for a given $u \in \mathcal{U}$ it is decidable whether $S_u \neq \emptyset$ then we say that the *$\mathcal{T}$-unification problem in $\mathcal{U}$ is solvable.* (4) A substitution σ will be called a *simultaneous $\mathcal{T}$-unifier* of a set U of multi-sets of literals if and only if $u\sigma$ is $\mathcal{T}$-complementary for every $u \in U$.

Example 5.7 (1) The complete set of theory connections given in Example 5.4 for theory $\mathcal{T}_1$ from Figure 1 has a solvable unification problem. It just reduces to syntactical unification of the terms occurring in the $\mathcal{T}_1$-connections. (2) The unification problem of the set of theory connections $\mathcal{U}_A$ discussed in Example 5.4 is solvable with oracle. The oracle is able to test whether two elements are equal.

For proving a theorem we have to find a *simultaneous $\mathcal{T}$-unifier* of a spanning $\mathcal{T}$-mating incrementally. By a straightforward calculation the reader may prove that the solvability of the $\mathcal{T}$-unification problem in a set of theory connections is sufficient for this purpose.

5.3 Theory Reasoning Calculi

Two of the three key capabilities of a theory reasoning calculus, which have been mentioned at the start of this section, have been described above — the capabilities of finding theory connections and of finding unifiers for them. Those capabilities may be seen as parameters of a calculus which is able to manage amplifications and to represent unsolved paths as the third key capability of a theorem reasoning calculus. For the technical details of such a calculus we refer the reader to [Petermann, 1998] or to Chapter I.2.6 in [Bibel and Schmitt, 1998]. The calculus may be simulated by a large range of theory reasoning calculi [Baumgartner and Furbach, 1993]. The completeness result cited below carries over to those calculi.

The so called theory pool calculus uses a pool[4] of so-called hooks $(p \perp \Gamma)$ to represent all paths in an amplification M which are not spanned by the $\mathcal{T}$-mating currently found. In particular, the hook $(p \perp \Gamma)$ represents all paths through M which continue the partial path p via one of the literals of the partial clause Γ. If Γ is empty the hook may be discarded. Otherwise it has the form $(p \perp L, \Gamma')$ and L will be chosen in order to perform an inference step. In the more desirable case a unifiable $\mathcal{T}$-connection u with $L \in u \subseteq \{L\} \cup p$ can be found. The current hook has to be replaced by $(p \perp \Gamma')$. In the other case clause copies $L_1, \Gamma_1, \ldots, L_n, \Gamma_n$ of the original matrix and a unifiable $\mathcal{T}$-connection u with $L \in u \subseteq \{L\} \cup p \cup \{L_1, \ldots, L_n\}$ can be found. Now, besides $(p \perp \Gamma')$ also $(p, L \perp \Gamma_1), \ldots, (p, L \perp \Gamma_n)$ have to be adjoined to the new pool. Different strategies are possible for the treatment of unifiers. Either it will always be applied to the whole resulting pool or the superposition of all substitutions found up to a current state of a derivation will be kept beside the pool. Moreover, a constraint reasoning variant of the treatment of the current goal is possible. More than one literal can be chosen at once and they may be replaced by simpler ones due to rewrite rules.

Example 5.8 Consider a successful derivation which as the first inference detects the connection 2 in Figure 4. The initial hook may be constructed from the first clause in Figure 4. The 3 resulting hooks have the clause fragments consisting of the literals marked by 2a, 2b (first), 2c, 2d (second) 2e, 2f, 2g (third hook). Each of the constraint problems is solvable.

Theorem 5.9 (Soundness and Completeness) *Suppose that for an open first-order theory $\mathcal{T}$ and a query language $\mathcal{Q}$ has been given a decidable set $\mathcal{U}$ of $\mathcal{T}$-connections which is $\mathcal{T}$-complete with regard to $\mathcal{Q}$ and the $\mathcal{T}$-unification problem in $\mathcal{U}$ is solvable. Then for every query M from $\mathcal{Q}$*

[4] I.e. a set.

holds: M is $\mathcal{T}$-unsatisfiable if and only if there exists a clause $\Gamma \in M$ and a successful derivation starting from the initial pool $\{(\emptyset \;\; \bot \;\; \Gamma)\}$ such that in each $\mathcal{T}$-connection inference for the chosen connection u holds $u \in \mathcal{U}$ and the chosen $\mathcal{T}$-unifier σ is a minimal $\mathcal{T}$-unifier out of the complete set of $\mathcal{T}$-unifiers S_u for u.

5.4 Application to a Certain Class of Hybrid Theories

Now we consider the special case of theories which are given by certain sub-theories — so called hybrid theories. We formulate a sufficient criterion which allows us to construct a complete set of theory connections with a solvable unification problem (possibly with oracle) for the hybrid theory from complete sets with solvable unification problem given for the sub-theories.

This criterion will be applied to theories constructed in Section 6.1. Using the ideas developed in Section 4, those theories transform semantic information into syntactic information. We therefore assume in this sub-section that both $\mathcal{T}$ and $\mathfrak{R}$ are given by (admittedly uncountable) sets of axioms.

Definition 5.10 Let a theory be given by its sub-theories $\mathcal{T}$ and $\mathfrak{R}$ which are formulated within the signatures Σ and Δ respectively. Then we say that $\mathcal{T}$ and $\mathfrak{R}$ *form a hybrid theory* in the union $\Sigma \cup \Delta$ of both signatures.

Definition 5.11 Let the theories $\mathcal{T}$ and $\mathfrak{R}$ form a hybrid theory in the union $\Sigma \cup \Delta$ of their signatures and let $\mathcal{Q}$ be a query language formulated in a signature which contains $\Sigma \cup \Delta$.

Every clause C in a matrix $M \in \mathcal{Q}$ contains two sub-clauses $C_{\mathcal{T}}$ and $C_{\mathfrak{R}}$, consisting of literals L expressed in signature Σ (respectively L' expressed in signature Δ). The set of nonempty sub-clauses $C_{\mathcal{T}}$ of M will be called the $\mathcal{T}$-layer of M. Analogously will be defined the $\mathfrak{R}$-layer of M. By $\mathcal{Q}_{\mathcal{T}}$ (analogously $\mathcal{Q}_{\mathfrak{R}}$) will be denoted the set of all matrices, being the $\mathcal{T}$-layer (respectively the $\mathfrak{R}$-layer) of a query from $\mathcal{Q}$. $\mathcal{Q}_{\mathcal{T}}$ (analogously $\mathcal{Q}_{\mathfrak{R}}$) will be called the $\mathcal{T}$-*layer* (respectively the $\mathfrak{R}$-*-layer*) of $\mathcal{Q}$.

The following proposition 5.12 gives a sufficient criterion for the theory completeness of the union of sets of theory connections that are theory complete with respect to the constituent sub-theories of a hybrid theory (expressed in the assumptions 4 and 5 of that proposition). Assumption 1 of Proposition 5.12 corresponds to the well known restriction that foreground and constraint language have no common predicate symbols. Assumption 2 is not really a restriction as we will see in Section 6.1. Perhaps restriction 3 could be relaxed.

Proposition 5.12 *Let theories $\mathcal{T}$ and $\mathfrak{R}$, which are expressed in the signatures Σ and Δ respectively, form a hybrid theory such that $\mathcal{T} \cup \mathfrak{R}$ is consistent. The query language $\mathcal{Q}$ is formulated in the union $\Sigma \cup \Delta$ of signatures. E denotes the set of ground equations which are $\mathfrak{R}$-valid. Moreover suppose that:*

(1) $\mathcal{T}$ and $\Re$ have no common predicate symbols.

(2) $\mathcal{U}_\Re$ consists only of singletons.

(3) $\mathcal{T}$ does not contain equality.

(4) The set of $\mathcal{T}$-connections $\mathcal{U}_\mathcal{T}$ is $\mathcal{T} \cup E$-complete with oracle with respect to $\mathcal{Q}_\mathcal{T}$.

(5) The set of $\Re$-connections $\mathcal{U}_\Re$ is $\Re$-complete with oracle with respect to $\mathcal{Q}_\Re$.

A) Then the set of $\mathcal{T} \cup \Re$-connections $\mathcal{U}_\mathcal{T} \cup \mathcal{U}_\Re$ is $\mathcal{T}, \Re$-complete with oracle with respect to $\mathcal{Q}$.

B) $\mathcal{U}_\mathcal{T} \cup \mathcal{U}_\Re$ is even $\mathcal{T}, \Re$-complete with respect to $\mathcal{Q}$ if both $\mathcal{U}_\mathcal{T}$ and $\mathcal{U}_\Re$ are theory complete.

Proof. Let us suppose that theories $\mathcal{T}$, $\Re$ and E, signatures Σ and Δ, query language $\mathcal{Q}$ and the sets of $\mathcal{T}$-connections $\mathcal{U}_\mathcal{T}$ and of $\Re$-connections $\mathcal{U}_\Re$ satisfy the assumptions of the proposition.

In order to show that $\mathcal{U}_\mathcal{T} \cup \mathcal{U}_\Re$ is $\mathcal{T}, \Re$-complete with oracle with respect to $\mathcal{Q}$ we show first of all that $\mathcal{U}_\mathcal{T} \cup \mathcal{U}_\Re$ has property (2.1) formulated in Definition 5.3. Let p be a $\mathcal{T}, \Re$-complementary ground path. We have to show that there exists a sub-path u of p such that $u \in \mathcal{U}_\mathcal{T} \cup \mathcal{U}_\Re$. We consider p as a set of unit clauses. By the compactness theorem for first-order logic there exists a finite set M of instances of clauses of $\mathcal{T}$ and of $\Re$ and a minimal mating U spanning $M \cup p$. Let u be the multi-set of all literals of p which are elements of a connection in U. Then u is not empty because of the consistency of $\mathcal{T} \cup \Re$. Because the sets of predicate symbols occurring in $\mathcal{T} \cup \mathcal{Q}_\mathcal{T}$ and $\Re \cup \mathcal{Q}_\Re$ are disjoint, we either have for every connection $u' \in U$ that $u \in \mathcal{Q}_\mathcal{T}$ or for every connection $u' \in U$ that $u \in \mathcal{Q}_\Delta$. Therefore, u is either an element of $\mathcal{Q}_\mathcal{T}$ or of $\mathcal{Q}_\Re$. If $u \in \mathcal{Q}_\mathcal{T}$ (the case $u \in \mathcal{Q}_\Re$ may be treated analogously) then there exists $u'' \in \mathcal{U}_\mathcal{T}$ such that $u'' \subseteq u$, and therefore $u'' \subseteq p$, because $\mathcal{U}_\mathcal{T}$ is $\mathcal{T}$-complete with respect to $\mathcal{Q}_\mathcal{T}$. Both $\mathcal{U}_\mathcal{T}$ and $\mathcal{U}_\Re$ satisfy condition (2.2) of Definition 5.3. Therefore also $\mathcal{U}_\mathcal{T} \cup \mathcal{U}_\Re$ has this property. In order to prove the thesis B) observe that if both $\mathcal{U}_\mathcal{T}$ and $\mathcal{U}_\Re$ are decidable then so is $\mathcal{U}_\mathcal{T} \cup \mathcal{U}_\Re$. **q.e.d.**

We now turn to a brief discussion of the unification problem in sets of hybrid theory connections. We restrict our attention to the case that for given theories $\mathcal{T}$ and $\Re$ a complete set of theory connections is given by the union of sets of theory connections that are complete with respect to the respective theories. It should be sufficient to apply the $\mathcal{T}$-unification procedure to a $\mathcal{T}$-connection and the $\Re$-unification procedure to a $\Re$-connection. This leads to the notion of non-interfering unification problems.

Definition 5.13 Let $\mathcal{U}_\Re$ and $\mathcal{U}_\mathcal{T}$ be sets of theory connections for the components of a hybrid theory $\mathcal{T}, \Re$. We say that *the unification problems in $\mathcal{U}_\Re$ and $\mathcal{U}_\mathcal{T}$ do not interfere* if and only if

(1) For every $u \in \mathcal{U}_\mathcal{T}$ and for every substitution σ holds: σ is a $\mathcal{T}$-unifier of u if and only if σ is $\mathcal{T}, \mathfrak{R}$-unifier of u and

(2) for every $u \in \mathcal{U}_\mathfrak{R}$ and for every substitution σ holds: σ is a $\mathfrak{R}$-unifier of u if and only if σ is $\mathcal{T}, \mathfrak{R}$-unifier of u.

Proposition 5.14 *Let theories $\mathcal{T}$ and $\mathfrak{R}$, which are expressed in the signatures Σ and Δ respectively, form a hybrid theory, such that $\mathcal{T} \cup \mathfrak{R}$ is consistent. The query language $\mathcal{Q}$ is formulated in the union $\Sigma \cup \Delta$ of signatures. Moreover suppose that the assumptions 1 ... 5 of Proposition 5.12 are satisfied. Then the unification problems in $\mathcal{U}_\mathcal{T}$ and $\mathcal{U}_\mathfrak{R}$ do not interfere.*

Proof. In the non-trivial direction of the equivalence to be proved we have to show that every $\mathcal{T} \cup \mathfrak{R}$-unifier of a $\mathcal{T}$-connection $u \in \mathcal{U}_\mathcal{T}$ is a $\mathcal{T}$-unifier of u and that every $\mathcal{T} \cup \mathfrak{R}$-unifier of a $\mathfrak{R}$-connection $u \in \mathcal{U}_\mathfrak{R}$ is a $\mathfrak{R}$-unifier of u. The latter claim is satisfied because $\mathcal{T}$ and $\mathfrak{R}$ have no common predicate symbols and $\mathfrak{R}$ does not contain the equality sign. The former claim follows from assumption 5. **q.e.d.**

6 APPLYING THE GENERIC APPROACH IN ORDER TO COMBINE SEMANTIC AND SYNTACTIC REASONING

In this section we apply the previously introduced formal framework for constructing complete theory reasoning calculi to our case of hybrid theories, i.e. we define the respective complete sets of theory connections and show how to solve the corresponding unification problems. Throughout this section let us suppose the following conditions. Σ and Δ are disjoint signatures. Σ does not contain function symbols. Δ contains equality. $\mathcal{T}$ is a theory expressed in Σ. $\mathcal{Q}_\mathcal{T}$ is a query language formulated in the signature of $\mathcal{T}$, and $\mathcal{U}_\mathcal{T}$ is a set of $\mathcal{T}$-connections complete with respect to $\mathcal{Q}_\mathcal{T}$. Let $\mathfrak{R}$ be a class of models of a signature Δ.

6.1 Complete Sets of Theory Connections for Sub-Theories

Definition 6.1 (Sets of $\mathfrak{R}, \mathfrak{A}$-connections $\mathcal{U}_{\mathfrak{R},\mathfrak{A}}$ and of $\mathcal{T}, th(\mathfrak{A})$-connections $\mathcal{U}_{\mathcal{T},th(\mathfrak{A})}$)

Let $\mathfrak{A} \in \mathfrak{R}$. By A we denote the carrier of $\mathfrak{A}$[5] and by $\Delta, \mathfrak{A}$ the extension of signature Δ by the elements of A treated as new constants. Formulas over $\Delta, \mathfrak{A}$ will be called $\Delta, \mathfrak{A}$-formulas. Let $th(\mathfrak{A})$ be the set of all $\Delta, \mathfrak{A}$-formulas valid in $\mathfrak{A}$, and $th_0(\mathfrak{A})$ the set of all $\Delta, \mathfrak{A}$-literals valid in $\mathfrak{A}$. By $E(\mathfrak{A})$ let us denote the set of all ground equations which are valid in $\mathfrak{A}$.

[5]Without restriction of generality we may assume that the carrier A is disjoint with the set of variables.

We define $\mathcal{U}_{\Re,\mathfrak{A}}$ as the set of all singleton $th(\mathfrak{A})$-connections. Let $\mathcal{Q}_{\Re,\mathfrak{A}}$ be the set of all disjunctions of $\Delta, \mathfrak{A}$-literals.

Moreover, we define $\mathcal{U}_{\mathcal{T},\mathfrak{A}}$ as the set of all $\mathcal{T}$, $th(\mathfrak{A})$-connections of the form $u\sigma$ where $u \in \mathcal{U}_\mathcal{T}$ and σ is a substitution assigning an element of the carrier of $\mathfrak{A}$ to each element of its domain. We construct $\mathcal{Q}_{\mathcal{T},\mathfrak{A}}$ in a similar way from $\mathcal{Q}_\mathcal{T}$.

Proposition 6.2 *For all $F \in \mathcal{Q}_{\Re,\mathfrak{A}}$: $th(\mathfrak{A}) \models F$ if and only if $th_0(\mathfrak{A}) \models F$.*

Proof. For the non-trivial direction suppose $th(\mathfrak{A}) \models F$ for some $F \in \mathcal{Q}_{\Re,\mathfrak{A}}$. By definition of $th(\mathfrak{A})$ holds $F \in th(\mathfrak{A})$ and by definition of $\mathcal{Q}_{\Re,\mathfrak{A}}$ also $F \in \mathcal{T}_0(\mathfrak{A})$. **q.e.d.**

Now the following proposition may be obtained by straightforward verification of the definitions. The reader should observe that the elements of $\mathcal{U}_{\Re,\mathfrak{A}}$ are singletons.

Proposition 6.3 *$\mathcal{U}_{\Re,\mathfrak{A}}$ is $th(\mathfrak{A})$-complete with oracle with respect to $\mathcal{Q}_{\Re,\mathfrak{A}}$.*

Now $E(\mathfrak{A})$ comes into play. We need those ground equations in order to know if two terms over the signature $\Delta, \mathfrak{A}$ denote the same element of $\mathfrak{A}$. Recall that Δ contains equality.

Proposition 6.4 *For every $u \in \mathcal{U}_{\mathcal{T},\mathfrak{A}}$ and every substitution σ holds:*

$$\mathcal{T}, th(\mathfrak{A}) \models \bigvee \overline{u}\sigma \quad \textit{if and only if} \quad \mathcal{T}, E(\mathfrak{A}) \models \bigvee \overline{u}\sigma$$

Proof. The main argument is that the common symbols of any $u \in \mathcal{U}_{\mathcal{T},\mathfrak{A}}$ and of $\mathcal{T}$, $th(\mathfrak{A})$ are the elements of the carrier of $\mathfrak{A}$ (as constants) and the equality symbol. **q.e.d.**

Proposition 6.5 *$\mathcal{U}_{\mathcal{T},\mathfrak{A}}$ is $\mathcal{T}, th(\mathfrak{A})$-complete with oracle with respect to query language $\mathcal{Q}_{\mathcal{T},\mathfrak{A}}$.*

By propositions 6.3 and 6.5 we have complete sets of theory connections for both sub-languages of the hybrid theory constructed from $\mathcal{T}, E(\mathfrak{A})$ and $\mathcal{T}_0(\mathfrak{A})$.

6.2 Solving the Unification Problems for the Sub-Theories

It remains to prove that the unification problems related to the sets of theory connections $\mathcal{U}_{\mathcal{T},\mathfrak{A}}$ and $\mathcal{U}_{\Re,\mathfrak{A}}$ are solvable — at least with an oracle concerning $\mathfrak{A}$. Solving the unification problem in $\mathcal{U}_{\Re,\mathfrak{A}}$ just means using the oracle related with model $\mathfrak{A}$. On the other hand, $\mathcal{U}_{\mathcal{T},\mathfrak{A}}$ is obtained from $\mathcal{U}_\mathcal{T}$ introducing free constant symbols. Thus an algorithm for solving the unification problem in $\mathcal{U}_{\mathcal{T},\mathfrak{A}}$ can be constructed from an algorithm given for $\mathcal{U}_\mathcal{T}$. We obtain:

Proposition 6.6 *The unification problem in $\mathcal{U}_{\mathcal{T},\mathfrak{A}}$ and the unification problem in $\mathcal{U}_{\Re,\mathfrak{A}}$ are both solvable with oracle.*

6.3 Putting Things Together

Under the assumptions made at the beginning of this section we may observe the following: Theories $\mathcal{T}$, $E(\mathfrak{A})$ and $th_0(\mathfrak{A})$ together with the query languages $\mathcal{Q}_{\mathcal{T},\mathfrak{A}}$ and $\mathcal{Q}_{\mathfrak{R},\mathfrak{A}}$ and the sets of theory connections $\mathcal{U}_{\mathcal{T},\mathfrak{A}}$ and $\mathcal{U}_{\mathfrak{R},\mathfrak{A}}$ satisfy the assumptions of Propositions 5.12 and 5.14. Thus, we have the following proposition.

Proposition 6.7 *$\mathcal{U}_{\mathcal{T},\mathfrak{A}} \cup \mathcal{U}_{\mathfrak{R},\mathfrak{A}}$ is complete (with oracle) with respect to the query language combined from $\mathcal{Q}_{\mathcal{T},\mathfrak{A}}$ and $\mathcal{Q}_{\mathfrak{R},\mathfrak{A}}$. Moreover, the unification problems in $\mathcal{U}_{\mathcal{T},\mathfrak{A}}$ and $\mathcal{U}_{\mathfrak{R},\mathfrak{A}}$ do not interfere.*

Proposition 6.7 has the important consequence that from the complete calculi which exist for both theories $\mathcal{T}, E(\mathfrak{A})$ and $th_0(\mathfrak{A})$ complete calculi may be constructed if a generalized Herbrand theorem (cf. Section 4) has been proved.

7 IMPLEMENTATION ISSUES, SUMMARY AND OUTLOOK

The more application fields that are opened for formal reasoning devices the greater the need for combining different formalisms. The results reported here meet this need.

The CAlculi PRogramming Interface (CaPrl) (see Part II.1.4 in [Bibel and Schmitt, 1998] or [Neugebauer and Petermann, 1995]) is based on a specification language for expressing inference rules by so called descriptors. A set of descriptors steers the translation process of a Prolog Technology Prover [Neugebauer and Petermann, 1995]. An important feature of CaPrl is the possibility to call Prolog-procedures as part of the execution of inference rules. This gives the possibility of combining syntactical and semantical information. Those features have been exploited in a case study by Gerd Neugebauer concerning proof problems involving data structures like natural numbers or lists. The (unpublished) study gives some evidence of the benefits of the substitution of deduction by computation (whenever this is possible).

In a case study [Rigó, 1995] provers for multi-modal logic relying on the algebraic translation [Debart *et al.*, 1990] have been implemented. In one of the implemented versions, two theories $\mathcal{T}$ and $\mathfrak{R}$ are combined. $\mathfrak{R}$ is given by the model of a definite theory. In another case study a further combination of deduction and computation has been explored. Examples using data structures, like lists and different representations of natural number, have been explored.

With this paper we developed a common formal framework for experimenting with combinations of formalisms. In further work the limitations of the present approach should be relaxed. In particular the restriction that theory $\mathcal{T}$ does not contain function symbols seems to be unnecessary. However, if it were dropped, problems of combining unification procedure appear. Therefore results like [Baader and Schulz, 1998] should be used to gain further progress.

Acknowledgment. This work has partially been funded by the Deutsche Forschungsgemeinschaft under grant Pe 480/6-1.

REFERENCES

[Baader and Schulz, 1998] Franz Baader and Klaus U. Schulz. Unification theory. In Wolfgang Bibel and Peter H. Schmitt, editors, *Automated Deduction. A Basis for Applications*, volume I, chapter 7, pages 225–263. Kluwer Academic Publishers, 1998.

[Baumgartner and Furbach, 1993] Peter Baumgartner and Ulrich Furbach. Consolution as a Framework for Comparing Calculi. *Journal of Symbolic Computation*, 16(5):445–477, 1993.

[Baumgartner and Stolzenburg, 1995] Peter Baumgartner and Frieder Stolzenburg. Constraint model elimination and a PTTP-implementation. In *Proceedings Workshop on Theorem Proving with Analytic Tableaux and Related Methods*, Lecture Notes in Artificial Intelligence, pages 201–216. Springer, 1995.

[Baumgartner, 1992] P. Baumgartner. Theory Model Elimination. In H. J. Ohlbach, editor, *Proc. GWAI 92*, 1992. MP-I-Inf.

[Bibel and Schmitt, 1998] Wolfgang Bibel and Peter Schmitt. *Deduction — A Base for Application.* Kluwer Academic Publishers, 1998.

[Bibel, 1982] W. Bibel. *Automated Theorem Proving.* Vieweg Verlag, Braunschweig, 1982.

[Bürckert, 1994] Hans-Jürgen Bürckert. A resolution principle for constrained logics. *Artificial Intelligence*, 66:235–271, 1994.

[Debart *et al.*, 1990] Francoise Debart, Patrice Enjalbert, and Madeleine Lescot. Multi Modal Logic Programming Using Equational and Order-Sorted Logic. In M. Okada and S. Kaplan, editors, *Proc. 2nd Conf. on Conditional and Typed Rewriting Systems.* Springer, 1990. LNCS.

[Frisch, 1991] A. M. Frisch. The Substitutional Framework for Sorted Deduction: Fundamental Results on Hybrid Reasoning. *Artificial Intelligence*, 1991.

[Miller, 1983] Dale A. Miller. *Proofs in Higher-Order Logic.* PhD thesis, Carnegie Mellon University, Pittsburg Pa., 1983.

[Moser and Steinbach, 97] Max Moser and Joachim Steinbach. STE-modification revisited. Technical Report 3, TU München, Institut für Informatik, 97.

[Murray and Rosenthal, 1987] N. Murray and E. Rosenthal. Theory Links: Applications to Automated Theorem Proving. *J. of Symbolic Computation*, 4:173–190, 1987.

[Neugebauer and Petermann, 1995] Gerd Neugebauer and Uwe Petermann. Specifications of inference rules and their automatic translation. In *Proceedings Workshop on Theorem Proving with Analytic Tableaux and Related Methods*, Lecture Notes in Artificial Intelligence, pages 185–200. Springer, 1995.

[Ohlbach and Schmidt-Schauß, 1985] H. J. Ohlbach and M. Schmidt-Schauß. The lion and the unicorn. *Journal of Automated Reasoning*, 1(3):327–332, 1985.

[Petermann, 1990] U. Petermann. Towards a connection procedure with built in theories. In Jan van Eijck, editor, *Logic in Artificial Intelligence, European Workshop, JELIA 90*, volume 478 of *LNCS*, pages 444–453. Springer, 1990.

[Petermann, 1993] Uwe Petermann. Completeness of the pool calculus with an open built in theory. In Georg Gottlob, Alexander Leitsch, and Daniele Mundici, editors, *3rd Kurt Gödel Colloquium '93*, volume 713 of *Lecture Notes in Computer Science*. Springer-Verlag, 1993.

[Petermann, 1996] Uwe Petermann. Multi-modal reasoning as reasoning in hybrid theories. In *Proceedings Workshop on Theorem Proving with Analytic Tableaux and Related Methods - Short Papers*. Dipartimento di Scienze dell'Informazione, Universita' degli Studi di Milano, 1996.

[Petermann, 1997] Uwe Petermann. Building-in hybrid theories. In *Proceedings Workshop on First-Order Theorem Proving*. RISC Linz, 1997.

[Petermann, 1998] Uwe Petermann. Theorem proving with built-in hybrid theories. Research Report FITL–98–1, Forschungsinstitut für Informations-Technologien Leipzig, April 1998.

[Rigó, 1995] Zoltán Rigó. Untersuchungen zum automatischen Beweisen in Modallogiken. Master's thesis, Universität Leipzig, 1995.

[Smullyan, 1978] Raymond M. Smullyan. *What is the name of this book? The riddle of Dracula and other puzzles.* Prentice Hall, 1978.

[Stickel, 1985] M.E. Stickel. Automated Deduction by Theory Resolution. *Journal of Automated Reasoning*, 1:333–355, 1985.

A Generic Approach to Combining Stochastic Algorithms With Systematic Constraint Solvers

Steven Prestwich

1 INTRODUCTION

Constraint satisfaction and optimization problems arise in many Artificial Intelligence applications, and a variety of algorithms have been proposed for their solution. Two quite different approaches are currently in competition: *systematic* algorithms such as forward checking with tree-search, and *stochastic* algorithms such as hill climbing with penalty functions. Neither has been shown to be consistently better than the other, and combining features of both is an active research area [Freuder et al., 1995].

This paper investigates a generic approach to creating hybrids of algorithms from the two classes. Our method performs stochastic search in the space of search strategies for a systematic constraint solver; the solver is used to probe the constrained space in order to compute an objective function. The idea is straightforward and clearly very flexible, but at first glance it appears prohibitively expensive; it is also not obvious how much effort should be spent evaluating a strategy. We describe a cheap evaluation technique, and show that for sufficiently large problems the hybrids may outperform their systematic and stochastic counterparts. There is a significant overhead which degrades the hybrids' performance on smaller problems in particular, but we describe ways of greatly reducing this overhead.

The paper is structured as follows: Section 2 surveys existing approaches to solving constraint problems; Section 3 describes our method; Section 4 evaluates its performance; Section 5 describes its extension to constrained optimization problems; Section 6 mentions related work; and Section 7 draws some conclusions and discusses future work.

2 BACKGROUND

We shall consider two kinds of constraint problem: Constraint Satisfaction Problems (CSPs) and Constrained Optimization Problems (COPs). In a CSP we are given a set of variables, a domain for each variable, each consisting of a finite set of values, and a set of constraints. Each constraint is defined over a subset of the variables, and restricts the combinations of values these variables may take. The problem is to assign a value to each variable so that all constraints are satisfied. In some applications all such solutions are required, but we restrict ourselves to *exemplification* CSPs in which only one solution is required. Moreover, if a CSP is unsolvable then some applications require proof that this is so; we assume that this is not required. A COP is a CSP in which each solution has a defined cost, specified by a *cost function*; the problem may be to find the solution with least cost (an optimal solution), or to find one with reasonably low cost (a near-optimal solution). There may also be a requirement to prove that the solution is indeed optimal (has least cost). In this paper we shall assume that the goal is to find a near-optimal solution. We choose these versions of the CSP and COP because we are interested in applications in which it is impractical to find all solutions, or to prove unsolvability or optimality — either because the problems are very large, or because we require a solution within a limited time.

Systematic constraint solvers based on tree search proceed by assigning a domain value to each variable, and checking after each assignment that no constraints are violated. If a violation occurs, a solver backtracks to a previous assignment and tries an alternative value. It may also perform additional reasoning, such as forward checking, to anticipate constraint violations. A typical performance problem associated with backtrack search is that an early poor decision cannot be undone until all its consequences have been explored. Various refinements to this basic *chronological backtracking* scheme have been suggested, which are sometimes collectively called *intelligent backtracking*. See, for example, [Kumar, 1992; Nadel, 1989] for surveys on these and other systematic algorithms.

Stochastic algorithms are essentially hill climbers, many of which use some mechanism for escape from local optima. Examples are simulated annealing and genetic algorithms. A stochastic algorithm has an *objective function* which it tries to minimize or maximize, depending upon the chosen metaphor: for a genetic algorithm the objective function is the *fitness function* (to be maximized); for simulated annealing it is the *energy function* (to be minimized). In this paper we shall assume that minimization is the goal; however, we use the common term "hill climbing" even though it seems to imply maximization. A common way of applying stochastic algorithms to CSPs is to generate candidate solutions which may violate constraints. A *penalty function* which penalizes violations is used as the objective function for the stochastic search. For COPs the same method is often used, by combining the

COP cost function (to be minimized) with the penalty function to form the objective function [Richardson et al., 1989]. Penalty functions can be defined in various ways [Eiben et al, 1994] and have been used many times (for example [Burke et al., 1995; Chew et al., 1992; Corne et al., 1994; Ghedira, 1994; Michalewicz and Janikow, 1991; Paredis, 1994; Thornton, 1994]). Some algorithms [Burke et al., 1995; Eiben et al, 1994] use modified genetic operators to preserve the most important constraints, and use a penalty function on the rest. Others modify stochastically generated solutions so that they satisfy all constraints; this works well when it can be achieved [Davis, 1985; Michalewicz and Janikow, 1991] but is usually rather problem-specific or limited to certain types of constraint. The *repair* algorithms also generate candidate solutions and attempt to remove constraint violations by reassigning values. One of the best-known is the Min-Conflicts Hill Climbing algorithm [Minton et al., 1994], which only changes the values of variables involved in constraint violations and selects a new value which minimizes the number of violations. Other refinements may also be used to allow escape from local minima [Morris, 1993; Yugami et al., 1994]. Another well-known stochastic algorithm is GSAT [Selman et al., 1992], which was designed to solve Boolean satisfaction problems but has been successfully applied to several other CSPs.

Both systematic and stochastic algorithms may be augmented by various forms of learning [Bowen and Dozier, 1996; Frost and Dechter, 1994; Ginsberg, 1993; Ginsberg and McAllester, 1994; Richards and Richards, 1997; Stallman and Sussman, 1977; Yokoo, 1994], thereby accumulating knowledge during search. This knowledge is used to avoid searching the same part of the space more than once. Some stochastic algorithms collect sufficient knowledge to ensure completeness, in which case they may be thought of as unusually flexible systematic algorithms.

3 THE HYBRIDIZATION METHOD

In this section we describe our new class of hybrid algorithms. First we define a simple stochastic algorithm: a hill climber. Next we show how this hill climber can be applied to a CSP via a penalty function, giving a repair-style algorithm. Then we show how it can also be applied to a space of search strategies. Finally we describe techniques for reducing overheads incurred by the hybrids.

3.1 Hill Climbing in Solution Space

The stochastic algorithm we use in this paper is a simple hill climber: given a search space consisting of *states*, and an *objective function* on states which we wish to minimize, start from a random state and apply random *actions* (small changes) to it; the action which leads to the greatest improvement (lowest value under the objective function) is accepted; on reaching a state from which no improvement can be made (a local optimum), restart from another random state; continue this process until a state is reached with an objective function value of zero (which is assumed to exist). This hill climber can be used in the space of unconstrained solutions via a penalty function:

- take a *state* to be an assignment of values to all variables;
- take the *objective function* to be the number of constraint violations among these values — a penalty function; [1]
- take an *action* to be the reassignment of a variable to a random value.

We shall use this algorithm (which we call HCR) for comparison with the hybrid algorithms.

3.2 Hill Climbing in Strategy Space

First we formalize what we mean by a search strategy for a systematic constraint solver. A *variable ordering* can be specified by a list of n integers, denoting the order in which the variables are to be assigned values. A simple specification for a *value ordering* for each variable is a starting position s in its domain, implying a value ordering $(s \ldots m, 1 \ldots s\text{-}1)$. Hence a *search strategy* can be specified by a variable ordering (a permutation of the integers $\{1 \ldots n\}$) and an integer in the range $1 \ldots m$ associated with each variable. For example if n=3 and m=4 then ((3,1),(1,4),(2,1)) is a strategy which assigns a value to variable 3, first starting with the first value in its domain; then it assigns a value to variable 1 starting with its fourth domain value; and so on. We could allow all permutations of the domain values, but this scheme seems sufficiently flexible. Now to use the hill climber in a space of search strategies:

- take a *state* to be a search strategy for the given constraint solver;
- take the *objective function* to be the number of variables which can be assigned values by the given constraint solver under the given strategy without backtracking;
- take an *action* to be a random change to a randomly chosen variable's value ordering.

An example of such an action is ((3,1),(1,4),(2,1)) $\longrightarrow$ ((3,1),(1,3),(2,1)). To evaluate a strategy, the values in each domain are tried until one is found which is permitted by the constraint solver; the meaning of *permitted* depends upon the solver: for a simple backtracking solver it means *consistent with previous assignments*; for a forward checker it also means *does not lead to*

[1] We experimented with alternative penalty functions, but this made little difference to the results below.

empty variable domains. Once a variable is assigned a value then this is fixed for the rest of the evaluation — hence the execution is backtrack-free. One reason we disallow backtracking is that in stochastic search the objective function must be cheap to calculate (the success of such methods depends partly upon many states being sampled in a reasonable time). Another reason is that any amount of backtracking we might allow would be arbitrary, and we would like to exclude arbitrary parameters from the hybrids.

3.3 Delta Evaluation

The performance of HCR (in solution space) can be improved by reusing results between actions, a technique sometimes called *delta evaluation* [Corne et al., 1994]. The idea is to compute the new objective function value $F+\Delta F$ from the known value F (computed for the previous state) and the more cheaply computed ΔF. The success of this technique depends upon strong coherence between successive states. In this case ΔF is the number of constraint violations caused by the new value minus the number caused by the old value.

Delta evaluation is more complex for the hybrids, because a small change in search strategy may change several values, and also change how the variable domains are pruned during forward checking. However, it is possible to reduce the overheads of restarting executions by evaluating several strategies together. This is best described by a simple example. Say that n=3, m=4, that we have the search strategy ((1,2),(2,3),(3,1)) and that we wish to find the best action. Each possible action leads to a neighbouring strategy in our search space. The set of these strategies is:

{ ((1,2),(2,3),(3,2)), ((1,2),(2,3),(3,3)), ((1,2),(2,3),(3,4)),
((1,2),(2,4),(3,1)), ((1,2),(2,1),(3,1)), ((1,2),(2,2),(3,1)),
((1,3),(2,3),(3,1)), ((1,4),(2,3),(3,1)), ((1,1),(2,3),(3,1)) }

Without delta evaluation we would generate each of these strategies and evaluate them independently. However, the set can be generated in the given order by a restricted form of backtracking: each variable cycles through all its possible values, starting with the specified value; but when the value of variable v_i is *reassigned*, backtracking is suppressed on variables $v_{i+1} \dots v_n$. When a dead-end is reached (where some variable cannot be assigned a value without violating a constraint) the current assignments are recorded. After enumerating all the possibilities, we then recover the strategy which led to the maximal assignment of values (breaking ties by random selection).

This technique makes two kinds of saving. Firstly, when reassigning the value of variable v_i there is no need to recompute the values of (or forward checking incurred by) the variables $v_1 \dots v_{i-1}$. Secondly, constraints may cause two strategies to yield the same partial assignment; using this restricted backtracking scheme, those assignments are enumerated once only.

3.4 Lamarckism

Overheads can be further reduced by using a technique called *Lamarckism*. A current area of research is the synergy between evolution and learning, and recent results in this area are collected in [Turney et al., 1993]. Although we are using a simple hill climber instead of a genetic algorithm, we can borrow techniques from this field. In our hybrids the search strategy corresponds to the *genotype*, the partial solution found during strategy evaluation to the *organism* (or its *phenotype*), and the number of bound variables to its *fitness*. [2] Lamarckism involves the direct encoding by an organism of acquired knowledge into its genotype. There is no known mechanism for this in biology, but it can be a useful computational technique and can be applied to our algorithms as follows. Say that the best neighbouring strategy is ((1,1),(2,3),(3,1)) and that this yields a partial assignment [v_1=3,v_2=4]. Then we can immediately derive another strategy which yields the same partial assignment more efficiently (rejecting fewer values): ((1,3),(2,4),(3,1)). Note that this strategy was *not* in the neighbourhood set; Lamarckism aids the discovery of larger leaps through the search space.

4 EXPERIMENTS

In this section we test the feasibility of our approach empirically. To compare the performance of various algorithms we shall count the number of constraint checks they incur. This is generally considered to be a reasonable performance metric: CPU time depends largely upon implementation details, and metrics such as the number of hill climbing steps or branch points do not permit the direct comparison of different types of algorithm. Unfortunately there does not seem to be a universally accepted definition of what constitutes a constraint check. We shall increment our constraint check counter whenever all of a constraint's variables are assigned values (even if only tentatively during forward checking) so that it can be checked. This definition may be unfair to stochastic algorithms, because it does not take into account the overhead incurred by a constraint solver of testing whether a constraint is ready to be checked. Nevertheless we shall use it because it seems to be the simplest definition.

4.1 A 3-Colouring Problem

We start with a simple map colouring problem, shown in Figure 1. It consists of n regions (numbered $1 \ldots n$ in the diagram) and we need to colour each region in such a way that no two adjacent regions have the same colour. We have a finite set $\{c_1 \ldots c_m\}$ of colours available. This problem can be cast as a CSP: each region corresponds to a domain variable $v_1 \ldots v_n$, each variable

[2] If we were to use a genetic algorithm instead of a hill climber then this correspondence would be exact.

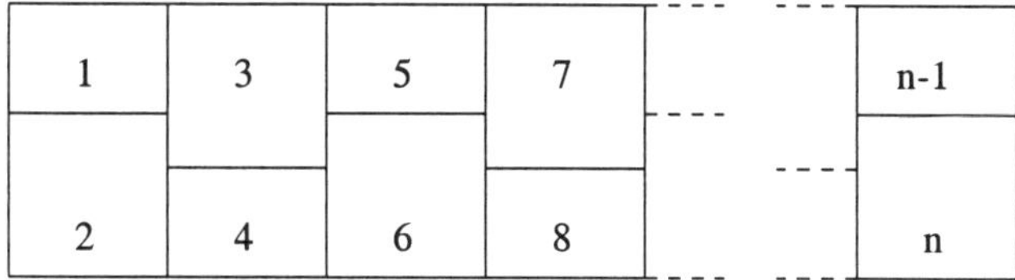

Figure 1: A map colouring problem

has a domain of m values (colours), and pairs of variables corresponding to adjacent regions are constrained to have different values. For m=3 and $n \geq 3$ this problem always has exactly 6 solutions of the form:

$$v_{4i+1} = v_{4i+4} = c_1 \qquad v_{4i+2} = c_2 \qquad v_{4i+3} = c_3 \qquad (i = 0, 1, 2 \ldots)$$

(The 6 solutions are formed by permuting the 3 colours.) Hence for large n the solutions are sparse. It is a highly artificial problem, and can be solved without backtracking simply by colouring regions in the order $1 \ldots n$. However, we shall assume that the variables are randomly assigned to regions so that this colouring order is not discovered. It could be discovered by appropriate heuristics, but the constraint solvers we test do not use such heuristics. A challenging feature of the problem is that its search space has deep local minima: if we colour from each end of the map, we may find that the two consistent colourings are incompatible when they meet in the middle. However, this is not a *hard* problem in the sense of being near the graph colouring phase transition; it is designed mainly to compare the scalability of our hybrids with that of systematic and stochastic algorithms. We consider hard CSPs below.

4.2 A Chronological Backtracker

As a first experiment, we take a simple finite domain constraint solver based on *chronological backtracking* (CB). CB takes each variable in turn and assigns to it a value which is consistent with all current variable assignments. If any variable cannot be assigned a value without violating a constraint, the solver backtracks to the previous assignment and tries another value. We compare the performance of three algorithms: HCR, CB (with random search strategy), and HCR[CB] (the hybrid of CB and HCR). We use the notation HCR[CB] because the HCR algorithm calls the CB constraint solver to compute its objective function.

The graph in Figure 2 plots the number of constraint checks required to solve the CSP against the problem size n, each point averaged over 100 runs. The graph shows that HCR performs quite poorly on this problem — though for sufficiently large n it may outperform CB (we were unable to run the programs long enough to verify this). HCR[CB] is also less efficient than CB for small n but becomes more efficient for larger n. Its poor performance on

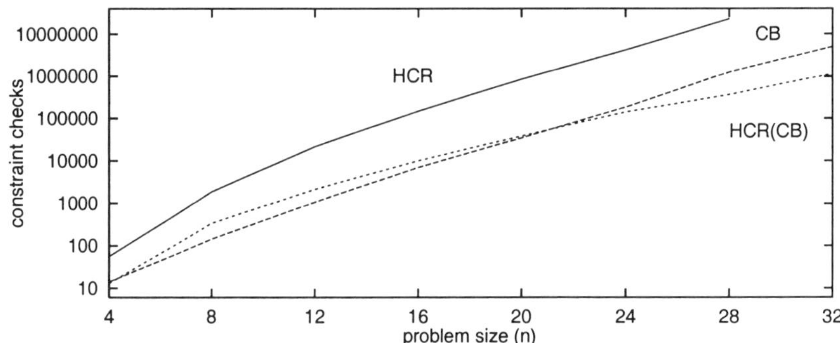

Figure 2: Map colouring results (chronological backtracking)

small problems is caused by overheads which become less significant on larger problems. Hence HCR[CB] is more scalable than CB.

4.3 A Forward Checker

CB is rather a trivial constraint solver, and easily beaten. We now show that hybridization can be successfully applied to a nontrivial solver, based on *forward checking* (FC). FC takes each variable in turn and assigns to it a value from its domain, then removes any inconsistent values from the domains of variables which currently have no value. If any domain becomes empty then backtracking occurs. FC is generally considered to be a powerful technique.

Again we compare the performance of three algorithms: HCR, FC (with random search strategy) and the hybrid HCR[FC]. The graph in Figure 3 shows the results, which are similar to those for CB except that (i) the stochastic algorithm is poorer in comparison with the other two, and (ii) the problem size n needs to be larger before the hybrid emerges as the best. Hence the hybridization again improves the scalability of the solver, and this result does not depend upon the triviality of CB.

4.4 A Forward Checker with Unit Propagation

Some constraint solvers use dynamic heuristics to modify their variable and value orderings, and this may greatly improve their performance. Can stochastically generated search strategies be reconciled with dynamic heuristics, and do similar results hold?

To explore this, we add a well-known dynamic variable ordering heuristic to the forward checker: when any domain is reduced to a single value, the corresponding variable is immediately bound to that value; otherwise variables are assigned values in the usual order. This may lead to further domain pruning, and hence earlier detection of inconsistency. The heuristic is sometimes called *unit propagation* and is a repeated application of the inference rule:

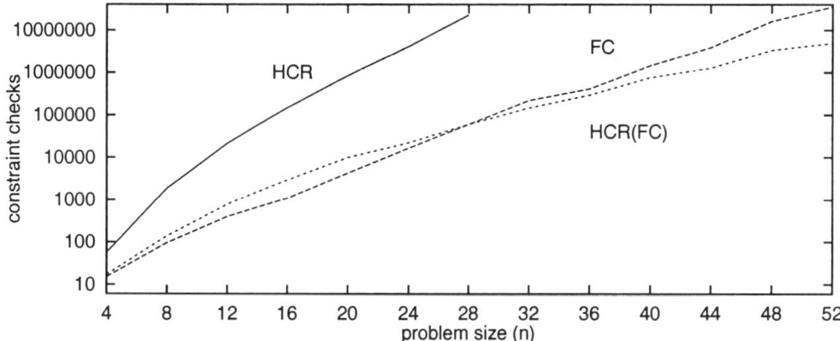

Figure 3: Map colouring results (forward checking)

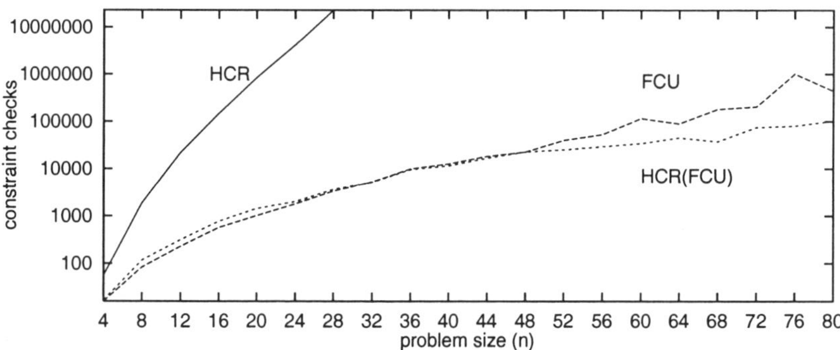

Figure 4: Map colouring results (forward checking with unit propagation)

$$\frac{\begin{array}{l} x \\ \neg x \vee y_1 \vee \ldots \vee y_n \end{array}}{y_1 \vee \ldots \vee y_n}$$

To integrate the stochastically-generated search strategy and the dynamic variable ordering heuristic, we simply take the former as the default and allow it to be superseded by the latter. We hope that the effect of the heuristic will be to judge more accurately the quality of a search strategy, by detecting failure earlier and assigning a lower objective function value to the strategy.

Again we compare the performance of three algorithms: HCR, FCU (the modified forward checker, with random search strategy) and HCR[FCU]. The graph in Figure 4 shows the results. Again the results are similar to those for FC and CB, except that (i) the stochastic search is poorer still in comparison with the other two, and (ii) for low values of n the systematic and hybrid algorithms have very similar performance — this is because FCU is sufficiently powerful to need very little backtracking to solve the problem for small n, so HCR[FCU] often succeeds with its first strategy.

```
procedure GSAT(Σ)
    for i := 1 to MaxTries
        T := random truth assignment
        for j := 1 to MaxFlips
            if T satisfies Σ then
                return T
            else
                PossFlips := {vars which increase satisfiability most}
                V := a random element of PossFlips
                T := T with V's truth assignment flipped
    return "no satisfying assignment found"
```

(Σ is a set of clauses)

Figure 5: The GSAT algorithm

4.5 Hard Random 3SAT Problems

It is not enough to demonstrate our method on a single problem, so we now turn to another. Hard random 3SAT problems are generally considered to be among the most challenging CSPs to solve and have been the subject of much recent research. They are also sufficiently different to our map colouring problem, which was highly structured. A further reason for our interest in 3SAT problems is that they have, roughly speaking, two leading classes of algorithm: stochastic hill climbers such as GSAT [Selman et al., 1992] and variations on the Davis-Putnam procedure such as Tableau [Crawford and Auton, 1986]. The latter are systematic constraint solving algorithms using backtracking, forward checking and unit propagation (plus other heuristics): essentially our FCU algorithm. It is certainly possible to create hybrids such as GSAT[FCU] or GSAT[Tableau] using our method, but do they outperform known algorithms?

Random 3SAT problems are generated as follows. Out of the n variables, random subsets of size 3 are selected to form clauses. Each variable in a clause is randomly made positive or negative, with equal probability. Each clause is interpreted as a disjunction of literals, and the set of clauses as a conjunction: that is a propositional statement in conjunctive normal form. The problem is to find a truth assignment for each variable such that the conjunction is true. Besides n, another parameter to the problem is the number of clauses ℓ. A phase transition occurs approximately at the value ℓ/n=4.3: above it problems are mostly unsolvable; below it they are mostly solvable; near it they are classed as *hard*. Following several other researchers, we shall compare algorithms on hard problems. To do this we generated 10 random solvable problems and applied each algorithm to them 10 times, for various values of n.

GSAT is described in Figure 5; it repeatedly flips (reassigns) variable values, allowing flips which either increase or leave unchanged the number of violated constraints: an important feature which allows random walks over plateaus in the search space. We set MaxTries to ∞ and MaxFlips to $5n$. The hybrid GSAT[FCU] flips *value orderings* instead of actual assignments. The results for FCU, GSAT, HCR, GSAT[FCU] and HCR[FCU] are shown in Figure 6.

It turns out that GSAT[FCU] performs less well than HCR[FCU]. In fact unit propagation has already been added to GSAT [Selman and Kautz, 1993] but did not improve it. The suggested explanation was that unit propagation does not match well with GSAT's randomized local search strategy. Since HCR[FCU] *does* perform well, we conjecture that the search space induced by unit propagation has smaller plateaus and/or a higher incidence of local minima, making GSAT[FCU]'s plateau traversal wasteful and favouring HCR[FCU]'s frequent restart strategy. This view is supported by the fact that GSAT[FCU]'s performance is improved by reducing the MaxFlips parameter.

It also turns out that GSAT is faster than HCR — unsurprisingly, since GSAT is known to be a good stochastic algorithm on this class of problems. A surprise is that GSAT appears to perform poorly compared to FCU and HCR[FCU]. This may be an artefact of the performance metric used (constraint checks). However, comparing the number of flips between HCR[FCU] and GSAT the results are still surprising: HCR[FCU] performs far fewer flips than GSAT, though admittedly its flips are more expensive. To investigate this further we intend to implement an efficient version of HCR[FCU] and compare CPU times on larger problems. For now, we note only that the results are very encouraging. Another satisfying result is that HCR[FCU] outperforms FCU on the larger problems. This agrees with the 3-colouring results: replacing backtracking by an appropriate stochastic algorithm appears to have improved FCU's scalability — and we hope that the same will apply to HCR[Tableau], which we intend to implement. Finally, note that the advantage of HCR[FCU] over FCU is more pronounced in the mean results than in the median; this indicates that its improvement is at least partly due to its avoidance of pathological behaviour.

4.6 Discussion of the Results

Stochastic algorithms search an unconstrained space, which may be much larger than the constrained space searched by systematic algorithms; however, a stochastic search problem only has the same number of optimal solutions as the corresponding CSP has solutions. To compensate for this, stochastic algorithms have superior navigational abilities such as the ability to follow local gradients. There is therefore a trade-off between the size of the search space and the navigation through it, and we should perhaps expect stochastic algorithms to be generally worse on more constrained (harder) problems. There is indeed some evidence of this [Michalewicz and Janikow, 1991], though stochastic algorithms can also solve hard problems efficiently [Selman et al.,

MEAN RESULTS: CONSTRAINT CHECKS

n	FCU	GSAT	HCR	GSAT[FCU]	HCR[FCU]
25	1,076	79,361	120,754	35,675	1,867
50	14,812	580,574	15,015,388	884,797	11,843
75	320,856	5,446,901	†	†	60,410
100	†	44,714,802	†	†	527,548

MEDIAN RESULTS: CONSTRAINT CHECKS

n	FCU	GSAT	HCR	GSAT[FCU]	HCR[FCU]
25	655	37,931	46,630	1,120	1,005
50	9,345	238,645	4,057,022	10,010	7,421
75	122,143	1,833,726	†	†	38,464
100	†	9,244,370	†	†	220,714

MEAN RESULTS: FLIPS

n	GSAT	HCR	GSAT[FCU]	HCR[FCU]
25	119	156	60	2
50	447	10,530	553	4
75	2,818	†	†	13
100	17,325	†	†	78

MEDIAN RESULTS: FLIPS

n	GSAT	HCR	GSAT[FCU]	HCR[FCU]
25	59	61	2	1
50	185	2,859	4	3
75	949	†	†	9
100	3,582	†	†	31

† *Failed to terminate in a reasonable time.*

Figure 6: Results for hard random 3SAT problems

1992]. However, the space explored by our hybrid algorithms is that of possible execution strategies, which (in our scheme) is exactly the same size as the stochastic search space — so why are the hybrids ever better than their stochastic counterparts?

First, note that for each solution to a CSP, there is at least one strategy which will find it without backtracking; we shall call such a strategy *optimal* for that solution. For example if n=3, m=4 and [v_3=1,v_1=4,v_2=1] is a solution, then ((3,1),(1,4),(2,1)) is clearly an optimal search strategy for that solution. Now suppose that the pair of assignments [v_3=1,v_1=3] is forbidden by a binary constraint. Then another optimal strategy for the same solution is ((3,1),(1,$\underline{3}$),(2,1)), because the assignment v_1=3 is rejected during execution under this strategy and the next assignment v_1=4 is discovered. Hence the more constrained a CSP is, the more optimal strategies are likely to exist for each solution. A hybrid's search space is therefore at least as dense in optimal solutions as the corresponding stochastic search space, and for a hard problem it is likely to be far more dense. So although our hybridization method does not actually preserve the space pruning abilities of systematic solvers, it does something almost as useful: it increases the solution density of the stochastic search space.

5 DYNAMIC PROBLEMS AND OPTIMIZATION

So far we have only considered CSPs. We now show that the hybrid algorithms can also be applied to COPs and dynamic CSPs.

Stochastic algorithms are often applied to CSPs by using a penalty function as the objective function. A common way of extending this method to COPs is to use instead a linear combination of the penalty and cost functions [Richardson et al., 1989]. However, this approach has the drawback that we must find an appropriate linear combination: a high coefficient of the cost function may cause the stochastic algorithm to follow false leads, which appear to lead to very good solutions but which actually lead to no solutions at all; conversely, a high coefficient of the penalty function may prevent the stochastic algorithm from finding good solutions. Choosing appropriate coefficients is rather arbitrary and problem-dependent.

Another approach is that of Tsang & Warwick [Tsang and Warwick, 1990]. Genetic algorithms are applied to COPs without using a penalty function. Solutions are generated which may violate constraints, and systematic (forward checking) search is used to find the first (under a random search strategy) true solution from this starting point. This is done during reproduction and for the initial population. The fitness function for the GA is derived from the cost of the solutions found. This approach is suitable for fairly unconstrained problems, but for harder problems too much time may be spent in systematic search. A related approach is that of Paredis [Paredis, 1993]. A GA is used to generate partial search states for a forward checking constraint solver with random search strategy. The fitness function is derived from several solutions found during bounded local search from this state.

Systematic constraint solvers can be used to solve COPs by *branch-and-bound* algorithms. First any solution is found and its cost noted; then another solution is found under the additional constraint that it must have lower cost; and so on, generating a sequence of solutions with decreasing cost. When the search fails then the last solution is guaranteed to be optimal, as long as the search procedure was complete. The additional cost constraint can be exploited by the solver to avoid subspaces which contain only solutions no better than those found so far. This algorithm effectively recasts a COP as a *dynamic* CSP. A dynamic CSP is one where constraints may be added or relaxed, and this algorithm adds cost constraints dynamically. Since the constraint solver is frequently restarted, it is simple to modify the set of constraints as often as necessary. Hence hybrid algorithms inherit the stochastic ability to solve dynamic CSPs. Now to solve a COP we can simply apply hybridization to a constraint solver which uses branch-and-bound. Such a hybrid works as follows: apply hybrid search to find any solution to the COP; add a cost constraint as in branch-and-bound; apply hybrid search under the new cost constraint; and continue this process until a sufficiently good solution is found. This approach requires no arbitrary coefficients because the cost function is not part of the objective function. Preliminary results using this technique on graph colouring problems are promising.

6 RELATED WORK

The combination of systematic and stochastic techniques is currently an active research area. Ours seems to be the first methodology for creating families of hybrids, but several specific hybrids already exist. The ones we derived in this paper have much in common with Yokoo's [Yokoo, 1994] Weak Commitment (WC): they incrementally complete partial solutions, and restart when a dead-end is reached. However, there are several differences. WC accumulates knowledge during search to ensure completeness, but our hybrids do not. WC uses conflict detection to select a variable for binding, whereas we use whatever variable ordering heuristics are used by the underlying systematic algorithm. Finally, WC uses the min-conflicts heuristic to select a new value while we select a value which maximizes the number of bound variables.

Another interesting hybrid is that of Ginsberg & McAllester [Ginsberg and McAllester, 1994]. Their systematic algorithm is Dynamic Backtracking [Ginsberg, 1993] which has a flexible backtracking strategy, and also accumulates knowledge during search, making it complete. Their stochastic algorithm is GSAT [Selman et al., 1992], which is highly efficient on many problems but may get stuck in local optima. The two are combined via reasoning on the accumulated information, to achieve great freedom of movement in the search space while preserving completeness. The algorithm requires exponential space in the worst case, but Selman et al. also describe a slightly restricted version (Partial-Order Dynamic Backtracking) which sacrifices some freedom of movement to achieve polynomial memory usage.

Another approach is Paredis' Genetic State-Space Search (GSSS) [Paredis, 1993] described earlier. However, there are significant differences between GSSS and our method: GSSS stochastically generates a partially defined *search state* for a constraint solver with random search strategy (we generate *search strategies*); it then evaluates the fitness of the state by performing a predetermined amount of constrained local search from the state, and evaluating any solution it finds (we use a very different evaluation).

7 CONCLUSIONS AND FUTURE WORK

This paper described a generic method for hybridizing systematic constraint solvers and stochastic algorithms. The method was first defined on exemplification CSPs then extended to COPs and dynamic CSPs. Unlike many algorithms, our hybrids do not apply stochastic algorithms to (total or partial) assignments of variables: they apply them to *search strategies*. A stochastic algorithm S is used to generate search strategies for a systematic constraint solver C, which probes the constrained problem space; the hybrid algorithm is denoted by S[C]. We demonstrated the method on specific algorithms, using delta evaluation and an evolutionary computation technique (Lamarckism) to improve efficiency. We found that S[C] often combines the scalability of S with the space-pruning ability of C, outperforming both on large, hard problems. However, the best systematic and stochastic algorithms do not always combine to give the best hybrids, and some hybrids perform poorly.

Although we believe that the hybrid HCR[FCU] (or HCR[FC+DVO] for some more powerful dynamic variable ordering heuristic DVO) has potential, the contribution of this work is not a single, definitive algorithm; it is rather a feasibility study of a new method for creating hybrid algorithms. We have not pinpointed exactly what classes of algorithms may be combined by our method, but the systematic algorithms include backtracking, domain pruning techniques such as forward checking and arc consistency, and dynamic variable ordering heuristics. On the stochastic side we can at least use variations on hill climbing, simulated annealing and genetic algorithms. For large, hard problems a particularly interesting systematic candidate is the MAC algorithm, a modified version of which has been shown to be very effective on such problems [Sabin and Freuder, 1997]. Another interesting candidate is the Tableau algorithm [Crawford and Auton, 1986], a systematic constraint solver which is already highly efficient on propositional satisfiability problems. Because it is similar to our systematic constraint solver (though with added heuristics) we believe that it could profitably be hybridized by our method.

Our hybridization method has three immediate applications:

- From the viewpoint of existing constraint solvers, which may already be highly efficient for a given class of problems (and which may describe the problems in a natural way), the method may be used as a "tweak" which sacrifices completeness to improve scalability.

- From the viewpoint of applying stochastic algorithms (such as simulated annealing) to constrained problems the method is an alternative to penalty functions, and seems better suited to hard problems.

- It allows rapid prototyping of new hybrid algorithms. In basic form (though not when optimized as in this paper) the hybrids have a neat software architecture: the stochastic and systematic components can be implemented separately, and connected via a simple interface.

Completeness is of course an important issue. Our hybrids are incomplete, but it is possible to execute a constraint solver in two modes: first as a hybrid to quickly find a CSP solution, if one exists, then systematically to prove unsolvability. Similarly for optimization problems, a solver could first be executed as a hybrid to quickly find a near-optimal solution, then systematically to refine this to an optimal solution and to prove it optimal. Alternatively some form of learning could be added to make the hybrids complete. However, incomplete search algorithms are useful in many applications, particularly where a limited amount of time is available for solving a problem.

REFERENCES

[Bowen and Dozier, 1996] J. Bowen and G. Dozier. Constraint Satisfaction Using A Hybrid Evolutionary Hill-Climbing Algorithm That Performs Opportunistic Arc and Path Revision. *Proceedings of the Thirteenth National Conference on Artificial Intelligence and Eighth Innovative Applications of Artificial Intelligence Conference* **1**, AAAI Press / The MIT Press, 1996.

[Burke et al., 1995] E. K. Burke, D. G. Elliman, R. F. Weare. A Hybrid Genetic Algorithm for Highly Constrained Timetabling Problems. *Proceedings of the Sixth International Conference on Genetic Algorithms*, Pittsburgh 1995, Morgan Kaufmann, pp. 605–610.

[Chew et al., 1992] T.-L. Chew, J.-M. David, A. Nguyen, and Y. Tourbier. Solving Constraint Satisfaction Problems with Simulated Annealing: the Car Sequencing Problem Revisited. *Proceedings of the 12th International Conference on Artificial Intelligence, Expert Systems and Natural Languages*, 1992, pp. 405–416.

[Corne et al., 1994] D. Corne, P. Ross, H.-L. Fang. Fast Practical Evolutionary Timetabling. *Proceedings of Evolutionary Computing, AISB Workshop*, Springer-Verlag 1994, pp. 250–263.

[Crawford and Auton, 1986] J. Crawford and L. Auton. Experimental Results on the Crossover Point in Random 3SAT. *Artificial Intelligence* **81**(1,2) pp. 31–57.

[Davis, 1985] L. Davis. Applying Adaptive Algorithms to Epistatic Domains. *Proceedings of the 9th International Joint Conference on Artificial Intelligence (IJCAI-85)*, Kaufmann 1985, pp. 162–164.

[Eiben et al, 1994] A. E. Eiben, P.-E. Raué, and Zs. Ruttkay. GA-Easy and GA-Hard Constraint Satisfaction Problems. *Proceedings of ECAI-94 Workshop W6: Constraint Processing*, 1994, pp. 87–96.

[Freuder et al., 1995] E. C. Freuder, R. Dechter, M. L. Ginsberg, B. Selman, and E. Tsang. Systematic Versus Stochastic Constraint Satisfaction. *Proceedings of IJCAI-95* **1**, 1995, pp. 2027–2032.

[Frost and Dechter, 1994] D. Frost and R. Dechter. Dead-End Driven Learning. *Proceedings of the 12th National Conference on Artificial Intelligence* **1**, AAAI Press 1994, pp. 294–300.

[Ghedira, 1994] K. Ghedira. Dynamic Partial Constraint Satisfaction by a Multi-Agent Simulated Annealing Approach. *Proceedings of ECAI-94 Workshop 19: Constraint Satisfaction Issues Raised by Practical Applications*, 1994, pp. 103–110.

[Ginsberg, 1993] M. L. Ginsberg. Dynamic Backtracking. *Journal of Artificial Intelligence Research* **1**, AI Access Foundation and Morgan Kaufmann 1993, pp. 25–46.

[Ginsberg and McAllester, 1994] M. L. Ginsberg and D. A. McAllester. GSAT and Dynamic Backtracking. *Proceedings of the 4th International Conference on Principles of Knowledge Representation and Reasoning (KR'94)*, Morgan Kaufmann, 1994, pp. 226–237.

[Kumar, 1992] V. Kumar. Algorithms for Constraint Satisfaction Problems: a Survey. *AI Magazine* **13**(1), 1992, pp. 32–44.

[Michalewicz and Janikow, 1991] Z. Michalewicz and C. Z. Janikow. Handling Constraints in Genetic Algorithms. *Proceedings of the International Conference on Genetic Algorithms*, Morgan Kaufmann 1991, pp. 151–157.

[Minton et al., 1994] S. Minton, M. D. Johnston, A. B. Philips, and P. Laird. Minimizing Conflicts: A Heuristic Repair Method For Constraint Satisfaction and Scheduling Problems. *Constraint-Based Reasoning*, Freuder & Mackworth (eds.), 1994.

[Morris, 1993] P. Morris. The Breakout Method for Escaping from Local Minima. *Proceedings of the 11th National Conference on Artificial Intelligence*, AAAI Press, 1993, pp. 40–45.

[Nadel, 1989] B. A. Nadel. Constraint Satisfaction Algorithms. *Computational Intelligence* **5**(4), 1989, pp. 188–224.

[Paredis, 1993] J. Paredis. Genetic State-Space Search for Constrained Optimization Problems. *Proceedings of the Thirteenth International Joint Conference on Artificial Intelligence*, Morgan Kaufmann, 1993.

[Paredis, 1994] J. Paredis. Co-Evolutionary Constraint Satisfaction. *Proceedings of the Third Conference on Parallel Problem Solving from Nature*, Springer-Verlag, 1994, pp. 46–55.

[Prestwich, 1998] S. Prestwich. Quasi-Systematic Constraint Satisfaction. (*Submitted to CP'98.*)

[Richards and Richards, 1997] E. T. Richards and B. Richards. Restart-Repair and Learning: An Empirical Study of Single Solution 3-SAT Problems. *Proceedings of CP97 Workshop on the Theory and Practice of Dynamic Constraint Satisfaction*, Linz, Austria, 1997.

[Richardson et al., 1989] J. T. Richardson, M. R. Palmer, G. Liepins, and M. Hilliard. Some Guidelines for Genetic Algorithms with Penalty Functions. *Proceedings of the 3rd International Conference on Genetic Algorithms*. Morgan Kaufmann, 1989, pp. 191–195.

[Sabin and Freuder, 1997] D. Sabin and E. C. Freuder. Understanding and Improving the MAC Algorithm. *Proceedings of the 3rd International Conference on Principles and Practice of Constraint Programming*, Lecture Notes in Computer Science **1330**, Springer-Verlag 1997, pp. 167–181.

[Selman and Kautz, 1993] B. Selman and H. Kautz. Domain-Independent Extension to GSAT: Solving Large Structured Satisfiability Problems. *Proceedings of the International Joint Conference on Artificial Intelligence 1993*, pp. 290–295.

[Selman et al., 1992] B. Selman, H. Levesque, and D. Mitchell. A New Method for Solving Hard Satisfiability Problems. *Proceedings of the 10th National Conference on Artificial Intelligence*, MIT Press, 1992, pp. 440–446.

[Stallman and Sussman, 1977] R. M. Stallman and G. J. Sussman. Forward Reasoning and Dependency-Directed Backtracking in a System for Computer-Aided Circuit Analysis. *Artificial Intelligence* **9**(2), 1977, pp. 135–196.

[Thornton, 1994] A. C. Thornton. Genetic Algorithms versus Simulated Annealing: Satisfaction of Large Sets of Algebraic Mechanical Design Constraints. *Proceedings of Artificial Intelligence in Design'94*, Kluwer Academic Publishers, 1994, pp. 381–398.

[Tsang and Warwick, 1990] E. P. K. Tsang and T. Warwick. Applying Genetic Algorithms to Constraint Satisfaction Optimization Problems. *Proceedings of the European Conference on Artificial Intelligence*, Pitman 1990.

[Turney et al., 1993] P. Turney, D. Whitley, and R. Anderson (editors). *Evolutionary Computation* **4**(3), Special Issue: The Baldwin Effect, MIT Press, Spring 1993.

[Yokoo, 1994] M. Yokoo. Weak-Commitment Search for Solving Constraint Satisfaction Problems. *Proceedings of the 12th National Conference on Artificial Intelligence* **1**, AAAI Press, 1994, pp. 313–318.

[Yugami et al., 1994] N. Yugami, Y. Ohta, and H. Hara. Improving Repair-based Constraint Satisfaction Methods by Value Propagation. *Proceedings of the 12th National Conference on Artificial Intelligence*, Volume 1, AAAI Press 1994, pp. 344–349.

Categorial Fibring of Logics with Terms and Binding Operators

Amílcar Sernadas, Cristina Sernadas, Carlos Caleiro, and Till Mossakowski

1 INTRODUCTION

Combination of logics has been attracting much attention [Blackburn and de Rijke, 1997]. Among the most interesting mechanisms, fibring deserves close attention [Gabbay, 1996a; Gabbay, 1996b; Gabbay, 1998]. Herein, we extend the work in [Sernadas et al., 1999] to logics with variables, terms and binding operators. For example, one can obtain first order temporal logic by fibring first order logic and propositional temporal logic. Our main goal is to provide a categorial characterization of both proof and model theoretic fibring of such logics.

Following the approach proposed in [Sernadas et al., 1997a; Sernadas et al., 1997b; Sernadas et al., 1999] we define free unconstrained fibring as a coproduct and constrained fibring (by sharing of symbols) as a cocartesian lifting in the appropriate categories (of Hilbert calculi and interpretation systems). Furthermore, we establish that soundness is preserved by the construction and that completeness is not always preserved. The extension of the approach to logics with variables, terms and binding operators is by no means trivial, at both proof and model theoretic levels.

At the level of the Hilbert calculi, we need an abstract solution to deal with the meta requirements in inference rules related to binding (such as term t is free for variable x in formula φ) having in mind that inference rules should be preserved by morphisms of Hilbert calculi. We are able to address this need by dealing with binding on the terminal signature where binding operators of the same kind are confused. Indeed, we consider two quite different kinds of binding operators: those that bind variables (like quantifiers and λ-abstraction) and those that bind flexible symbols (like modal operators).

At the level of the interpretation systems, the main problem is to cope with the different semantics of rigid and flexible symbols at the envisaged degree of generality. In this paper, we assume that domains are constant and values of variables are rigid. This assumption makes the semantics simpler but is the key reason for not achieving preservation of completeness. Indeed, by fibring first order logic and propositional temporal logic we obtain a logic where the Barcan formulae are valid but are not obtained from the axioms.

The proposed categorial approach was essential in finding a suitably abstract treatment of binding. Furthermore, as already shown in [Sernadas et al., 1999] for the simpler case of propositional logics, the categorial imperative was very helpful in fine tuning the envisaged semantics of fibring with explicit models and satisfaction and in establishing the technical lemmata for preservation of soundness.

In section 2 we present the category of signatures and some of its properties. In section 3 we develop the notion of Hilbert calculus including binding predicates used for stating requirements in schema rules. We conclude section 3 with the category of Hilbert calculi and give categorial characterizations of unconstrained and constrained fibring. In section 4 we deal with semantics. Section 5 is aimed at some preliminary preservation results. In section 6 we compare the notions and results presented in this paper with the work on model theoretic parchments [Mossakowski et al., 1998].

2 SIGNATURES

For the purpose of this paper, a logic system is composed of a signature, a Hilbert calculus and an interpretation system. In this section we develop the syntactic components of the logic system: signature and language. We also define the category of signatures and indicate some of its properties used in the sequel. We assume two disjoint sets X of *object variables* and Ξ of *schema variables* (to be used in inference schemata) are given. When convenient we shall use Z for $X \cup \Xi$: the set of all variables.

Definition 2.1 A *signature* is a tuple $\Sigma = \langle S, R, F, B, O, \zeta \rangle$ where: S is a pointed set with selected element a; $R = \{R_{\vec{g}s}\}_{\vec{g} \in S^*, s \in S}$ where each $R_{\vec{g}s}$ is a set; $F = \{F_{\vec{g}s}\}_{\vec{g} \in S^*, s \in S}$ where each $F_{\vec{g}s}$ is a set; $B = \{B_{v\vec{g}s}\}_{v \in S, \vec{g} \in S^*, s \in S}$ where each $B_{v\vec{g}s}$ is a set; $O = \{O_{\vec{g}s}\}_{\vec{g} \in S^*, s \in S}$ where each $O_{\vec{g}s}$ is a set; and $\zeta \subseteq Z \times S$ such that if $\langle z, s_1 \rangle, \langle z, s_2 \rangle \in \zeta$ then $s_1 = s_2$. We assume that all the sets are disjoint among them and from X and Ξ.

The elements of S are called *sorts* with a being the *assertions sort*. The elements of $R_{\vec{g}s}$ are called *rigid constructors* of type $\vec{g}s$. The elements of $F_{\vec{g}s}$ are called *flexible constructors* of type $\vec{g}s$. The elements of $B_{v\vec{g}s}$ are called *variable binding operators* of type $v\vec{g}s$. The elements of $O_{\vec{g}s}$ are called *modal binding operators* of type $\vec{g}s$. Relation ζ establishes which variables may be used with which sorts. Each variable may be used at most with one sort. Clearly, ζ is a partial map from Z to S. When setting up a signature for a given logic, we have to decide which symbols are to be considered rigid and which should be taken as flexible. Even if this partition does not affect the logic as given, it still plays a crucial role in any subsequent fibring of that logic with another one, as we shall see.

Example 2.2 *Propositional modal logic.* Given a set Π of propositional symbols, plus a partition of this set in rigid and flexible symbols, the corresponding signature is as follows (omitting the empty sets):

- $S = \{a\}$;
- $R_{\epsilon a} = \{\pi : \pi \in \Pi, \pi \text{ rigid}\}$; $R_{aa} = \{\neg\}$; $R_{aaa} = \{\Rightarrow\}$;
- $F_{\epsilon a} = \{\pi : \pi \in \Pi, \pi \text{ flexible}\}$;
- $O_{aa} = \{\Box\}$;
- $\zeta = \{\langle \xi, a\rangle : \xi \in \Xi\}$.

Example 2.3 *First order linear temporal logic.* Let a family $F = \{F_n\}_{n \in I\!N}$ of sets of function symbols be given, together with a family $P = \{P_n\}_{n \in I\!N}$ of sets of predicate symbols, plus a partition of each of these sets in rigid and flexible symbols; the corresponding signature is as follows (omitting the empty sets):

- $S = \{i, a\}$, i being the sort of individuals;
- $R_{i^n i} = \{f : f \in F_n, f \text{ rigid}\}$; $R_{i^n a} = \{p : p \in P_n, p \text{ rigid}\}$;
  $R_{aa} = \{\neg\}$; $R_{aaa} = \{\Rightarrow\}$;
- $F_{i^n i} = \{f : f \in F_n, f \text{ flexible}\}$; $F_{i^n a} = \{p : p \in P_n, p \text{ flexible}\}$;
- $B_{iaa} = \{\forall\}$;
- $O_{aaa} = \{\mathsf{U}\}$;
- $\zeta = \{\langle z, i\rangle : z \in Z\} \cup \{\langle \xi, a\rangle : \xi \in \Xi\}$.

Other examples are easily given, such as the λ-calculus, second order logic and equational logic. In the sequel we use $X_s^\Sigma = \{x \in X : \langle x, s\rangle \in \zeta\}$ plus $X^\Sigma = \{x \in X : \exists s \langle x, s\rangle \in \zeta\}$ and likewise for the other sets of variables. We also use the following *constructors*:

- $C_{\epsilon s} = R_{\epsilon s} \cup F_{\epsilon s} \cup \{bx : b \in B_{v \epsilon s}, x \in X_v^\Sigma\} \cup O_{\epsilon s} \cup X_s^\Sigma$;
- $C_{\vec{g}s} = R_{\vec{g}s} \cup F_{\vec{g}s} \cup \{bx : b \in B_{v\vec{g}s}, x \in X_v^\Sigma\} \cup O_{\vec{g}s}$ for $\vec{g} \neq \epsilon$.

The family of sets of *schema terms* $L(\Sigma, \Xi) = \{L(\Sigma, \Xi)_s\}_{s \in S}$ generated by a signature Σ is inductively defined as follows:

- $\xi \in L(\Sigma, \Xi)_s$ provided that $\xi \in \Xi_s$;

- $c(\tau_1, \ldots, \tau_n) \in L(\Sigma, \Xi)_s$ provided that $c \in C_{g_1 \ldots g_n\, s}$ and $\tau_i \in L(\Sigma, \Xi)_{g_i}$ for each $i = 1, ..., n$ with $n \geq 0$.

The family of sets of *ground terms* $L(\Sigma) = \{L(\Sigma)_s\}_{s \in S}$ is defined similarly without using schema variables. Terms of sort a are said to be *formulae.*

The notion of *signature morphism* $h : \Sigma \to \Sigma'$ is straightforward: sorts are mapped to sorts respecting the assertion sort and symbols are mapped to symbols of the same kind taking into account the change of sorts. Furthermore, if $\langle z, s\rangle \in \zeta$ then $\langle z, h(s)\rangle \in \zeta'$. Clearly, such a morphism induces a *translation map* between terms, also denoted by h.

Signatures and their morphisms constitute the category *Sig*. In this category there is a *terminal signature* Σ_T that we shall use for defining the binding mechanism. The terminal signature contains a unique sort and a single operator of each kind. We denote by 1_Σ the unique signature morphism from Σ to Σ_T. We shall also use coproducts, coequalizers and pushouts when combining signatures. Coproducts do not always exist. We say that two signatures Σ', Σ'' are *variable independent* iff $X^{\Sigma'} \cap X^{\Sigma''} = \emptyset$. Coproducts of variable independent signatures always exist. Otherwise, the coproduct of the signatures may not exist since the resulting ζ may not be functional. One may ask why we do not relax our definition of signature allowing non functional sorting of variables. This would guarantee the existence of coproduct of signatures in all cases, but it would bring undesired side effects when confusing variables of two sorts collapsed by a morphism (e.g., when sharing sorts in constrained fibring).

We shall also need the concept of substitution. A *schema substitution* on $L(\Sigma, \Xi)$ is a family $\sigma = \{\sigma_s\}_{s \in S}$ where each $\sigma_s : \Xi_s \to L(\Sigma, \Xi)_s$. We denote by $Sub(\Sigma, \Xi)$ the set of all schema substitutions on $L(\Sigma, \Xi)$. The *instance* of a schema term τ by a schema substitution σ, denoted by $\tau\sigma$, is the schema term obtained from τ by simultaneously replacing each occurrence of ξ in τ by $\sigma_s(\xi)$ for every $\xi \in \Xi_s$. We denote the composition of substitution σ with substitution σ' by $\sigma\sigma'$. Clearly, $\tau(\sigma\sigma') = (\tau\sigma)\sigma'$. We extend the notion of instance to sets of schema terms: $T\sigma$ denotes the set $\{\tau\sigma : \tau \in T\}$. Furthermore, a *ground substitution* on $L(\Sigma)$ is a family $\rho = \{\rho_s\}_{s \in S}$ where each $\rho_s : \Xi_s \to L(\Sigma)_s$. We denote by $Sub(\Sigma)$ the set of all ground substitutions on $L(\Sigma)$.

3 HILBERT CALCULI

A Hilbert calculus is composed of a signature and inference rules. Before saying what an inference rule is, we need to be able to express meta requirements related to binding. For instance, in first order logic, we have the rule (actually an axiom since the set of premises is empty): $((\forall x\, \xi_1) \Rightarrow \xi_1|^x_{\xi_2})$ provided that ξ_2 of type i is free for x in ξ_1 of type a. Such meta requirements are written using meta predicates:

Definition 3.1 A (meta) *binding predicate* of arity $\langle i, j\rangle$ with $i, j \in I\!N$ is a pair $\pi = \langle \mathrm{dom}\,\pi, \mathrm{val}\,\pi\rangle$ where:

- $\mathrm{dom}\,\pi = \{\mathrm{dom}_\Sigma\,\pi\}_{\Sigma\in|Sig|}$ with $\mathrm{dom}_\Sigma\,\pi \subseteq S^i \times S^j$;
- $\mathrm{val}\,\pi$ provides a value $\mathrm{val}\,\pi(\vec{v}, \vec{s}, \vec{x}, \vec{\theta})\in\{0,1\}$ for every $\langle\vec{v}, \vec{s}\rangle$ in $\mathrm{dom}_{\Sigma_T}\,\pi$, sequence $\vec{x}$ with type $\vec{v}$ of object variables, and sequence $\vec{\theta}$ with type $\vec{s}$ of ground terms of the terminal signature Σ_T;

such that, for every signature morphism $h : \Sigma \to \Sigma'$, $\langle h(\vec{v}), h(\vec{s})\rangle \in \mathrm{dom}_{\Sigma'}\,\pi$ whenever $\langle\vec{v}, \vec{s}\rangle \in \mathrm{dom}_\Sigma\,\pi$.

So far, a binding predicate is evaluated only over the terminal signature. We shall see below how that value can be used for evaluating the predicate over any signature. The advantage of evaluating binding predicates in the terminal signature should be obvious: binding mechanisms are universal since they always work in the same way, independent of any particular binding operator.

Example 3.2 *Term free for variable in term:* $\triangleright$ with arity $\langle 1, 2\rangle$:

- $\mathrm{dom}_\Sigma \triangleright = \{\langle v, vs\rangle : v, s \in S\}$;
- $\mathrm{val} \triangleright (v, vs, x, \theta_1\theta_2) = 1$ iff $\theta_1 \triangleright_{\Sigma_T} x : \theta_2$, that is, iff θ_1 is free for x in θ_2.

We refrain from spelling out the details of the definition of the values of this binding predicate over the terminal signature, since it coincides with the standard definition in any logic with variable binding operators.

In the sequel, we shall also use other binding predicates, such as an object variable being free in a term, as well as the notion of substitution of a term for an object variable, again without providing any defining details.

Definition 3.3 The set $Req(\Sigma, \Xi)$ of *requirements* over Σ is composed of expressions of the form $\pi(\vec{v}, \vec{s}, \vec{x}, \vec{\tau})$ where π is a binding predicate, $\langle\vec{v}, \vec{s}\rangle$ belongs to $\mathrm{dom}_\Sigma\,\pi$, $\vec{x}$ is a sequence with type $\vec{v}$ of object variables, and $\vec{\tau}$ is a sequence with type $\vec{s}$ of schema terms over Σ.

Example 3.4 *Term free for variable in term* – $\triangleright(v, vs, x, \tau_1\tau_2)$: τ_1 of sort v is free for x of sort v in τ_2 of sort s.

Given $q = \pi(\vec{v}, \vec{s}, \vec{x}, \vec{\tau}) \in Req(\Sigma, \Xi)$ and a schema substitution σ, the *instance of requirement* q by σ is the requirement $q\sigma = \pi(\vec{v}, \vec{s}, \vec{x}, \vec{\tau}\sigma)$. Furthermore, given $h : \Sigma \to \Sigma'$, the *image of requirement* q by h is the requirement $h(q) = \pi(h(\vec{v}), h(\vec{s}), \vec{x}, h(\vec{\tau}))$ over Σ'.

Definition 3.5 The *denotation* of a requirement $\pi(\vec{v}, \vec{s}, \vec{x}, \vec{\tau})$ over the terminal signature Σ_T is the partial map $[\![\pi(\vec{v}, \vec{s}, \vec{x}, \vec{\tau})]\!]_{\Sigma_T} : Sub(\Sigma_T, \Xi) \rightharpoonup \{0,1\}$ defined as follows:

$$[\![\pi(\vec{v}, \vec{s}, \vec{x}, \vec{\tau})]\!]_{\Sigma_T}(\sigma) = \begin{cases} 1 & \text{if } \mathrm{val}\,\pi(\vec{v}, \vec{s}, \vec{x}, \vec{\tau}\sigma\rho) = 1 \ \ \forall\rho \in Sub(\Sigma_T) \\ 0 & \text{if } \mathrm{val}\,\pi(\vec{v}, \vec{s}, \vec{x}, \vec{\tau}\sigma\rho) = 0 \ \ \forall\rho \in Sub(\Sigma_T) \\ \bot & \text{otherwise} \end{cases}.$$

We can easily extend this notion to any signature:

Definition 3.6 The *denotation* of a requirement $\pi(\vec{v},\vec{s},\vec{x},\vec{\tau})$ over Σ is the partial map $[\![\pi(\vec{v},\vec{s},\vec{x},\vec{\tau})]\!]_\Sigma : Sub(\Sigma,\Xi) \rightharpoonup \{0,1\}$ such that:

$$[\![\pi(\vec{v},\vec{s},\vec{x},\vec{\tau})]\!]_\Sigma(\sigma) = [\![1_\Sigma(\pi(\vec{v},\vec{s},\vec{x},\vec{\tau}))]\!]_{\Sigma_T}(1_\Sigma \circ \sigma).$$

It is easy to establish the following technical lemma:

Proposition 3.7 *Let $h : \Sigma \to \Sigma'$ be a signature morphism and $q = \pi(\vec{v}, \vec{s}, \vec{x}, \vec{\tau})$ a requirement over Σ and Ξ. Then, $[\![q]\!]_\Sigma(\sigma) = [\![h(q)]\!]_{\Sigma'}(h \circ \sigma)$.*

We are finally ready to define inference rules and Hilbert calculi:

Definition 3.8 A *rule schema* over a signature Σ is a triple

$$r = \langle Prem(r), Conc(r), Req(r)\rangle$$

where $Prem(r) \in \wp_{\text{fin}}(L(\Sigma,\Xi)_a)$ – the finite set of premises, $Conc(r) \in L(\Sigma,\Xi)_a$ – the conclusion, $Req(r) \in \wp_{\text{fin}}(Req(\Sigma,\Xi))$ – the finite set of meta requirements, and such that $[\![q]\!]_\Sigma(id_\Sigma)\uparrow$ for every $q \in Req(r)$.

Translation of rules along signature morphisms is defined componentwise.

Example 3.9 *Some first order rules.*

- $\langle\{\xi_1,(\xi_1 \Rightarrow \xi_2)\},\xi_2,\emptyset\rangle$;
- $\langle\emptyset,((\forall x\xi_1) \Rightarrow \xi_1|^x_{\xi_2}),\{\triangleright(i,ia,x,\xi_2\xi_1)\}\rangle$.

Definition 3.10 A *Hilbert calculus* is a pair $\langle\Sigma, R\rangle$ where Σ is a signature and R is a set of rule schemata.

For the sake of simplicity, we do not distinguish here between proof and derivation rules, contrarily to what we did in [Sernadas et al., 1999].

Example 3.11 *Propositional modal logic.*
A propositional (uni)modal Hilbert calculus over Π is a pair $\langle\Sigma, R\rangle$ such that Σ is as defined before and:

- $R \supseteq \{\langle\emptyset,(\xi_1 \Rightarrow (\xi_2 \Rightarrow \xi_1)),\emptyset\rangle,$
  $\langle\emptyset,((\xi_1 \Rightarrow (\xi_2 \Rightarrow \xi_3)) \Rightarrow ((\xi_1 \Rightarrow \xi_2) \Rightarrow (\xi_1 \Rightarrow \xi_3))),\emptyset\rangle,$
  $\langle\emptyset,(((\neg\,\xi_1) \Rightarrow (\neg\,\xi_2)) \Rightarrow (\xi_2 \Rightarrow \xi_1)),\emptyset\rangle,$
  $\langle\emptyset,((\Box(\xi_1 \Rightarrow \xi_2)) \Rightarrow ((\Box\xi_1) \Rightarrow (\Box\xi_2))),\emptyset\rangle,$
  $\langle\{\xi_1,(\xi_1 \Rightarrow \xi_2)\},\xi_2,\emptyset\rangle,$
  $\langle\{\xi_1\},(\Box\xi_1),\emptyset\rangle\}$.

Definition 3.12 Within the context of a Hilbert calculus $\langle\Sigma, R\rangle$, given $Q \in \wp_{\text{fin}}(Req(\Sigma,\Xi))$, we say that $\delta \in L(\Sigma,\Xi)_a$ is *Q-provable* from $\Gamma \subseteq L(\Sigma,\Xi)_a$ iff there are sequences $\gamma_1,\ldots,\gamma_m \in (L(\Sigma,\Xi)_a)^+$ and $Q_0,\ldots,Q_m \in Req(\Sigma,\Xi)^+$ such that $Q_0 = \emptyset$, γ_m is δ and $Q_m = Q$, and for each $i = 1,\ldots,m$ either $\gamma_i \in \Gamma$ and $Q_i = Q_{i-1}$, or there are a substitution σ and a rule $r \in R$ such that γ_i and Q_i result by r via σ from the previous entries, i.e.:

- γ_i is $Conc(r)\sigma$;
- every element of $Prem(r)\sigma$ occurs in $\gamma_1,\ldots,\gamma_{i-1}$;
- $[\![q]\!]_\Sigma(\sigma) \neq 0$ for every $q \in Req(r)$;
- $Q_i = Q_{i-1} \cup \{q\sigma : q \in Req(r), [\![q]\!]_\Sigma(\sigma)\uparrow\}$.

When $Q = \emptyset$ we say that δ is *provable* from Γ. When $\Gamma = \emptyset$ we say that δ is *Q-provable*, or just provable if in addition $Q = \emptyset$.

We are now ready to put forward the envisaged notion of Hilbert calculus morphism: a signature morphism that preserves the inference rules.

Definition 3.13 A *Hilbert calculus morphism* $h : \langle\Sigma, R\rangle \to \langle\Sigma', R'\rangle$ is a signature morphism $h : \Sigma \to \Sigma'$ such that $h(Conc(r))$ is $h(Req(r))$-provable from $h(Prem(r))$ in $\langle\Sigma', R'\rangle$ for every $r \in R$.

Proposition 3.14 *Hilbert calculi morphisms preserve provability: if δ is Q-provable from Γ then $h(\delta)$ is $h(Q)$-provable from $h(\Gamma)$.*

Proof: We first note that if δ is Q-provable from Γ and $\sigma \in Sub(\Sigma,\Xi)$ is such that $[\![q]\!]_\Sigma(\sigma) \neq 0$ for every $q \in Q$ then $\delta\sigma$ is $\{q\sigma : q \in Q, [\![q]\!]_\Sigma(\sigma)\uparrow\}$*-provable* from $\Gamma\sigma$. From this fact it easily follows that $h(\delta)$ is $h(Q)$-provable from $h(\Gamma)$ using induction on the length of the Q-proof of δ from Γ. □

The preservation lemma above is needed to establish the category *Hil* of Hilbert calculi and it will also be useful later on.

When defining constrained fibring, we shall need a mechanism for constructing a new Hilbert calculus from a given one along a given signature morphism. To this end we need the (forgetful) functor $N : Hil \to Sig$ such that:

- $N(\langle\Sigma, R\rangle) = \Sigma$;
- $N(h : \langle\Sigma, R\rangle \to \langle\Sigma', R'\rangle) = h$.

Proposition 3.15 *For each Hilbert calculus $\langle\Sigma, R\rangle$ and each signature morphism $h : \Sigma \to \Sigma'$, the morphism $h : \langle\Sigma, R\rangle \to \langle\Sigma', h(R)\rangle$ is cocartesian by N for h on $\langle\Sigma, R\rangle$. We denote the codomain of the cocartesian morphism above by $h(\langle\Sigma, R\rangle)$.*

Proof: Clearly, $\langle\Sigma', h(R)\rangle$ is a Hilbert calculus and $h : \langle\Sigma, R\rangle \to \langle\Sigma', h(R)\rangle$

a *Hil* morphism. As far as the universal property is concerned, suppose that $f : \langle \Sigma, R \rangle \to \langle \Sigma'', R'' \rangle$ is a *Hil* morphism and $g : \Sigma' \to \Sigma''$ is a *Sig* morphism such that $f = g \circ h$. It is straightforward to verify that $g : \langle \Sigma', h(R) \rangle \to \langle \Sigma'', R'' \rangle$ is a *Hil* morphism. $\square$

Unconstrained Fibring

Intuitively, in the unconstrained fibring of two Hilbert calculi we import the constructors and the inference rules from both calculi. The schema variables are essential in order to allow the application of the rules to new terms and formulae.

Definition 3.16 Let $\langle \Sigma', R' \rangle$ and $\langle \Sigma'', R'' \rangle$ be Hilbert calculi such that their signatures are variable independent. Then, their *unconstrained fibring* is

$$\langle \Sigma', R' \rangle \oplus \langle \Sigma'', R'' \rangle = \langle \Sigma' \oplus \Sigma'', i'(R') \cup i''(R'') \rangle$$

where i', i'' are the injections of the coproduct $\Sigma' \oplus \Sigma''$.

Unconstrained fibring is therefore defined only in the case of variable independent signatures. But no generality is lost since it is very simple and always possible to prepare two logics for fibring: just make sure beforehand that no variable is used by both.

Proposition 3.17 *Every unconstrained fibring is a coproduct in Hil.*

Proof: Clearly, $H' \oplus H''$ is a Hilbert calculus, and $i' : H' \to H' \oplus H''$ and $i'' : H'' \to H' \oplus H''$ are *Hil* morphisms. Given any other two *Hil* morphisms $h' : H' \to H$ and $h' : H'' \to H$, the unique morphism $h : H' \oplus H'' \to H$ such that $h' = h \circ i'$ and $h'' = h \circ i''$ is given by the corresponding universal property in *Sig*. $\square$

The coproduct construction captures the intuitive idea that fibring should extend the given logics in a minimal and conservative way. Indeed, the injections i', i'' being morphisms preserve proofs as envisaged.

Constrained Fibring

In order to constrain the fibring by imposing some interaction between the two given Hilbert calculi we have two approaches that can be used together: sharing of constructors and addition of new (mixed) rules. The technique of cocartesian lifting provides the means for sharing constructors: it provides a canonical Hilbert calculus guided by the sharing defined at the signature level.

Definition 3.18 Let $\langle \Sigma', R' \rangle$ and $\langle \Sigma'', R'' \rangle$ be Hilbert calculi such that their signatures are variable independent and $f' : \Sigma \to \Sigma'$, $f'' : \Sigma \to \Sigma''$ be injective signature morphisms. Then, their *constrained fibring by sharing* the symbols in Σ is:

$$\langle \Sigma', R' \rangle \overset{f'\Sigma f''}{\oplus} \langle \Sigma'', R'' \rangle = q(\langle \Sigma', R' \rangle \oplus \langle \Sigma'', R'' \rangle)$$

where $q : \Sigma' \oplus \Sigma'' \to \Sigma' \overset{f'\Sigma f''}{\oplus} \Sigma''$ is the coequalizer in *Sig* of $i' \circ f' : \Sigma \to \Sigma' \oplus \Sigma''$ and $i'' \circ f'' : \Sigma \to \Sigma' \oplus \Sigma''$.

Note that we allow sharing of any symbols of the same kind. For instance, if we are fibring two modal logics we may impose that the two boxes are identified, thereby obtaining a box which inherits the properties of the two original ones.

As an interesting example of constrained fibring, consider the problem of obtaining the calculus of first order predicate linear temporal logic (FOTL) by combining the calculi of first order predicate logic (FOL) and propositional linear temporal logic (TL). By constrained fibring of the given calculi (imposing the sharing of the propositional connectives) we obtain an approximation of FOTL where the Barcan formulae are not provable.

4 INTERPRETATION SYSTEMS

Intuitively, an interpretation system provides for a given signature a class of models and the means for interpreting the constructors in each model over a domain of points. Then, terms can be inductively evaluated. We should stress that for each model we provide two independent parts: one for interpreting rigid symbols and another for interpreting flexible symbols.

Definition 4.1 Given a signature Σ and a family $D = \{D_s\}_{s \in S}$ of carrier sets, a $\langle \Sigma, D \rangle$-*rigid structure* is a family:

- $\kappa = \{\kappa_{\vec{g}s}\}_{\vec{g} \in S^*, s \in S}$
  with each $\kappa_{g_1 \dots g_n s}$: $R_{g_1 \dots g_n s} \to [(D_{g_1}) \times \dots \times (D_{g_n}) \to (D_s)]$;

and a $\langle \Sigma, D \rangle$-*flexible structure* is a triple $\langle W, \eta, \mu \rangle$ where:

- W is a non empty set;
- $\eta = \{\eta_{\vec{g}s}\}_{\vec{g} \in S^*, s \in S}$
  with each $\eta_{g_1 \dots g_n s}$: $F_{g_1 \dots g_n s} \to [(D_{g_1}) \times \dots \times (D_{g_n}) \to (D_s)]^W$;
- $\mu = \{\mu_{\vec{g}s}\}_{\vec{g} \in S^*, s \in S}$
  with each $\mu_{g_1 \dots g_n s}$: $O_{g_1 \dots g_n s} \to [(D_{g_1})^W \times \dots \times (D_{g_n})^W \to (D_s)^W]$.

We say that D is the (sorted) *domain*, κ is the family of *rigid denotation maps*, W is the *world space*, η is the family of *flexible denotation maps* and μ is the family of *modal binding denotation maps*. We denote by $Str^R(\Sigma, D)$ and $Str^F(\Sigma, D)$ the classes of all rigid and flexible structures, respectively.

Given a signature $\Sigma = \langle S, R, F, B, O, \zeta\rangle$, we need to refer to the family $\{\overline{X}_{\vec{g}s}\}_{\vec{g}\in S^*, s\in S}$ of sets of symbols:

- $\overline{X}_{\epsilon s} = X_s^\Sigma$;
- $\overline{X}_{\vec{g}s} = \{bx : b \in B_{v\vec{g}s}, x \in X_v^\Sigma\}$ for $\vec{g} \neq \epsilon$.

Definition 4.2 A *pre-interpretation system* is a triple $\langle\Sigma, \mathbb{D}, \mathbb{I}\rangle$ where Σ is a signature, $\mathbb{D}$ is a class of families of carrier sets over Σ such that $D_a = \{0,1\}$ for every $D \in \mathbb{D}$, and $\mathbb{I}$ maps each $D \in \mathbb{D}$ to a tuple $\langle U_D$, ν_D, M_D^R, A_D^R, M_D^F, $A_D^F\rangle$ where:

- U_D is a non empty set;
- $\nu_D = \{\nu_{D\vec{g}s}\}_{\vec{g}\in S^*, s\in S}$
  with each $\nu_{Dg_1\ldots g_n s}$: $\overline{X}_{g_1\ldots g_n s} \to [(D_{g_1})^{U_D} \times \ldots \times (D_{g_n})^{U_D} \to (D_s)^{U_D}]$,
- M_D^R is a class;
- $A_D^R : M_D^R \to Str^R(\Sigma, D)$;
- M_D^F is a class;
- $A_D^F : M_D^F \to Str^F(\Sigma, D)$;

and such that for every $b \in B_{v\vec{g}s}$ and $x \in X_v^\Sigma$:

BC1 $\nu_{D\vec{g}s}(bx)(\vec{V})(u_1) = \nu_{D\vec{g}s}(bx)(\vec{V})(u_2)$ provided that $u_1 \equiv_x u_2$;

BC2 $\nu_{D\vec{g}s}(bx)(\vec{V_1})(u_2) = \nu_{D\vec{g}s}(bx)(\vec{V_2})(u_2)$ provided that $\vec{V_1}(u_1) = \vec{V_2}(u_1)$ for every $u_1 \in U_D$ such that $u_1 \equiv_x u_2$;

where $u_1 \equiv_x u_2$ iff $\nu_{D\epsilon v}(y)(u_1) = \nu_{D\epsilon v}(y)(u_2)$ for every $y \in X_v^\Sigma \setminus \{x\}$.

We say that U_D is the *assignment space*, ν_D is the family of *variable binding denotation maps*, M_D^R is the class of *models for rigid constructors* and M_D^F is the class of *models for flexible constructors*. When $u_1 \equiv_x u_2$ we say that the assignments u_1 and u_2 are x-equivalent. Conditions BC1 and BC2 are known as *binding conditions*. BC1 states that the result of applying bx is a term with a value not dependent on the value of x. BC2 states that the value of $bx\tau$ for a given assignment u is only dependent on the values of τ for assignments x-equivalent to u.

In the sequel, taking D from the context, we use the notations $A_D^R(m) = \kappa_m$ for each $m \in M_D^R$, and $A_D^F(m) = \langle W_m, \eta_m, \mu_m\rangle$ for each $m \in M_D^F$.

Example 4.3 *First order linear temporal logic.* Let Σ be a first order linear temporal signature as introduced before. Then, the envisaged pre-interpretation system over Σ is as follows:

- $\mathbb{D}$ is the class of all carrier sets D with D_i non empty and $D_a = \{0,1\}$;
- $\mathbb{I}(D) = \langle D_i^{X_i^\Sigma}, \nu, M^R, A^R, M^F, A^F \rangle$ where:
  - $\nu_{\epsilon i}(x)(u) = u(x)$;
    $\nu_{iaa}(\forall x)(V)(u) = \Pi_{u' \equiv_x u} V(u')$;
  - M^R is the class of all denotations I such that:
    - $f_I : D_i^n \to D_i$ for rigid function symbol $f \in F_n$;
    - $p_I : D_i^n \to D_a$ for rigid predicate symbol $p \in P_n$;
  - $A^R(I) = \kappa$ such that:
    - $\kappa_{i^n i}(f) = f_I$;
      $\kappa_{i^n a}(p) = p_I$;
      $\kappa_{aa}(\neg)(b) = 1 - b$;
      $\kappa_{aaa}(\Rightarrow)(b_1, b_2) = b_1 \leq b_2$;
  - M^F is the class of all denotations J such that:
    - $f_J : [D_i^n \to D_i]^{\mathbb{N}}$ for flexible function symbol $f \in F_n$;
    - $p_J : [D_i^n \to D_a]^{\mathbb{N}}$ for flexible predicate symbol $p \in P_n$;
  - $A^F(J) = \langle \mathbb{N}, \eta, \mu \rangle$ where:
    - $\eta_{i^n i}(f) = f_J$;
      $\eta_{i^n a}(p) = p_J$;
    - $\mu_{aaa}(\mathsf{U})(V_1, V_2)(w) = sg(\Sigma_{w<w'}[V_2(w') * \Pi_{w \leq w'' < w'} V_1(w'')])$.

Definition 4.4 A pre-interpretation system $\langle \Sigma, \mathbb{D}, \mathbb{I} \rangle$ is an *interpretation system* provided that for every $D \in \mathbb{D}$, $m_1 \in M_D^F$ and bijection $j : W_{m_1} \cong W$ there is $m_2 \in M_D^F$ such that:

- $W_{m_2} = W$;
- $\eta_{m_2 \vec{g} s}(f) \circ j = \eta_{m_1 \vec{g} s}(f)$;
- $\mu_{m_2 \vec{g} s}(o)(\vec{V}) \circ j = \mu_{m_1 \vec{g} s}(o)(\vec{V} \circ j)$.

Clearly, given any pre-interpretation system it is always possible to establish an interpretation system. This enrichment operation does not affect semantic entailment. It is straightforward to extract the (contextual and floating) *satisfaction* relations and, from them, the corresponding *entailments*.

Definition 4.5 Given a pre-interpretation system I, a family of carrier sets $D \in \mathbb{D}$ and a pair $m = \langle m_R, m_F \rangle \in M_D^R \times M_D^F$, the family $[\![.]\!]_s^{Dm}$ of *interpretation* maps $\{[\![.]\!]_s^{Dm} : L(\Sigma)_s \to D_s^{U_D \times W_{m_F}}\}_{s \in S}$ is inductively defined by:

- $[\![x]\!]_s^{Dm}(u, w) = \nu_{D\epsilon s}(x)(u)$ provided that $x \in X_s^\Sigma$;

- $[\![r(\theta_1,\ldots,\theta_n)]\!]_s^{Dm}(u,w) =$
  $\kappa_{m_R g_1\ldots g_n s}(r)([\![\theta_1]\!]_{g_1}^{Dm}(u,w),\ldots,[\![\theta_n]\!]_{g_n}^{Dm}(u,w))$
  provided that $r \in R_{g_1\ldots g_n s}$ and each $\theta_i \in L(\Sigma)_{g_i}$;

- $[\![f(\theta_1,\ldots,\theta_n)]\!]_s^{Dm}(u,w) =$
  $\eta_{m_F g_1\ldots g_n s}(f)(w)([\![\theta_1]\!]_{g_1}^{Dm}(u,w),\ldots,[\![\theta_n]\!]_{g_n}^{Dm}(u,w))$
  provided that $f \in F_{g_1\ldots g_n s}$ and each $\theta_i \in L(\Sigma)_{g_i}$;

- $[\![bx(\theta_1,\ldots,\theta_n)]\!]_s^{Dm}(u,w) =$
  $\nu_{D g_1\ldots g_n s}(bx)(\lambda u'.[\![\theta_1]\!]_{g_1}^{Dm}(u',w),\ldots,\lambda u'.[\![\theta_n]\!]_{g_n}^{Dm}(u',w))(u)$
  provided that $b \in B_{v g_1\ldots g_n s}$, $x \in X_v^\Sigma$ and each $\theta_i \in L(\Sigma)_{g_i}$;

- $[\![o(\theta_1,\ldots,\theta_n)]\!]_s^{Dm}(u,w) =$
  $\mu_{m_F g_1\ldots g_n s}(o)(\lambda w'.[\![\theta_1]\!]_{g_1}^{Dm}(u,w'),\ldots,\lambda w'.[\![\theta_n]\!]_{g_n}^{Dm}(u,w'))(w)$
  provided that $o \in O_{g_1\ldots g_n s}$ and each $\theta_i \in L(\Sigma)_{g_i}$.

Fixing an assignment $u \in U_D$ and a world $w \in W_{m_F}$, the *contextual satisfaction relation* for formulae $\varphi \in L(\Sigma)_a$ is defined by

$$Dmuw \Vdash \varphi \;\text{ iff }\; [\![\varphi]\!]_a^{Dm}(u,w) = 1.$$

Furthermore, the *floating satisfaction relation* is defined by

$$Dm \Vdash \varphi \;\text{ iff }\; Dmuw \Vdash \varphi \text{ for every } u \in U_D, w \in W_{m_F}.$$

We extend the above defined satisfaction relations to sets of formulae in the usual way by requiring the joint satisfaction of all the formulae in the set.

The *contextual entailment* operator $.^{\vDash} : 2^{L(\Sigma)_a} \to 2^{L(\Sigma)_a}$ is defined, as expected, by $\Phi^{\vDash} = \{\varphi \in L(\Sigma)_a : Dmuw \Vdash \varphi \text{ whenever } Dmuw \Vdash \Phi\}$. And the *floating entailment* operator $.^{\overline{\vDash}} : 2^{L(\Sigma)_a} \to 2^{L(\Sigma)_a}$ is defined by $\Phi^{\overline{\vDash}} = \{\varphi \in L(\Sigma)_a : Dm \Vdash \varphi \text{ whenever } Dm \Vdash \Phi\}$.

The binding conditons BC1 and BC2 allow us to obtain the following semantic lemmata (proved by straightforward induction):

Proposition 4.6 *Independence of non-free variables lemma:*

$$[\![\theta]\!]_s^{Dm}(u_1,w) = [\![\theta]\!]_s^{Dm}(u_2,w)$$

whenever $\nu_{\epsilon v}(y)(u_1) = \nu_{\epsilon v}(y)(u_2)$ *for every object variable* y *free in* θ.

Proposition 4.7 *Substitution lemma:*

$$[\![\theta_1]\!]_s^{Dm}(u_1,w) = [\![\theta_1|_{\theta_2}^{x}]\!]_s^{Dm}(u_2,w)$$

whenever θ_2 *free for* x *in* θ_1, $u_1 \equiv_x u_2$, *and* $\nu_{\epsilon v}(x)(u_1) = [\![\theta_2]\!]_s^{Dm}(u_2,w)$.

We proceed now with the envisaged notion of interpretation system morphism. Clearly, such a morphism must relate the constructors and the models in a contravariant way, with the same domains modulo change of sorts. Given a signature morphism $h : \Sigma \to \Sigma'$ and a sorted domain D' over Σ', we denote by $D'|_h$ the sorted domain over Σ such that $(D'|_h)_s = D'_{h(s)}$. Each rigid target model is mapped back to a rigid source model. Each flexible target model is associated with a collection of flexible source models. Note that, contrarily to assignments, we do not map back the worlds, since we are considering rich systems and, therefore, the change of worlds is not essential.

Definition 4.8 An *interpretation system morphism*

$$h : \langle \Sigma, \mathbb{D}, \mathbb{I} \rangle \to \langle \Sigma', \mathbb{D}', \mathbb{I}' \rangle$$

is a tuple $\langle \overline{h}, \underline{h}^{\bullet}, \underline{h}^{R}, \underline{h}^{F} \rangle$ where:

- $\overline{h} : \Sigma \to \Sigma'$ is a signature morphism such that $D'|_{\overline{h}} \in \mathbb{D}$ whenever $D' \in \mathbb{D}'$;
- $\underline{h}^{\bullet} = \{\underline{h}^{\bullet}_{D'}\}_{D' \in \mathbb{D}'}$ where each $\underline{h}^{\bullet}_{D'} : U_{D'} \to U_{D'|_{\overline{h}}}$;
- $\underline{h}^{R} = \{\underline{h}^{R}_{D'}\}_{D' \in \mathbb{D}'}$ where each $\underline{h}^{R}_{D'} : M^{R}_{D'} \to M^{R}_{D'|_{\overline{h}}}$;
- $\underline{h}^{F} = \{\underline{h}^{F}_{D'm'w'}\}_{D' \in \mathbb{D}', m' \in M^{F}_{D'}, w' \in W_{m'}}$ where each $\underline{h}^{F}_{D'm'w'} \in M^{F}_{D'|_{\overline{h}}}$;

such that for every $\vec{g} \in S^*$, $s \in S$, $D' \in \mathbb{D}'$, $n' \in M^{R}_{D'}$, $m' \in M^{F}_{D'}$, $u' \in U_{D'}$, $w' \in W_{m'}$, $\overline{x} \in \overline{X}_{\vec{g}s}$, $\vec{V} \in D_{\vec{g}}^{U_{D'|_{\overline{h}}}}$, $\vec{V}' \in D_{h(\vec{g})}^{W_{m'}}$

- $\nu_{D'\overline{h}(\vec{g}s)}(\overline{h}(\overline{x}))(\vec{V} \circ \underline{h}^{\bullet}_{D'}) = \nu_{D'|_{\overline{h}}\vec{g}s}(\overline{x})(\vec{V}) \circ \underline{h}^{\bullet}_{D'}$;
- $\kappa_{n'\overline{h}(\vec{g}s)} \circ \overline{h} = \kappa_{\underline{h}^{R}_{D'}(n')\vec{g}s}$;
- $W_{\underline{h}^{F}_{D'm'w'}} = \{y' \in W_{m'} : \underline{h}^{F}_{D'm'y'} = \underline{h}^{F}_{D'm'w'}\}$;
- $\eta_{m'\overline{h}(\vec{g}s)}(\overline{h}(f))(w') = \eta_{\underline{h}^{F}_{D'm'w'}\vec{g}s}(f)(w')$;
- $\mu_{m'\overline{h}(\vec{g}s)}(\overline{h}(o))(\vec{V}')(w') = \mu_{\underline{h}^{F}_{D'm'w'}\vec{g}s}(o)(\vec{V}' \circ \iota W^{h}_{D'm'w'})(w')$

  where $\iota W^{h}_{D'm'w'} : W_{\underline{h}^{F}_{D'm'w'}} \hookrightarrow W_{m'}$ is the inclusion map.

Besides the obvious conditions on the preservation of the interpretation of each kind of constructor, the condition on the world spaces imposes that the world space of each flexible target model is partitioned by the world spaces of the corresponding collection of flexible source models [Sernadas et al., 1999].

The following result reflects the logical nature of the proposed notion of morphism: morphisms preserve contextual entailment, and even floating entailment under a reasonable assumption.

Proposition 4.9 *Interpretation system morphisms preserve contextual entailment. Namely, if $\varphi \in \Phi^{\models}$ then $\overline{h}(\varphi) \in \overline{h}(\Phi)^{\models'}$. Moreover, if each map in $\underline{h}^\bullet$ is surjective, floating entailment is also preserved i.e., if $\varphi \in \Phi^{\overline{\models}}$ then $\overline{h}(\varphi) \in \overline{h}(\Phi)^{\overline{\models}'}$.*

Proof: We start by noting that we can prove that

$$[\![\varphi]\!]_a^{Dm}(\underline{h}^\bullet_{D'}(u'), w') = [\![\overline{h}(\varphi)]\!]_{a'}^{D'm'}(u', w')$$

by induction on the structure of φ. Then, the result on the preservation of contextual entailment follows directly. The surjectivity of each map in $\underline{h}^\bullet$ is required for floating entailment so that we can infer $D\langle \underline{h}^R_{D'}(m'), \underline{h}^F_{D'm'w'}\rangle \Vdash \Phi$ from $D'm' \Vdash \overline{h}(\Phi)$. □

Interpretation systems and their morphisms constitute the category *Int*. We need a mechanism for constructing a new interpretation system from a given one along a given signature morphism. To this end we need the (forgetful) functor $N : Int \to Sig$ such that:

- $N(\langle \Sigma, \mathbb{D}, \mathbb{I}\rangle) = \Sigma$;
- $N(h : \langle \Sigma, \mathbb{D}, \mathbb{I}\rangle \to \langle \Sigma', \mathbb{D}', \mathbb{I}'\rangle) = \overline{h}$.

Proposition 4.10 *For each $\langle \Sigma, \mathbb{D}, \mathbb{I}\rangle$ in Int and each morphism $h : \Sigma \to \Sigma'$ in Sig such that each component map of h is surjective, the morphism*

$$\langle h\rangle : \langle \Sigma, \mathbb{D}, \mathbb{I}\rangle \to \langle \Sigma', \mathbb{D}', \mathbb{I}'\rangle$$

where:

- *$\mathbb{D}'$ is the class of all families $h(D)$ with D in $\mathbb{D}$ such that:*
  - *$D_{s_1} = D_{s_2}$ whenever $h(s_1) = h(s_2)$;*
  - *$\nu_{D\vec{g}_1 s_1}(\overline{x}_1) = \nu_{D\vec{g}_2 s_2}(\overline{x}_2)$ if $h(\vec{g}_1 s_1) = h(\vec{g}_2 s_2)$, $h(\overline{x}_1) = h(\overline{x}_2)$;*

  where:

  - *$h(D)_{h(s)} = D_s$;*
- *$\mathbb{I}'(h(D)) = \langle U_D, \nu_{h(D)}, M^R_{h(D)}, A^R_{h(D)}, M^F_{h(D)}, A^F_{h(D)}\rangle$ where:*
  - *$\nu_{h(D)h(\vec{g}s)}(h(\overline{x})) = \nu_{D\vec{g}s}(\overline{x})$;*
  - *$M^R_{h(D)}$ is the subclass of all elements m of M^R_D such that:*
    - *$\kappa_{D\vec{g}_1 s_1}(r_1) = \kappa_{D\vec{g}_2 s_2}(r_2)$ if $h(\vec{g}_1 s_1) = h(\vec{g}_2 s_2)$, $h(r_1) = h(r_2)$;*
  - *$A^R_{h(D)}(m) = \kappa'_m$ where:*
    - *$\kappa'_{mh(\vec{g}s)}(h(r)) = \kappa_{m\vec{g}s}(r)$;*
  - *$M^F_{h(D)}$ is the subclass of all elements m of M^F_D such that:*

  * $\eta_{m\vec{g}_1 s_1}(f_1) = \eta_{m\vec{g}_2 s_2}(f_2)$ *if* $h(\vec{g}_1 s_1) = h(\vec{g}_2 s_2)$, $h(f_1) = h(f_2)$;
  * $\mu_{m\vec{g}_1 s_1}(o_1) = \mu_{m\vec{g}_2 s_2}(o_2)$ *if* $h(\vec{g}_1 s_1) = h(\vec{g}_2 s_2)$, $h(o_1) = h(o_2)$;
- $A^F_{h(D)}(m) = \langle W_m, \eta'_m, \mu'_m \rangle$ *where:*
  * $\eta'_{mh(\vec{g}s)}(h(f)) = \eta_{m\vec{g}s}(f)$;
  * $\mu'_{mh(\vec{g}s)}(h(o)) = \mu_{m\vec{g}s}(o)$;

- $\overline{\langle h \rangle} = h$;
- $\underline{\langle h \rangle}^{\bullet}_{h(D)} = id_{U_D}$;
- $\underline{\langle h \rangle}^{R}_{h(D)} : M^R_{h(D)} \hookrightarrow M^R_D$;
- $\underline{\langle h \rangle}^{F}_{h(D)mw} = m$;

is cocartesian by N *for* h *on* $\langle \Sigma, \mathbb{D}, \mathbb{I} \rangle$. *We denote the codomain of the cocartesian morphism defined above by* $h(\langle \Sigma, \mathbb{D}, \mathbb{I} \rangle)$.

Proof: Clearly, I' is a pre-interpretation system. That it is an interpretation system follows from the facts that any two isomorphic flexible models are, by definition, either both kept or excluded, and the binding conditions hold by construction. The fact that $\langle h \rangle : I \to I'$ is a *Int* morphism is an immediate consequence of the definition. As far as the universal property is concerned, suppose that $f : I \to I''$ is a *Int* morphism and $\overline{g} : \Sigma' \to \Sigma''$ a *Sig* morphism such that $\overline{f} = \overline{g} \circ h$. We refrain from describing here the unique morphism $g : I' \to I''$ such that $f = g \circ \langle h \rangle$. It is easily adapted from [Sernadas et al., 1999] with the necessary extensions. □

Unconstrained Fibring

Intuitively, in the unconstrained fibring of two interpretation systems we obtain models by pairing rigid models and by fibring flexible models.

Prop/Definition 4.11 *Let both* $\langle \Sigma', \mathbb{D}', \mathbb{I}' \rangle$ *and* $\langle \Sigma'', \mathbb{D}'', \mathbb{I}'' \rangle$ *be interpretation systems such that their signatures are variable independent. Then, their unconstrained fibring* $\langle \Sigma', \mathbb{D}', \mathbb{I}' \rangle \oplus \langle \Sigma'', \mathbb{D}'', \mathbb{I}'' \rangle$ *is the interpretation system* $\langle \Sigma' \oplus \Sigma'', \mathbb{D}, \mathbb{I} \rangle$ *defined as follows, using the injections* $i' : \Sigma' \to \Sigma' \oplus \Sigma''$ *and* $i'' : \Sigma'' \to \Sigma' \oplus \Sigma''$ *in Sig:*

- $\mathbb{D}$ *is the class of all sorted domains* $D' \bullet D''$ *such that* $(D' \bullet D'')_{i'(s')} = D'_{s'}$ *and* $(D' \bullet D'')_{i''(s'')} = D''_{s''}$ *where* $D' \in \mathbb{D}'$, $D'' \in \mathbb{D}''$;
- $\mathbb{I}(D' \bullet D'') = \langle U, \nu, M^R, A^R, M^F, A^F \rangle$ *where:*
  - $U = U_{D'} \times U_{D''}$ *with projections* $\pi'_{D' \bullet D''}$ *and* $\pi''_{D' \bullet D''}$;

- $\nu_{i'(\vec{g}'s')}(i'(\overline{x}'))(\vec{V}' \circ \pi'_{D'\bullet D''}) = \nu_{D'\vec{g}'s'}(\overline{x}')(\vec{V}') \circ \pi'_{D'\bullet D''}$;
  $\nu_{i''(\vec{g}''s'')}(i''(\overline{x}''))(\vec{V}'' \circ \pi''_{D'\bullet D''}) = \nu_{D''\vec{g}''s''}(\overline{x}'')(\vec{V}'') \circ \pi''_{D'\bullet D''}$;
- $M^R = M^R_{D'} \times M^R_{D''}$;
- $A^R(\langle m', m''\rangle) = \kappa$ *where:*
  * $\kappa_{i'(\vec{g}'s')} \circ i' = \kappa_{m'\vec{g}'s'}$;
    $\kappa_{i''(\vec{g}''s'')} \circ i'' = \kappa_{m''\vec{g}''s''}$;
- M^F *is the class of all pairs* $m = \langle W, \tau\rangle$ *such that:*
  * W *is a non empty set;*
  * $\tau = \{\langle \tau'_w, \tau''_w\rangle\}_{w\in W}$ *with each* $\tau'_w \in M^F_{D'}$ *and* $\tau''_w \in M^F_{D''}$;
  * $W_{\tau'_w} = \{y \in W : \tau'_y = \tau'_w\}$ and $W_{\tau''_w} = \{y \in W : \tau''_y = \tau''_w\}$ *for each* $w \in W$;
  * $\mathrm{lfp}(O, \{w\}) = W$ *for every* $w \in W$*, where* $O : 2^W \to 2^W$ *is such that*
    $$O(Y) = \bigcup_{y\in Y} (W_{\tau'_y} \cup W_{\tau''_y});$$
- $A(\langle W, \tau\rangle) = \langle W, \eta, \mu\rangle$ *where:*
  * $\eta_{i'(\vec{g}'s')}(i'(f'))(w) = \eta_{\tau'_w\vec{g}'s'}(f')(w)$;
    $\eta_{i''(\vec{g}''s'')}(i''(f''))(w) = \eta_{\tau''_w\vec{g}''s''}(f'')(w)$;
  * $\mu_{i'(\vec{g}'s')}(i'(f'))(\vec{V}' \circ \iota' W_w)(w) = \mu_{\tau'_w\vec{g}'s'}(f')(\vec{V}')(w)$;
    $\mu_{i''(\vec{g}''s'')}(i''(f''))(\vec{V}'' \circ \iota'' W_w)(w) = \mu_{\tau''_w\vec{g}''s''}(f'')(\vec{V}'')(w)$
    with $\iota' W_w : W_{\tau'_w} \hookrightarrow W$ *and* $\iota'' W_w : W_{\tau''_w} \hookrightarrow W$.

Proof: It is straightforward to verify that the proposed unconstrained fibring is an interpretation system. □

Proposition 4.12 *Unconstrained fibrings are coproducts in Int.*

Proof: Easily adapted from [Sernadas et al., 1999] with the necessary extensions. □

Again, now at the semantic level, the coproduct construction captures the intuitive idea that fibring should extend the given logics in a minimal and conservative way. Indeed, the injections of the coproduct, being morphisms with the necessary surjective components, preserve the entailments as envisaged.

Constrained Fibring

It is straightforward to repeat what we did before for Hilbert calculi now for interpretation systems. In order to obtain the constrained (by sharing some constructors) fibring of two interpretation systems with variable independent signatures, we first calculate their unconstrained fibring and then we get the result by cocartesian lifting along the coequalizer of signatures.

As an illustration, consider now the problem of obtaining the semantics of FOTL by combining the semantics of FOL and TL. The envisaged interpretation system is indeed obtained by constrained fibring of the given systems (imposing the sharing of the propositional connectives). Note that the Barcan formulae are valid in the resulting system, although they were not proved as theorems in the fibring of the corresponding Hilbert calculi (recall the example at the end of section 3).

5 TRANSFER RESULTS

It is straightforward to make precise the notion of *logic system presentation* as a tuple $\langle \Sigma, \mathbb{D}, \mathbb{I}, R \rangle$ containing an interpretation system $\langle \Sigma, \mathbb{D}, \mathbb{I} \rangle$ and a Hilbert calculus $\langle \Sigma, R \rangle$, both over the signature Σ. The morphisms are obvious. The unconstrained and constrained fibrings of logic system presentations with variable independent signatures are easily obtained from the fibrings of their model and proof theoretic components. The study of transference results (like preservation of soundness and completeness) should of course be carried out within this context.

We start by analysing the preservation of soundness. To this end, we need to define carefully what we mean by a sound rule in a given logic system presentation. Note that, contrarily to what was possible in the simpler case of propositional based logics considered in [Sernadas et al., 1999], we have to consider here that schema variables are to be instantiated by substitutions at the symbolic level, in order to be able to take into account requirements. Therefore, we are led to the following definition.

Definition 5.1 Let $\mathcal{L} = \langle \Sigma, \mathbb{D}, \mathbb{I}, R \rangle$ be a logic system presentation. A rule $r \in R$ is said to be *sound* (in $\mathcal{L}$) iff, for every interpretation system morphism $h : \langle \Sigma, \mathbb{D}, \mathbb{I} \rangle \to \langle \Sigma', \mathbb{D}', \mathbb{I}' \rangle$ such that each map in $\underline{h}^\bullet$ is surjective, and ground substitution $\sigma' \in Sub(\Sigma')$:

- $\overline{h}(Conc(r))\sigma' \in \overline{h}(Prem(r))\sigma'^{\models_{\Sigma'}}$ whenever $[\![\overline{h}(q)]\!]_{\Sigma'}(\sigma') = 1$ for every $q \in Req(r)$.

This definition seems to require too much for a rule to be sound in a particular $\mathcal{L}$ by looking also at substitutions elsewhere. However, it is not so. Indeed, we must take into account that a schema variable may be instantiated within the context of a much richer language (over Σ') with any semantics for the new symbols! Otherwise, for instance, in a FOL-like logic but without quantifiers, the rule "from φ infer φ^x_τ" would be sound thanks to the paucity of the language. Of course, the soundness of such a rule would not be preserved in general. Unfortunately, the proposed "cautious" definition imposes a heavy burden when verifying soundness of a Hilbert calculus: how can we hope to verify the soundness of a rule?

Example 5.2

Consider the problem of verifying the soundness of the rule in FOL: "infer $((\forall x(\varphi \Rightarrow \psi)) \Rightarrow (\varphi \Rightarrow (\forall x \psi)))$ provided that x is not free in φ". The verification follows the structure of the classical verification of the "local" soundness of the rule in FOL using the semantic lemma of independence of non-free variables. Clearly, we have to consider the possible presence of other symbols with unknown semantics and we can use the fact that the semantics of the FOL symbols is preserved by h. This approach should be feasible in all cases. For instance, when verifying the soundness of the rule "from $(\forall x \varphi)$ infer $\varphi|_t^x$ provided that t is free for x in φ", we again follow the steps of the classical "local" verification using both semantic lemmata (of independence of non-free variables and of susbstitution).

On the other hand, the proposed definition of soundness of rules pays off when we look at preservation, since it is trivial to obtain the following result:

Proposition 5.3 *Let $h : \mathcal{L} \to \mathcal{L}'$ be a logic system presentation morphism such that each map in $\underline{h}^\bullet$ is surjective. If $\mathcal{L}$ has sound rules, then so does $\mathcal{L}'$.*

As immediate corollaries we obtain that both unconstrained and constrained fibring preserve soundness of the rules. Clearly, a system with sound rules is sound in the usual sense, that is, provability implies floating entailment.

On the other hand, it is easy to see that completeness is not always preserved. For instance, when fibring FOL and TL we obtain a logic system presentation where the Barcan formulae are valid but are not provable. We might relax our notion of interpretation system (and morphism) so that by fibring we would get models with changing domains and flexible object variables, therefore leading to invalid Barcan formulae. However, that is not so desirable given the applications in sight. In any case, we are always free to add the Barcan formulae as axioms after the combination when applicable and so hopefully recover completeness.

6 RELATED WORK

This work capitalizes on the previous paper [Sernadas et al., 1999] extending the results therein to the more complex case of logics with variables, terms and binding operators. We believe that our proof and model theoretic accounts of fibring follow the concepts in [Gabbay, 1996a; Gabbay, 1996b; Gabbay, 1998] even in this more complex case.

The categorial approach to the study of combination of logics was already advocated in [Sernadas et al., 1997a] and it has been a key feature in the literature on institutions [Goguen and Burstall, 1992; Tarlecki, 1996] and parchments [Mossakowski, 1996]. One difference to these works is that, for the sake of simplicity, we do not work at the signature indexed level. Working at the signature indexed level would mean to consider a logic system to be a category of object signatures (not to be confused with the meta signatures introduced in Section 2) and a functor from this category into the category of logic system presentations (or interpretation systems, or Hilbert calculi). We expect that some form of fibring can also be done at the signature indexed level using general results about indexed categories similar to those in [Mossakowski et al., 1998].

Concerning proof theory, the non-indexed constituents of entailment systems [Meseguer, 1989] or Π-institutions [Fiadeiro and Sernadas, 1988] are consequence systems [Sernadas et al., 1997a], and there is an obivous functor from the category of Hilbert calculi to the category of consequence systems. At the model-theoretic side, the non-indexed constituents of institutions are satisfaction systems [Sernadas et al., 1997a], called twisted relations in [Goguen and Burstall, 1992]. There are two obvious functors from the category of interpretation systems to the category of satisfaction systems, one using contextual and one using floating satisfaction.

It should not be difficult to factor these functors through the category of rooms for model-theoretic parchments (which, like interpretation systems, are algebraic presentations of satisfaction systems, but much simpler than interpretation systems). Indeed, the notion of interpretation system can be seen as an extension of the notion of room for model theoretic parchments [Mossakowski et al., 1998]. Roughly speaking, the latter are interpretation systems with sorts and rigid constructors only. Variables and quantifiers are added when moving to context parchments [Pawłowski, 1998], but with our approach, we can also model binding operators like λ-abstraction. The most important contributions of our approach are the flexible constructors and modal binding operators. Though model theoretic parchments allow one to model these within a much simpler formalism (an example is given in [Mossakowski et al., 1998]), the extra complexity of our approach is justified by the possibility to model fibring [Gabbay, 1996a; Gabbay, 1996b]. For example, when combining a modality of time with a modality of space, in the fibred models, different points in space may have different time lines associated with them. This is not possible in the parchment approach.

On the other hand, model theoretic parchments are more general than our interpretations systems in the respect that they cover also multi-valued logics. Moreover, domains can be changed along morphisms. Both features are useful when adding partial functions to a given logic—even if the resulting logic is two-valued [Mossakowski et al., 1998]. The generalization of interpretation systems in this respect is straightforward—but it is unclear whether the results carry over. Of course, two-valued partial first-order logic [Burmeister, 1982] can be modelled already with the present approach.

7 CONCLUDING REMARKS

We were able to provide categorial characterizations of both unconstrained and constrained fibring of Hibert calculi and interpretation systems for languages with variables, terms, variable binding operators and modal like operators. We also established some preliminary transference results and briefly compared our notions and results to those on model theoretic parchments.

The categorial approach helped in fine tuning the envisaged semantics of fibring with explicit models. It was also particularly helpful when defining the mechanism of binding at the level of Hibert calculi. By defining the binding predicates over the terminal signature we were able to treat in the same way all variable binding operators as it was envisaged, and we obtained for free the desired structurality (consistent change of values with changes of signatures).

Further work should be directed at obtaining additional preservation results. However, preservation of completeness is out of question within the proposed framework since Barcan formulae are valid but not provable. A more relaxed notion of interpretation system seems to be the most feasible way to recover the preservation of completeness.

Another line of research should be aimed at generalizing our notion of interpretation system in order to be able to cope with multi-valued and partial logics, adapting the results on model theoretic parchments, while keeping the link to Hilbert calculi. In this way we might achieve the mentioned advantages of the theory of parchments together with those of our theory of interpretation sytems where we are able to deal with fibring of modal structures.

We should also mention the importance of looking again at propositional based logics with the machinery for defining inference rules with requirements. Indeed, requirements seem essential to a correct understanding of propositional intuitionistic logic and its fibring with classical logics.

Acknowledgments. This research was partially supported by FCT, the PRAXIS XXI Projects 2/2.1/ MAT/262/94 SitCalc, PCEX/P/MAT/46/96 ACL and 2/2.1/TIT/1658/95 LogComp, as well as by the ESPRIT IV Working Groups 22704 ASPIRE and 23531 FIREworks.

REFERENCES

[Blackburn and de Rijke, 1997] P. Blackburn and M. de Rijke. Why combine logics? *Studia Logica*, 59(1):5–27, 1997.

[Burmeister, 1982] P. Burmeister. Partial algebras — survey of a unifying approach towards a two-valued model theory for partial algebras. *Algebra Universalis*, 15:306–358, 1982.

[Fiadeiro and Sernadas, 1988] J. Fiadeiro and A. Sernadas. Structuring theories on consequence. In D. Sannella and A. Tarlecki, editors, *Recent Trends in Data Type Specification*, volume 332 of *Lecture Notes in Computer Science*, pages 44–72. Springer-Verlag, 1988.

[Gabbay, 1996a] D. Gabbay. Fibred semantics and the weaving of logics: part 1. *Journal of Symbolic Logic*, 61(4):1057–1120, 1996.

[Gabbay, 1996b] D. Gabbay. An overview of fibred semantics and the combination of logics. In F. Baader and K. Schulz, editors, *Frontiers of Combining Systems*, pages 1–55. Kluwer Academic Publishers, 1996.

[Gabbay, 1998] D. Gabbay. *Fibring logics.* Oxford University Press, 1998.

[Goguen and Burstall, 1992] J. Goguen and R. Burstall. Institutions: Abstract model theory for specification and programming. *Journal of the ACM*, 39(1):95–146, 1992.

[Meseguer, 1989] J. Meseguer. General logics. In H.-D. Ebbinghaus et al, editor, *Proceedings of the Logic Colloquium, 1987*, pages 275–329. North-Holland, 1989.

[Mossakowski, 1996] T. Mossakowski. Using limits of parchments to systematically construct institutions of partial algebras. In M. Haveraaen, O. Owe, and O.-J. Dahl, editors, *Recent Trends in Data Type Specifications. 11th Workshop on Specification of Abstract Data Types*, volume 1130 of *Lecture Notes in Computer Science*, pages 379–393. Springer Verlag, 1996.

[Mossakowski et al., 1998] T. Mossakowski, A. Tarlecki, and W. Pawłowski. Combining and representing logical systems using model-theoretic parchments. In F. Parisi Presicce, editor, *Recent Trends in Algebraic Development Techniques. 12th Workshop on Algebraic Development Techniques, Tarquinia 1997*, volume 1376 of *Lecture Notes in Computer Science*, pages 349–364. Springer Verlag, 1998.

[Pawłowski, 1998] W. Pawłowski. Context parchments. In F. Parisi Presicce, editor, *Recent Trends in Algebraic Development Techniques. 12th Workshop on Algebraic Development Techniques, Tarquinia 1997*, volume 1376 of *Lecture Notes in Computer Science*, pages 381–401. Springer Verlag, 1998.

[Sernadas et al., 1997a] A. Sernadas, C. Sernadas, and C. Caleiro. Synchronization of logics. *Studia Logica*, 59(2):217–247, 1997a.

[Sernadas et al., 1997b] A. Sernadas, C. Sernadas, and C. Caleiro. Synchronization of logics with mixed rules: Completeness preservation. In M. Johnson, editor, *AMAST'97 - 6th International Conference on Algebraic Methodology and Software Technology*, volume 1349 of *Lecture Notes in Computer Science*, pages 465–478. Springer-Verlag, 1997b.

[Sernadas et al., 1999] A. Sernadas, C. Sernadas, and C. Caleiro. Fibring of logics as a categorial construction. *Journal of Logic and Computation*, 9(2):149–179, 1999.

[Tarlecki, 1996] A. Tarlecki. Moving between logical systems. In M. Haveraaen, O. Owe, and O.-J. Dahl, editors, *Recent Trends in Data Type Specifications. 11th Workshop on Specification of Abstract Data Types*, volume 1130 of *Lecture Notes in Computer Science*, pages 478–502. Springer Verlag, 1996.

Iterative Dialogues and Automated Proof

Konrad Slind and Richard Boulton

1 INTRODUCTION

Proof planning [Bundy, 1991] uses artificial intelligence planning techniques to discover proofs of theorems in mathematics and logic. CLAM is a proof planner, implemented in Prolog, for Oyster, an implementation of Martin-Löf Type Theory. Its main area of application is in finding proofs by mathematical induction but it has been used for other kinds of proof as well. The paper [Boulton *et al.*, 1998] describes an interface between CLAM and HOL90, a Standard ML implementation of the HOL interactive theorem prover [Gordon and Melham, 1993]. The HOL system is an LCF-style [Gordon *et al.*, 1979] theorem prover for a classical higher-order logic. The aim of the interface is to make CLAM's expertise in proof by induction available in HOL; in the combined system, CLAM is treated as a tactic, an invocation of which either completely solves the current goal, or fails to return a proof. The main issues in the integration are translating between the different logics of the two systems and obtaining HOL tactics from CLAM plans.

Here we improve the combined system by introducing an *extended interaction* protocol in which the two systems can take part in an unfolding dialogue as they search for a proof. A particular novelty of the approach is the use of iterative dialogues as opposed to recursive dialogues. We believe this to be an important way of combining systems that have substantial amounts of internal state. In the protocol, CLAM may return a *partial* proof plan. Included in the plan are conjectures that CLAM believes HOL may be able to prove more easily than it can itself. HOL may choose to return these conjectures to the user as new subgoals, but more likely it will attempt to prove them automatically, e.g., using its implementation of Nelson-Oppen co-operating decision procedures [Boulton, 1995], and then re-invoke CLAM, stating whether it proved, refuted, or failed to prove each conjecture. CLAM then attempts to advance the proof using the new information. We believe this to be the first time two mechanized reasoning systems have been combined in this way.

In the following sections we outline the original 'single-shot' combination of the two systems, argue the merits of the recursive and iterative approaches to extended dialogues, and describe the iterative approach in abstract terms. We then go on to describe relevant details of the implementation and demonstrate the system on some examples.

2 SINGLE-SHOT INTERACTION

Our approach to combining the two systems is based on the following principles:

1. The CLAM system should not be modified to suit the classical logic used by the HOL system. (This is possible since the area where CLAM is planning fits comfortably into the intersection of the two logics.) As a consequence, custom HOL tactics corresponding to CLAM's basic methods were written.

2. The HOL system should not take the plans resulting from CLAM on faith. In effect, HOL is checking the work done by CLAM when it executes the tactic derived from a plan. In every case we have encountered so far, the time spent to execute a tactic is much less than that required to find the plan.

The combination is achieved with the help of three translations [Boulton *et al.*, 1998]: a couple which map back and forth between expressions in the two object logics; and one mapping plans—which CLAM creates in the belief that they are to be used by Oyster—into tactics usable by HOL. This treatment allows the systems to interact while remaining independent, in several senses. For example, the consistency of HOL is not threatened by its use of CLAM as a tactic (invocation of a tactic cannot introduce inconsistency); similarly, the theory and practice of proof planning in CLAM can be developed without regard to the impact on HOL.

In our new design, we have made a few simple additions to CLAM; we therefore give a brief introduction to proof plans in order to facilitate the reader's understanding of the sequel. CLAM works by using formalized pre- and post-conditions of Oyster tactics as the basis of plan search. These high-level specifications of tactics are called *methods*. A *plan* is a hierarchy of methods, with the high-level methods representing steps in a proof that a human mathematician might make, e.g., induction, while low-level methods correspond to more detailed (and directed) syntactic manipulations. When a plan for a goal is found, the expectation is that the resulting tactic will solve the goal. CLAM uses a wide variety of methods as it searches for solutions, e.g., induction, generalization, *etc.* A significant feature of CLAM is its *rippling* method, which is typically used after an induction step in order to transform the goal so that the induction hypothesis can be used. This transformation happens via the application of *wave rules*, a form of rewrite rule in which meta-level annotations are used to direct rewriting and in particular to ensure termination.

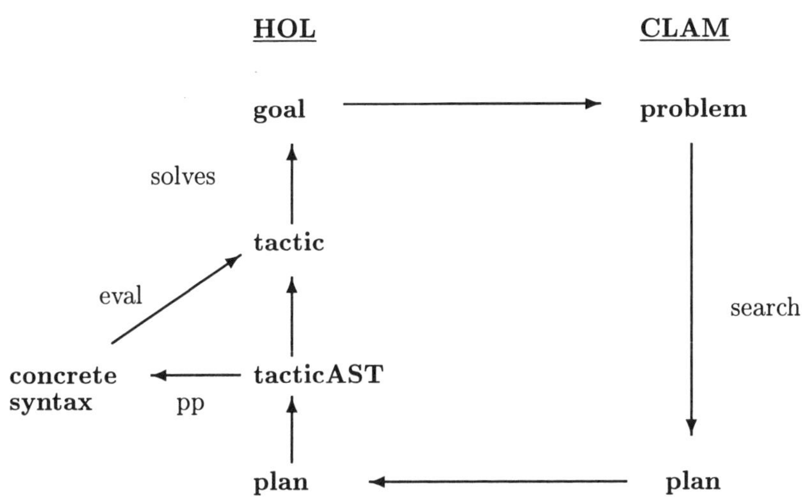

Figure 1: Original System Structure

A diagram for the system is given in Fig. 1. The data flow is represented by arrows. The HOL formula (goal) to be proved is translated into the abstract syntax of Oyster's logic and, together with supporting definitions, wave rules, and induction schemes, is passed to the CLAM process. CLAM then attempts to plan the goal. If a plan is found, it is passed back to the HOL process. The plan is then translated to an intermediate form from which both tactics and (equivalent) concrete syntax can be generated.[1] Finally, the tactic is applied to the HOL goal.

At the system level, HOL and CLAM are connected via Unix sockets, and are thus able to interact in a machine independent and distributed fashion over the Internet (for example, the system has been operated with a HOL process in Pennsylvania connected to a CLAM process in Scotland). Another significant point is that the bulk of the translation to and from CLAM has been automatically generated by Boulton's ML-Syn tool [Boulton, 1996] from a single high-level source document. This enabled us to quickly construct translations between the syntaxes of the two object languages. Such translations are often required when combining systems in an 'arms-length' fashion.

The following theorems, which are lemmas in the correctness of an 'insertion' sort function, are an example of what the combined system can automatically prove:

[1]The generation of concrete syntax is used to produce stand-alone proof scripts which can be run in the absence of CLAM.

$$\forall x\ l.\ \texttt{MEMBER}\ x\ (\texttt{SORT}\ l) \supset \texttt{MEMBER}\ x\ l$$
$$\forall x\ l.\ \texttt{MEMBER}\ x\ l\ \supset \texttt{MEMBER}\ x\ (\texttt{SORT}\ l)$$

These examples use the following definitions/rules:[2]

$$\begin{array}{rcl}
\forall x.\ \texttt{LESS}\ x\ 0 & = & \texttt{F} \\
\forall y.\ \texttt{LESS}\ 0\ (\texttt{SUC}\ y) & = & \texttt{T} \\
\forall x\ y.\ \texttt{LESS}\ (\texttt{SUC}\ x)\ (\texttt{SUC}\ y) & = & \texttt{LESS}\ x\ y \\
\forall m.\ \texttt{MEMBER}\ m\ [\,] & = & \texttt{F} \\
\forall el\ h\ l.\ (el = h) \supset (\texttt{MEMBER}\ el\ (\texttt{CONS}\ h\ l) & = & \texttt{T}) \\
\forall el\ h\ l.\ \neg(el = h) \supset (\texttt{MEMBER}\ el\ (\texttt{CONS}\ h\ l) & = & \texttt{MEMBER}\ el\ l) \\
\forall n.\ \texttt{INSERT}\ n\ [\,] & = & [n] \\
\forall n\ h\ t.\ \texttt{LESS}\ n\ h \supset (\texttt{INSERT}\ n\ (\texttt{CONS}\ h\ t) & = & \texttt{CONS}\ n\ (\texttt{CONS}\ h\ t)) \\
\forall n\ h\ t.\ \neg(\texttt{LESS}\ n\ h) \supset (\texttt{INSERT}\ n\ (\texttt{CONS}\ h\ t) & = & \texttt{CONS}\ h\ (\texttt{INSERT}\ n\ t)) \\
\texttt{SORT}\ [\,] & = & [\,] \\
\texttt{SORT}(\texttt{CONS}\ h\ t) & = & \texttt{INSERT}\ h\ (\texttt{SORT}\ t)
\end{array}$$

and the induction scheme

$$\forall P.\ (\forall y.\ P\ 0\ y) \supset (\forall x.\ P\ x\ 0) \supset (\forall x\ y.\ P\ x\ y \supset P\ (\texttt{SUC}\ x)\ (\texttt{SUC}\ y)) \\ \supset \forall x\ y.\ P\ x\ y$$

along with the standard induction schemes for lists and natural numbers.

A detailed description of the combined system can be found in [Boulton *et al.*, 1998].

2.1 Drawbacks

The single-shot approach provides a tool which considerably increases the automation of HOL on inductive proofs. However, it also has several limitations:

- CLAM can only be used as a final step in a branch of a HOL proof: it cannot partly plan a proof.
- CLAM plans proofs to a very low level. CLAM has high-level methods, composed of lower level methods. In principle, HOL tactics could be written to implement the high-level methods without explicitly implementing the lower level methods. This approach is not taken because we want CLAM, not HOL, to search for the proof. Thus, HOL tactics corresponding to the lower level methods of CLAM are used. However,

[2]The constant F stands for falsity in the HOL logic; T stands for truth; and SUC stands for the successor operation on natural numbers.

in order to know with high probability that a corresponding tactic will succeed, CLAM has to plan proofs to a yet lower level. Often this results in the detailed planning of steps that HOL can do with ease (without search).

- There are only very weak means of influencing the behaviour of the planner as it works. Wave rules, definitions and induction schemes may be given, and time limits on planning may be set, but there is no way, from HOL, to steer CLAM away from fruitless searches.
- Lack of modularity. Sub-problems that are solved during planning may represent general lemmas of wider utility. Currently, these lemmas are hidden inside the returned proof plan.
- Requiring all wave rules to be given before planning starts means that the HOL user needs to know — ahead of time — the wave rules that will be required in the proof plan (so that they can be supplied to CLAM). Since this is not possible in general, the possibilities of automation are limited.
- Currently, planning in CLAM is not *logic-independent*: some aspects of its operation, e.g., transitivity reasoning, are not reflected in the final proof plan. In such cases, CLAM may successfully plan the proof of a theorem, but the generated HOL tactic will fail.

3 EXTENDED INTERACTION

A solution to many (and possibly all) of the limitations described above is to permit more extended dialogues between the two systems, wherein, part way through planning, CLAM can ask HOL for information (which can lead to an embedded invocation of CLAM by HOL, *etc*). How this addresses the limitations is described in Section 3.3. But what form should the extended interaction take? Perhaps the most obvious approach is for CLAM to be able to 'call' HOL during planning. However, this may be difficult to achieve, as we now argue.

3.1 Recursion *versus* Iteration

(Although the discussion in this section concerns CLAM and HOL, it seems to apply quite generally.) During a simple interaction, the HOL process is blocked waiting on a socket for CLAM to return with a result. Thus HOL is not ready to be called on by CLAM to perform some other task. The CLAM process seems to have two choices if it wants to appeal to HOL:

1. Start a new HOL process to perform the task. This is unsatisfactory for two reasons. First, HOL processes are large, so there may be an

unreasonable demand on computational resources, especially if the dialogue goes further, spawning multiple processes. Second, a fresh HOL process may not have the same context as the original HOL process. For deciding a subproblem in some standard logical theory this may not be a problem, but usually problems exist in user-defined extensions of standard theories, in which case ensuring that the same context exists in the new process is a difficult job.

2. Pass control to the original HOL process, with a request to perform the task. This solves the problem of HOL waiting for CLAM to return and of having access to a suitable context. However, once HOL has serviced the new request it must wait again for CLAM to return with the answer to the original request and it must do so in the state it was in originally. Likewise, the CLAM process may get called a second time while HOL is servicing its request from CLAM and similar arguments apply. This all points to the need for both HOL and CLAM to maintain (in some sense) a stack of states, a new stack frame being invoked each time a new request is received. We call this approach the *recursive* option. Since both HOL and CLAM have substantial amounts of context associated with a proof attempt, the recursive option would seem to require major modifications to both systems and large amounts of memory.

An alternative way for CLAM to transfer control to HOL proceeds, not by recursion, but by the following expedient: rather than attempting to build a complete plan by making intermediate recursive invocations, CLAM's requests for help become embedded in a partial plan. In this scenario, once CLAM has returned a partial plan, it has—as far as it is concerned—finished its work. HOL detects the requests within the partial plan, attempts to service them, and then invokes CLAM again, on the same goal, but with strictly more information than in the original invocation. (HOL may also call CLAM during its attempts to service CLAM's requests. CLAM will be in a position to receive such requests because it believes it has finished its previous business.) In this approach, which we call the *iterative* option, neither system need maintain a stack of states, though it may be necessary for them to encapsulate the extra information to be passed around in each round of the dialogue. The obvious disadvantage of the iterative option is that CLAM will repeat (at least part of) the planning process on each iteration. We now examine how the iterative approach works when used to extend the single-shot interaction scheme of CLAM and HOL.

3.2 Iterative Dialogues in CLAM/HOL

Starting from a goal, the two systems will iterate through a number of rounds, attempting to converge on a proof plan. The CLAM process functions as a server, and HOL is the client. The intent of the design, in order to help the server find a proof plan, is that the client should supply it with more and more information about sub-problems that occur in the planning process. To support this, CLAM needs to be altered so that it accepts a database of arbitrary *facts* as input. These facts will be used to help control the planning process; their main function is to close off search spaces (build leaf nodes in the plan). A fact may be classified as an element of the following ML datatype:[3]

```
fact = Proved    of thm
     | Refuted   of term list * term
     | CantProve of term list * term
```

3.2.1 The Server Side

CLAM accepts the given fact database and the problem (along with, as usual, wave rules, induction schemes, and definitions) and starts planning. It continues to a point where a specific lemma M could be used. (The crucial point of how M is determined is discussed in Section 3.4.) Instead of attempting to plan a proof of M, which is the normal way that CLAM operates, CLAM will attempt to match M against the given facts. There are 4 possible results of the match operation: `NotFound`, `Proved`, `Refuted`, and `CantProve`.

`NotFound`. Information pertaining to M is not in the database. This means that sub-problem M has not been previously encountered. CLAM *assumes* M is true, logs it in the proof plan, and continues planning.

`Proved`. This means that sub-problem M had been encountered and proved by the client in an earlier round. The effect is that planning of M succeeds; the server augments the plan with the name of the proved theorem and continues.

`Refuted`. This means that sub-problem M had been encountered and assumed by CLAM in an earlier round. However, the client had subsequently proven $\neg M$. The effect is that planning of M fails and the planner will backtrack.

`CantProve`. This means that sub-problem M had been encountered and assumed by CLAM in an earlier round. However, the client was subsequently unable to either prove or refute M. At this point, CLAM can conclude that appealing to the client for help on M was a waste of time, and should apply any of its remaining methods in attempting to prove M.

CLAM continues planning in this way until it succeeds or fails. If it fails, that means that neither CLAM nor HOL could prove the goal. If planning

[3] The ML type `thm` represents theorems in the HOL logic; the type `term` represents the terms and formulas of the HOL logic. The infix '`*`' denotes pairing.

succeeds, CLAM has found a partial plan, one that is contingent upon the truth of the assumed conjectures. CLAM has now finished its work in round n and it returns the partial plan $\mathcal{P}$ and the lemmas $\mathcal{L}$ to the client and waits for another planning request. Thus the interface to CLAM, from the point of view of HOL, would essentially have the type

$$definitions * wave\_rules * schemes * facts * goal \longrightarrow partial\_plan * goal\ list.$$

This is basically the representing type for LCF-style *tactics* [Gordon *et al.*, 1979] because it takes a goal g and returns a list of goals plus a means of combining their proofs into a proof for g.

3.2.2 The Client Side

If the server failed to find a plan, the client will likewise fail. Otherwise, the client, on receipt of $(\mathcal{P}, \mathcal{L})$, attempts to prove each lemma in $\mathcal{L}$. The client is allowed to use its native proof procedures or other external proof tools. If all lemmas in $\mathcal{L}$ are proved, denoted $\vdash \mathcal{L}$, the extended interaction is finished, and the pair $(\mathcal{P}, \vdash \mathcal{L})$ is translated into a tactic that is applied to the initial goal. This mapping is a simple modification of the original translation of proof plans to tactics. Otherwise, each member of $\mathcal{L}$ is tagged whether it was proved (`Proved`) or refuted (`Refuted`) or neither (`CantProve`) and added to the facts for round $n + 1$. Thus CLAM enters round $n + 1$ with a more extensive collection of facts than in any of the previous rounds. Indeed, this collection of facts is the only thing that changes in the input to CLAM from round to round: we repeatedly try to plan the same problem, adding more information to the input of the planner each time, in response to the output of the planner in the previous round.

Note. $\mathcal{P}$ does not persist between rounds: in the client, the partial plan from round n is erased when round $n + 1$ starts.

3.2.3 The Combined System

Now we focus on round $n + 1$ in the server. First, note that the plan from the previous round is discarded: planning starts afresh and may proceed as in round n until the first point where CLAM assumed a proposition M in round n.[4] For every conjecture in $\mathcal{L}$ from the server in round n, the client has classified it as `Proved`, `Refuted`, or `CantProve`, and entered it into the database for round $n + 1$. Therefore M has been classified. If M had been proved by the client, then planning of the current sub-problem can cease. If M had been refuted by the client, then the server can backtrack. If `CantProve`(M) was the judgement of the client, then we only know that the client couldn't

[4] The planner may also proceed along a different path because of new information from the client: it is still trying to solve the same initial problem (with more information than it had in the last round).

prove M. Therefore, CLAM can invoke the remaining applicable methods on M. If this succeeds, then planning continues; otherwise, neither the client nor CLAM can prove M, and thus we are at a dead-end and can backtrack.

Remarks

- Facts are added incrementally: in round $n+1$, the server only gets new facts about conjectures assumed in round n and whose classification as a `fact` has just been performed by the client.
- In all cases, each round gives more information, thus constraining the search space.
- The current behaviour of the CLAM/HOL system can be regained by having the client fail to prove every conjecture sent its way. This forces the original behaviour of the system, where the planner has to do all the work.

3.3 Drawbacks Revisited

Now we return to our original list of limitations:

CLAM cannot partly plan a proof. This is no longer true: by single-stepping the extended interaction, CLAM can now behave as a tactic that produces subgoals for HOL, and not simply as an all-or-nothing prover. Thus the automation offered by proof planning can be neatly combined with interactive proof.

CLAM has to plan proofs to a very low level. This is no longer true: the database of facts, plus the access to client proof procedures will allow proofs to get cut-off at an arbitrarily high level.

Hard to influence planning. This is no longer true, for the same reason supplied in the previous point. However, notice that although successful proofs can be shortcut, failing ones—where the planner has blundered into a black hole—remain uncontrollable in the new scheme.

Lack of modularity. Sub-problems that are assumed by CLAM get registered in the client, and thus all intermediate results obtained by asking the client for an answer are retained.

All wave rules need to be given before planning starts. The collection of wave rules used by the planner should be lazily augmented as planning progresses. In the extended interaction scheme, the planner could assume wave rules as they are needed (and have the client attempt to prove them in the ensuing round). This approach seems to require CLAM to dynamically synthesize wave rules. A *lemma discovery critic* [Ireland and Bundy, 1996] has been proposed as a solution to this problem.

Planning is not logic-independent. If we employ our scheme to have client proof procedures prove the logic-*dependent* sub-problems that arise in planning, then a degree of logic-independence might be attainable. The planner would have to recognize formulas (e.g., propositional formulas) whose validity may depend on the logic used by the client and ask the client to say whether they are valid or not. Recall that it is not sufficient for the planner to simply assume the validity of a formula. In the iterative approach this assumption is made *speculatively* and may be reversed on the next iteration due to new information from the client.

3.4 Methods for External Facts

The preceding discussion skirts a fundamental question: how does the planner know when to ask the client for help? In our design, there are two facets to this functionality: (1) checking if the current planning problem is an instance of a `Proved` fact in the database (and, if so, placing the name of the theorem in the proof plan); and (2) checking if the current planning problem lies in the domain of known client proof procedures (and, if so, placing the conjecture in the proof plan). If neither of these cases hold, then the planner should continue with the next applicable method.

This functionality — actually a more general version — has been implemented as a method named `external`, which has been added to CLAM's stock of methods. Since adding methods is the standard 'user-level' way of extending the planning capabilities of CLAM, we have achieved an important software engineering victory: the basic infrastructure of CLAM has not been meddled with in order to implement the server side of the extended interaction, as we would have been forced to do had we taken the recursive option.

Now we discuss when `external` is called, i.e., its priority with respect to other methods. The `external` method has guards to control its applicability. These may be constraints on the goal, or more general constraints on the applicability. In its pure form, `external` has no guards; hence it will always succeed (and ask HOL to prove the goal) on any goal that is not already in the database. If this pure form of `external` were made the highest priority method, CLAM would do very little planning in each iteration and HOL would be asked to prove goals it had little chance of handling automatically. By this reasoning, the pure form of the method should be given the lowest priority or not be included at all in the list of top-level methods.

On the other hand, giving `external` a low priority is not ideal either, since CLAM sometimes gets into a non-terminating branch of the search space. In such a situation a low-priority `external` may not be able to help because it may never be reached. For example, induction (which is often the lowest priority method) can prevent a lower priority method from being reached. (The situation is not always so dire: although it is always possible to do an induction if there is a variable in the current goal, CLAM only tries inductions that it has reason to believe may succeed, typically those having a suitable induction scheme and wave rules that allow some progress in rippling.) In any case, it seems a bad use of resources for `external` to be of low priority: for example, if the current problem fits into the syntax of a decision procedure in the client, then the planner should not use its resources in attempting to plan the problem.

We decided to use an instance of `external` called `external_decision`. This method is made the highest priority method in CLAM; its guard tests whether the current problem is a formula of the theory of linear arithmetic. If not, the next method is tried; otherwise, if the problem is an instance of a `Proved` fact in the database, the method places the name of the theorem in the proof plan; if the problem is not in the database it is placed in the proof plan as a conjecture instead.

Other instances of the `external` method are intended to be included in the planner in the future.

4 EXAMPLES

Now we come to examples. The first example is standard: prove the commutativity of multiplication, $\forall m\ n.\ m * n = n * m$, given only the recursion equations for addition and multiplication in Peano arithmetic:

$$\begin{aligned} 0 + n &= n \\ \mathtt{SUC}\ m + n &= \mathtt{SUC}(m + n) \\ 0 * n &= 0 \\ \mathtt{SUC}\ m * n &= m * n + n \end{aligned}$$

This apparently simple task is made much more difficult by restricting to definitions of addition and multiplication: no other lemmas are given. This forces the system to prove its required intermediate results by repeated induction and generalization: the original proof takes seven inductions and two generalizations. In contrast, the proof via extended interaction takes two inductions and one generalization. The plan returned by CLAM is as follows:

```
ind_strat([(m:holnum)-holSUC(v0)]) then
  [external_decision(...),
   ind_strat([(n:holnum)-holSUC(v2)]) then
     generalise(...) then
       external_decision(...)
  ]
```

The generated tactic (an ML function) can be found in the Appendix to this paper. Reading it, one can see that the proof plan inducts on m and then simplifies with the definition of multiplication. The base case is then solved by the decision procedure with the fact $\vdash \forall n.\ 0 = n * 0$. In the step case, the induction hypothesis is used and then induction on n is invoked, followed by simplification with the definitions. After this, the second base case is immediate; the second step case proceeds by using the second induction hypothesis and then generalizing the term $v_2 * v_0$. After this, the second base case is immediate; it only remains to prove

$$(v_2 * v_0 + v_0) + \mathsf{SUC}\ v_2 = v_2 * \mathsf{SUC}\ v_0 + \mathsf{SUC}\ v_0$$

under the (second) inductive hypothesis $v_2 * v_0 + v_2 = v_2 * \mathsf{SUC}\ v_0$. The proof proceeds by using this hypothesis (from right to left) and then generalizing the term $v_2 * v_0$. After this, the decision procedure again kicks in, to prove

$$\vdash \forall v_4.\ (v_4 + v_0) + \mathsf{SUC}\ v_2 = (v_4 + v_2) + \mathsf{SUC}\ v_0$$

and the proof is finished. In the example, it is clear that the decision procedure cut down the magnitude of the planning obligation; instead of CLAM having to plan to a very low level, now the system is planning down to the point that the decision procedure can fire.

In another example, the original system simply fails, while the combination manages to find a plan:

$$\forall p\ q\ n.\ n^{p+q} = n^p * n^q$$

The HOL tactic corresponding to the plan can be found in the Appendix.

Finally, we tested our claim that the original behaviour of the system can be recovered by having the client fail to prove any assumed lemmas in partial plans. In a proof of the associativity of multiplication,

$$\forall m\ n\ p.\ (m * n) * p = m * (n * p)$$

the protocol iterates through four rounds before finding a plan that doesn't use the external decision procedure.

5 RELATED WORK

This work is related to Boyer and Moore's integration of a linear arithmetic decision procedure [Boyer and Moore, 1988] into Nqthm. They noticed that a black-box decision procedure is rarely used in inductive proofs and concluded that only if there is closer integration is the co-operation effective. Our approach is more general than that of [Boyer and Moore, 1988], but the specific instantiation we describe is closely related, since we are using a linear arithmetic decision procedure.

More recently, Armando and Ranise [Armando and Ranise, 1998] have studied the Boyer-Moore linear arithmetic procedure, specifying its logical services as an open mechanized reasoning system (OMRS) [Giunchiglia *et al.*, 1996]. There seem to be close connections between OMRS and our work more generally. It would, for example, be interesting to try to specify our interface in the OMRS framework.

PVS [Shankar, 1996] is an interactive proof system augmented with various tightly-integrated decision procedures, for example BDD-based propositional simplification and model checking, equality, and linear arithmetic.

The ΩMEGA system [Benzmüller *et al.*, 1997] is intended to provide theorem proving support for mathematics. It includes a proof planner together with various other theorem proving components and proof presentation tools. Most of these components appear to have been designed to be part of ΩMEGA rather than as systems in their own right, but there has been some work on integrating provers such as Otter, TPS, constraint solvers, and also computer algebra systems.

In contrast to PVS and the Boyer-Moore system, our combination system is loosely integrated: since our two systems need not share an address space, or the same machine, or even a file system, one might think this strategy is almost forced on us. However, that overlooks the existence of so-called binary-compatible distributed object architecures such as CORBA [OMG Group, 1998] and DCOM [Eddon and Eddon, 1997]. In an infrastructure such as these, components written in different languages and running on different machines can be transparently accessed through a chosen host programming language. Such an infrastructure could be used to implement recursive dialogues running over a network.

At the time we performed this research we were unaware of the existence of distributed object architectures, and moreover, had we been, it would have not helped, since CORBA (or DCOM) interfaces for Standard ML do not yet exist. Even if they did, it is not clear to us that the recursive approach is better than the iterative approach.

The work of Lowe et al. on providing a graphical user interface to CLAM [Lowe *et al.*, 1998] resembles our work, with a human user taking the role of the client. Their system (BARNACLE) allows the user to initiate planning on a goal, stop it at any point, undo steps, and influence the planning process by modifying CLAM's knowledge of wave rules, etc., and by intervening at choice points in the planning search space. Apart from the difference in the nature of the client (computer program vs human), the work differs in aspects of control. In the BARNACLE work, CLAM does not normally request additional information on its own initiative, except when it gets stuck. The user may, however, select certain methods as ones for which CLAM must stop and ask for user confirmation before it applies them. In our work, CLAM decides for itself when it could usefully ask the client (HOL) to intervene. On the other hand, BARNACLE does not begin planning from scratch after being given new information; it is able to continue planning from where it stopped. An interesting question is how might the two lines of research be merged to produce a system in which a proof planner, an object-level theorem prover, and a human user all co-operate to produce a proof. Perhaps the biggest difficulty would be that HOL already has a user interface; somehow this would have to be merged with the BARNACLE interface to present the human user with a coherent view.

Other related work is that of Felty and Howe [Felty and Howe, 1997] on using HOL theories and definitional packages within a Nuprl proof development. CLAM was designed to use the same logic as Nuprl, and hence there are related issues concerning the translation between logics.

6 FUTURE WORK

First of all, more extensive testing should reveal whether our approach scales up on larger examples. The extended interaction protocol is nearly stateless, and its iterative nature requires that the server keep rebuilding the plan. For big plans, this might be unacceptable, in which case we could investigate techniques for storing and restoring snapshots of the state of the planner. Doing so would not vitiate our argument against recursive dialogues; if only the planner saves state, the iterative style is still simple.

Another point is that, in addition to using syntactic forms to control the application of the `external` method, e.g., looking to see if the formula belongs to the theory of linear arithmetic, the planner could also use a simple counter-example finder to increase the expectation that the current subproblem is provable.

Proof critics [Ireland and Bundy, 1996] allow CLAM to dynamically synthesize wave rules while planning. Currently, CLAM would attempt to plan a proof for such a synthesized rule. In the extended interaction scheme, CLAM would synthesize the rule and then use the protocol to ask HOL if it knows of any theorems that match it. HOL would then search its database of previously proved theorems. The work on critics by Ireland and Bundy may also assist in determining when to apply an external method (i.e., make a request to the client) in other parts of the planning process.

Right now, the tactics resulting from CLAM plans are made up of 'small' proof steps that don't have to do any search; however, one could easily provide a planner that returned plans containing some 'high-level' steps. Confronted with such, the client would either have to re-perform the search that validates such steps or, alternatively, immediately accept the 'big step' as a theorem. The latter choice may be unacceptable to those who worry about such issues as: the "goodness of fit" between the two logics; the heuristic nature of plans *versus* the exact requirements of a theorem prover; and becoming dependent on the existence of a separate system (or even a particular implementation!). In the case that some trustworthy third party oracle is asserting the validity of a 'big-step', then (subject to concerns over differences in logics) it may be acceptable for CLAM to use the oracle directly and for HOL to then trust the result from CLAM, i.e., for both HOL and CLAM to trust the third party.

Such decisions become important when proof systems are combined: however, it seems reasonable to ask the question *what is the notion of proof that is being implemented in such a combination*? Using our solution, that question is easy to answer, since the notion of proof is just that of HOL, which is well-understood [Gordon and Melham, 1993].

Finally, Boyer and Moore's view—that close integration of decision procedures with other proof procedures is required for higher automation—is an interesting challenge. We intend to see if an extended interaction protocol can enable planning, loosely integrated with decision procedures, to be as effective as Boyer and Moore's tight combination.

7 CONCLUSIONS

The main contribution of this paper is a novel protocol for extended interaction between two mechanized proof systems, where one can be regarded as a client and the other as a server. In the protocol, the server can appeal to the client for help; we argue that supporting such a facility by recursion, especially when the systems are only connected by a network, is a daunting task (in the absence of the support offered by a distributed object architecture). In contrast, our protocol iterates through a sequence of rounds, in a fashion somewhat similar to an iterative deepening search strategy. We also argued that our design remedies many of the limitations of the original combined system.

We have implemented this protocol, and tried it out on some standard examples. These first results are quite encouraging, since they achieve a smooth and nearly effortless integration of an inductive prover with a linear arithmetic package. This integration is not as tight as the hand-crafted one of [Boyer and Moore, 1988], but we feel that the comparative ease of assembling the combined system speaks highly for our approach: it took less than a week of work and required very little modification of the component systems. Moreover, in an iterative dialogue, the user has a natural place in the scheme of things; if desired, the user can easily single step the protocol by tactic invocations in HOL. By manually adding facts into the fact database, the user can also help delimit the search space of the planner. Such functionality may be difficult to support in a multi-lingual and recursive protocol.

From a system development point of view, we think our choice is also highly satisfactory: there is a marked tendency for research software development to fragment into many parallel branches that never again converge. We were anxious to avoid this, as we want to be able to exploit new developments in both CLAM and HOL. We are hopeful that we will be able to do so because our approach does not involve major modifications to the two systems.

Acknowledgements. The attendees of the 1998 CLAM/INKA/OMRS meeting (held in the Edinburgh University Department of AI) offered useful advice on this work. In particular, we thank Alan Bundy, Erica Melis, Andrew Stevens, and Toby Walsh. Silvio Ranise also offered helpful comments on this paper.

This work was supported by the Engineering and Physical Sciences Research Council of Great Britain under grants GR/L03071 and GR/L14381.

REFERENCES

[Armando and Ranise, 1998] A. Armando and S. Ranise. From integrated reasoning specialists to "plug-and-play" reasoning components. In *Proceedings of the Fourth International Conference on Artificial Intelligence And Symbolic Computation (AISC'98)*, Plattsburgh, NY, USA, September 1998.

[Benzmüller *et al.*, 1997] C. Benzmüller, L. Cheikhrouhou, D. Fehrer, A. Fiedler, X. Huang, M. Kerber, M. Kohlhase, K. Konrad, E. Melis, A. Meier, W. Schaarschmidt, J. Siekmann, and V. Sorge. ΩMEGA: Towards a mathematical assistant. In McCune [1997], pages 252–255.

[Boulton *et al.*, 1998] R. Boulton, K. Slind, A. Bundy, and M. Gordon. An interface between CLAM and HOL. In J. Grundy and M. Newey, editors, *Proceedings of the 11th International Conference on Theorem Proving in Higher Order Logics (TPHOLs'98)*, Canberra, Australia, September/October 1998. *Lecture Notes in Computer Science*, Springer.

[Boulton, 1995] R. J. Boulton. Combining decision procedures in the HOL system. In E. T. Schubert, P. J. Windley, and J. Alves-Foss, editors, *Proceedings of the 8th International Workshop on Higher Order Logic Theorem Proving and Its Applications*, volume 971 of *Lecture Notes in Computer Science*, pages 75–89, Aspen Grove, UT, USA, September 1995. Springer-Verlag.

[Boulton, 1996] R. J. Boulton. Syn: A single language for specifying abstract syntax trees, lexical analysis, parsing and pretty-printing. Technical Report 390, University of Cambridge Computer Laboratory, March 1996.

[Boyer and Moore, 1988] R. S. Boyer and J S. Moore. Integrating decision procedures into heuristic theorem provers: A case study of linear arithmetic. In J. E. Hayes, D. Michie, and J. Richards, editors, *Machine Intelligence 11*, chapter 5, pages 83–124. Oxford University Press, 1988.

[Bundy *et al.*, 1990] A. Bundy, F. van Harmelen, C. Horn, and A. Smaill. The OYSTER-CLAM system. In M. E. Stickel, editor, *Proceedings of the 10th International Conference on Automated Deduction*, volume 449 of *Lecture Notes in Artificial Intelligence*, pages 647–648, Kaiserslautern, FRG, July 1990. Springer-Verlag. Also: Research Paper 507, Department of Artificial Intelligence, University of Edinburgh, 1990.

[Bundy, 1991] A. Bundy. A science of reasoning. In J.-L. Lassez and G. Plotkin, editors, *Computational Logic: Essays in Honor of Alan Robinson*, pages 178–198. MIT Press, 1991. Also: Research Paper 445, Department of Artificial Intelligence, University of Edinburgh.

[Eddon and Eddon, 1997] G. Eddon and H. Eddon. *Inside Distributed COM.* Microsoft Press. ISBN 1-57231-849-X.

[Felty and Howe, 1997] A. P. Felty and D. J. Howe. Hybrid interactive theorem proving using Nuprl and HOL. In McCune [1997], pages 351–365.

[Giunchiglia *et al.*, 1996] F. Giunchiglia, P. Pecchiari, and C. Talcott. Reasoning theories — towards an architecture for open mechanized reasoning systems. In F. Baader and K. U. Schulz, editors, *Proceedings of the First International Workshop on Frontiers of Combining Systems (FroCoS'96)*, Munich, Germany, March 1996. Volume 3 of *Applied Logic Series*, pages 157–174. Kluwer Academic Publishers.

[Gordon and Melham, 1993] M. J. C. Gordon and T. F. Melham, editors. *Introduction to HOL: A theorem proving environment for higher order logic.* Cambridge University Press, 1993.

[Gordon *et al.*, 1979] M. J. Gordon, A. J. Milner, and C. P. Wadsworth. *Edinburgh LCF: A Mechanised Logic of Computation*, volume 78 of *Lecture Notes in Computer Science.* Springer-Verlag, 1979.

[Ireland and Bundy, 1996] A. Ireland and A. Bundy. Productive use of failure in inductive proof. *Journal of Automated Reasoning*, 16(1–2):79–111, 1996.

[Lowe *et al.*, 1998] H. Lowe, A. Bundy, and D. McLean. The use of proof planning for co-operative theorem proving. *Journal of Symbolic Computation*, 25(2):239–261, February 1998. Special Issue on Graphical User Interfaces and Protocols.

[McCune, 1997] W. McCune, editor. *Proceedings of the 14th International Conference on Automated Deduction (CADE-14)*, volume 1249 of *Lecture Notes in Artificial Intelligence*, Townsville, North Queensland, Australia, July 1997. Springer.

[OMG Group, 1998] The Common Object Request Broker: Architecture and Specification. Object Management Group Inc., OMG Headquarters, 492 Old Connecticut Path, Framingham, Massachusetts 01701, USA. Also available at `http://www.omg.org`.

[Shankar, 1996] N. Shankar. PVS: Combining specification, proof checking, and model checking. In M. Srivas and A. Camilleri, editors, *Proceedings of the First International Conference on Formal Methods in Computer-Aided Design (FMCAD'96)*, volume 1166 of *Lecture Notes in Computer Science*, pages 257–264, Palo Alto, CA, USA, November 1996. Springer.

APPENDIX

Goal. $\forall m\, n.\, m * n = n * m.$

Proof Plan.

```
ind_strat([(m:holnum)-holSUC(v0)]) then
  [external_decision(...),
   ind_strat([(n:holnum)-holSUC(v2)]) then
     generalise(...) then
       external_decision(...)
  ]
```

Tactic.

```
let
val th0 = VisibleTactic.DECIDE `!n. 0 = n * 0`
val th1 =
  VisibleTactic.DECIDE
    `!(v4 :num).
       (v4 + (v0 :num)) + SUC (v2 :num) = (v4 + (v2 :num)) + SUC (v0 :num)`
in
  QIND_TAC (RENAME (scheme "hol_num_INDUCTION") [["v0"]]) [`m:num`] THENL
  [OCC_RW_TAC "hol_MULT1" [1,1] LEFT,
   OCC_RW_TAC "hol_MULT2" [1,1] LEFT THEN
   ALL_TAC THEN WFERT_EQ_TAC([1, 1, 1],LEFT)] THENL
  [PROVED_TAC th0,
   QIND_TAC (RENAME (scheme "hol_num_INDUCTION") [["v2"]]) [`n:num`] THENL
   [OCC_RW_TAC "hol_MULT1" [1,1,1] LEFT THEN
    OCC_RW_TAC "hol_ADD1" [1,1] LEFT THEN OCC_RW_TAC "hol_MULT1" [2,1] LEFT
    THEN ELEM_TAC,
    OCC_RW_TAC"hol_MULT2" [2,1] LEFT THEN OCC_RW_TAC"hol_MULT2" [1,1,1] LEFT
    THEN ALL_TAC THEN WFERT_EQ_TAC([1, 2, 1],RIGHT)] THEN
   QSPEC_TAC (`(v2 :num) * (v0 :num)`,`(v4 :num)`) THEN PROVED_TAC th1]
end
```

Goal. $\forall p\ q\ n.\ n^{p+q} = n^p * n^q$.

Proof Plan.

```
ind_strat([(p:holnum)-holSUC(v0)]) then
  [generalise(...) then
     external_decision(...),
   generalise(...) then
     generalise(...) then
       ind_strat([(n:holnum)-holSUC(v4)]) then
         generalise(...) then
           ind_strat([(v6:holnum)-holSUC(v7)]) then
             generalise(...) then
               generalise(...) then
                 external_decision(...) ]
```

Tactic.

```
let
val th0 =
  VisibleTactic.DECIDE `!(v0 :num) (q :num) (n :num). v0 = 1 * v0`
val th1 =
  VisibleTactic.DECIDE
    `!(v10 :num) (v9 :num) (v3 :num) (v2 :num) (q :num).
       (v10 + v2) + v9 = (v10 + v9) + v2`
in
  QIND_TAC (RENAME (scheme "hol_num_INDUCTION") [["v0"]]) [`p:num`] THENL
  [OCC_RW_TAC"hol_ADD1"[2,1,1] LEFT THEN OCC_RW_TAC "hol_EXP1"[1,2,1] LEFT,
   OCC_RW_TAC"hol_EXP2" [1,2,1] LEFT THEN
   OCC_RW_TAC"hol_ADD2" [2,1,1] LEFT THEN OCC_RW_TAC"hol_EXP2" [1,1] LEFT
   THEN ALL_TAC THEN WFERT_EQ_TAC([2, 1, 1],LEFT)] THENL
  [QSPEC_TAC (`(n :num) EXP (q :num)`,`(v0 :num)`) THEN PROVED_TAC th0,
   QSPEC_TAC (`(n :num) EXP (q :num)`,`(v2 :num)`) THEN
   QSPEC_TAC (`(n :num) EXP (v0 :num)`,`(v3 :num)`) THEN
   QIND_TAC (RENAME (scheme "hol_num_INDUCTION") [["v4"]]) [`n:num`] THENL
   [OCC_RW_TAC"hol_MULT1" [1,1] LEFT THEN
    OCC_RW_TAC"hol_MULT1" [1,2,1] LEFT THEN OCC_RW_TAC"hol_MULT1" [2,1] LEFT
    THEN ELEM_TAC,
    OCC_RW_TAC"hol_MULT2" [1,1] LEFT THEN OCC_RW_TAC"hol_MULT2" [1,2,1] LEFT
    THEN ALL_TAC THEN WFERT_EQ_TAC([1, 1, 1],LEFT)] THEN
   QSPEC_TAC (`(v4 :num) * (v3 :num)`,`(v6 :num)`) THEN
   QIND_TAC (RENAME (scheme "hol_num_INDUCTION") [["v7"]]) [`v6:num`] THENL
   [OCC_RW_TAC "hol_MULT1" [1,1,1] LEFT THEN
    OCC_RW_TAC "hol_ADD1" [1,1] LEFT THEN OCC_RW_TAC "hol_ADD1" [1,2,1] LEFT
    THEN ELEM_TAC,
    OCC_RW_TAC"hol_ADD2" [1,2,1] LEFT THEN
    OCC_RW_TAC"hol_MULT2" [2,1] LEFT THEN OCC_RW_TAC"hol_MULT2" [1,1,1] LEFT
    THEN ALL_TAC THEN WFERT_EQ_TAC([1, 2, 1],RIGHT)] THEN
   QSPEC_TAC (`(v3 :num) * (v2 :num)`,`(v9 :num)`) THEN
   QSPEC_TAC (`(v7 :num) * (v2 :num)`,`(v10 :num)`) THEN PROVED_TAC th1]
end
```

Towards Heterogeneous Specifications

Andrzej Tarlecki

There is a population explosion among the logical systems used in computer science...

[Goguen and Burstall, 1984]

1 INTRODUCTION

The evident multitude of logical systems in use in various areas of computer science is in principle a great advantage for potential applications: it is so much more likely to find a logic that best suits the needs of each particular project. It is also a research challenge to enable proper use of the possibilities this multitude offers and to cope with this without generating a mess of (often repetitive) problems.

The very first task is to present some formal concept of what a logical system is. We follow the model-theoretic tradition, and adopt a view of logic based on the satisfaction relation between models and sentences, as formally captured in the notion of *institution* introduced by Goguen and Burstall [1984; 1992]. Given this, we can attempt to work as much as possible independently from the institution chosen, and provide ideas, concepts, results and tools that work in an *arbitrary* institution, and so can potentially be used in any project, whatever logical system it involves. The original motivation behind the notion of institution was to use this strategy in the design of specification languages. Indeed, largely institution-independent foundations for software specification and development emerged, see [Sannella and Tarlecki, 1997].

The multitude of logical systems offers another exciting possibility. A number of logical systems can be used in the same project, for instance to describe various parts or various features of the system at various stages of its specification and development. A necessary prerequisite for sensible use of many logical systems together is that they are linked in some way with each other. In the work on institutions, this is captured by notions of an "arrow" between institutions: institution morphisms [Goguen and Burstall, 1992], maps of institutions [Meseguer, 1989], institution semi-morphisms [Sannella and Tarlecki, 1988b], simulations [Astesiano and Cerioli, 1993], representations [Tarlecki, 1987], pre-institution transformations [Salibra and Scollo, 1993]. [Tarlecki, 1996] gives a systematic presentation of the main concepts and problems in the area together with a possible intuition behind and the possible role for some of these concepts. That paper and some following work by Mossakowski *et al.* [1997; 1998] deal mainly with the issues of combining institutions linked by "arrows", providing rudimentary results for systematic construction of complex logical systems by putting together simpler logics and their extensions.

In the present paper we discuss in some detail what happens with specifications in the presence of a number of institutions linked by arrows of various kinds — this topic is treated only marginally in [Tarlecki, 1996].

We recall the basic concept of *institution* in Section 2. Section 3 follows [Sannella and Tarlecki, 1988a] to illustrate that working in an arbitrary institution can be quite fruitful — *structured specifications* together with a related proof system are introduced in this way. To allow a natural construction of specifications spanning a number of institutions, we introduce the concepts of *institution semi-morphism* and *semi-representation* in Section 4, and discuss their use for moving specifications from one institution to another and for building *heterogeneous specifications* in this way. The expected possibility of translating specifications from one institution to another provides an alternative motivation for *institution representations* — in Section 5 we show that representations allow for syntactic translation of specifications between institutions. We consider combining families of compatible representations into some "universal institution" (Section 6). An important result is that we can turn arbitrary heterogeneous specifications over institutions so represented into usual specifications built entirely within the "universal institution". Finally, Section 7 introduces yet another kind of arrows between institutions, *institution encodings*, motivated by examples from the work on behavioural interpretation of specifications [Hennicker and Bidoit, 1998]. Conclusions, summary of results and plans for future work are given in Section 8.

2 INSTITUTIONS

The theory of *institutions* [Goguen and Burstall, 1992] takes a predominantly model-theoretic view of logic, with the satisfaction relation between models and logical sentences adopted as a primary notion. Somewhat unlike the classical model-theory though, a family of such relations is considered at once, indexed by a category of signatures.

Definition 2.1 An *institution* $\mathbf{I}$ consists of:

- a category $\mathbf{Sign_I}$ of *signatures*;
- a functor $\mathbf{Sen_I}\colon \mathbf{Sign_I} \to \mathbf{Set}$, giving a set $\mathbf{Sen}(\Sigma)$ of Σ-*sentences* for each signature $\Sigma \in |\mathbf{Sign_I}|$;

- a functor $\mathbf{Mod_I}: \mathbf{Sign_I}^{op} \to \mathbf{DCat}^1$, giving a class $\mathbf{Mod}(\Sigma)$ of Σ-*models* for each signature $\Sigma \in |\mathbf{Sign_I}|$; and

- for $\Sigma \in |\mathbf{Sign_I}|$, a *satisfaction relation* $\models_{\mathbf{I},\Sigma} \subseteq \mathbf{Mod_I}(\Sigma) \times \mathbf{Sen_I}(\Sigma)$

such that for any signature morphism $\sigma: \Sigma \to \Sigma'$, Σ-sentence $\varphi \in \mathbf{Sen_I}(\Sigma)$ and Σ'-model $M' \in \mathbf{Mod_I}(\Sigma')$ the following *satisfaction condition* holds:

$$M' \models_{\mathbf{I},\Sigma'} \mathbf{Sen_I}(\sigma)(\varphi) \iff \mathbf{Mod_I}(\sigma)(M') \models_{\mathbf{I},\Sigma} \varphi$$

When working within an institution, we freely use the standard logical terminology, and, for example, say that a Σ-model $M \in \mathbf{Mod}(\Sigma)$ *satisfies* a Σ-sentence $\varphi \in \mathbf{Sen}(\Sigma)$, where $\Sigma \in |\mathbf{Sign}|$, or that φ *holds* in M, whenever $M \models_{\mathbf{I},\Sigma} \varphi$. When there is no danger of confusion, we omit the subscript $\mathbf{I}$ when referring to the components of an institution $\mathbf{I}$. Similarly, the subscript Σ on the satisfaction relations is often omitted. Moreover, symbols **Sign**, **Sen**, **Mod** and $\models$ are always used for the corresponding components of an institution. We omit obvious expansions of institutions to such components if some natural decorations are used to make the relationship evident. So, for instance, $\mathbf{Sign}'$ and $\mathbf{Sign}_1$ denote the categories of signatures of institutions $\mathbf{I}'$ and $\mathbf{I}_1$, respectively.

For any signature morphism $\sigma: \Sigma \to \Sigma'$, the function $\mathbf{Sen}(\sigma): \mathbf{Sen}(\Sigma) \to \mathbf{Sen}(\Sigma')$ is denoted simply by σ and referred to as the σ-*translation* of sentences. The function $\mathbf{Mod}(\sigma): \mathbf{Mod}(\Sigma') \to \mathbf{Mod}(\Sigma)$ is denoted by $\_|_\sigma$ and referred to as the σ-*reduct* (or σ-*translation*) of models.

The satisfaction relation extends naturally to sets of sentences and classes of models: for any $\Sigma \in |\mathbf{Sign}|$, set $\Phi \subseteq \mathbf{Sen}(\Sigma)$ of Σ-sentences, Σ-sentence $\varphi \in \mathbf{Sen}(\Sigma)$, Σ-model $M \in \mathbf{Mod}(\Sigma)$, and class $\mathcal{M} \subseteq \mathbf{Mod}(\Sigma)$ of Σ-models, we write $M \models \Phi$, $\mathcal{M} \models \varphi$ and $\mathcal{M} \models \Phi$ with the obvious meaning (for instance, $M \models \Phi$ means $M \models \psi$ for all $\psi \in \Phi$).

For any signature Σ, the satisfaction relation induces a *semantic* (or *model-theoretic*) *consequence*: a Σ-sentence $\varphi \in \mathbf{Sen}(\Sigma)$ is a consequence of a set $\Phi \subseteq \mathbf{Sen}(\Sigma)$ of Σ-sentences, written $\Phi \models_\Sigma \varphi$ (with the index Σ omitted when clear), if for all models $M \in \mathbf{Mod}(\Sigma)$, $M \models \Phi$ implies $M \models \varphi$. Semantic consequence is preserved by translation along signature morphisms:

> for any signature morphism $\sigma: \Sigma \to \Sigma'$, if $\Phi \models_\Sigma \varphi$ then $\sigma(\Phi) \models_{\Sigma'} \sigma(\varphi)$.

The semantic consequence is often approximated by a proof-theoretic component of the logical system. A *logical entailment* (or *consequence relation*, see [Avron, 1991] for a survey) on a set of sentences $\mathcal{S}$ is a relation $\vdash \subseteq \wp(\mathcal{S}) \times \mathcal{S}$

[1]**DCat** is the category of all discrete categories (classes). For simplicity, we disregard in this paper morphisms between models. Hence, classes of models, rather than model categories are considered.

that satisfies the usual reflexivity, transitivity (cut) and weakening properties. Given an institution **I**, an *entailment system* for **I** is a family

$$\langle \vdash_\Sigma \rangle_{\Sigma \in |\mathbf{Sign}|}$$

of entailment relations on the corresponding sets of Σ-sentences that is closed under translation along signature morphisms. The family of semantic consequence relations $\langle \models_\Sigma \rangle_{\Sigma \in |\mathbf{Sign}|}$ is an entailment system for **I** in this sense. Typically, however, we expect that a logical system comes equipped with an entailment system given in a proof-theoretic way, for instance by a set of inference rules. An entailment system $\langle \vdash_\Sigma \rangle_{\Sigma \in |\mathbf{Sign}|}$ is *sound* (respectively, *complete*) for **I** if for each $\Sigma \in |\mathbf{Sign}|$, $\vdash_\Sigma \subseteq \models_\Sigma$ (respectively, $\vdash_\Sigma \supseteq \models_\Sigma$). Thus, $\langle \vdash_\Sigma \rangle_{\Sigma \in |\mathbf{Sign}|}$ is sound and complete for **I** if the entailment $\vdash$ coincides with the model-theoretic consequence $\models$. See [Meseguer, 1989] for a notion of a *general logic* which extends institutions with sound entailment systems.

Examples of institutions abound. Essentially, any standard logical system, such as many-sorted equational logic, first-order logic with and without equality, various infinitary and higher-order logics, logical systems for partial, regular, continuous, order-sorted algebras, etc, may be presented as institutions and typically equipped with standard entailment systems. We refer to [Tarlecki, 1999] for details of a few standard examples and to [Sannella and Tarlecki, 1999] for many more, including non-standard examples presenting for instance a semantic view of a simple programming language as an institution.

2.1 Institutions with Extra Structure

The notion of institution is very general and imposes only very light requirements on the logical system to be presented as an institution (see [Tarlecki, 1999] for some discussion). To restrict the range of logical systems considered, further requirements may be imposed. Examples of properties of institutions used in the rest of the paper are given below.

An institution **I** has *negation* if for each signature $\Sigma \in |\mathbf{Sign}|$ and Σ-sentence $\varphi \in \mathbf{Sen}(\Sigma)$, there exists a Σ-sentence $\neg\varphi \in \mathbf{Sen}(\Sigma)$ such that for all models $M \in \mathbf{Mod}(\Sigma)$, $M \models \neg\varphi$ if and only if $M \not\models \varphi$.

I has (*finite*) *conjunction* if for each signature $\Sigma \in |\mathbf{Sign}|$ and (finite) set $\Phi \subseteq \mathbf{Sen}(\Sigma)$, there exists a Σ-sentence $\bigwedge \Phi \in \mathbf{Sen}(\Sigma)$ such that for all models $M \in \mathbf{Mod}(\Sigma)$, $M \models \bigwedge \Phi$ if and only if $M \models \Phi$.

An institution **I** has (*finitely*) *composable signatures* if its category of signatures **Sign** is (finitely) cocomplete and its model functor $\mathbf{Mod}\colon \mathbf{Sign}^{op} \to \mathbf{DCat}$ is (finitely) continuous, mapping (finite) colimits in **Sign** to limits in **DCat**.

Lemma 2.2 (Amalgamation Property) *Assume that an institution* **I** *has finitely composable signatures. Consider a pushout in the category* **Sign** *of signatures:*

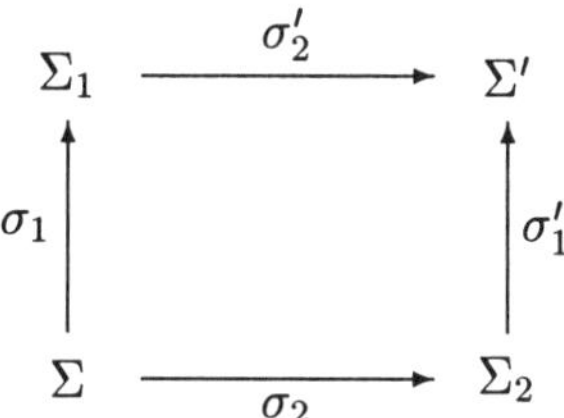

Then, for any models $M_1 \in \mathbf{Mod}(\Sigma_1)$ *and* $M_2 \in \mathbf{Mod}(\Sigma_2)$ *such that* $M_1|_{\sigma_1} = M_2|_{\sigma_2}$*, there exists a unique model* $M' \in \mathbf{Mod}(\Sigma')$ *such that* $M'|_{\sigma'_2} = M_1$ *and* $M'|_{\sigma'_1} = M_2$.

The above Amalgamation Property is equivalent to the requirement that the model functor **Mod** of **I** maps pushouts in the category **Sign** of signatures to pullbacks in the category **DCat** of all discrete categories. We will refer to institutions that satisfy this property as institutions *with semi-composable signatures*[2]. For instance, the pushout in **Sign** as above is mapped to the following pullback in **DCat**:

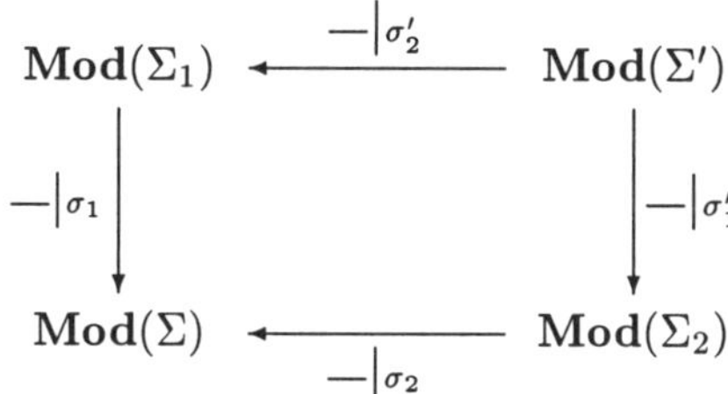

An institution **I** has *interpolation* if for any pushout in **Sign**

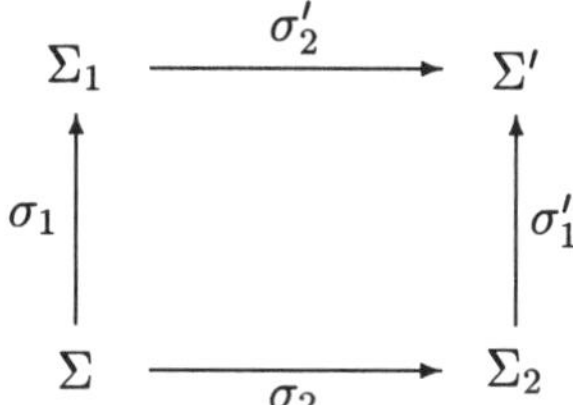

and sentences $\varphi_1 \in \mathbf{Sen}(\Sigma_1)$ and $\varphi_2 \in \mathbf{Sen}(\Sigma_2)$ such that $\sigma'_2(\varphi_1) \models_{\Sigma'} \sigma'_1(\varphi_2)$, there exists a sentence $\psi \in \mathbf{Sen}(\Sigma)$ such that $\varphi_1 \models_{\Sigma_1} \sigma_1(\psi)$ and $\sigma_2(\psi) \models_{\Sigma_2} \varphi_2$.

3 STRUCTURED SPECIFICATIONS

The notion of institution emerged as a formalisation of the concept of logical system underlying algebraic specification formalisms. Given an institution **I**,

[2]The importance of this property for the work on specifications in an arbitrary institution was stressed in [Sannella and Tarlecki, 1988a]; see also [Diaconescu *et al.*, 1993] where institutions with (semi-)composable signatures were called (semi-)exact.

the rough idea is that signatures determine the syntax of the systems to be specified, models of the institution represent (the semantics of) the systems themselves, and sentences are used to specify the desired system properties. Whatever the specifications are, they should ultimately describe a signature and a class of models over this signature, called *models of the specifications* to capture the admissible realizations of the specified system. Consequently, any specification formalism over **I** determines a class of specifications, and then, for any specification SP, its signature $Sig[SP] \in |\mathbf{Sign}|$ and the collection of its models $Mod[SP] \subseteq \mathbf{Mod}(Sig[SP])$. If $Sig[SP] = \Sigma$, we refer to SP as Σ-specification.

Following [Sannella and Tarlecki, 1988a], we consider the following rudimentary ways of building specifications in an arbitrary institution **I**:

presentations: For any signature $\Sigma \in |\mathbf{Sign}|$ and set $\Phi \subseteq \mathbf{Sen}(\Sigma)$ of Σ-sentences, the *presentation* $\langle \Sigma, \Phi \rangle$ is a specification with:

$$\begin{aligned} Sig[\langle \Sigma, \Phi \rangle] &= \Sigma \\ Mod[\langle \Sigma, \Phi \rangle] &= \{M \in \mathbf{Mod}(\Sigma) \mid M \models \Phi\} \end{aligned}$$

union: For any signature $\Sigma \in |\mathbf{Sign}|$, given Σ-specifications SP_1 and SP_2, their *union* $SP_1 \cup SP_2$ is a specification with:

$$\begin{aligned} Sig[SP_1 \cup SP_2] &= \Sigma \\ Mod[SP_1 \cup SP_2] &= Mod[SP_1] \cap Mod[SP_2] \end{aligned}$$

translation: For any signature morphism $\sigma: \Sigma \to \Sigma'$ and Σ-specification SP, **translate** SP **by** σ is a specification with:

$$\begin{aligned} Sig[\textbf{translate } SP \textbf{ by } \sigma] &= \Sigma' \\ Mod[\textbf{translate } SP \textbf{ by } \sigma] &= \{M' \in \mathbf{Mod}(\Sigma') \mid M'|_\sigma \in Mod[SP]\} \end{aligned}$$

hiding: For any signature morphism $\sigma: \Sigma \to \Sigma'$ and Σ'-specification SP', **derive from** SP' **by** σ is a specification with:

$$\begin{aligned} Sig[\textbf{derive from } SP' \textbf{ by } \sigma] &= \Sigma \\ Mod[\textbf{derive from } SP' \textbf{ by } \sigma] &= \{M'|_\sigma \mid M' \in Mod[SP']\} \end{aligned}$$

The above *specification-building operations*, although extremely simple, already provide flexible mechanisms for expressing basic ways of putting specifications together and so for building specifications in a structured manner. Further, perhaps more convenient-to-use operations may be built on this basis. See for instance [Sannella and Tarlecki, 1988a] for examples, definitions of other similarly abstract operations, and further discussion.

We denote by $Spec_{\mathbf{I}}$ the class of all specifications built in **I** using the above operations. For any signature $\Sigma \in |\mathbf{Sign}|$, $Spec_{\mathbf{I}}(\Sigma)$ denotes the class of all such Σ-specifications.

The semantic consequence introduced in Section 2 for sets of sentences readily extends to specifications. Given any Σ-specification SP, a Σ-sentence

$\varphi \in \mathbf{Sen}(\Sigma)$ is a *semantics consequence* of SP, written $SP \models \varphi$, if $Mod[SP] \models \varphi$.

The obvious next problem is how to build a proof-theoretic entailment that would approximate the semantic consequence for structured specifications. Interestingly, given an entailment system $\langle \vdash_\Sigma \rangle_{\Sigma \in |\mathbf{Sign}|}$ for $\mathbf{I}$, the following rules provide its compositional extension to structured specifications (as usual, we omit the signature index on entailment relations):

$$\frac{\text{for } i \in \mathcal{I}, SP \vdash \varphi_i \qquad \{\varphi_i\}_{i \in \mathcal{I}} \vdash \varphi}{SP \vdash \varphi}$$

$$\frac{\varphi \in \Phi}{\langle \Sigma, \Phi \rangle \vdash \varphi} \qquad \frac{SP_1 \vdash \varphi}{SP_1 \cup SP_2 \vdash \varphi} \qquad \frac{SP_2 \vdash \varphi}{SP_1 \cup SP_2 \vdash \varphi}$$

$$\frac{SP \vdash \varphi}{\textbf{translate } SP \textbf{ by } \sigma \vdash \sigma(\varphi)} \qquad \frac{SP \vdash \sigma(\varphi)}{\textbf{derive from } SP \textbf{ by } \sigma \vdash \varphi}$$

It can be easily checked that the above rules preserve soundness of the entailment system. Completeness is preserved only for institutions with a sufficiently rich logical structure [Borzyszkowski, 1998a]:

Theorem 3.1 *Let* $\mathbf{I}$ *be an institution with negation and conjunction, semi-composable signatures and interpolation, and let* $\langle \vdash_\Sigma \rangle_{\Sigma \in |\mathbf{Sign}|}$ *be a sound and complete entailment system for* $\mathbf{I}$.

Then the above proof rules extend $\langle \vdash_\Sigma \rangle_{\Sigma \in |\mathbf{Sign}|}$ *to a sound and complete proof system for consequences of structured specifications in* $Spec_{\mathbf{I}}$*: for any specification* $SP \in Spec_{\mathbf{I}}(\Sigma)$ *and* Σ*-sentence* $\varphi \in \mathbf{Sen}(\Sigma)$, $SP \vdash \varphi$ *if and only if* $SP \models \varphi$.

The detailed proof and its corollaries in [Borzyszkowski, 1998a] show a rather delicate balance between the exact form of conjunction, signature composability, interpolation and compactness of the institution required for the completeness of the above proof system. In any case though, amalgamation and interpolation (in perhaps a weaker form) are essential.

Two specifications SP_1 and SP_2 are *equivalent*, written $SP_1 \equiv SP_2$, if they have the same signatures and classes of models. This can also be expressed using the refinement relation between specifications: SP_2 is a *refinement* of SP_1, written $SP_1 \rightsquigarrow SP_2$ if $Sig[SP_2] = Sig[SP_1]$ and $Mod[SP_2] \subseteq Mod[SP_1]$. This notion captures the basic steps in the process of system development via a sequence of consecutive refinements (see [Sannella and Tarlecki, 1997]). Again, a sound compositional proof system for specification refinement in an arbitrary institution can be provided, and its completeness requires some form of amalgamation and interpolation.

4 HETEROGENEOUS SPECIFICATIONS

In Section 3 we have discussed specifications in an arbitrary (but fixed) institution. Suppose now that we want to construct specifications that combine components built in two different institutions $\mathbf{I}$ and $\mathbf{I}'$, or that we want to capture how specifications in $\mathbf{I}$ may be refined to specifications in $\mathbf{I}'$. Of course, in general there can be no meaningful way to do this if the institutions considered are completely unrelated. Since the ultimate meaning of a specification in an institution is a class of its models (as given by the semantics of the specification formalism) the minimal requirement on how the institutions considered should be related is that their classes of models are related to each other. Here is one way to capture this:

Definition 4.1 An *institution semi-morphism* $\mu: \mathbf{I}' \to_{semi} \mathbf{I}$ between institutions $\mathbf{I}'$ and $\mathbf{I}$ consists of:

- a functor $\mu^{Sig}: \mathbf{Sign}' \to \mathbf{Sign}$, and
- a natural transformation $\mu^{Mod}: \mathbf{Mod}' \to (\mu^{Sig})^{op};\mathbf{Mod}$,[3] i.e., a family of functions $\mu^{Mod}_{\Sigma'}: \mathbf{Mod}'(\Sigma') \to \mathbf{Mod}(\mu^{Sig}(\Sigma'))$ natural in $\Sigma' \in |\mathbf{Sign}'|$:

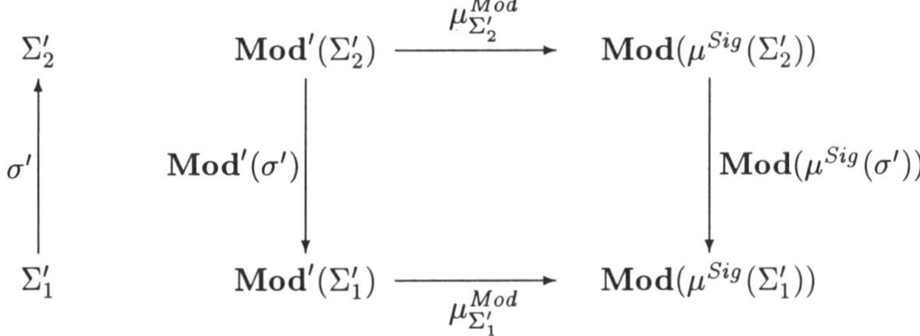

The idea behind the above definition is that $\mathbf{I}'$ is an institution with a "richer" structure of models, which typically contains more components and more detailed information about their semantic interpretation than more "poor" models of the institution $\mathbf{I}$. An institution semi-morphism $\mu: \mathbf{I}' \to_{semi} \mathbf{I}$ "extracts" simpler $\mathbf{I}$-signatures out of more complex $\mathbf{I}'$-signatures and simpler $\mathbf{I}$-models out of more complex $\mathbf{I}'$-models.

We do not provide any recipe for defining an institution semi-morphism between two arbitrary institutions; in each case this has to reflect the way in which models of the institutions considered capture the relevant properties of systems they represent. Examples include a semi-morphism from the institution of first-order logic to the institution of equational logic, where the semi-morphism simply "forgets" about predicate symbols and their interpretation in first-order structures. Another example is a semi-morphism from an institution for a programming language to an institution for partial algebras

[3] Functor $(\mu^{Sig})^{op}: (\mathbf{Sign}')^{op} \to \mathbf{Sign}^{op}$ is the "same" as $\mu^{Sig}: \mathbf{Sign}' \to \mathbf{Sign}$ but considered between the opposite categories.

that builds a partial algebra from (the semantics of) a program by extracting the data types on which the program operates and partial functions that the program procedures and functions provide.

Once an institution semi-morphism $\mu: \mathbf{I}' \to_{semi} \mathbf{I}$ is given, it can be used to move specifications from $\mathbf{I}'$ to $\mathbf{I}$:

Deriving specifications: If SP' is a Σ'-specification in $\mathbf{I}'$ (with models $Mod[SP'] \subseteq \mathbf{Mod}'(\Sigma')$) then **derive from** SP' **via** μ is a specification in $\mathbf{I}$ with the following semantics:

$$Sig[\textbf{derive from } SP' \textbf{ via } \mu] = \mu^{Sig}(\Sigma')$$
$$Mod[\textbf{derive from } SP' \textbf{ via } \mu] = \{\mu^{Mod}_{\Sigma'}(M') \mid M' \in Mod[SP']\}$$

In principle it is also possible to move specifications from $\mathbf{I}$ to $\mathbf{I}'$, in the direction opposite to the institution semi-morphism $\mu: \mathbf{I}' \to_{semi} \mathbf{I}$. This is a bit awkward though, as for a Σ-specification in $\mathbf{I}$ we would have to identify at least a signature $\Sigma' \in |\mathbf{Sign}'|$ in $\mathbf{I}'$ to which the specification is to be moved, satisfying $\mu^{Sig}(\Sigma') = \Sigma$. Since such a signature need not always exist, and rarely is unique, a different way of linking models of $\mathbf{I}'$ and $\mathbf{I}$ seems more appropriate here:

Definition 4.2 An *institution semi-representation* $\rho: \mathbf{I} \to_{semi} \mathbf{I}'$ between institutions $\mathbf{I}$ and $\mathbf{I}'$ consists of:

- a functor $\rho^{Sig}: \mathbf{Sign} \to \mathbf{Sign}'$, and
- a natural transformation $\rho^{Mod}: (\rho^{Sig})^{op};\mathbf{Mod}' \to \mathbf{Mod}$, i.e., a family of functions $\rho^{Mod}_{\Sigma}: \mathbf{Mod}'(\rho^{Sig}(\Sigma)) \to \mathbf{Mod}(\Sigma)$ natural in $\Sigma \in |\mathbf{Sign}|$:

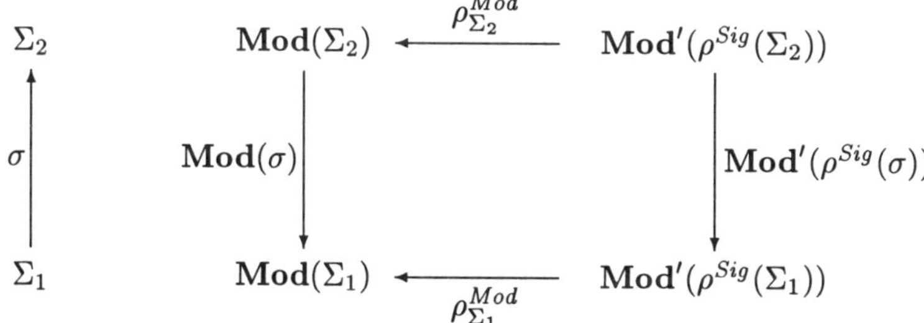

Even though the difference with respect to the definition of a semi-morphism seems very slight, institution semi-representations capture quite a different intuition, at least in typical examples. An institution semi-representation $\rho: \mathbf{I} \to_{semi} \mathbf{I}'$ indicates how to find for each signature Σ in the "poorer" institution $\mathbf{I}$, a signature $\rho^{Sig}(\Sigma)$ in the "richer" institution $\mathbf{I}'$ that represents Σ (at least, as far as model classes are considered).

An archetypical example is a semi-representation of equational logic in first-order logic, where each algebraic signature is represented by (or just: is in the institution of first-order logic) the same signature.

Any institution semi-representation $\rho: \mathbf{I} \to_{semi} \mathbf{I}'$ can be naturally used to move specifications from $\mathbf{I}$ to $\mathbf{I}'$:

Translating specifications: If SP is a Σ-specification in $\mathbf{I}$ (with models $Mod[SP] \subseteq \mathbf{Mod}(\Sigma)$) then **translate** SP' **via** ρ is a specification in $\mathbf{I}'$ with the following semantics:

$$Sig[\textbf{translate } SP \textbf{ via } \rho] = \rho^{Sig}(\Sigma)$$
$$Mod[\textbf{translate } SP \textbf{ via } \rho] = \{M' \in \mathbf{Mod}'(\rho^{Sig}(\Sigma)) \mid \rho_\Sigma^{Mod}(M') \in Mod[SP]\}$$

Once we have extended the repertoire of specification-building operations by the above ways of moving specifications between institutions linked by semi-morphisms and semi-representations, there is no reason to restrict the way in which they are used and combined with other specification-building operations. For instance, given a specification $SP' \in Spec_{\mathbf{I}'}$ and a semi-morphisms $\mu: \mathbf{I}' \rightarrow_{semi} \mathbf{I}$, the specification **derive from** SP' **via** μ is a specification in $\mathbf{I}$ which can be used exactly as any other specification, whether built entirely within $\mathbf{I}$ or not. In particular, it may be combined with other $\mathbf{I}$-specifications using the operations introduced in Section 3 and moved further to another institution using a semi-representation or a semi-morphism from $\mathbf{I}$. Quite purposefully, in the definitions of specification-building operations (both above and earlier in Section 3) we have not assumed that the argument specifications are given within any fixed specification formalism. When necessary, we always do so explicitly below, by indicating the class of specifications considered — otherwise the specifications are quite arbitrary (but always come equipped with their model-theoretic semantics, as above).

Given a family $\mathcal{H}$ of institutions linked by semi-morphisms and semi-representations, we can thus build structured *heterogeneous specifications* that may involve component specifications from a number of institutions in $\mathcal{H}$ moved around and combined using the operations introduced above (both here and in Section 3). We denote by $Spec^{\mathcal{H}}$ the class of all specifications built using these operations. For each institution $\mathbf{I}$ in $\mathcal{H}$, generalising the notation for structured specifications built entirely over $\mathbf{I}$, the class of such heterogeneous specifications in $\mathbf{I}$ (that is, informally, those that "end up" with the semantics given by a signature and a class of models in $\mathbf{I}$) is denoted by $Spec_{\mathbf{I}}^{\mathcal{H}}$.

Moving specifications along an institution semi-morphism seems especially important in the process of software development. Namely, consider an institution semi-morphism $\mu: \mathbf{I}' \rightarrow_{semi} \mathbf{I}$. Informally, think of $\mathbf{I}'$ as the institution of a programming language used for implementation and $\mathbf{I}$ as the institution of a logic used for specification. Then, given a specification SP in $\mathbf{I}$ and a specification SP' in $\mathbf{I}'$ (think of the former as a requirements specification and of the latter as a possibly "unfinished" program), refinement $SP \rightsquigarrow$ **derive from** SP' **via** μ provides an abstract way to capture how (the "program") SP' implements (the "requirements") SP.

Not much can be said about proving consequences of heterogeneous specifications built using institution semi-morphisms and semi-representations in general. The trouble is that semi-morphisms and semi-representations offer no way to link consequences of specifications in the institutions considered, as sentences of the institutions remain unrelated. This provides other motivations for the "full" notions, covering sentences as well, as originally introduced in [Goguen and Burstall, 1992] and [Tarlecki, 1996], respectively.

Definition 4.3 An *institution morphism* $\mu: \mathbf{I}' \to \mathbf{I}$ extends an institution semi-morphism consisting of a functor $\mu^{Sig}: \mathbf{Sign}' \to \mathbf{Sign}$ and a natural transformation $\mu^{Mod}: \mathbf{Mod}' \to (\mu^{Sig})^{op};\mathbf{Mod}$ by:

- a natural transformation $\mu^{Sen}: \mu^{Sig};\mathbf{Sen} \to \mathbf{Sen}'$, i.e., a family of functions $\mu^{Sen}_{\Sigma'}: \mathbf{Sen}(\mu^{Sig}(\Sigma')) \to \mathbf{Sen}'(\Sigma')$ natural in $\Sigma' \in |\mathbf{Sign}'|$:

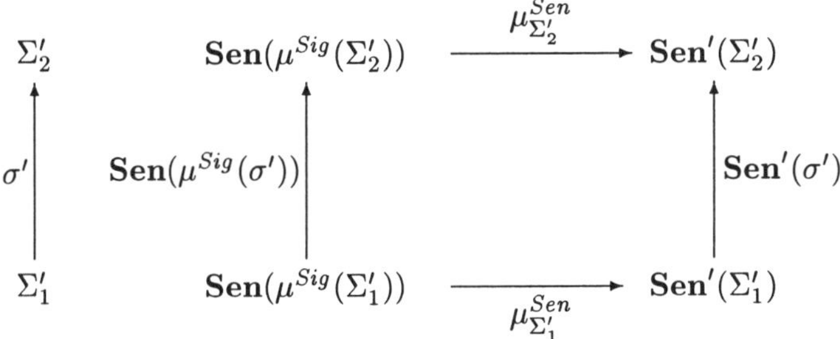

such that for any $\Sigma' \in |\mathbf{Sign}'|$, translations $\mu^{Sen}_{\Sigma'}: \mathbf{Sen}(\mu^{Sig}(\Sigma')) \to \mathbf{Sen}'(\Sigma')$ and $\mu^{Mod}_{\Sigma'}: \mathbf{Mod}'(\Sigma') \to \mathbf{Mod}(\mu^{Sig}(\Sigma'))$ preserve the satisfaction relation, that is, for any $\varphi \in \mathbf{Sen}(\mu^{Sig}(\Sigma'))$ and $M' \in \mathbf{Mod}'(\Sigma')$ the following *satisfaction condition* holds:

$$M' \models'_{\Sigma'} \mu^{Sen}_{\Sigma'}(\varphi) \iff \mu^{Mod}_{\Sigma'}(M') \models_{\mu^{Sig}(\Sigma')} \varphi$$

Definition 4.4 An *institution representation* $\rho: \mathbf{I} \to \mathbf{I}'$ extends an institution semi-representation consisting of a functor $\rho^{Sig}: \mathbf{Sign} \to \mathbf{Sign}'$ and a natural transformation $\rho^{Mod}: (\rho^{Sig})^{op};\mathbf{Mod}' \to \mathbf{Mod}$ by:

- a natural transformation $\rho^{Sen}: \mathbf{Sen} \to \rho^{Sig};\mathbf{Sen}'$, that is, a family of functions $\rho^{Sen}_{\Sigma}: \mathbf{Sen}(\Sigma) \to \mathbf{Sen}'(\rho^{Sig}(\Sigma))$ natural in $\Sigma \in |\mathbf{Sign}|$:

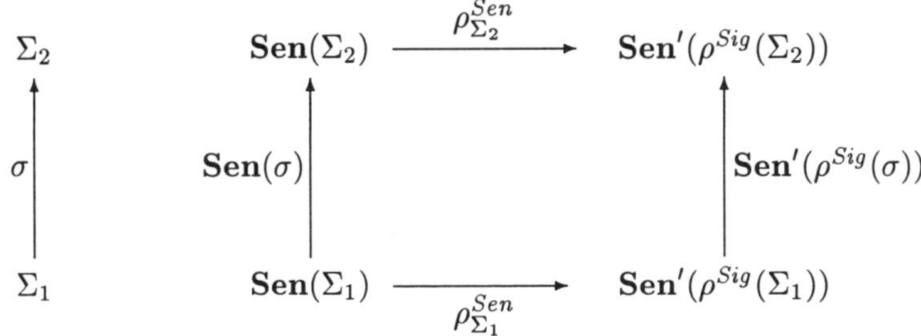

such that for any $\Sigma \in |\mathbf{Sign}|$, translations $\rho^{Sen}_{\Sigma}: \mathbf{Sen}(\Sigma) \to \mathbf{Sen}'(\rho^{Sig}(\Sigma))$ and $\rho^{Mod}_{\Sigma}: \mathbf{Mod}'(\rho^{Sig}(\Sigma)) \to \mathbf{Mod}(\Sigma)$ preserve the satisfaction relation, that is, for any $\varphi \in \mathbf{Sen}(\Sigma)$ and $M' \in \mathbf{Mod}'(\rho^{Sig}(\Sigma))$ the following *representation condition* holds:

$$M' \models'_{\rho^{Sig}(\Sigma)} \rho^{Sen}_{\Sigma}(\varphi) \iff \rho^{Mod}_{\Sigma}(M') \models_{\Sigma} \varphi$$

We will use institution morphisms and representations to move specifications between institutions in exactly the same way as the semi-morphisms and semi-representations, respectively, they extend.

Given an institution morphism μ and an institution representation ρ, the following proof rules may be added to the proof system presented in Section 3:

$$\frac{SP \vdash \varphi}{\textbf{translate } SP \textbf{ via } \rho \vdash \rho^{Sen}(\varphi)} \qquad \frac{SP \vdash \mu^{Sen}(\varphi)}{\textbf{derive from } SP \textbf{ via } \mu \vdash \varphi}$$

Soundness of these rules, and hence of the resulting proof system for heterogeneous specifications, can be checked directly. Unfortunately, in general, completeness is not preserved. Formulation of useful sufficient conditions that would ensure a proper generalisation of Theorem 3.1 to $Spec^{\mathcal{H}}$ is an interesting open problem.

5 TRANSLATING SPECIFICATIONS

Consider an institution representation $\rho: \mathbf{I} \to \mathbf{I}'$.

Theorem 5.1 *Let $\Sigma \in |\mathbf{Sign}|$ be a signature, $\Sigma' = \rho^{Sig}(\Sigma)$, and suppose that for some sets of sentences $\Gamma \subseteq \mathbf{Sen}(\Sigma)$ and $\Gamma' \subseteq \mathbf{Sen}'(\Sigma')$ we have that*

$$\rho^{Mod}_{\Sigma}(Mod[\langle \Sigma', \Gamma' \rangle]) = Mod[\langle \Sigma, \Gamma \rangle]$$

(where $\rho^{Mod}_{\Sigma}(Mod[\langle \Sigma', \Gamma' \rangle])$ is the image of $Mod[\langle \Sigma', \Gamma' \rangle]$ under ρ^{Mod}_{Σ}).

Then for any set $\Phi \subseteq \mathbf{Sen}(\Sigma)$ and sentence $\varphi \in \mathbf{Sen}(\Sigma)$,

$$\Phi \cup \Gamma \models_{\Sigma} \varphi \iff \rho^{Sen}_{\Sigma}(\Phi) \cup \Gamma' \models'_{\Sigma'} \rho^{Sen}_{\Sigma}(\varphi)$$

This allows one to move entailments between the "richer" institution $\mathbf{I}'$ and the "poorer" institution $\mathbf{I}$. It becomes particularly important when Γ' is empty (and so Γ characterizes the image of the model-translation function) or even better, when Γ can be taken to be empty as well.

Definition 5.2 An institution (semi-)representation $\rho: \mathbf{I} \to \mathbf{I}'$ is *total* if for each signature $\Sigma \in |\mathbf{Sign}|$, the model translation $\rho^{Mod}_{\Sigma}: \mathbf{Mod}'(\rho^{Sig}(\Sigma)) \to \mathbf{Mod}(\Sigma)$ is a surjective function.

Corollary 5.3 *Given a total institution representation $\rho: \mathbf{I} \to \mathbf{I}'$, for any signature $\Sigma \in |\mathbf{Sign}|$, set $\Phi \subseteq \mathbf{Sen}(\Sigma)$ and sentence $\varphi \in \mathbf{Sen}(\Sigma)$,*

$$\Phi \models \varphi \iff \rho^{Sen}_{\Sigma}(\Phi) \models' \rho^{Sen}_{\Sigma}(\varphi)$$

In the rest of this section we briefly present a view of the recent work by Borzyszkowski [1998b] that provides a generalisation of the above corollary to arbitrary (homogeneous) structured specifications.

The key property is the relationship between the presentations $\langle\Sigma', \Gamma'\rangle$ and $\langle\Sigma, \Gamma\rangle$ that was the basic assumption in Theorem 5.1. Translation of specifications along any total institution (semi-)representation enjoys a similar property:

Proposition 5.4 *For any total institution semi-representation $\rho\colon \mathbf{I} \to_{semi} \mathbf{I}'$, signature $\Sigma \in |\mathbf{Sign}|$ and Σ-specification SP in* **I**,

$$\rho_\Sigma^{Mod}(Mod[\textbf{\textit{translate}}\ SP\ \textbf{\textit{via}}\ \rho]) = Mod[SP]$$

Consequently, the translation along a total semi-representation $\rho\colon \mathbf{I} \to_{semi} \mathbf{I}'$ may be used to faithfully represent in the "richer" institution $\mathbf{I}'$ any specification from the "poorer" institution **I**. Then, informally, all the work from **I** may be moved to $\mathbf{I}'$. For instance:

Corollary 5.5 *For any total institution representation $\rho\colon \mathbf{I} \to \mathbf{I}'$, $\Sigma \in |\mathbf{Sign}|$, Σ-specifications SP, SP_1, SP_2 in* **I**, *and Σ-sentence $\varphi \in \mathbf{Sen}(\Sigma)$,*

- $SP \models \varphi \iff \textbf{\textit{translate}}\ SP\ \textbf{\textit{via}}\ \rho \models' \rho_\Sigma^{Sen}(\varphi)$
- $SP_1 \rightsquigarrow SP_2 \iff \textbf{\textit{translate}}\ SP_1\ \textbf{\textit{via}}\ \rho \rightsquigarrow \textbf{\textit{translate}}\ SP_2\ \textbf{\textit{via}}\ \rho$
- $SP_1 \equiv SP_2 \iff \textbf{\textit{translate}}\ SP_1\ \textbf{\textit{via}}\ \rho \equiv \textbf{\textit{translate}}\ SP_2\ \textbf{\textit{via}}\ \rho$

The trick is to represent specifications of the form **translate** *SP* **via** ρ syntactically within $\mathbf{I}'$, without referring to the semantic translation given by the representation. For homogeneous specifications, built entirely using the operations introduced in Section 3, this seems rather straightforward:

Definition 5.6 Let $\rho\colon \mathbf{I} \to \mathbf{I}'$ be an institution representation. Given a (homogeneous) structured specification $SP \in Spec_{\mathbf{I}}$, its *representation under* ρ, written $\hat{\rho}(SP)$, is defined as follows:

- $\hat{\rho}(\langle\Sigma, \Phi\rangle)$ is $\langle\rho^{Sig}(\Sigma), \rho_\Sigma^{Sen}(\Phi)\rangle$,
- $\hat{\rho}(SP_1 \cup SP_2)$ is $\hat{\rho}(SP_1) \cup \hat{\rho}(SP_2)$,
- $\hat{\rho}($**translate** SP **by** $\sigma)$ is **translate** $\hat{\rho}(SP)$ **by** $\rho^{Sig}(\sigma)$,
- $\hat{\rho}($**derive from** SP **by** $\sigma)$ is **derive from** $\hat{\rho}(SP)$ **by** $\rho^{Sig}(\sigma)$.

Lemma 5.7 *Let $\rho\colon \mathbf{I} \to \mathbf{I}'$ be an institution representation. For any $\Sigma \in |\mathbf{Sign}|$ and structured Σ-specification $SP \in Spec_{\mathbf{I}}(\Sigma)$, $\hat{\rho}(SP)$ is a well-formed structured $\rho^{Sig}(\Sigma)$-specification over $\mathbf{I}'$, that is, $\hat{\rho}(SP) \in Spec_{\mathbf{I}'}(\rho^{Sig}(\Sigma))$.*

Moreover, $Mod[\hat{\rho}(SP)] \subseteq Mod[\textbf{\textit{translate}}\ SP\ \textbf{\textit{via}}\ \rho]$.

The opposite inclusion between model classes does not hold in general. One way to ensure it does hold, is by imposing an inter-institutional analogue of semi-composability of signatures (cf. Section 2.1).

Definition 5.8 An institution (semi-)representation $\rho: \mathbf{I} \to \mathbf{I}'$ has *amalgamation* if the natural transformation $\rho^{Mod}: (\rho^{Sig})^{op};\mathbf{Mod}' \to \mathbf{Mod}$ is *cartesian* [Jay, 1995], i.e., for each signature morphism $\sigma: \Sigma_1 \to \Sigma_2$ in **Sign**, the following diagram is a pullback in **DCat**:

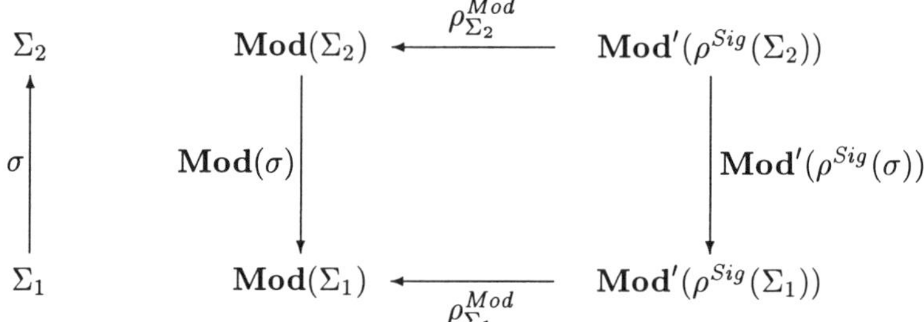

That is, for all models $M_2 \in \mathbf{Mod}(\Sigma_2)$ and $M'_1 \in \mathbf{Mod}'(\rho^{Sig}(\Sigma_1))$ such that $M_2|_\sigma = \rho^{Mod}_{\Sigma_1}(M'_1)$, there exists a unique model $M'_2 \in \mathbf{Mod}'(\rho^{Sig}(\Sigma_2))$ such that $\rho^{Mod}_{\Sigma_2}(M'_2) = M_2$ and $M'_2|_{\rho^{Sig}(\sigma)} = M'_1$.

Theorem 5.9 *For any institution representation $\rho: \mathbf{I} \to \mathbf{I}'$ with amalgamation and (homogeneous) structured specification $SP \in Spec_{\mathbf{I}}$,*

$$\hat{\rho}(SP) \equiv \textbf{\textit{translate}}\ SP\ \textbf{\textit{via}}\ \rho$$

Corollary 5.10 *For any total institution representation $\rho: \mathbf{I} \to \mathbf{I}'$ with amalgamation, signature $\Sigma \in |\mathbf{Sign}|$, specifications $SP, SP_1, SP_2 \in Spec_{\mathbf{I}}(\Sigma)$, and Σ-sentence $\varphi \in \mathbf{Sen}(\Sigma)$,*

- $SP \models \varphi \iff \hat{\rho}(SP) \models' \rho^{Sen}_{\Sigma}(\varphi)$
- $SP_1 \rightsquigarrow SP_2 \iff \hat{\rho}(SP_1) \rightsquigarrow \hat{\rho}(SP_2)$
- $SP_1 \equiv SP_2 \iff \hat{\rho}(SP_1) \equiv \hat{\rho}(SP_2)$

The above facts still hold if totality and amalgamation are assumed in weaker versions; in some form they are essential though, see [Borzyszkowski, 1998b] for details.

The importance of Corollary 5.10 is worth stressing, as it justifies the following scenario. Suppose that we have built a nice, expressive, "universal" institution **UI** with composable signatures, interpolation, a (sound and complete) proof system, supported by efficient tools like a powerful theorem prover, etc. Then, when we want to work in some institution **I**, instead of building all the utilities for it from scratch (or not being able to do this at all) we can provide a "good" (total and with amalgamation) representation $\rho: \mathbf{I} \to \mathbf{UI}$ and move all the technical work from **I** to **UI** using ρ.

For instance, consider the task of proving consequences of structured specifications over **I**. First, direct proofs may be difficult, since in general even developing an entailment system for **I** need not be easy. Even if we have a sound and complete entailment system for **I**, the logic here may lack interpolation, and so the compositional extension of the entailment system for **I** to consequences of structured specifications cannot be complete (cf. Section 3). On the other hand, given all the nice properties for **UI**, we have the sound and complete extension of the entailment system for **UI** to consequences of structured specifications built over **UI** (Theorem 3.1). Assuming then that ρ is total and has amalgamation, Corollary 5.10 justifies the following proof rule (let us index the entailment symbol by institutions for clarity):

$$\frac{\hat{\rho}(SP) \vdash^{\mathbf{UI}} \rho(\varphi)}{SP \vdash^{\mathbf{I}} \varphi}$$

Together with the proof system for consequences of specifications in $Spec_{\mathbf{UI}}$, this yields a sound and complete proof system for consequences of structured specifications in $Spec_{\mathbf{I}}$ (homogeneous specifications built over **I**).

Of course, it is not very realistic to expect all the nice features of **UI** mentioned above to be immediately available — but with the development of "universal logical frameworks" (see e.g. [Pfenning, 1996] for an excellent annotated bibliography) this way of proceeding and this line of research becomes more and more important.

The two assumptions about representations considered (totality and amalgamation) turn out not to be major problems in practice. Moreover, some standard ways of building institution (semi-)representations from simpler components may be provided so that the totatily and amalgamation properties of the resulting semi-representations follow from more elementary and quite natural properties of these components.

6 COMPATIBLE REPRESENTATIONS

In the previous section we have presented the possibilities of "internalizing" homogeneous specifications built over an institution **I** within a "universal" institution **UI** along any adequate (total and with amalgamation) institution representation $\rho: \mathbf{I} \to \mathbf{UI}$. Here, we generalize this further to heterogeneous specifications, as introduced in Section 4.

Let us fix an institution $\mathbf{UI} = \langle \mathbf{USign}, \mathbf{USen}, \mathbf{UMod}, \langle \models_\Sigma \rangle_{\Sigma \in |\mathbf{USign}|} \rangle$, viewed informally as a sufficiently rich, "universal" logic — throughout this section we consider institutions represented in **UI**.

Consider two institutions **I** and **I**′ and institution representations $\rho: \mathbf{I} \to \mathbf{I}'$ and $\rho': \mathbf{I}' \to \mathbf{UI}$. Their *composition*[4] $\rho;\rho': \mathbf{I} \to \mathbf{UI}$ is defined in the obvious, componentwise manner.

[4]We write composition in the diagrammatic order and denote it by ";" (semicolon).

Proposition 6.1 *Let $\rho: \mathbf{I} \to \mathbf{I}'$ and $\rho': \mathbf{I}' \to \mathbf{UI}$ be institution representations. Then for any specification SP in* $\mathbf{I}$,

$$\textbf{translate}\ (\textbf{translate}\ SP\ \textbf{via}\ \rho)\ \textbf{via}\ \rho' \equiv \textbf{translate}\ SP\ \textbf{via}\ (\rho;\rho')$$

We say that two representations $\rho_{\mathbf{I}}: \mathbf{I} \to \mathbf{UI}$ and $\rho_{\mathbf{I}'}: \mathbf{I}' \to \mathbf{UI}$ are *compatible* with a representation $\rho: \mathbf{I} \to \mathbf{I}'$ if $\rho;\rho_{\mathbf{I}'} = \rho_{\mathbf{I}}$. The above proposition provides an easy basis for eliminating translations along institution representations from heterogeneous specifications built in institutions represented in **UI** in a compatible way.

Let's turn now to specifications of the form **derive from** *SP* **via** μ built using institution semi-morphisms. Compatibility of the semi-morphisms used with representations of the institutions they link is a bit more tricky:

Definition 6.2 Consider two institution representations $\rho: \mathbf{I} \to \mathbf{UI}$ and $\rho': \mathbf{I}' \to \mathbf{UI}$. A *representation semi-map* from ρ to ρ' consists of:

- an institution semi-morphism $\mu: \mathbf{I}' \to_{semi} \mathbf{I}$, and
- a natural transformation $\theta: \mu^{Sig};\rho^{Sig} \to (\rho')^{Sig}$, i.e., a family of morphisms $\theta_{\Sigma'}: \rho^{Sig}(\mu^{Sig}(\Sigma')) \to (\rho')^{Sig}(\Sigma')$ in **USign** natural in $\Sigma' \in |\mathbf{Sign}'|$,

such that $(\rho')^{Mod};\mu^{Mod} = (\theta^{op}\cdot\mathbf{UMod});((\mu^{Sig})^{op}\cdot\rho^{Mod})^5$, i.e., such that for each signature $\Sigma' \in |\mathbf{Sign}'|$ with $\Sigma = \mu^{Sig}(\Sigma')$, the following diagram commutes:

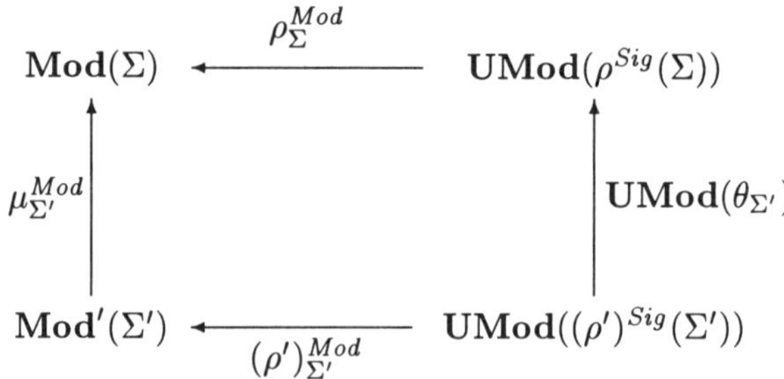

We say that such a representation map is *based on* μ. It has *amalgamation* if for each $\Sigma' \in |\mathbf{Sign}'|$, the above diagram is a pullback in **DCat**, that is, for all models $M' \in \mathbf{Mod}'(\Sigma')$ and $UM \in \mathbf{UMod}(\rho^{Sig}(\Sigma))$ such that $\mu^{Mod}_{\Sigma'}(M') = \rho^{Mod}_{\Sigma}(UM)$ there is a unique model $UM' \in \mathbf{UMod}((\rho')^{Sig}(\Sigma'))$ such that $UM'|_{\theta_{\Sigma'}} = UM$ and $(\rho')^{Mod}_{\Sigma'}(UM') = M'$.

[5] We write "$\cdot$" for the vertical, "diagonal" composition of natural transformations as well as for the multiplication of natural transformations by functors; ";" denotes the horizontal, "sequential" composition of natural transformations. $\theta^{op}: ((\rho')^{Sig})^{op} \to (\mu^{Sig};\rho^{Sig})^{op}$ is the "same" natural transformation as $\theta: \mu^{Sig};\rho^{Sig} \to (\rho')^{Sig}$ between the "same" functors but considered between the opposite categories.

The idea behind this definition is that given a richer institution $\mathbf{I}'$ built over $\mathbf{I}$, as witnessed for models by an institution semi-morphism $\mu: \mathbf{I}' \to_{semi} \mathbf{I}$, we formally state that the model part of representation $\rho': \mathbf{I}' \to \mathbf{UI}$ is built over the model part of representation $\rho: \mathbf{I} \to \mathbf{UI}$. For each $\mathbf{I}'$-signature $\Sigma' \in |\mathbf{Sign}'|$, its representation $(\rho')^{Sig}(\Sigma') \in |\mathbf{USign}|$ "includes" the representation $\rho^{Sig}(\Sigma) \in |\mathbf{USign}|$ of the $\mathbf{I}$-signature $\Sigma = \mu^{Sig}(\Sigma') \in |\mathbf{Sign}|$ over which Σ' is built. These "inclusions" are given by morphisms $\theta_{\Sigma'}: \rho^{Sig}(\Sigma)) \to (\rho')^{Sig}(\Sigma')$ in $\mathbf{USign}$, required to be natural in $\Sigma' \in |\mathbf{Sign}'|$. The extra "commutativity" requirements express the fact that "inclusions" $\theta_{\Sigma'}$ induce in $\mathbf{UI}$ reducts of models that are consistent with μ-translations of models they represent.

We refer to [Tarlecki, 1996] for the notion of representation map based on institution morphisms (rather than semi-morphisms) and so covering also sentences of represented institutions. Let us only mention in passing that institution representations in **UI** with representation maps (and, respectively, semi-maps) as morphisms form a category. Colimits in the category of representations and their maps provide a rudimentary tool for systematic combination of institutions represented in **UI**.

Theorem 6.3 *Consider institution representations $\rho: \mathbf{I} \to \mathbf{UI}$ and $\rho': \mathbf{I}' \to \mathbf{UI}$ and a representation semi-map $\langle \mu, \theta \rangle$ from ρ to ρ'. For any specification SP' in $\mathbf{I}'$,*

$$Mod[\textbf{derive from } (\textbf{translate } SP' \textbf{ via } \rho') \textbf{ by } \theta_{Sig[SP']}] \\ \subseteq Mod[\textbf{translate } (\textbf{derive from } SP' \textbf{ via } \mu) \textbf{ via } \rho]$$

Moreover, if the semi-map $\langle \mu, \theta \rangle$ has amalgamation then

$$\textbf{translate } (\textbf{derive from } SP' \textbf{ via } \mu) \textbf{ via } \rho \\ \equiv \textbf{derive from } (\textbf{translate } SP' \textbf{ via } \rho') \textbf{ by } \theta_{Sig[SP']}$$

Representations $\rho_{\mathbf{I}}: \mathbf{I} \to \mathbf{UI}$ and $\rho_{\mathbf{I}'}: \mathbf{I}' \to \mathbf{UI}$ are *compatible* with an institution semi-morphisms $\mu: \mathbf{I}' \to_{semi} \mathbf{I}$ if there exists a representation semi-map from $\rho_{\mathbf{I}}$ to $\rho_{\mathbf{I}'}$ based on μ that has amalgamation. The above theorem allows us to represent the operation of moving specifications along an institution semi-morphism "internally" within the universal institution **UI** as the usual hiding operation, provided that the institution semi-morphism is compatible with the representations of the institutions involved in **UI**.

Theorem 5.9, Proposition 6.1 and Theorem 6.3 together yield the main conclusion of this section:

Corollary 6.4 *Let $\mathcal{H}$ be a family of institutions linked by semi-morphisms and representations. Let $\mathcal{R} = \langle \rho_{\mathbf{I}}: \mathbf{I} \to \mathbf{UI} \rangle_{\mathbf{I} \in |\mathcal{H}|}$ be a family of representations in **UI** of the institutions in $\mathcal{H}$. Assume that the representations in $\mathcal{R}$ are compatible with the semi-morphisms and representations in $\mathcal{H}$, and that all of them have amalgamation.*

For any heterogeneous specification $SP \in Spec_{\mathbf{I}}^{\mathcal{H}}$ (built entirely using the operations of Sections 3 and 4) there exists a homogeneous structured specification $\hat{\mathcal{R}}(SP) \in Spec_{\mathbf{UI}}$ over **UI** *(built using the operations of Section 3 only) such that*

$$\textbf{\textit{translate}}\ SP\ \textbf{\textit{via}}\ \rho_{\mathbf{I}} \equiv \hat{\mathcal{R}}(SP).$$

Moreover, if $\rho_{\mathbf{I}}: \mathbf{I} \to \mathbf{UI}$ is total, then for any $\Sigma \in |\mathbf{Sign}|$, specifications $SP, SP_1, SP_2 \in Spec_{\mathbf{I}}^{\mathcal{H}}(\Sigma)$, and Σ-sentence $\varphi \in \mathbf{Sen}(\Sigma)$,

- $SP \models \varphi \Longleftrightarrow \hat{\mathcal{R}}(SP) \models^{\mathbf{UI}} \rho_{\Sigma}^{Sen}(\varphi)$
- $SP_1 \leadsto SP_2 \Longleftrightarrow \hat{\mathcal{R}}(SP_1) \leadsto \hat{\mathcal{R}}(SP_2)$
- $SP_1 \equiv SP_2 \Longleftrightarrow \hat{\mathcal{R}}(SP_1) \equiv \hat{\mathcal{R}}(SP_2)$

7 INSTITUTION ENCODINGS

A characteristic feature of institution morphisms and representations is that models and sentences are translated contravariantly with respect to each other. [Tarlecki, 1996] mentions yet another kind of arrows between institutions, where these two components are translated covariantly, and disregards this possibility in view of the lack of convincing examples.[6] However, it turns out that such a notion may be quite useful in a study of behavioural interpretation of specifications, for instance to define or characterize the notion of behavioural satisfaction and its related proof strategies (see [Hennicker and Bidoit, 1998] for some hints in this direction).

This section is an addendum both to [Tarlecki, 1996] and to the current paper. We briefly introduce a third basic concept of arrow between institutions, and sketch results similar to those developed in the previous sections.

Definition 7.1 An *institution encoding* $\varepsilon: \mathbf{I} \to \mathbf{I}'$ between institutions **I** and **I**$'$ consists of:

- a functor $\varepsilon^{Sig}: \mathbf{Sign} \to \mathbf{Sign}'$,
- a natural transformation $\varepsilon^{Sen}: \mathbf{Sen} \to \varepsilon^{Sig};\mathbf{Sen}'$, and
- a natural transformation $\varepsilon^{Mod}: \mathbf{Mod} \to (\varepsilon^{Sig})^{op};\mathbf{Mod}'$

such that for any $\Sigma \in |\mathbf{Sign}|$, translations $\varepsilon_{\Sigma}^{Sen}: \mathbf{Sen}(\Sigma) \to \mathbf{Sen}'(\varepsilon^{Sig}(\Sigma))$ and $\varepsilon_{\Sigma}^{Mod}: \mathbf{Mod}(\Sigma) \to \mathbf{Mod}'(\varepsilon^{Sig}(\Sigma))$ preserve the satisfaction relation, that is, for any $\varphi \in \mathbf{Sen}(\Sigma)$ and $M \in \mathbf{Mod}(\Sigma)$ the following *encoding condition* holds:

$$M \models_{\Sigma} \varphi \Longleftrightarrow \varepsilon_{\Sigma}^{Mod}(M') \models_{\varepsilon^{Sig}(\Sigma)} \varepsilon_{\Sigma}^{Sen}(\varphi)$$

[6] Pre-institution transformations of [Salibra and Scollo, 1993] seem to have this property. However, models are mapped there in the same direction as sentences, but to classes of models, not to individual models. This is often (though not always) given as the pre-image function for an underlying, more basic construction on models mapping them in the opposite direction.

Although an institution encoding $\varepsilon\colon \mathbf{I} \to \mathbf{I}'$ extends an institution semi-morphism $\langle \varepsilon^{Sig}, \varepsilon^{Mod} \rangle \colon \mathbf{I} \to_{semi} \mathbf{I}'$, the intuitions here (and certainly in typical examples) correspond rather to those for institution representations. We might think of $\mathbf{I}'$ as a richer institution, where we pick signatures, models and sentences to encode the signatures, models and sentences of $\mathbf{I}$.

The following fact corresponds to Theorem 5.1 for representations:

Proposition 7.2 *Consider an institution encoding $\varepsilon\colon \mathbf{I} \to \mathbf{I}'$. Let $\Sigma \in |\mathbf{Sign}|$ be a signature, $\Sigma' = \varepsilon^{Sig}(\Sigma)$, and suppose that for some sets of sentences $\Gamma \subseteq \mathbf{Sen}(\Sigma)$ and $\Gamma' \subseteq \mathbf{Sen}'(\Sigma')$ we have that*

$$\varepsilon^{Mod}_{\Sigma}(Mod[\langle \Sigma, \Gamma \rangle]) = Mod[\langle \Sigma', \Gamma' \rangle]$$

Then for any set $\Phi \subseteq \mathbf{Sen}(\Sigma)$ and sentence $\varphi \in \mathbf{Sen}(\Sigma)$,

$$\Phi \cup \Gamma \models_{\Sigma} \varphi \iff \varepsilon^{Sen}_{\Sigma}(\Phi) \cup \Gamma' \models'_{\Sigma'} \varepsilon^{Sen}_{\Sigma}(\varphi)$$

Unlike for institution representations, it is rarely possible to eliminate both Γ and Γ' in the above theorem. However, the situations when we can eliminate Γ are of particular interest and importance, and in fact can be achieved (albeit sometimes at the expense of the use of a logic stronger than expected, e.g., infinitary logic as used in [Hennicker and Bidoit, 1998] to characterize the image of "behaviour functor").

Definition 7.3 An institution encoding $\varepsilon\colon \mathbf{I} \to \mathbf{I}'$ is a *logical encoding characterized by* $\langle \Gamma'_{\Sigma} \subseteq \mathbf{Sen}'(\varepsilon^{Sig}(\Sigma)) \rangle_{\Sigma \in |\mathbf{Sign}|}$ if for each $\Sigma \in |\mathbf{Sign}|$, Γ'_{Σ} defines the image of the Σ-model encoding, that is:

$$\varepsilon^{Mod}_{\Sigma}(\mathbf{Mod}(\Sigma)) = Mod[\langle \varepsilon^{Sig}(\Sigma), \Gamma'_{\Sigma} \rangle]$$

Corollary 7.4 *If $\varepsilon\colon \mathbf{I} \to \mathbf{I}'$ is a logical encoding characterized by $\langle \Gamma'_{\Sigma} \rangle_{\Sigma \in |\mathbf{Sign}|}$ then for any signature $\Sigma \in |\mathbf{Sign}|$, set $\Phi \subseteq \mathbf{Sen}(\Sigma)$ of sentences and sentence $\varphi \in \mathbf{Sen}(\Sigma)$,*

$$\Phi \models_{\Sigma} \varphi \iff \varepsilon^{Sen}_{\Sigma}(\Phi) \cup \Gamma' \models'_{\Sigma'} \varepsilon^{Sen}_{\Sigma}(\varphi)$$

As in Section 5, a syntactic translation of structured specifications under an institution encoding is rather obvious — but note that we have to explicitly limit the models of the translation to those given by the encoding:

Definition 7.5 Let $\varepsilon\colon \mathbf{I} \to \mathbf{I}'$ be a logical encoding characterized by a family $\langle \Gamma'_{\Sigma} \rangle_{\Sigma \in |\mathbf{Sign}|}$. Given a (homogeneous) structured specification $SP \in Spec_{\mathbf{I}}$, its *encoding under* ε, written $\hat{\varepsilon}(SP)$, is defined as follows:

- $\hat{\varepsilon}(\langle \Sigma, \Phi \rangle)$ is $\langle \varepsilon^{Sig}(\Sigma), \varepsilon^{Sen}_{\Sigma}(\Phi) \cup \Gamma'_{\Sigma} \rangle$,
- $\hat{\varepsilon}(SP_1 \cup SP_2)$ is $\hat{\varepsilon}(SP_1) \cup \hat{\varepsilon}(SP_2)$,
- let $\sigma\colon Sig[SP] \to \Sigma$; then $\hat{\varepsilon}$(**translate** SP **by** σ) is (**translate** $\hat{\varepsilon}(SP)$ **by** $\varepsilon^{Sig}(\sigma)$) $\cup \langle \varepsilon^{Sig}(\Sigma), \Gamma'_{\Sigma} \rangle$,

- $\hat{\varepsilon}$(**derive from** *SP* **by** σ) is **derive from** $\hat{\varepsilon}(SP)$ **by** $\varepsilon^{Sig}(\sigma)$.

Lemma 7.6 *Let* $\varepsilon\colon \mathbf{I} \to \mathbf{I}'$ *be a logical institution encoding. For any* $\Sigma \in |\mathbf{Sign}|$ *and* Σ*-specification* $SP \in Spec_{\mathbf{I}}(\Sigma)$, $\hat{\varepsilon}(SP)$ *is a well-formed* $\varepsilon^{Sig}(\Sigma)$*-specification over* $\mathbf{I}'$*, that is,* $\hat{\varepsilon}(SP) \in Spec_{\mathbf{I}'}(\varepsilon^{Sig}(\Sigma))$*. Moreover:*

- $Mod[\hat{\varepsilon}(SP)] \subseteq \varepsilon^{Mod}_{\Sigma}(\mathbf{Mod}(\Sigma))$*, and*
- $\varepsilon^{Mod}_{\Sigma}(Mod[SP]) \subseteq Mod[\hat{\varepsilon}(SP)]$.

To obtain the identity of the two model classes in the last inclusion, some form of amalgamation is required again.

We say that an institution encoding has *image amalgamation* if for each signature morphism $\sigma\colon \Sigma_1 \to \Sigma_2$ in **Sign**, for all models $M_1 \in \mathbf{Mod}(\Sigma_1)$ and $M_2' \in \varepsilon^{Mod}_{\Sigma_2}(\mathbf{Mod}(\Sigma_2))$ such that $\varepsilon^{Mod}_{\Sigma_1}(M_1) = M_2'|_{\varepsilon^{Sig}(\sigma)}$, there exists a unique model $M_2 \in \mathbf{Mod}(\Sigma_2)$ such that $M_2|_\sigma = M_1$ and $\varepsilon^{Mod}_{\Sigma_2}(M_2) = M_2'$:

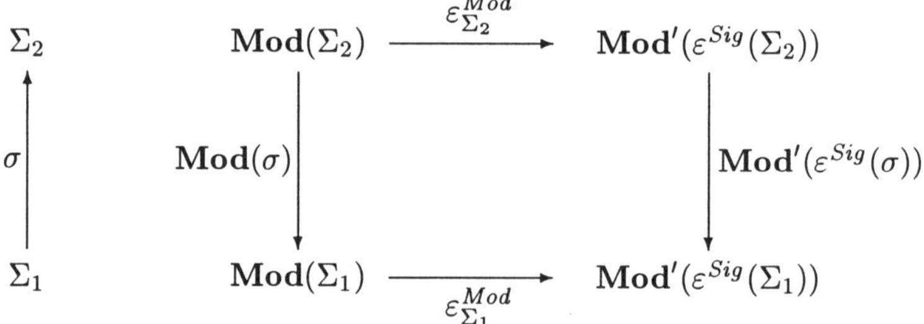

Theorem 7.7 *Let* $\varepsilon\colon \mathbf{I} \to \mathbf{I}'$ *be a logical institution encoding with image amalgamation. Then for any* $\Sigma \in |\mathbf{Sign}|$ *and structured* Σ*-specification* $SP \in Spec_{\mathbf{I}}(\Sigma)$, $Mod[\hat{\varepsilon}(SP)] = \varepsilon^{Mod}_{\Sigma}(Mod[SP])$.

Moreover, for any $\Sigma \in |\mathbf{Sign}|$*, specifications* $SP, SP_1, SP_2 \in Spec_{\mathbf{I}}(\Sigma)$*, and* Σ*-sentence* $\varphi \in \mathbf{Sen}(\Sigma)$*,*

- $SP \models \varphi \iff \hat{\varepsilon}(SP) \models' \varepsilon^{Sen}_{\Sigma}(\varphi)$
- $SP_1 \rightsquigarrow SP_2 \iff \hat{\varepsilon}(SP_1) \rightsquigarrow \hat{\varepsilon}(SP_2)$
- $SP_1 \equiv SP_2 \iff \hat{\varepsilon}(SP_1) \equiv \hat{\varepsilon}(SP_2)$

The benefits of this theorem are quite analogous to those for Corollary 5.10, as discussed at the end of Section 5. In fact, institution encodings seem to provide a more expected and potentially more useful way of representing logics than institution representations, where totality of the representation considered often has to be imposed, and all models of the richer institution $\mathbf{I}'$ have to be linked to the models of $\mathbf{I}$. In a way, in a (logical) institution encoding for each Σ-model M in $\mathbf{I}$, we pick its single "canonical" encoding $\varepsilon^{Mod}_{\Sigma}(M)$ in $\mathbf{I}'$. The typical trouble is that this choice cannot be natural in Σ, and therefore institution encodings seem even more difficult to build than institution representations. Once this is possible though, as in the examples based on [Hennicker and Bidoit, 1998], the results are nice!

8 CONCLUSIONS

This paper presents some ideas, concepts and results related to the problems of structured specifications built over a number of institutions linked by institution (semi) morphisms, representations and (newly introduced here) encodings. We discuss how heterogeneous specifications that span a number of institutions can be constructed. The main moral here is that to move specifications between logical systems in a meaningful way, the models of the logics involved must be related to each other. Extension of this to logical sentences is useful if a good compositional proof system for such specifications is needed. We also considered representations (or encodings) of logical systems in a "universal" logic. Then, under some technical conditions about representations (and encodings) structured specifications can be faithfully represented internally within the universal logic. The same holds for heterogeneous specifications provided that the representations are compatible with the morphisms used to move specifications between institutions. As concluded in Corollary 6.4, the main new technical result in the paper, this allows one to work entirely within the universal logic where the other institutions have been represented.

It should be stressed though that whilst promising, this covers only preliminary work in the area. In fact:

I have cheated a bit in this paper.

Of course, this is not to mean that the technical results as stated are incorrect. However, some of the technical definitions and assumptions have to be considerably refined before the results presented become practically important.

A minor point is that amalgamation is used throughout the paper in an unnecessarily strong version: less familiar *weak* amalgamation is sufficient for the results (uniqueness of the "amalgamated" model need not be required).

More importantly, the technical assumptions about representations and their compatibility must be checked for typical representations (or encodings) of logical systems in the existing logical frameworks. In fact, as already pointed out in [Tarlecki, 1987; Tarlecki, 1996] the representation condition used here seems too strong for practical applications — its weaker version has to be considered. For many purposes, for instance for logic combination, it also turns out to be more appropriate to work with some form of institution presentations (*parchments* [Goguen and Burstall, 1986] in various versions), providing more internal structure of the logic than institutions. Although very little of this has been worked out in detail yet, I believe that the ideas presented here should lead to more adequate formulations in the future, much as the general ideas presented in [Tarlecki, 1996] for institutions resulted in more practically relevant work on combinations of parchments and their representations by Mossakowski *et al.* [1997, 1998].

Acknowledgments. The work presented here has roots in joint work with Don Sannella, Bob Harper, Till Mossakowski, Wiesiek Pawłowski and Tomek Borzyszkowski (Theorems 3.1 and 5.9 are due to him). Special thanks to Rod Burstall and Joseph Goguen who introduced me to the theory of institutions and to other friends of institutions, including Maura Cerioli, Jose Fiadeiro, Jose Meseguer, Antonino Salibra, Giuseppe Scollo, Amílcar and Cristina Sernadas, Martin Wirsing, Uwe Wolter. Michel Bidoit pointed out the need for Section 7 — many thanks to him and Rolf Hennicker for the work that provided motivating examples there.

This research has been partially supported by KBN grant 8 T11C 018 11 and the LoSSeD workpackage of CRIT-2 funded by ESPRIT and INCO.

REFERENCES

[Astesiano and Cerioli, 1993] Egidio Astesiano and Maura Cerioli. Relationships between logical frames. In Michel Bidoit and Christine Choppy, editors, *Recent Trends in Data Type Specification, Selected Papers, 8th Workshop on Specification of Abstract Data Types ADT'91*, volume 655 of *Lecture Notes in Computer Science*, pages 126–143. Springer Verlag, 1993.

[Avron, 1991] Arnon Avron. Simple consequence relations. *Information and Computation*, 92(1):105–139, 1991.

[Borzyszkowski, 1998a] Tomasz Borzyszkowski. Completeness of a logical system for structured specifications. In F. Parisi-Presicce, editor, *Recent Trends in Data Type Specifications. Selected Papers, 12th Workshop on Specification of Abstract Data Types WADT'97*, volume 1376 of *Lecture Notes in Computer Science*, pages 107–121. Springer Verlag, 1998.

[Borzyszkowski, 1998b] Tomasz Borzyszkowski. Moving specification structures between logical systems. Institute of Mathematics and Informatics, Gdańsk University, submitted for publication, 1998.

[Diaconescu *et al.*, 1993] Ražvan Diaconescu, Joseph Goguen, and Petros Stefaneas. Logical support for modularisation. In Gerard Huet and Gordon Plotkin, editors, *Logical Environments*, pages 83–130. Cambridge University Press, 1993.

[Goguen and Burstall, 1984] Joseph Goguen and Rod Burstall. Introducing institutions. In Edmund Clarke and Dexter Kozen, editors, *Proceedings, Logics of Programming Workshop*, volume 164 of *Lecture Notes in Computer Science*, pages 221–256. Springer Verlag, 1984.

[Goguen and Burstall, 1986] Joseph Goguen and Rod Burstall. A study in the foundations of programming methodology: Specifications, institutions, charters and parchments. In David Pitt, Samson Abramsky, Axel Poigné, and David Rydeheard, editors, *Proceedings, Conference on Category Theory and Computer Programming CTCS'86*, number 240 in Lecture Notes in Computer Science, pages 313–333. Springer Verlag, 1986.

[Goguen and Burstall, 1992] Joseph Goguen and Rod Burstall. Institutions: Abstract model theory for specification and programming. *Journal of the Association for Computing Machinery*, 39(1):95–146, January 1992.

[Hennicker and Bidoit, 1998] Rolf Hennicker and Michel Bidoit. Observational logic. Technical Report LSV-98-6, Lab. Specification et Verification, ENS de Cachan, Cachan, France, 1998.

[Jay, 1995] C. Barry Jay. A semantics for shape. *Science of Computer Programming*, 25:251–283, 1995.

[Meseguer, 1989] José Meseguer. General Logics. In H.-D. Ebbinghaus, J. Fernández-Prida, M. Garrido, D. Lascar, and M. Rodríguez Artalejo, editors, *Logic Colloquium '87*, pages 275–329. North-Holland, 1989.

[Mossakowski *et al.*, 1997] Till Mossakowski, Wiesław Pawłowski, and Andrzej Tarlecki. Combining and representing logical systems. In Eugenio Moggi and Giuseppe Rosolini, editors, *Proc. 7th Intl. Conference on Category Theory and Computer Science CTCS'97*, volume 1290 of *Lecture Notes in Computer Science*, pages 177–198. Springer Verlag, 1997.

[Mossakowski *et al.*, 1998] Till Mossakowski, Wiesław Pawłowski, and Andrzej Tarlecki. Combining and representing logical systems using model-theoretic parchments. In F. Parisi-Presicce, editor, *Recent Trends in Data Type Specifications. Selected Papers, 12th Workshop on Specification of Abstract Data Types WADT'97*, volume 1376 of *Lecture Notes in Computer Science*, pages 349–364. Springer Verlag, 1998.

[Pfenning, 1996] Frank Pfenning. The practice of logical frameworks. In H. Kirchner, editor, *Proc. 20th Coll. on Trees in Algebra and Computing CAAP'96*, volume 1059 of *Lecture Notes in Computer Science*, pages 119–134. Springer Verlag, 1996.

[Salibra and Scollo, 1993] Antonio Salibra and Giuseppe Scollo. A soft stairway to institutions. In Michel Bidoit and Christine Choppy, editors, *Recent Trends in Data Type Specification, Selected Papers, 8th Workshop on Specification of Abstract Data Types ADT'91*, volume 655 of *Lecture Notes in Computer Science*, pages 310–329. Springer Verlag, 1993.

[Sannella and Tarlecki, 1988a] Donald Sannella and Andrzej Tarlecki. Specifications in an arbitrary institution. *Information and Computation*, 76:165–210, 1988.

[Sannella and Tarlecki, 1988b] Donald Sannella and Andrzej Tarlecki. Toward formal development of programs from algebraic specifications: Implementations revisited. *Acta Informatica*, 25:233–281, 1988.

[Sannella and Tarlecki, 1997] Donald Sannella and Andrzej Tarlecki. Essential concepts of algebraic specification and program development. *Formal Aspects of Computing*, 9:229–269, 1997.

[Sannella and Tarlecki, 1999] Donald Sannella and Andrzej Tarlecki. *Foundations of Algebraic Specifications and Formal Program Development*. Cambridge University Press, 1999.

[Tarlecki, 1987] Andrzej Tarlecki. Institution representation. Unpublished note, Dept. of Computer Science, University of Edinburgh, 1987.

[Tarlecki, 1996] Andrzej Tarlecki. Moving between logical systems. In M. Haveraaen, O.-J. Dahl, and O. Owe, editors, *Recent Trends in Data Type Specifications. Selected Papers, 11th Workshop on Specification of Abstract Data Types ADT'95*, volume 1130 of *Lecture Notes in Computer Science*, pages 478–502. Springer Verlag, 1996.

[Tarlecki, 1999] Andrzej Tarlecki. Institutions: An abstract framework for formal specifications. In Egidio Astesiano, Hans-Jörg Kreowski, and Bernd Krieg-Brückner, editors, *Algebraic Foundations of Systems Specification*, chapter 4. Springer Verlag, 1999.

Integration of Linear Arithmetic and Goal-Oriented Resolution for Software Reasoning

Tie-Chung Wang and Allen Goldberg

1 INTRODUCTION

This paper describes a tight integration approach to incorporate an integer linear arithmetic solver into a resolution-based general deduction procedure. Our approach is motivated by improving the efficiency of an existing inference system, that was developed to support formal software analysis in a formal testing project. Many relations abstracted from program constructs that involve enumerated types, array indexing, counters, partial ordering, and integers are in the language of linear arithmetic. Since applying general deduction to reason about these relations often contributes a major portion of the inference cost, the incorporation of a linear arithmetic (LA) decision procedure is important.

However, a naïve incorporation of an LA procedure into a general deduction system may not improve efficiency. This is because most inference problems from practical application are not presented as theorems of pure linear arithmetic. To make the combination effective, there are two fundamental problems that must be solved. One problem is how to deal with those relations which are embedded in or mixed with the general relations. Another problem is how to preserve the existing refinements of general deduction, such as subsumption, ordering constraints, set of supports, etc., so that the combined system is efficient with regard to both general deduction and special reasoning.

To achieve a satisfactory solution of these problems, we have pursued a tight integration approach, and developed two special inference rules for promoting the use of decision procedures. One is called linear arithmetic resolution (LA resolution) and the other is called linear condition resolution (LC resolution). LA resolution allows the general deduction procedure to make a productive use of the decision procedure when reasoning about those subgoals that are linear relations. LC resolution is used for extracting linear relations that are embedded in general relations without restricting the application of normal resolution on general relations. We employ as the general deduction procedure a goal-oriented refinement of resolution, called hierarchical deduction [Wang & Bledsoe, 1987].

Resolution-based refinements remain some of the most successful techniques for automated theorem proving for first-order logic. Goal-oriented refinements, such as Set of Support Strategy [Wos et. al., 1992], Model Elimination [Loveland, 1968], natural deduction [Bledsoe, 1977; Plaisted, 1988], and the hierarchical deduction [Wang & Bledsoe, 1987] considered in this paper, are often used by systems for software reasoning [Smith, 1990; Jüllig, 1993]. Proving theorems for formal analysis of real-world software requires the use of a large rule base of data type theories, semantics of program constructs, and hypotheses from the problem domains. In order to avoid the search explosion caused by a large rule base, the goal-oriented methods and strategies are often used to help the system focus its search on the conclusion of the theorem to be proved and the closely related hypotheses.

Our approach follows the tight integration approach made by Bledsoe with his SUP-INF method [Bledsoe, 1975], and Boyer and Moore's integration of a linear arithmetic decision procedure into the NQTHM system [Boyer & Moore, 1988]. The SUP-INF method is a tight combination of LA with natural deduction. It extends the LA decision procedure with certain natural deduction mechanisms for extracting the embedded linear relations. Boyer-Moore's method promotes the use of LA by means of a sophisticated interaction between the term-rewriting procedure of the NQTHM system and an LA prover.

In contrast, our approach focuses on how to integrate LA into goal-oriented resolution. In order to keep the goal-oriented nature of the deductive procedure, we employ LC resolution to extract the embedded linear relations, and LA resolution to incorporate the results from an integer linear arithmetic solver. This is different to Boyer-Moore's method, which uses the decision procedure mainly for checking the validity of the set of linear relations. We incorporate, more generally, the solutions produced by the LA solver when the set is consistent. Consider the simple example of [Boyer & Moore, 1988]. Suppose one needs to prove that (here A_0, l_0, k_0 are constants)

$$l_0 \leq min(A_0) \wedge 0 < k_0 \rightarrow l_0 < max(A_0) + k_0,$$

under the hypothesis

$$\forall(x)(min(x) \leq max(x)).$$

With NQTHM, the system will first obtain a ground instance $min(A_0) \leq max(A_0)$ from $min(x) \leq max(x)$ by means of an instantiation process. Then a refutation proof can be obtained by calling the LA prover with the inconsistent linear set,

$$\{l_0 \leq min(A_0), 0 < k_0, \neg(l_0 < max(A_0) + k_0), min(A_0) \leq max(A_0)\}$$

With our resolution-based prover, we first obtain a set of clauses by normalizing the hypothesis and the negation of the conjecture:

A1. $min(x) \leq max(x)$
A2. $l_0 \leq min(A_0)$
A3. $1 \leq k_0$
B1. $max(A_0) + k_0 \leq l_0$ Negation of the conclusion.

Then we call the hierarchical deduction procedure with B1, the negation of the conclusion, as the top goal. Since the first subgoal is a linear relation, the LA solver will be invoked with the input set of ground linear relations, $\{l_0 \leq min(A_0), 1 \leq k_0, max(A_0) + k_0 \leq l_0\}$. From this set, the solver will produce a solution $\{max(A_0) < min(A_0)\}$. With this solution, the prover will generate a so called LA-resolvent,

B2. $\neg(min(A_0) \leq max(A_0))$

Then a refutation is produced from a resolution of B2 against A1.

Broadly speaking, our method is related to previous research in incorporating specialized reasoning to improve the efficiency of resolution-based procedures. For instance, the LA resolution rule is based on an incorporation of a specialized version of theory resolution [Stickel, 1985] with hierarchical deduction. The LC resolution can trace back to constraint resolution [Bürckert, 1990], constraint logic program(CLP) [Jaffar & Lassez, 1987], RUE [Digricoli, 1986], and E-resolution [Morris, 1969]. We shall give more comments on these relations below. Besides resolution-based methods, other results include the Nelson-Oppen combination procedure [Nelson & Oppen, 1979], which combines decision procedures for a class of first-order theories, the incorporation of specialized theories and algorithms into unification [Baader & Schulz, 1992], and the incorporation of decision procedures into higher-order or temporal deduction systems [Björner, et. al., 1997].

We shall first introduce the general deduction procedure and the LA decision procedure that are considered by our integration approach. Then we describe linear arithmetic resolution and its incorporation with the general procedure. After that, we describe linear condition resolution, and compare these special rules with existing methods. Finally, we present our combined procedure, and discuss instantiation strategies, the overall control, its completeness, and its implementation as an integrated inference component in a formal testing system.

2 THE GENERAL DEDUCTION PROCEDURE

Our general deduction procedure is a goal-oriented resolution procedure, called hierarchical deduction (HD). HD accepts as input a set S of clauses and a subset GS of S as the initial goals. It attempts to deduce the empty clause $\Box$, and thus prove that S is unsatisfiable.

The procedure carries out the deduction by traversing a tree of nodes. Each node is a record of several fields. Among them, $LABEL$ is a natural number used for the node label, $GOALS$ a sequence of goal clauses, and $RULES$ a sequence of rule clauses. A clause is a sequence of literals. Each literal is attached by a node label, called the index of the literal. At the beginning of the deduction, the tree contains only the root node: $< 1, GS, S >$, where 1 is the node label, S the given set of clauses for $RULES$, and GS a subset of S for $GOALS$ of the root node. All literals of GS are indexed by the root node label.

Figure 1 below is a version of HD with a depth-first searching strategy. It is called initially with a root node as the input. The functions $FIRST$ and $EMPTY$ are standard operators on a sequence. *GET-NODE(k)* retrieves the node labeled by k from the node tree being generated by the HD procedure, and *GEN-LABEL()* returns the next available natural number as the label of the new node. For the current goal clause g, the procedure retrieves a set rs of rules from the node labeled by the index of the first literal l_1 of g. An H-resolvent of g and rs with *label* is a resolvent of HD by applying resolution or paramodulation upon the first literal of g against a rule clause in rs. For each of the resolvents produced, the indices and the order of the literals inherited from the goal clauses are retained. But the indices of the literals inherited from the rule clause are then replaced by the value of *label*.

```
procedure HD(node nd)
  gs := GOALS(nd);
  enumerate g over gs do
    if EMPTY(g) then return □
    else begin
      l1 := FIRST(g);
      rs := RULES(GET-NODE(INDEX(l1)));
      label := GEN-LABEL();
      new-gs := {all legal H-resolvents of g and rs upon l1 with label};
      new-nd := < label, new-gs, new-gs ∪ rs >;
      if HD(new-nd) = □ then return □;
      end;
  return fail
```

Figure 1: An Hierarchical Deduction Procedure HD.

A legal H-resolvent is an H-resolvent satisfying the set of constraints of HD, which includes local subsumption, constraints on common tails, proper reduction, global subsumption, resource restrictions, etc. The labels and indices are used to retrieve rules, and also to implement these constraints. Incorporating these constraints and implementing the node-tree of HD are beyond the scope of this paper, and so we shall include no further description about these constraints, nor labels or indices in the illustration examples to be given. The interested reader can refer to [Wang & Bledsoe, 1987] for a full description about these constraints and the related labeling and indexing mechanism.

As indicated by the generation of *new-gs* in the above algorithm, HD deduces and retains all possible legal H-resolvents by applying rules that can be reached along the hierarchical deduction structure. This feature is important in the incorporation of the LA solver into HD, because the LA solver may possibly produce multiple solutions, each of which must be explored.

3 THE LINEAR ARITHMETIC SOLVER

The LA solver will be used by our combined system for two purposes. One is to check the consistency of a set of linear literals. One is to deduce logical consequences for constructing LA-resolvents when the set is consistent. Since the LA solver is treated as a black box, we need only to describe the interface.

The LA solver accepts a set of ground linear relations as input and produces a set of canonical linear relations. A linear relation is a form t_1 *op* t_2, where t_1 and t_2 are polynomial terms, and *op* is a linear relation predicate. A polynomial term is a form $c_1 * t_1 + \ldots + c_n * t_n$, where c_i is an integer, and t_i is an atomic symbol or a term whose operator differs from '+' and '−'. Each t_i is called a *monomial term.* The set of linear relation predicates include '$=$,' '$<$', '$\leq$', '$>$', '$\geq$'. Note that we allow the input of the LA solver to include all ground linear relations. This is an extension to the language used by a typical LA solver. A typical LA solver accepts only those linear relations that contain no terms other than constant symbols. But the translation for this extension is trivial.

The basic notion supported by the interface is called projection. Let S_0 be a set of linear relations on a set of ground terms U_0, and let U' be a sequence $[t_1, \ldots, t_n]$, that specifies an ordering of a subset of U_0. U' is called a projecting sequence. The solver will produce a projection of S_0 onto U', denoted by $LP_{U'}(S_0)$. A *projection* of S_0 onto U' is a set of canonical linear relations on U' that has the same integer linear solutions over the terms of U' as the original problem. A canonical linear relation is a linear relation whose polynomial terms are ordered with the ordering of elements in the sequence U'.

For our combined system, the purpose of using different projecting sequences is to deduce different logical consequences from the input set. For example, given the set $\{0 \leq a \leq 4b, a < b \leq 4\}$, if we choose $[b]$ as the projecting sequence, then the output will be $\{1 \leq b, b \leq 4\}$. If we choose $[a]$, then the output will be $\{0 \leq a, a \leq 3\}$.

Without going into the details of how to choose projecting sequences we shall use $LP(S_0)$ to denote the entire set of linear relations that can be deduced by the LA solver from S_0. Since there is only a finite number of distinct projecting sequences, and for each projecting sequence the solver produces a finite set which consists of one or more linear relation, then $LP(S_0)$ is always finite. However, if S_0 is determined to be inconsistent, then $LP(S_0)$ must be empty.

Another extension that we made to the LA solver is the inclusion of linear inequalities into linear relations. As usual, we shall handle a linear inequality relation by splitting. That is, we treat $t1 \neq t2$ by a disjunction $t1 < t2 \vee t2 < t1$. Let $S_1 \equiv \{t1 \neq t2\} \cup S_0$ be a set of linear relations. Let S_{1L} be $\{t1 < t2\} \cup S_0$ and let S_{1R} be $\{t2 < t1\} \cup S_0$. Then

$$LP(S_1) \equiv \begin{cases} LP(S_{1R}) & \text{if } LP(S_{1L}) \equiv \{\}, \\ LP(S_{1L}) & \text{if } LP(S_{1R}) \equiv \{\}, \\ \{r1 \vee r2 \mid r1 \in LP(S_{1L}) \wedge r2 \in LP(S_{1R})\} & \text{otherwise.} \end{cases}$$

4 LINEAR ARITHMETIC RESOLUTION

We now turn to the combination of the LA solver and the hierarchical deduction procedure. First, we describe the special inference rule, LA resolution. The LA resolution can be viewed as a specialized version of theory resolution [Stickel, 1985]. It builds the integer linear arithmetic theory of the LA solver into the resolution rule employed by the hierarchical deduction. For our combined system, this rule will be applied when the current subgoal is a linear relation.

Definition 4.1 (Linear Arithmetic Resolution) Let $G \equiv L_g \vee G_1$ be a goal clause whose first literal L_g is a linear relation. Let $RS \equiv \{L_1 \vee C_1, \ldots, L_m \vee C_m\}$ be a set of rule clauses, such that for each j, $1 \leq j \leq m$, $L_j \vee C_j$ can be used as a rule clause for G in the hierarchical deduction, L_j can be resolved upon, and L_j is a linear relation.

Let $G' \equiv L'_g \vee G'_1$ be an instance of G, such that L'_g is a ground linear relation. Let RS' be $\{L'_1 \vee C'_1, \ldots, L'_n \vee C'_n\}$, such that for each $k, 1 \leq k \leq n$, $L'_k \vee C'_k$ is an instance of $L_j \vee C_j$, where $1 \leq j \leq m$, and L'_k is a ground literal.

Let RS'_{la} be $\{L'_g, L'_1, \ldots, L'_n\}$, and let $LP(RS'_{la})$ be the set of linear relations deduced from RS'_{la} by the LA solver. If $LP(RS'_{la})$ is empty, then RS'_{la} is LA unsatisfiable. Then the set S'_{la} of LA-resolvents of G' against RS' is the singleton set,

$$S_{la} \equiv \{C'_1 \vee \ldots \vee C'_n \vee G'_1\}.$$

If $LP(RS'_{la})$ is not empty, then

$$S_{la} \equiv \{p_i \vee C'_1 \vee \ldots \vee C'_n \vee G'_1 \mid p_i \in LP(RS'_{la})\}.$$

We mention that this is a partial definition concerning only the specific features of LA resolution. The other portion for a complete definition will be similar to a resolution rule employed by the hierarchical deduction, and use the similar constraints for checking the "legal" LA-resolvents. Only "legal" LA-resolvents will be retained. For instance, the common tails of the residue clauses $C_1, \ldots, C_m, G_1$ are required to be mergable.

Theorem 4.2 *LA resolution is sound.*

Proof. If $LP(RS'_{la})$ given in the definition is empty, then RS'_{la} is LA unsatisfiable. Then the proof is obvious. If $LP(RS'_{la})$ is not empty, we need only to show that $p_i \vee C'_1 \vee \ldots \vee C'_n \vee G'_1$ is a consequence of $RS \cup \{G\}$ in the LA theory for every member p_i of $LP(RS'_{la})$. Suppose that is not the case. Then there exists an LA model M, such that M is a model of RS and G, but there exists a p_j of $LP(RS'_{la})$, such that $p_j \vee C'_1 \vee \ldots \vee C'_n \vee G'_1$ is *false* in M. Since M is a model of RS and G, but $RS \equiv \{L_1 \vee C_1, \ldots, L_m \vee C_m\}$ and $G' \equiv L'_g \vee G'_1$, each member of RS'_{la} must be *true* in M. Because the LA solver is sound, then p_i must be *true* in M. This is a contradiction. $\dashv$

Similar to the proper reduction defined for the hierarchical deduction, there is an LA reduction associated with LA resolution.

Definition 4.3 **(LA reduction)** Let $L_g \vee G_1$ be a goal. If G_1 is an LA-resolvent of $L_g \vee G_1$, then G_1 is called an LA reduction of $L_g \vee G_1$. Recursively, G' is called an LA reduction of $L_1 \vee \ldots \vee L_n \vee G'$ if for each $i, 1 \leq i < n$, $L_{i+1} \vee \ldots \vee L_n \vee G'$ is an LA reduction of $L_i \vee \ldots \vee L_n \vee G'$, and G' is an LA reduction of $L_n \vee G'$.

If an LA reduction is obtained from the current goal G, then all other resolvents deduced from the same goal can be discarded without introducing incompleteness.

Instantiation. Since a standard LA decision procedure can only accept ground linear relations, LA resolution is defined by using ground instances RS'_{la}. But we have not specified how G and each member of RS involved in an LA resolution are instantiated for producing the ground instances. There may be many plausible instantiations. For instance, a naïve method is to enumerate a finite subset of the Herbrand's universe limited by certain complexity constraints. However, such a breadth-first style approach may lead to the generation of many useless LA-resolvents, specially when the rule base contains integer boundary conditions, such as $lb \leq x$, $x \leq ub$.

Instead, we will employ a number of distinct instantiation strategies with a global control on the use of these strategies. For instance, by default or initially, the system may choose the null instantiation strategy, that is, to apply LA resolution only to those existing ground linear-relations. This strategy is sufficient for the case for the example given in the introduction, and most of examples to be given below. We shall describe other possible instantiation strategies below.

We now present an example to illustrate LA resolution and the incorporation of it in the general deduction process. We assume that our combined procedure uses the null instantiation strategy. Note that our system canonicalizes all integer inequality relations to use '$\leq$' (i.e., less or equal).

Example 4.4 A1 specifies that A is a sorted array of 100 elements. The theorem to be proved is that if all program variables I, J, N satisfy the constraints A2-A6, then $A(J) \leq A(N)$ implies $A(J) < A(I+N)$, where

A1. $\forall(x,y)(0 \leq x \leq 99 \wedge 0 \leq y \leq 99 \Rightarrow (x \leq y \Leftrightarrow A(x) \leq A(y)))$.

To simplify the presentation, we do not present the clauses transformed from A1.

A2. $1 \leq I$
A3. $0 \leq N \leq 50$
A4. $0 \leq J \leq 50$
A5. $A(J) \leq A(N)$
A6. $I + N \leq 99$
B1. $A(I+N) \leq A(J)$ — Negation of the conclusion
B2. $\neg(0 \leq I+N) \vee \neg(I+N \leq 99) \vee \neg(0 \leq J) \vee \neg(J \leq 99) \vee I+N \leq J$
H-resolvent of B1, A1
B3. $I + N \leq J$ — LA reduction of B2, A2 – A6
B4. $\neg(J \leq N)$ — LA-resolvent of B3, A2 – A6
B5. $\neg(0 \leq J) \vee \neg(J \leq 99) \vee \neg(0 \leq N) \vee \neg(N \leq 99) \vee \neg(A(J) \leq A(N))$
H-resolvent of B4, A1
B6. $\square$ — LA reduction of B5, A2 – A6

Note the LA redutions obtained by using LA resolution. For instance, B3 is obtained from B2 by eliminating $\neg(0 \leq I+N)$, $\neg(I+N \leq 99)$, $\neg(0 \leq J)$, and $\neg(J \leq 99)$, consecutively with LA reduction. Similarly, B6 is the result of a sequence of LA reductions from B5.

This example demonstrates how the solution of the LA solver is incorporated in the goal-oriented general procedure. For instance, B4 is an LA-resolvent of B3 and A2 – A6 with a projection to the sequence $[J, N]$ by the LA solver. If the decision procedure were used only as a validity checker, one would first have to find a suitable instantiation of (the clauses of) A1, before generating the needed resolvent.

The combined use of the general deduction is helpful. For instance, B2 and B5 are deduced as ordinary hierarchical resolvents. One may note that, since the parent goals of these resolvents are linear relations, they may also be deduced by LA resolution using results of the LA solver. However, such a derivation needs to first generate some ground instances from A1.

The next example illustrates another feature of the LA resolution: the subsequent use of the results of the LA solver by the general deduction procedure is a necessary condition for obtaining a goal-oriented refutation.

Example 4.5 Assume that x, y, z are variables, and a, b, c are constants. Prove that A1–A7 implies $R(b, c)$.

A1. $P(0, 1)$	
A2. $Q(x, x)$	
A3. $\neg P(x, b) \vee \neg Q(x, c)$	
A4. $R(y, z) \vee 0 \leq y + 3z$	
A5. $3a - 4c = 0$	
A6. $\neg Q(c, c) \vee c \leq 0$	
A7. $a + b = 1$	
B1. $\neg R(b, c)$	The top goal
B2. $0 \leq b + 3c$	H-resolvent of B1, A4
B3. $a = 0 \vee \neg Q(c, c)$	LA-resolvent of B2, A7, A6, A5
B4. $b = 1 \vee \neg Q(c, c)$	LA-resolvent of B2, A7, A6, A5
B5. $c = 0 \vee \neg Q(c, c)$	LA-resolvent of B2, A7, A6, A5
B6. $\neg P(x, b) \vee \neg Q(x, 0) \vee \neg Q(c, c)$	H-resolvent of B5, A3
B7. $\neg P(x, 1) \vee \neg Q(x, 0) \vee \neg Q(c, c)$	H-resolvent of B6, B4
B8. $\neg Q(0, 0) \vee \neg Q(c, c)$	H-resolvent of B7, A1
B9. $\neg Q(c, c)$	H-resolvent of B8, A2
B10. $\square$	H-resolvent of B9, A2

As shown by the above proof tree, the LA-resolvents B3, B4, and B5 are deduced by LA resolution, which uses the linear relations obtained by the LA solver from a (consistent) set of linear constraints $\{0 \leq b + 3c, 3a - 4c = 0, a + b = 1, c \leq 0\}$. This use of the LA solver is necessary to the proof, because there exists no other hierarchical proof path that can be found, unless the LA axioms are used directly by the general procedure.

5 LINEAR CONDITION RESOLUTION

The linear arithmetic resolution introduced in the preceding section cannot be directly applied to those linear relations which are embedded in the term structures of the given clauses. For instance, consider the set of clauses $S_k = \{1 \leq k, k \leq 3, p(a+1), p(a+2), p(a+3), \neg p(a+k)\}$. Although S_k is unsatisfiable in the LA theory, there is no refutation that can be deduced if the last clause $\neg p(a + k)$ is used as the top goal. In order to fix this problem, we need to use a specialized resolution rule. It is called a linear condition resolution. We first introduce a linear disagreement formula.

Definition 5.1 (Linear disagreement formula) Let s and t be two terms. The linear disagreement formula of s and t is denoted by $LDS(s, t)$. If s is identical to t, then $LDS(s, t)$ is the empty clause $\square$. If s and t are both LA terms, then $LDS(s, t)$ is $s \neq t$. If s and t have the forms $f(a_1, \ldots, a_n)$ and $f(b_1, \ldots, b_n)$, respectively, then $LDS(s, t)$ is $LDS(a_1, b_1) \vee \ldots \vee LDS(a_n, b_n)$. Otherwise $LDS(s, t)$ is the tautology $True$.

The linear disagreement formula of two literals reflects their difference. If $LDS(s,t) \equiv \square$, then s and t must be identical. If $LDS(s,t) \equiv True$, then they must be different on some general terms. Otherwise, they differ only in LA terms contained in $LDS(s,t)$.

Definition 5.2 **(Linear Condition Resolution)** Let $G \equiv L_g \vee G_1$ be a goal clause, and let $R \equiv L_r \vee R_1$ be a rule clause with no variables in common with G. If there is a most general partial unifier (*mgpu*) σ, such that $LDS(\neg L_g\sigma, L_r\sigma)$ is not the tautology *True*, then $R_1\sigma \vee LDS(\neg L_g\sigma, L_r\sigma) \vee G_1\sigma$ is an LC-resolvent of G and R.

Similar to the generation of normal hierarchical resolvents, there are some constraints attached to the "legal" LC-resolvents, and only "legal" LC-resolvents will be used and maintained by the procedure.

There exist different versions of partial unification [Morris, 1969; Digricoli, 1986]. We give an outline of the one used in this paper. Given terms t_1 and t_2, the partial unification procedure attempts to unify t_1 and t_2 by a top-down traversal of the term structure, in the same way as a standard unification algorithm. However, for each pair of different subterms t_{11} and t_{22} that are both of integer type, unless t_{11} or t_{22} is a variable, the procedure will not try to unify them, but instead construct an inequality $t_{11} \neq t_{22}$.

Another case in which linear constraints may need to be extracted occurs when the leading subgoal of G is an inequality $t \neq s$. To include this case, we make an extension to the above definition:

Definition 5.3 **(LC Resolution extended)** Let $G \equiv L_g \vee G_1$ be a goal clause. If L_g has a form $s \neq t$, and if there is a *mgpu* σ of s and t, such that $LDS(s\sigma, t\sigma)$ is not the tautology *True*, then $LDS(s\sigma, t\sigma) \vee G_1\sigma$ is an LC-resolvent of G.

Extracting a linear disagreement formula by partial unification makes an implicit use of the equality substitution axioms. However, if the literals to be unified contain no LA subterms, then LC resolution is exactly the same as the resolution employed by a hierarchical deduction procedure.

We now present an example to illustrate LC resolution, and the combination of LA resolution, LC resolution, and the hierarchical deduction procedure.

Example 5.4 This example is the clausal form extracted from a theorem proving problem from software analysis of an Ada program. An Ada array is represented by a term $Ary(x, lb, ub, e)$, which means that x is an array with lower bound lb, upper bound ub, and a content term e (whose structure is not relevant here). *Aeq* denotes array equality. $Ael(A, n)$ denotes the value of array A at the entry n. Two arrays are equal if they have the same shape and the corresponding elements are equal.

A52. $Aeq(Ary(x, l1, u1, e1), Ary(y, l2, u2, e2)) \vee (l2 + u1 \neq l1 + u2) \vee$
$(l1 \leq Sk)$
A53: $Aeq(Ary(x, l1, u1, e1), Ary(y, l2, u2, e2)) \vee (l2 + u1 \neq l1 + u2) \vee$
$(Sk \leq u1)$
A54: $Aeq(Ary(x, l1, u1, e1), Ary(y, l2, u2, e2)) \vee (l2 + u1 \neq l1 + u2) \vee$
$Ael(Ary(x, l1, u1, e1), Sk) \neq Ael(Ary(y, l2, u2, e2), Sk + l2 - l1)$
A75. $Ael(Ary(A_3, 0, 2, Nl), 2) = Ael(Ary(A_1, 0, 2, Nl), 2)$
A77. $Ael(Ary(A_3, 0, 2, Nl), 1) = Ael(Ary(A_1, 0, 2, Nl), 1)$
A79. $Ael(Ary(A_3, 0, 2, Nl), 0) = Ael(Ary(A_1, 0, 2, Nl), 0)$
B1. $\neg Aeq(Ary(A_3, 0, 2, Nl), Ary(A_1, 0, 2, Nl)))$
Negation of the conclusion
B2. $0 + 2 \neq 0 + 2 \vee 0 \leq Sk$ H-resolvent of B1, A52
B3. $0 + 2 \neq 0 + 2 \vee Sk \leq 2$ H-resolvent of B1, A53
B4. $0 + 2 \neq 0 + 2 \vee Ael(Ary(A_3, 0, 2, Nl), Sk) \neq Ael(Ary(A_1, 0, 2, Nl), Sk)$
H-resolvent of B1, A54
B5. $Ael(Ary(A_3, 0, 2, Nl), Sk) \neq Ael(Ary(A_1, 0, 2, Nl), Sk)$
LA reduction of B4
B6. $Sk \neq 2$ LC-resolvent of B5, A75
B7. $Sk \neq 1$ LC-resolvent of B5, A77
B8. $Sk \neq 0$ LC-resolvent of B5, A79
B9. $0 + 2 \neq 0 + 2$ LA-resolvent of B8, B7, B6, B3, B2
B10. $\square$ LA reduction of B9

Related work. Our mechanism for extracting linear disagreement formulas follows a line of previous work that was originated by E-resolution [Morris, 1969], and continued by RUE [Digricoli, 1986], constraint linear programming (CLP) [Jaffar & Lassez, 1987], and constraint resolution (C-resolution) [Bürckert, 1990]. The basic mechanism of this approach is to resolve complementary literals involved in the resolution to a disjunction D of inequalities.

The difference is mainly in what kind of disjunction D that is to be extracted, and how D is processed in the remaining deduction. For instance, consider the following two clauses,

$\neg P(2x + a, f(b), d) \vee C1$
$P(2c + e, f(y), y) \vee C2$

Assume that all terms, except b, d, and y are of integer type. Then we can list the *mgpu* σ, the disagreement inequality D, and the final resolvent in the following tables.

Method	Mgpu (σ)	Disagreement formula (D)
LC Resolution	$\{d/y\}$	$2x + a \neq 2c + e \vee f(b) \neq f(d)$
E-resolution	$\{c/x, d/y\}$	$a \neq e \vee b \neq d$
RUE	$\{\}$	$2x + a \neq a * c + e \vee f(b) \neq f(y) \vee d \neq y$
C-Resolution	$\{d/y\}$	$2x + a \neq 2c + e \vee f(b) \neq f(d)$
CLP	$\{d/y\}$	$2x + a \neq 2c + e \vee f(b) \neq f(d)$

Method	Resolvent
LC Resolution	$D \vee C1\sigma \vee C2\sigma$
E-resolution	$C1\sigma \vee C2\sigma$
RUE	$D \vee C1 \vee C2$
C-Resolution	$C1\sigma \vee C2\sigma/\neg D$
CLP	$C1\sigma \vee C2\sigma/\neg D$

As shown by the tables, while the *mgpu* of our LC resolution is $\{d/y\}$, the *mgpu* of E-resolution and RUE are $\{c/x, d/y\}$ and $\{\}$, respectively. This is because E-resolution tries to unify two terms by an exhaustive traversal of the term structures, collecting only the irreducible terms into the disagreement formula D. Moreover, before the generation of the resolvent, each element of D must be eliminated by paramodulation with other input rules. For RUE, there are different options to collect disagreement formulas. The RUE result shown here uses the option of topmost disagreement formula. Both E-resolution and RUE assumes no special theory other than equality.

Regarding *mgpu* and the resulting disagreement formula, LC resolution is closer to constraint resolution and CLP. However, both constraint resolution and CLP transform D into a constraint $\neg D$, which is attached to the kernel clause K. Syntactically, this is represented as a constraint clause $K/\neg D$. This constraint will not be resolved upon by the general deduction procedure, but merely serves as a criteria for checking the consistency of the constraint resolvents to be deduced. In contrast, LC resolution maintains disagreement formula as a part of the hierarchical resolvent.

Another difference is in the nature of the general deduction. For instance, constraint resolution is established by using a binary resolution with a full saturation strategy. CLP is designed for reasoning about Horn sets. Our LC resolution rule extends the hierarchical refinement of resolution, and in particular keeps the goal-oriented feature of the deduction process. We can use the following example to illustrate this point.

Example 5.5 This example is from [Bledsoe75]. To improve the readability, we did not write all hypotheses in clausal form.

A1. $1 \leq n$
A2. $\forall(x)(2 \leq n \wedge 1 \leq x \wedge x \leq 1 \Rightarrow A(x) \leq A(2))$
A3. $\forall(x)(x+1 \leq n \wedge 2 \leq x \Rightarrow A(x) \leq A(x+1))$
A4. $k+1 \leq n$
A5. $1 \leq k$
B1. $\neg(A(k) \leq A(k+1))$ Neg. of the conclusion
B2. $\neg(2 \leq n) \vee \neg(1 \leq k) \vee \neg(k \leq 1) \vee 2 \neq k+1$
LC-resolvent of B1, A2
B3. $\neg(k+1 \leq n) \vee \neg(2 \leq k)$ H-resolvent of B1, A3
B4. $\neg(2 \leq k)$ H-resolvent of B3, A4

B5. $k = 1$ LA-resolvent of B4, A5, A4, A1
B6. $2 \leq n$ LA-resolvent of B4, A5, A4, A1
B7. $\neg(2 \leq n) \vee \neg(1 \leq 1) \vee \neg(1 \leq 1) \vee 2 \neq 1 + 1$
H-resolvent of B5, B2
B8. $\neg(1 \leq 1) \vee \neg(1 \leq 1) \vee 2 \neq 1 + 1$ H-resolvent of B7, B6
B9. $\Box$ LA reduction of B8

Note that the LC-resolvent B2 is used as a rule by the goal B5 for deducing B7. This deduction is possible due to the hierarchical deduction structure. However, this deduction and the use of LC resolution for extracting the disagreement formula $2 \neq k + 1$ in producing B2 are essential steps to the proof.

Finally, we mention another approach that is also related to the extraction. This approach is called separation [Nelson & Oppen, 1979]. Separation allows one to take out terms in a specified sublanguage before the deduction. If separation is applied to the LA theory, then for our example clauses, we shall obtain

$$(2x + a = u1 \wedge f(x) = v1) \Rightarrow \neg P(u1, v1, d) \vee C1$$
$$(2c + e = u2 \wedge f(y) = v2) \Rightarrow P(u2, v2, y) \vee C2$$

The interesting feature is that by using separation one can replace the LC resolution. One potential problem is that separation intendes to expand a literal into a clause. This will introduce additional complexity. So it is not always a good idea to use separation. For example, by separation, the unit clause $P(0, 1)$ of Example 2 will be $(x = 0 \wedge y = 1) \Rightarrow P(x, y)$. This is unnatural.

6 THE COMBINED PROOF PROCEDURE

We are now ready to make an overall description of the combined procedure.

6.1 Outline of the Combined Proof Procedure

Since the combined procedure is a modification and enhancement of the HD procedure, which has been reviewed in Section 2, we need only to describe the diferences.

- First the combined procedure will have a built-in LA theory employed by LA resolution. We assume that the knowledge-base (KB) of the procedure contains no declarative axioms from the LA theory. Given a set of clauses, the prover is to produce a refutation of S under the hypothesis of $LA \cup KB$.

- The root node is constructed in the same way as the old procedure, and starting from the root node, the combined procedure will repeat similar proof steps, and use the same set of rules *rs* for developing a goal *g*. However, the set *new-gs* of resolvents will be constructed differently. One difference is that the combined procedure uses LC resolution while the old hierarchical deduction uses ordinary resolution. Another difference is that, if the first literal of *g* is a linear relation, then in addition to LC resolution and paramodulation, linear arithmetic resolution is also applied to *g* by using linear relations contained in the set of rules *rs*.

- There is an instantiation algorithm assigned to the procedure that produces, for each application of LA resolution, the ground instances passed to the LA solver. We present below three unpolished strategies for these algorithms.

6.2 Instantiation Strategies

Let $RS_{la} \equiv \{L_g, L_1, \ldots, L_n\}$ be the set of linear relations that is considered in the definition of LA resolution (see Section 4). Let RS'_{la} be the set of ground linear relations that is to be produced by applying the strategy.

Strategy 1: RS'_{la} is the extraction of the ground elements of RS_{la}. This will be the default instantiation algorithm.

Strategy 2: Let GM_{la} be the entire set of ground monomial terms and subterms contained in RS_{la}. Let $S(RS_{la}, GM_{la})$ be the entire set of ground linear relations that can be obtained by enumerating all possible substitutions (which satisfy the sort restrictions). Let RS'_{la} be the subset of $S(RS_{la}, GM_{la})$, such that each member contains only those monomial terms that are instances of monomial terms of RS_{la}.

Strategy 3: Let HS_{la} be the subset of Herbrand's universe obtained with certain criteria. Let GM_{la} be the union of HS_{la} and the entire set of ground monomial terms and subterms contained in RS_{la}. The rest is the same as Strategy 2.

Typically, given a theorem to prove, the procedure will start with a less expensive instantiation strategy and the associated algorithm, and run until a proof is found, or until it is aborted due to the resource limitation. For the later case, the procedure may run again with a more expensive instantiation algorithm. Another possible control strategy is to execute in parallel with a different instantiation algorithm.

6.3 Completeness

Similar to the tight integration approach stated in [Boyer & Moore, 1988], our goal of combining the LA solver is to eliminate the need for direct access axioms from LA theory in the main goal-oriented deductive component.

Thus the completeness of our combined procedure is a theoretically interesting topic.

However, we cannot claim that our combined procedure is complete, but we have not found a counterexample that this procedure cannot handle.

6.4 Application of the Combined Procedure

The combined procedure described in this paper is not implemented as a stand-alone prover. Instead, it is a part of the backward inference component of a large inference system developed for supporting the feasible path analysis. This inference system has been described in [Goldberg et. al., 1994; Goldberg & Wang, 1995]. The work described in this paper is the result of a further improvement of the inference system. Besides the combination described here, we also incorporated the LA solver into our term-rewriting procedure to discharge the conditions of term rewriting rules. We also incorporated the solver with our context-dependent simplification procedure.

We wish to point out that this combined proof procedure is only one component of the entire inference system. Among them, there is a natural deduction controller (NDC), which is based on natural deduction [Bledsoe, 1977] and mating [Andrews, 1981; Bibel, 1981]. The task of NDC is to decompose the input task into small subtasks, and then assign the subtasks to other inference procedures. It is well-known that transforming a large formula into clauses required by resolution suffers possible explosion. This is particularly true for the problems from feasible path analysis. This explosion problem is avoided completely because a subtask sent to the resolution procedure usually contains only a conjunction of unit clauses. With these improvements, the main difficulty for the resolution procedure is how to efficiently reason about the conjecture with the large rule base.

7 CONCLUSIONS AND PERSPECTIVES

We described a practical proof procedure that combines a linear arithmetic solver with a goal-oriented resolution proof procedure. We introduced a linear arithmetic resolution rule, a linear constraint resolution rule, and special control strategies to achieve a tight integration of the decision procedure and the general deduction procedure. We described how these inference components are combined into a proof procedure, and demonstrated the advantages of our approach by solving problems that would be hard to solve by general methods. We have compared our work with many existing results in this area.

Our primary goal is to develop a tractable inference system for supporting feasible path analysis, and other applications of software reasoning. We have had particular success using this system to determine test coverage criteria for a large safety-critical control application. Also see [Jasper et. al., 1994; Chilenski & Newcomb, 1994] for related work extending our system. In particular they incorporate a constraint satisfaction system that generates test cases from satisfiable formulas resulting from symbolic evaluation.

In the future we will continue experimentation to enhance the speed and effectiveness of the system. A planned enhancement is to integrate our typed inference [Wang & Goldberg, 1994] resolution prover.

Acknowledgments. This work has been partly supported by NASA and Boeing Corporation. Thanks go to Alessandro Coglio and Cordell Green for proofreading the draft, and to the anonymous referees for valuable suggestions.

REFERENCES

[Andrews, 1981] P.B. Andrews. Theorem proving via general mating. *J. ACM* 28(2), pages 193-214, 1981.

[Baader & Schulz, 1992] F. Baader and K. Schulz. Unification in the union of disjoint equational theories: Combining decision procedures. In D. Kapur editor, *Proceedings CADE-11*, LNAI 607, pages 50-65, Springer-Verlag, 1992.

[Bibel, 1981] W.W. Bibel. On matrices with connections. *J. ACM* 28(4), pages 633-645, 1981.

[Björner, et. al., 1997] N.S. Björner, M. E. Stickel, T. E. Uribe. A practical integration of first-order reasoning and decision procedures. In B. McCune editor, *Proceedings CADE-97*, LINCS 1249, Springer-Verlag, 1997.

[Bledsoe, 1975] W. W. Bledsoe. A new method for proving certain Presburger formulas. In Advance Papers, *Proceedings of Fourth IJCAI*, pages 15-20, U.S.S.R. 1975.

[Bledsoe, 1977] W. W. Bledsoe. Non-resolution theorem proving. *Artificial Intelligence* 9, pages 1-35, 1977.

[Boyer & Moore, 1988] R. S. Boyer and J. S Moore. Integrating decision procedure into heuristic theorem provers: A case study with linear arithmetic. *Machine Intelligence 11,* Oxford University Press, 1988.

[Bürckert, 1990] H.-J. Bürckert. A resolution principle for clauses with constraints. In M.E. Stickel editor, *Proceedings of 10th International Conference on Automated Deduction*, LNAI 568, pages 178-202, 1990.

[Chilenski & Newcomb, 1994] J. J. Chilenski and P. H. Newcomb. Formal specification tools for test coverage analysis. In *KBSE'94: The Ninth Knowledge-Based Software Engineering Conference,* pages 59-68, IEEE press, 1994.

[Digricoli, 1986] V. J. Digricoli. Resolution by unification and equality. *J. ACM* 33, pages 253-289, 1986.

[Goldberg et. al., 1994] A. Goldberg, T.C. Wang, and D. Zimmerman. Application of feasible path analysis to program testing. In *International Symposium on Software Testing and Analysis*, pages 80-93, IEEE press, 1994.

[Goldberg & Wang, 1995] A. Goldberg and T.C. Wang. Integration of symbolic evaluation and specialized inference components for software analysis. *Kestrel Tech. Report KES.95.3.* 1995.

[Jasper et. al., 1994] R. Jasper, M. Brennan, K. Williamson, C. Currier, and D. Zimmerman. Test data generation and feasible path analysis. In *International Symposium on Software Testing and Analysis*, IEEE press, 1994.

[Jaffar & Lassez, 1987] J. Jaffar and J.-L. Lassez. Constraint logic programming. In *Proceedings of ACM Symposium on Principles of Programming Languages,* pages 111-120, 1987.

[Jüllig, 1993] R. K. Jüllig. Applying formal software synthesis. *IEEE Software,* 10(3), pages 11-22, 1993.

[Loveland, 1968] D. W. Loveland. Mechanical theorem-proving by model-elimination. *J. ACM* 15, pages 236-251, 1968.

[Morris, 1969] J. Morris. E-resolution: an extension of resolution to include the equality relation. *IJCAI*, 1969.

[Nelson & Oppen, 1979] G. Nelson and D. Oppen. Simplification by cooperating decision procedures. *ACM Transitions on Programming Languages and Systems*, No 1, pages 245-257, 1979.

[Plaisted, 1988] D. A. Plaisted. Non-Horn clause logic programming without contrapositives. *Journal of Automated Reasoning* 4(3), pages 287-325, 1988.

[Smith, 1990] D. R. Smith. KIDS – a semi-automatic program development system. *IEEE Transaction on Software Engineering*, Special Issue on Formal Methods in Software Engineering 16(9), pages 1024-1043, 1990.

[Stickel, 1985] M.E. Stickel. Automated deduction by theory resolution. *Journal of Automated Reasoning* 1(4), pages 333-355, 1985.

[Wang & Bledsoe, 1987] T.C. Wang and W. W. Bledsoe. Hierarchical deduction. *Journal of Automated Reasoning,* 3(1), pages 35-71, 1987.

[Wang & Goldberg, 1994] T. C. Wang and A. Goldberg. KITP-93: An automated inference system for program analysis. In A. Bundy editor, *Proceedings CADE-12*, LNAI 814, pages 831-835, Springer-Verlag, 1994.

[Wos et. al., 1992] L. Wos, R. Overbeek, E. Lusk, J. Boyle. *Automated Reasoning: Introduction and Applications,* Prentice Hall, 1992.

Temporalizing Description Logics

Frank Wolter and Michael Zakharyaschev

1 INTRODUCTION

Traditional first order predicate logic was designed for representing and manipulating *static* knowledge (e.g. mathematical theories). The same is true for many of its applications. Knowledge representation systems based on concept description logics are not exceptions.

In the framework of a description logic, one can represent an application domain in terms of concepts, roles, and object names. Concepts are understood as classes of objects, roles as binary relations between objects, and object names denote certain objects in the domain. The expressive power of the description logic depends on the concept and role constructs available in its language. Typical examples are conjunction, negation and restricted quantification of concepts, and composition, union, inversion, and reflexive transitive closure of roles. In general, description logics can be characterized as variable-free fragments of first order logic, sometimes augmented with fixpoint-operators (see [De Giacomo and Lenzerini, 1994]). Unlike first order logic, description logics are often decidable and, moreover, effectively implementable (see e.g. [Brachman and Schmolze, 1985; Borgida *et al.*, 1989; Baader and Hollunder, 1991]). Recently description logics have found numerous applications, in particular, to information systems [Catarci and Lenzerini, 1993], databases [Borgida, 1995], software engineering [Wright *et al.*, 1993]. They have also been advocated as a unifying framework for different types of databases and knowledge representation formalisms [Bergamaschi and Sartori, 1992].

To capture various *dynamic* features of application domains in computer science and artificial intelligence (such as program executions, information flows, temporal databases, multi-agent and distributed systems, etc.), first order logic is usually extended by explicit program, temporal, epistemic or some other kind of "modal" operators. However, this often results in logics of a higher degree of undecidability, for instance, recursively non-enumerable (see e.g. [Gabbay *et al.*, 1994; Kröger, 1990; Szalas and Holenderski, 1988]), which is the main reason why only the propositional fragments of temporal, dynamic and other logics of this sort have mostly been studied and used in practice.

On the other hand, having such a natural, well motivated and established knowledge representation formalism as description logics, it would be strange not to try to extend it by adding, say, a temporal dimension so that the underlying description logic would represent knowledge about states of a process while the temporal component describes the behaviour of the process in time, i.e., the resulting sequence of states.

The main aim of this paper is to show that by combining expressive decidable description logics and point-based temporal propositional logics we can obtain decidable hybrids. In a sense, our results can be regarded as an optimal compromise between expressive power and decidability: even harmless looking extensions of the constructed systems lead to undecidable logics.

We deal with three types of underlying description logics. First we consider the logic $\mathcal{CIQ}$ developed and investigated by De Giacomo and Lenzerini [1996] and De Giacomo [1995]. It has the usual concept constructs including qualified number restrictions and an extensive set of role constructs: union, chaining, transitive reflexive closure, inversion, and test. (Note that because of the transitive reflexive closure construct this logic is not a fragment of first order logic.) We allow not only TBox-reasoning but also object names and assertions of the form $a : C$ (object a is in concept C), aRb (objects a and b are in relation R). Two other description logics are $\mathcal{CIO}$ and $\mathcal{CQO}$ introduced by De Giacomo [1995].[1] In their languages one can form concepts $\{a\}$ for all object names a, which are interpreted as singletons and correspond to names or nominals known in the modal logic literature (see e.g. [Blackburn, 1993]). In these cases to obtain decidability either the construct of inverse roles or number restrictions have to be omitted.

In the temporal dimension, we consider the operators "Since" and "Until" over natural and integer numbers, and the operators "sometime in the future" and "sometime in the past" over arbitrary strict linear orders and rational numbers. The pure temporal part of our logics is also well known and investigated; see e.g. [Gabbay *et al.*, 1994].

In the variety of possible ways of combining the formalisms of description and temporal logics we follow one which was first proposed by Baader and Laux [1995], who integrated polymodal **K** with the description logic $\mathcal{ALC}$ by applying modal operators to both concepts and formulas. In our case, we also allow applications of the temporal operators to concepts and formulas. This way seems to be an optimal choice, for, as was shown by Baader and Ohlbach [1995], modal operators applicable to roles can ruin decidability.

Our attempt to combine description and temporal logics is not the first one. Schmiedel [1990] proposed a very expressive temporal description logic based on intervals as introduced by Halpern and Shoham [1991]; however, it turned out to be undecidable. Devanbu and Litman [1991], Weida and

[1]In [De Giacomo, 1995], $\mathcal{CQO}$ was called $\mathcal{CNO}$.

Litman [1992], Artale and Franconi [1994] continued this work by weakening Schmiedel's logic (they integrated constraint networks and fragments of Allen's interval calculus into description logics). Schild [1993] introduced a decidable point-based temporal description logic in which temporal operators can be applied only to concepts. On the other hand, a number of approaches to combining modal and temporal logics have been proposed. Finger and Gabbay [1992] studied temporal modal logics in which (speaking in terms of description logics) temporal operators are applied only to formulas. Both Schild's and Finger–Gabbay's constructions are covered by our approach. Fagin et al. [1995] considered a logic for modelling the behaviour of parallel processes on the basis of epistemic and temporal operators. Their system for one agent who does not forget, does not learn and knows time is a fragment of our logics based on natural numbers. Reynolds [1996] interpreted this system on arbitrary strict linear orders. This is also covered by our formalisms.

This paper is organized in the following way. Having defined (in Sections 2 and 3) the syntax and semantics of the temporal description logic $\mathcal{CIQ}_{\mathcal{US}}$, we introduce and investigate (in Section 4) our main tool for establishing decidability, the notion of a quasimodel. Unlike standard models, worlds in quasimodels are always finite; however, modulo a given formula, every model can be represented as a suitable quasimodel. In [Wolter and Zakharyaschev, 1998] we introduced the notion of a quasimodel for proving the decidability of other combinations of modal and description logics. In Sections 5 and 6 we use the idea of the mosaic technique of [Reynolds, 1996] to establish the decidability of the satisfiability problem for various temporal description logics based on $\mathcal{CIQ}$, and Section 7 extends the obtained results to temporal logics based on $\mathcal{CIO}$ and $\mathcal{CQO}$. The paper closes with a discussion of open problems.

2 BASIC DESCRIPTION LOGIC

The underlying concept description logic we deal with in the first part of the paper was introduced by De Giacomo and Lenzerini [1996] and De Giacomo [1995] under the name $\mathcal{CIQ}$.

Definition 1 (Language). The *language* of $\mathcal{CIQ}$ is based on a list of *concept names* $C_0, C_1, \ldots$, a list of *role names* $R_0, R_1, \ldots$, and a list of *object names* $a_0, a_1, \ldots$. Starting from these we can form compound roles, concepts, and formulas using the following constructs. First, by a *basic role* we mean any role name R_i as well as its "inversion" R_i^-. Now, if R, S are roles, B is a basic role, C, D are concepts (for the basis of our inductive definition we assume basic roles to be roles and concept names to be concepts) and $n < \omega$, then $R \vee S$, $R \circ S$, R^*, R^-, $C?$ are *roles* and $\top$, $C \wedge D$, $\neg C$, $\exists R.C$, $\exists_{\geq n} B.C$ are *concepts*. *Atomic formulas* are expressions of the form $\top$, $C = D$, $a : C$, aRb, where C and D are concepts, R is a role name and a, b are object names. If φ and ψ are formulas then so are $\varphi \wedge \psi$ and $\neg\varphi$. (The booleans $\rightarrow$, $\vee$, $\bot$ and the universal quantifier $\forall$ are defined as abbreviations in the usual (classical) way.)

The intended meaning of the introduced constructs will be clear from Definition 3 below.

Definition 2 (Model). A $\mathcal{CIQ}$-*model* is a structure of the form

$$I = \langle \Delta, R_0^I, \ldots, C_0^I, \ldots, a_0^I, \ldots \rangle,$$

where Δ is a non-empty set, the *domain* of the model, R_i^I $(i = 0, \ldots)$ are binary relations on Δ (interpreting the role names), C_i^I subsets of Δ (interpreting the concept names), and a_i^I are elements of Δ (interpreting the object names).

Definition 3 (Satisfaction). For a $\mathcal{CIQ}$-model I, the *value* C^I of a concept C, the *value* R^I of a role R, and the *truth-relation* $\models$ are defined inductively in the following way: $\top^I = \Delta$ and $C^I = C_i^I$, for $C = C_i$; $(C \wedge D)^I = C^I \cap D^I$; $(\neg C)^I = \Delta - C^I$; $x \in (\exists R.C)^I$ iff $\exists y \in C^I\; xR^I y$; $x \in (\exists_{\geq n} R.C)^I$ iff $\left|\{y \in C^I : xR^I y\}\right| \geq n$; $(R \vee S)^I = R^I \cup S^I$; $(R \circ S)^I = R^I \circ S^I$ (the composition of R^I and S^I); $(R^*)^I = (R^I)^*$ (the transitive and reflexive closure of R^I); $(R^-)^I = (R^I)^{-1}$ (the inversion of R^I); $(C?)^I = \{\langle x, x\rangle : x \in C^I\}$; $I \models \top$; $I \models C = D$ iff $C^I = D^I$; $I \models a : C$ iff $a^I \in C^I$; $I \models aRb$ iff $a^I R^I b^I$; $I \models \varphi \wedge \psi$ iff $I \models \varphi$ and $I \models \psi$; $I \models \neg\varphi$ iff $I \not\models \varphi$. (Here and below $|X|$ is the cardinality of X.) A formula φ is called *satisfiable* if there is a $\mathcal{CIQ}$-model I such that $I \models \varphi$.

As was shown by De Giacomo and Lenzerini [1996], the satisfiability problem for $\mathcal{CIQ}$ is decidable;[2] however, it becomes undecidable for the extended language in which one can construct concepts of the form $\exists_{\geq n} R.C$ for all (not only basic) roles R; see [De Giacomo and Lenzerini, 1996], where the reader can find also some examples illustrating the expressive power of $\mathcal{CIQ}$. Another important fact observed by De Giacomo and Lenzerini [1996] is that $\mathcal{CIQ}$ does not have the finite model property: there exists a formula satisfiable in an infinite model but not in finite ones.

3 TEMPORAL DESCRIPTION LOGIC

We now add to the static language $\mathcal{CIQ}$ a temporal dimension.

Definition 4 (Language). Let $\mathcal{CIQ}_{\mathcal{US}}$ be the extension of $\mathcal{CIQ}$ with the binary temporal operators $\mathcal{U}$ (Until) and $\mathcal{S}$ (Since) which may be applied

[2]The satisfiability problem for $\mathcal{CIQ}$-formulas can easily be reduced to the satisfiability problem for $\mathcal{CIQ}$ knowledge bases (consisting of inclusion assertions $C \subseteq D$ and instance assertions $a : C$, aRb) which was shown to be decidable in the cited paper; for instance, in the definition of the reduced form in [De Giacomo and Lenzerini, 1996], $\neg a_i R a_j$ will give rise to the inclusion assertion $A_i \subseteq \forall R.\neg A_j$.

to concepts and formulas, i.e., if C, D are concepts and φ, ψ formulas then $C\mathcal{U}D$, $C\mathcal{S}D$ are concepts and $\varphi\mathcal{U}\psi$, $\varphi\mathcal{S}\psi$ formulas. $\mathcal{CIQ}_{\mathcal{U}}$ is the extension of $\mathcal{CIQ}$ with only $\mathcal{U}$. And by $\mathcal{CIQ}_{\diamond}$ we denote the extension of $\mathcal{CIQ}$ with the operators $\diamond^{+}$ (sometime in the future) and $\diamond^{-}$ (sometime in the past) defined by $\diamond^{+}E = \top\mathcal{U}E$, $\diamond^{-}E = \top\mathcal{S}E$, where E is either a concept or a formula. As usual, $\Box^{-}$ and $\Box^{+}$ (always in the past and always in the future) are the duals of $\diamond^{-}$ and $\diamond^{+}$.

Below we define models and other semantic notions only for the full language $\mathcal{CIQ}_{\mathcal{US}}$; they are easily relativized to its fragments $\mathcal{CIQ}_{\mathcal{U}}$ and $\mathcal{CIQ}_{\diamond}$.

Definition 5 (Model). A *$\mathcal{CIQ}_{\mathcal{US}}$-model* with *domain* Δ is a pair of the form $\mathfrak{M} = \langle\langle W, <\rangle, I\rangle$ in which $\langle W, <\rangle$ is a strict linear order[3] and I a function associating with each $w \in W$ a $\mathcal{CIQ}$-model

$$I(w) = \left\langle \Delta, R_0^{I(w)}, \ldots, C_0^{I(w)}, \ldots, a_0^{I(w)}, \ldots \right\rangle$$

such that $a_i^{I(u)} = a_i^{I(v)}$ for all $u, v \in W$. Without loss of generality we may identify the objects $a_i^{I(w)}$ with the object names a_i.

It is worth emphasizing that all our models satisfy the *constant domain assumption*; as was shown in [Wolter and Zakharyaschev, 1998], the cases of expanding and varying domains are reducible to that of constant domains, at least as far as the decidability of the satisfiability problem is concerned.

Definition 6 (Satisfaction). Given a $\mathcal{CIQ}_{\mathcal{US}}$-model $\mathfrak{M} = \langle\langle W, <\rangle, I\rangle$ and a "world" w in it, the *values* $C^{I(w)}$ and $R^{I(w)}$ of a concept C and a role R in w, and the *truth-relation* $(\mathfrak{M}, w) \models \varphi$ (or simply $w \models \varphi$, if $\mathfrak{M}$ is understood) are computed inductively according to the rules of Definition 3 and the following clauses:

1. $x \in (C\mathcal{U}D)^{I(w)}$ iff there is $v > w$ such that $x \in D^{I(v)}$ and $x \in C^{I(u)}$ for every u in the interval $(w, v) = \{u \in W : w < u < v\}$;

2. $x \in (C\mathcal{S}D)^{I(w)}$ iff there is $v < w$ such that $x \in D^{I(v)}$ and $x \in C^{I(u)}$ for every $u \in (v, w)$;

3. $w \models \psi\mathcal{U}\chi$ iff there is $v > w$ such that $v \models \chi$ and $u \models \psi$ for every $u \in (w, v)$;

4. $w \models \psi\mathcal{S}\chi$ iff there is $v < w$ such that $v \models \chi$ and $u \models \psi$ for every $u \in (v, w)$.

A formula φ is *satisfiable* in the frame $\langle W, <\rangle$ if there are a $\mathcal{CIQ}_{\mathcal{US}}$-model based on $\langle W, <\rangle$ and a world w in it such that $w \models \varphi$.

[3] I.e., $<$ is an irreflexive transitive relation on W such that $u < v$ or $v < u$ for all $u \neq v$.

Example 7. The following formulas describing some properties of a mail system will help the reader to grasp the expressive capabilities of $\mathcal{CIQ}_{\mathcal{US}}$:

$$\Box(\mathsf{Received} = \mathsf{Mail} \wedge \Diamond^-_\bullet \exists\mathsf{is\_in}.\mathsf{Mailbox}),$$
$$\Box(\mathsf{Replied} = \mathsf{Received} \wedge \Diamond^-_\bullet \exists\mathsf{reply}.\mathsf{Mail}),$$
$$\Box(\mathsf{Saved} = \mathsf{Received} \wedge \exists\mathsf{copy}.\mathsf{File}),$$
$$\Box(\mathsf{Deleted} = \neg\exists\mathsf{is\_in}.\mathsf{Mailbox} \wedge (\neg\exists\mathsf{is\_in}.\mathsf{Mailbox})\mathcal{S}\mathsf{Received}),$$
$$\Box(\mathsf{Received} \wedge \neg\mathsf{Deleted} \rightarrow (\neg\mathsf{Deleted}\,\mathcal{U}\mathsf{Saved}) \vee \Box^+\neg\mathsf{Deleted} = \top),$$
$$\Box(\mathsf{Received} \rightarrow \Diamond^+\mathsf{Replied} = \top).$$

Here $\Diamond^-_\bullet C = C \vee \Diamond^- C$ and $\Box C = C \wedge \Box^- C \wedge \Box^+ C$. It is worth noting that the language $\mathcal{CIQ}_{\mathcal{US}}$ is able to express the dynamics of concepts (e.g. $\Box(\mathsf{Received} \rightarrow \Box^+\mathsf{Received} = \top)$), but it provides no direct way to describe the dynamics of roles, say, the development of the file system of [De Giacomo and Lenzerini, 1996] in time.

In this paper, our concern is only the satisfiability problem in various frames. Other standard reasoning tasks, say subsumption or instantiation, are known to be reducible to it. The entailment problems in both local and global formulations can also be reduced to the satisfiability problem:

- (*local consequence*) $\Gamma \models \varphi$, for a finite set of formulas Γ, iff for every model $\mathfrak{M} = \langle\langle W, <\rangle, I\rangle$ in a given class and every $w \in W$, we have $w \models \varphi$ whenever $w \models \Gamma$; it is easily seen that $\Gamma \models \varphi$ iff $\bigwedge \Gamma \wedge \neg\varphi$ is not satisfiable in the class;[4]

- (*global consequence*) $\Gamma \models^* \varphi$ iff, for every $\mathfrak{M}$ in a given class, we have $w \models \varphi$ for all $w \in W$ whenever $w \models \Gamma$ for all $w \in W$; in this case $\Gamma \models^* \varphi$ iff $\Gamma \cup \{\Box^+\psi : \psi \in \Gamma\} \cup \{\Box^-\psi : \psi \in \Gamma\} \models \varphi$.

In the semantics introduced above, the object names are interpreted globally, whereas the role and concept names are interpreted locally (in the AI literature locally interpreted terms are known as fluents). We can easily simulate global concepts with the help of the equation $C = \Box^+ C \wedge \Box^- C$: a concept C satisfying it in each world of a model is a global concept in the sense that $C^{I(w)}$ does not depend on w. On the other hand, global roles cannot be simulated by means of local ones, and this restriction is essential for the satisfiability problem to be decidable. Indeed, let us assume that the roles are interpreted globally, i.e., $R^{I(v)} = R^{I(w)}$ for all $v, w \in W$. Then the resulting temporal description logic would contain as its fragments products of modal logics (see [Gabbay and Shehtman, 1998; Marx and Venema, 1997]) interpreted in structures of the form $\langle W, <\rangle \times \langle \Delta, R\rangle$. As was shown by Spaan [1993] and Marx [1999], the global consequence problem for products of this form is mostly

[4] Note that in general $\models$ is not compact.

undecidable. From their results it follows immediately, for example, that the satisfiability problem for $\mathcal{ALC}_{\mathcal{U}}$ in $\langle \mathbb{N}, < \rangle$ with the global interpretation of roles is undecidable. Thus, to ensure decidability we are forced either to interpret the roles locally or to omit some of the boolean operators and the universal role quantification. The latter way was taken by Artale and Franconi [1994] who considered interval-based temporal description logics. In this paper we deal with only the former choice. Our main aim is to prove the following:

Theorem 8. *There are algorithms that are capable of deciding whether*

1. *a given $\mathcal{CIQ}_{\mathcal{US}}$-formula is satisfiable in $\langle \mathbb{Z}, < \rangle$ and in $\langle \mathbb{N}, < \rangle$ ($\mathbb{Z}$ and $\mathbb{N}$ are the sets of all integer and natural numbers, respectively)*

2. *a given $\mathcal{CIQ}_{\Diamond}$-formula is satisfiable in some (strictly linearly ordered) frame as well as in $\langle \mathbb{Q}, < \rangle$ ($\mathbb{Q}$ is the set of all rational numbers).*

As in [Wolter and Zakharyaschev, 1998], our first step is to represent $\mathcal{CIQ}_{\mathcal{US}}$-models in the form of quasimodels, sequences of certain finite structures called quasiworlds.

4 QUASIMODELS

Fix a $\mathcal{CIQ}_{\mathcal{US}}$-formula φ. Let $ob\varphi$ be the set of all object names in φ. And by $con\varphi$ and $sub\varphi$ we denote the closure under negation of, respectively, the set of all concepts in φ and the set of all subformulas in φ. Without loss of generality we may identify E and $\neg\neg E$, for every concept or formula E; so both $con\varphi$ and $sub\varphi$ are finite.

Definition 9 (Types). A *concept type* t for φ is a subset of $con\varphi$ such that

- $C \wedge D \in t$ iff $C, D \in t$, for every $C \wedge D \in con\varphi$;
- $\neg C \in t$ iff $C \notin t$, for every $C \in con\varphi$.

By a *named concept type* for φ we mean a pair $\langle a, t \rangle$ in which $a \in ob\varphi$ and t is a concept type for φ. We will denote $\langle a, t \rangle$ by t_a and write $C \in t_a$ instead of $C \in t$, for t in $\langle a, t \rangle$. A *formula type* Φ for φ is a subset of $sub\varphi$ such that

- $\psi \wedge \chi \in \Phi$ iff $\psi, \chi \in \Phi$, for every $\psi \wedge \chi \in sub\varphi$;
- $\neg\psi \in \Phi$ iff $\psi \notin \Phi$, for every $\psi \in sub\varphi$.

Definition 10 (Quasiworld candidate). Let T be a set of concept types for φ, T^o a set containing one named concept type t_a for every $a \in ob\varphi$, and let Φ be a formula type for φ. The triple $\langle T, T^o, \Phi \rangle$ is called a *quasiworld candidate* for φ if the following holds:

- $t \in T$ for every $\langle a, t \rangle \in T^o$;

- $(a : C) \in \Phi$ iff $C \in t_a$, for every $(a : C) \in sub\varphi$ and every $t_a \in T^o$;
- $(C = D) \in \Phi$ iff each $t \in T$ contains or does not contain simultaneously both C and D, for every $(C = D) \in sub\varphi$.

For every quasiworld candidate $\langle T, T^o, \Phi\rangle$ for φ we clearly have

$$|T| \leq 2^{|con\varphi|}, \ |T^o| = |ob\varphi|, \ |\Phi| \leq |sub\varphi|.$$

Also, it is not hard to see that, given a triple $\langle T, T^o, \Phi\rangle$ as described in the first sentence of Definition 10, one can effectively check whether it is a quasiworld candidate for φ or not.

Definition 11 (Extended $\mathcal{CIQ}$-model). By an *extended $\mathcal{CIQ}$-model* for φ we mean a $\mathcal{CIQ}$-model

$$(1) \qquad I = \langle \Delta, R_0^I, \dots, C_0^I, \dots, (C\mathcal{U}D)^I, \dots, (C'\mathcal{S}D')^I, \dots, a_0^I, \dots \rangle$$

in which all concepts of the form $C\mathcal{U}D$ and $C'\mathcal{S}D'$ occurring in φ are regarded as concept names. For every $x \in \Delta$ we put

$$t^I(x) = \{C \in con\varphi : x \in C^I\}, \ \ [x]^I = \{y \in \Delta : t^I(x) = t^I(y)\}.$$

Clearly, $t^I(x)$ is a concept type.

Definition 12 (Quasiworld). Say that an extended $\mathcal{CIQ}$-model I for φ of the form (1) *realizes* a quasiworld candidate $\mathfrak{w} = \langle T, T^o, \Phi\rangle$ for φ if the following conditions hold: $T = \{t^I(x) : x \in \Delta\}$; $t_a = \langle a, t^I(a)\rangle \in T^o$ for every $a \in ob\varphi$; $a^I R^I b^I$ iff $aRb \in \Phi$, for every $aRb \in sub\varphi$.

A realizable quasiworld candidate $\mathfrak{w}$ for φ is called a *quasiworld* for φ. Instead of $\psi \in \Phi$ we will often write $\mathfrak{w} \models \psi$ and say that ψ is *true* in $\mathfrak{w}$.

Lemma 13. *Given a quasiworld candidate for φ, one can effectively recognize whether it is quasiworld for φ.*

Proof. It is easy to see that a quasiworld candidate $\langle T, T^o, \Phi\rangle$ for φ is realizable iff the conjunction of the formulas $\bigvee\{\bigwedge t : t \in T\} = \top$, $\bigwedge t \neq \bot$ for $t \in T$, $a : \bigwedge t_a$ for $t_a \in T^o$, aRb for $aRb \in \Phi$, and $\neg(aRb)$ for $\neg(aRb) \in \Phi$ ($\bigwedge t$ is the conjunction of all concepts in t) is satisfied in an extended $\mathcal{CIQ}$-model for φ. It remains to recall that the satisfiability problem for $\mathcal{CIQ}$ is decidable. □

Observe that the number of distinct quasiworlds for φ does not exceed

$$\sharp(\varphi) = 2^{2^{|con\varphi|}} \cdot |ob\varphi| \cdot 2^{|con\varphi|} \cdot 2^{|sub\varphi|}.$$

Fix a strict linear order $\mathfrak{F} = \langle W, <\rangle$ and consider a sequence

$$(2) \qquad Q = \langle \mathfrak{w}_w : w \in W\rangle$$

of quasiworlds $\mathfrak{w}_w = \langle T_w, T_w^o, \Phi_w\rangle$ for φ. We will call it an *$\mathfrak{F}$-sequence* for φ. Concept types in T_w will be denoted by t_w, named concept types in T_w^o by t_a^w, $a \in ob\varphi$. $Q(w)$ is another name for $\mathfrak{w}_w$.

Definition 14 (Run). A *run* in Q is a sequence $r = \langle r(w) : w \in W \rangle$ such that

(a) $r(w) \in T_w$ for every $w \in W$;

(b) for every concept $C\mathcal{U}D \in con\varphi$ and every $w \in W$, $C\mathcal{U}D \in r(w)$ iff there exists $u > w$ such that $D \in r(u)$ and $C \in r(v)$ for all $v \in (w, u)$;

(c) for every concept $C\mathcal{S}D \in con\varphi$ and every $w \in W$, $C\mathcal{S}D \in r(w)$ iff there exists $u < w$ such that $D \in r(u)$ and $C \in r(v)$ for all $v \in (u,,w)$.

Definition 15 (Quasimodel). An $\mathfrak{F}$-sequence Q for φ of the form (2) is called a *quasimodel for φ based on $\mathfrak{F}$* if the following conditions hold:

(d) for every $a \in ob\varphi$, the sequence $r_a = \langle t_a^w : w \in W \rangle$ is a run in Q;

(e) for every $w \in W$ and every $t \in T_w$, there is a run r in Q such that $r(w) = t$;

(f) for every $w \in W$ and every $\psi\mathcal{U}\chi \in sub\varphi$, we have $Q(w) \models \psi\mathcal{U}\chi$ iff there exists $u > w$ such that $Q(u) \models \chi$ and $Q(v) \models \psi$ for all $v \in (w, u)$;

(g) for every $w \in W$ and every $\psi\mathcal{S}\chi \in sub\varphi$, we have $Q(w) \models \psi\mathcal{S}\chi$ iff there exists $u < w$ such that $Q(u) \models \chi$ and $Q(v) \models \psi$ for all $v \in (u, w)$.

A formula $\psi \in sub\varphi$ is said to be *satisfied* in Q if $Q(w) \models \psi$ for some $w \in W$.

Theorem 16. *A formula φ is satisfiable in a $\mathcal{CIQ}_{\mathcal{US}}$-model based on $\langle W, < \rangle$ iff it is satisfiable in a quasimodel for φ based on $\langle W, < \rangle$.*

Proof. ($\Rightarrow$) Suppose that φ is satisfied in a $\mathcal{CIQ}_{\mathcal{US}}$-model $\langle \langle W, < \rangle, I \rangle$ with domain Δ. For every $w \in W$, we define $\mathfrak{w}_w = \langle T_w, T_w^o, \Phi_w \rangle$ by taking

$$T_w = \{t^{I(w)}(x) : x \in \Delta\},\ T_w^o = \{t_a^w = \left\langle a, t^{I(w)}(a) \right\rangle : a \in ob\varphi\},$$
$$\Phi_w = \{\psi \in sub\varphi : w \models \psi\}.$$

It is not hard to see that $\mathfrak{w}_w$ is a quasiworld for φ (realized in $I(w)$ extended by the concepts $C\mathcal{U}D$ and $C'\mathcal{S}D'$ in φ) and $Q = \langle \mathfrak{w}_w : w \in W \rangle$ is a quasimodel on $\langle W, < \rangle$ satisfying φ (the sequence $\left\langle t^{I(w)}(x) : w \in W \right\rangle$ is a run through $t^{I(u)}(x)$, for every $u \in W$ and every $x \in \Delta$).

($\Leftarrow$) To show the converse we require the following lemma.

Lemma 17. *There is a cardinal $\kappa \geq \aleph_0$ such that, for any cardinal $\kappa' \geq \kappa$, every quasiworld $\mathfrak{w}$ for φ is realized in an extended $\mathcal{CIQ}$-model J in which $\left|[x]^J\right| = \kappa'$ for all x in the domain of J.*

Proof. For each quasiworld $\mathfrak{u}$ for φ fix an extended $\mathcal{CIQ}$-model $I_\mathfrak{u}$ realizing $\mathfrak{u}$. Let $\Delta_\mathfrak{u}$ be the domain of $I_\mathfrak{u}$. Then we define κ to be the supremum of $\aleph_0$ and $\left|[x]^{I_\mathfrak{u}}\right|$, for all quasiworlds $\mathfrak{u}$ for φ and all $x \in \Delta_\mathfrak{u}$, and show that it satisfies the required conditions.

Suppose $\mathfrak{w}$ is a quasiworld for φ and $\kappa' \geq \kappa$. Take an extended $\mathcal{CIQ}$-model

$$I = \left\langle \Delta, R_0^I, \ldots, C_0^I, \ldots, (C\mathcal{U}D)^I, \ldots, (C'\mathcal{S}D')^I, \ldots, a_0^I, \ldots \right\rangle$$

realizing $\mathfrak{w}$ and such that $\left|[x]^I\right| \leq \kappa$ for every $x \in \Delta$. Now we define

$$J = \left\langle \Delta', R_0^J, \ldots, C_0^J, \ldots, (C\mathcal{U}D)^J, \ldots, (C'\mathcal{S}D')^J, \ldots, a_0^J, \ldots \right\rangle$$

to be the disjoint union of κ' copies of I; more precisely, we put

$$\Delta' = \{\langle x, \xi \rangle : x \in \Delta, \xi < \kappa'\}, \ R_i^J = \{\langle \langle x, \xi \rangle, \langle y, \xi \rangle \rangle : \langle x, y \rangle \in R_i^I, \xi < \kappa'\},$$

$$C_i^J = \{\langle x, \xi \rangle : x \in C_i^I, \xi < \kappa'\}, \ a_i^J = \left\langle a_i^I, 0 \right\rangle.$$

Clearly, $\left|[x]^J\right| = \kappa'$ for every $x \in \Delta'$, and one can readily check by induction that J realizes $\mathfrak{w}$. □

Let us now return to the proof of our theorem. Suppose φ is satisfied in a quasimodel $Q = \langle \mathfrak{w}_w : w \in W \rangle$ with $\mathfrak{w}_w = \langle T_w, T_w^o, \Phi_w \rangle$. Assume also that κ' is a cardinal exceeding the cardinality of the set Ω of all runs in Q and the cardinal κ supplied by Lemma 17 as well. Let

$$\Delta = \{\langle r, \xi \rangle : r \in \Omega, \xi < \kappa'\}.$$

Notice that $|\{\langle r, \xi \rangle \in \Omega : r(w) = t\}| = \kappa'$, for every $w \in W$ and every $t \in T_w$. By Lemma 17, for every $w \in W$ there exists an extended $\mathcal{CIQ}$-model

$$I(w) = \langle \Delta, R_0^{I(w)}, \ldots, C_0^{I(w)}, \ldots, (C\mathcal{U}D)^{I(w)}, \ldots, (C'\mathcal{S}D')^{I(w)}, \ldots, a_0^{I(w)}, \ldots \rangle$$

such that $a^{I(w)} = \langle r_a, 0 \rangle$, for each $a \in ob\varphi$, and $t^{I(w)}(\langle r, \xi \rangle) = r(w)$, for every $r \in \Omega$ and every $\xi < \kappa'$. For $w \in W$ let

$$J(w) = \left\langle \Delta, R_0^{I(w)}, \ldots, C_0^{I(w)}, \ldots, a_0^{I(w)}, \ldots \right\rangle.$$

Consider the $\mathcal{CIQ}_{\mathcal{US}}$-model $\mathfrak{M} = \langle \langle W, < \rangle, J \rangle$ and show by induction on the construction of $\psi \in sub\varphi$ that

(3) $$\mathfrak{w}_w \models \psi \text{ iff } (\mathfrak{M}, w) \models \psi.$$

Observe first that for every $C \in con\varphi$, we have $C^{I(w)} = C^{J(w)}$. This is also proved by induction the only non-trivial step in which is to show

$$(C\mathcal{U}D)^{I(w)} = (C\mathcal{U}D)^{J(w)}, \ (C\mathcal{S}D)^{I(w)} = (C\mathcal{S}D)^{J(w)}$$

assuming that $C^{I(u)} = C^{J(u)}$ and $D^{I(u)} = D^{J(u)}$ for all $u \in W$.

Suppose $\langle r, \xi \rangle \in \Delta$. By the definition of $I(w)$, $\langle r, \xi \rangle \in (C\mathcal{U}D)^{I(w)}$ iff $r(w) \in C\mathcal{U}D$. By (b) of Definition 14, this means that there is $u > w$ such that $D \in r(u)$ and $C \in r(v)$ for all $v \in (w, u)$, which is equivalent to $\langle r, \xi \rangle \in D^{I(u)}$ and $\langle r, \xi \rangle \in C^{I(v)}$ for $v \in (w, u)$, and so, by IH, $\langle r, \xi \rangle \in (C\mathcal{U}D)^{J(w)}$. The concept $C\mathcal{S}D$ is treated analogously.

By the definition of a quasiworld, it follows that (3) holds for atomic ψ. The induction step for $\psi = \chi_1 \wedge \chi_2$ and $\psi = \neg\chi_1$ is trivial, and the cases $\psi = \chi_1 \mathcal{U} \chi_2$, $\psi = \chi_1 \mathcal{S} \chi_2$ follow from (f) and (g) in Definition 15. Thus $\mathfrak{M}$ satisfies φ. □

5 SATISFIABILITY PROBLEM FOR $\mathcal{CIQ}_{\mathcal{U}}$ AND $\mathcal{CIQ}_{\mathcal{US}}$

In this section we prove the first claim of Theorem 8. To make the idea of the proof more transparent, we develop a satisfiability checking algorithm for $\mathcal{CIQ}_{\mathcal{U}}$-formulas in the frame $\langle \mathbb{N}, < \rangle$. Fix a $\mathcal{CIQ}_{\mathcal{U}}$-formula φ. Given a sequence $s = s(0), s(1), \ldots$ and $i \geq 0$, we denote by $s^{\leq i}$ and $s^{>i}$ the head $s(0), \ldots, s(i)$ and the tail $s(i+1), s(i+2), \ldots$ of s, respectively; $s_1 * s_2$ is the concatenation of s_1 and s_2; $|s|$ denotes the length of s and $s^* = s * s * s * \ldots$

Lemma 18. *Let $Q = Q(0), Q(1), \ldots$ be a quasimodel for φ and $Q(n) = Q(m)$ for some $n < m$. Then $Q_{nm} = Q^{\leq n} * Q^{>m}$ is a quasimodel for φ too.*

Proof. It suffices to observe that if r_1 and r_2 are runs in Q with $r_1(n) = r_2(m)$ then $r_1^{\leq n} * r_2^{>m}$ is a run in Q_{nm}. □

Lemma 19. *Every quasimodel Q for φ contains a subquasimodel $Q' = Q_1 * Q_2$ such that $|Q_1| \leq \sharp(\varphi)$ and each quasiworld in Q_2 occurs in this sequence infinitely often.*

Proof. Let n be the maximal number such that $Q(n) \neq Q(m)$ for all $m > n$. If $n = 0$ then we take $Q' = Q = Q_2$ (Q_1 is empty). Otherwise we apply Lemma 18 to $Q = Q^{\leq n} * Q^{>n}$ deleting from its head $Q^{\leq n}$ all repeating quasiworlds, which yields the subquasimodel $Q' = Q_1 * Q^{>n}$ we need. □

Definition 20. Suppose that $Q = \langle \mathfrak{w}_i : i \in \mathbb{N} \rangle$ is a sequence of quasiworlds $\mathfrak{w}_i = \langle T_i, T_i^o, \Phi_i \rangle$ for φ and r is a sequence of elements from T_i, $i \in \mathbb{N}$, such that $r(i) \in T_i$. Say that r *realizes* a concept $C\mathcal{U}D \in r(n)$ *in m steps* if there is $l \leq m$ such that $D \in r(n + l)$ and $C \in r(n + k)$ for all $k \in (0, l)$. A formula $\psi \mathcal{U} \chi \in \Phi_n$ is *realized in m steps* if there is $l \leq m$ such that $\chi \in \Phi_{n+l}$ and $\psi \in \Phi_{n+k}$ for all $k \in (0, l)$.

Lemma 21. *Let $Q = Q_1 * Q_2$ be a quasimodel for φ (with quasiworlds of the form $\langle T_i, T_i^o, \Phi_i \rangle$ for $i \in \mathbb{N}$) satisfying the requirements of Lemma 19, let*

$n = |Q_1| + 1$ *and* $\flat(\varphi) = 2^{|con\varphi|} + |ob\varphi|$. *Then* Q *contains a subquasimodel of the form* $Q_1 * Q_0 * Q_2^{>l}$, *for some* $l \geq 0$, *such that*

(i) $|Q_0| \leq \flat^2(\varphi) \cdot |con\varphi| \cdot \sharp(\varphi) + |sub\varphi| \cdot \sharp(\varphi) + \sharp(\varphi)$;

(ii) *for every* $t \in T_n$ *there is a run* r *through* t *realizing all concepts of the form* $C\mathcal{U}D \in r(n)$ *in* $|Q_0|$ *steps (for* $t_a \in T_n^o$ *one can take* $r = r_a$*);*

(iii) *every formula* $\psi\mathcal{U}\chi \in \Phi_n$ *is realized in* $|Q_0|$ *steps;*

(iv) $Q_0(1) = Q_2^{>l}(1)$.

Proof. Suppose $t \in T_n$, $C\mathcal{U}D \in t$ and r is a run in Q through t, i.e., $r(n) = t$. Then there exists $m > 0$ such that $D \in r(n+m)$ and $C \in r(n+k)$ for all $k \in (0, m)$. Assume now that $0 < i < j < m$, $r(n+i) = r(n+j)$ and $Q(n+i) = Q(n+j)$. In view of Lemma 18, $Q_1 * Q_2^{\leq i} * Q_2^{>j}$ is a subquasimodel of Q and $r^{\leq n+i} * r^{>n+j}$ is a run through t. It follows that we can construct a subquasimodel $Q_1 * Q_2^{\leq 1} * Q_3$ of Q and a run r_1 in it which comes through t and realizes $C\mathcal{U}D$ in $m_1 \leq \flat(\varphi) \cdot \sharp(\varphi)$ steps. Then we consider another concept $C'\mathcal{U}D' \in t$ and assume that it is realized in $m_2 > m_1$ steps in r_1. Using Lemma 18 once again (and deleting repeating quasiworlds in the interval $Q_3(m_1), \ldots, Q_3(m_2)$) we select a subquasimodel $Q_1 * Q_2^{\leq 1} * Q_3^{\leq m_1} * Q_4$ of Q and a run r_2 through t which realizes both $C\mathcal{U}D$ and $C'\mathcal{U}D'$ in $2 \cdot \flat(\varphi) \cdot \sharp(\varphi)$ steps.

Having analyzed all distinct concepts of the form $C\mathcal{U}D \in t$ we obtain a subquasimodel $Q_1 * Q_2^{\leq 1} * Q'$ of Q and a run r' through t which realizes all those concepts in $m' \leq |con\varphi| \cdot \flat(\varphi) \cdot \sharp(\varphi)$ steps. After that we consider in the same manner another type $t' \in T_n$. However, this time we can delete quasiworlds only after $Q'(m')$, and so to realize in some run through t' the concepts $C\mathcal{U}D \in t'$ we need $\leq 2 \cdot |con\varphi| \cdot \flat(\varphi) \cdot \sharp(\varphi)$ steps. And so on. Since $|T_n| + |T_n^o| \leq \flat(\varphi)$, to satisfy (ii) at most $|con\varphi| \cdot \flat^2(\varphi) \cdot \sharp(\varphi)$ quasiworlds are required.

The formulas $\psi\mathcal{U}\chi \in sub\varphi$ that are true in $Q_2(1)$ are treated analogously. This may give us $\leq |sub\varphi| \cdot \sharp(\varphi)$ more quasiworlds. And $\leq \sharp(\varphi)$ quasiworlds may be required to comply with (iv). □

Definition 22 (Suitable pair). A pair t, t' of concept types for φ is called *suitable* if for every $C\mathcal{U}D \in con\varphi$, $C\mathcal{U}D \in t$ iff either $D \in t'$ or $C \in t'$ and $C\mathcal{U}D \in t'$.

Lemma 23. *Suppose* Q_1 *and* Q_2 *are finite sequences of quasiworlds for* φ *of length* l_1 *and* l_2*, respectively, and let* $Q = Q_1 * Q_2^*$ *with* $Q(n) = \langle T_n, T_n^o, \Phi_n \rangle$. *Then* Q *is a quasimodel for* φ *whenever the following conditions hold:*

1. *for every* $i \leq l_1 + l_2$ *and every* $t' \in T_{i+1}$*, there is* $t \in T_i$ *such that the pair* t, t' *is suitable;*

2. *for every* $i \leq l_1 + 1$ *and every* $t_i \in T_i$*, all concepts of the form* $C\mathcal{U}D \in t_i$ *are realized in* $l_1 + l_2 - i$ *steps in some sequence* $t_i, t_{i+1}, \ldots, t_{l_1+l_2}$ *in*

which $t_{i+j} \in T_{i+j}$ and every pair of adjacent elements is suitable (for $t_a^i \in T_i^o$ one can take the sequence $t_a^i, t_a^{i+1}, \ldots, t_a^{l_1+l_2}$, where $t_a^j \in T_j^o$);

3. *for every $i \leq l_1 + l_2$, and every $\psi \mathcal{U} \chi \in sub\varphi$, $Q(i) \models \psi \mathcal{U} \chi$ iff either $Q(i+1) \models \chi$ or $Q(i+1) \models \psi$ and $Q(i+1) \models \psi \mathcal{U} \chi$;*

4. *for every $i \leq l_1 + 1$, all formulas of the form $\psi \mathcal{U} \chi \in \Phi_i$ are realized in $l_1 + l_2 - i$ steps.*

Proof. To construct a run through $t_m \in T_m$, we first take types $t_i \in T_i$, for $i < m$, such that every pair of adjacent elements in the sequence $t_1, \ldots, t_m$ is suitable—this can indeed be done by item 1. Then using condition 2 we select a sequence $t_m, \ldots, t_{m+n}$, for some $n \leq l_1 + l_2$, such that every pair of adjacent elements in it is suitable and all concepts of the form $C \mathcal{U} D \in t_m$ are realized in it in n steps. After that we select such a sequence starting from t_{m+n} and so on. It is readily seen that the resulting sequence is a run in Q. This establishes (e) and (d). And condition (f) follows from 3 and 4. □

As a consequence of the two preceding lemmas we immediately obtain:

Theorem 24. *A $\mathcal{CIQ}_{\mathcal{U}}$-formula φ is satisfiable in $\langle \mathbb{N}, < \rangle$ iff there are two sequences Q_1 and Q_2 of quasiworlds for φ such that $Q = Q_1 * Q_2^*$ satisfies conditions 1–4 of Lemma 23, all quasiworlds in Q_1 are distinct (and so $|Q_1| \leq \sharp(\varphi)$), $Q(1) \models \varphi$ and $|Q_2| \leq \flat^2(\varphi) \cdot |con\varphi| \cdot \sharp(\varphi) + |sub(\varphi)| \cdot \sharp(\varphi) + \sharp(\varphi)$.*

This theorem provides us with an algorithm which is capable of deciding, given an arbitrary $\mathcal{CIQ}_{\mathcal{U}}$-formula, whether it is satisfiable in $\langle \mathbb{N}, < \rangle$. In a similar manner one can construct a satisfiability checking algorithm for $\mathcal{CIQ}_{\mathcal{US}}$-formulas in the frame $\langle \mathbb{Z}, < \rangle$.

6 SATISFIABILITY PROBLEM FOR $\mathcal{CIQ}_{\diamond}$

The aim of this section is to prove the second claim of Theorem 8. Now our frames are arbitrary strict linear orders. For $\mathcal{CIQ}_{\diamond}$ Definition 6 becomes somewhat simpler: its items 1–4 are reduced to

1. $x \in (\diamond^+ C)^{I(w)}$ iff there is $v > w$ such that $x \in C^{I(v)}$;
2. $x \in (\diamond^- C)^{I(w)}$ iff there is $v < w$ such that $x \in C^{I(v)}$;
3. $w \models \diamond^+ \psi$ iff there is $v > w$ such that $v \models \psi$;
4. $w \models \diamond^- \psi$ iff there is $v < w$ such that $v \models \psi$.

Fix an arbitrary $\mathcal{CIQ}_{\diamond}$-formula φ.

Definition 25 (Suitable triple). Let $\mathfrak{u} = \langle T_{\mathfrak{u}}, T_{\mathfrak{u}}^o, \Phi_{\mathfrak{u}} \rangle$ and $\mathfrak{v} = \langle T_{\mathfrak{v}}, T_{\mathfrak{v}}^o, \Phi_{\mathfrak{v}} \rangle$ be quasiworlds for φ and $\sigma \subseteq T_{\mathfrak{u}} \times T_{\mathfrak{v}}$. The triple $\langle \mathfrak{u}, \mathfrak{v}, \sigma \rangle$ is called *suitable* if it satisfies the conditions:

- $\forall t \in T_{\mathfrak{u}} \exists t' \in T_{\mathfrak{v}}\ t\sigma t'$, $\forall t' \in T_{\mathfrak{v}} \exists t \in T_{\mathfrak{u}}\ t\sigma t'$, $\forall a \in ob\varphi\ t_a^{\mathfrak{u}} \sigma t_a^{\mathfrak{v}}$;
- $\forall \Diamond^+ C \in con\varphi \forall t \in T_{\mathfrak{u}} \forall t' \in T_{\mathfrak{v}}\ (t\sigma t' \ \&\ \Diamond^+ C \notin t \Rightarrow C \notin t' \ \&\ \Diamond^+ C \notin t')$;
- $\forall \Diamond^- C \in con\varphi \forall t \in T_{\mathfrak{u}} \forall t' \in T_{\mathfrak{v}}\ (t\sigma t' \ \&\ \Diamond^- C \notin t' \Rightarrow C \notin t \ \&\ \Diamond^- C \notin t)$;
- $\forall \Diamond^+ \psi \in sub\varphi\ (\mathfrak{u} \not\models \Diamond^+ \psi \Rightarrow \mathfrak{v} \not\models \psi \ \&\ \mathfrak{v} \not\models \Diamond^+ \psi)$;
- $\forall \Diamond^- \psi \in sub\varphi\ (\mathfrak{v} \not\models \Diamond^- \psi \Rightarrow \mathfrak{u} \not\models \psi \ \&\ \mathfrak{u} \not\models \Diamond^- \psi)$.

The relation σ is called a *connection* between $\mathfrak{u}$ and $\mathfrak{v}$. Note that the same pair of quasiworlds may have several different connections.

It is easily checked that if $\langle \mathfrak{u}, \mathfrak{v}, \tau \rangle$ and $\langle \mathfrak{v}, \mathfrak{w}, \rho \rangle$ are suitable triples then $\langle \mathfrak{u}, \mathfrak{w}, \tau \circ \rho \rangle$ is a suitable triple as well.

Definition 26 (Satisfying set). Say that a set $\mathcal{S}$ of suitable triples for φ is a *satisfying set* for φ if the following conditions hold:

(S1) there is a triple in $\mathcal{S}$ which contains a quasiworld $\mathfrak{w}$ such that $\mathfrak{w} \models \varphi$;

(S2) if $\langle \mathfrak{u}, \mathfrak{v}, \sigma \rangle \in \mathcal{S}$ and $\mathfrak{v} \models \Diamond^+ \psi$, then there is $\langle \mathfrak{v}, \mathfrak{w}, \tau \rangle \in \mathcal{S}$ such that $\mathfrak{w} \models \psi$;

(S3) if $\langle \mathfrak{u}, \mathfrak{v}, \sigma \rangle \in \mathcal{S}$ and $\mathfrak{u} \models \Diamond^- \psi$, then there is $\langle \mathfrak{w}, \mathfrak{u}, \tau \rangle \in \mathcal{S}$ such that $\mathfrak{w} \models \psi$;

(S4) if $\langle \mathfrak{u}, \mathfrak{v}, \sigma \rangle \in \mathcal{S}$ and $\Diamond^+ C \in t \in T_{\mathfrak{v}}$, then there are $\langle \mathfrak{v}, \mathfrak{w}, \tau \rangle \in \mathcal{S}$ and $t' \in T_{\mathfrak{w}}$ such that $C \in t'$ and $t\tau t'$ (if $t = t_a$, for $a \in ob\varphi$, then one can take $t' = t'_a$);

(S5) if $\langle \mathfrak{u}, \mathfrak{v}, \sigma \rangle \in \mathcal{S}$ and $\Diamond^- C \in t \in T_{\mathfrak{u}}$, then there are $\langle \mathfrak{w}, \mathfrak{u}, \tau \rangle \in \mathcal{S}$ and $t' \in T_{\mathfrak{w}}$ such that $C \in t'$ and $t'\tau t$ (if $t = t_a$, for $a \in ob\varphi$, then one can take $t' = t'_a$);

(S6) if $\langle \mathfrak{u}, \mathfrak{v}, \sigma \rangle \in \mathcal{S}$, $\mathfrak{u} \models \Diamond^+ \psi$, $\mathfrak{v} \not\models \psi$ and $\mathfrak{v} \not\models \Diamond^+ \psi$, then there is a quasiworld $\mathfrak{w} \models \psi$ such that $\langle \mathfrak{u}, \mathfrak{w}, \tau \rangle \in \mathcal{S}$, $\langle \mathfrak{w}, \mathfrak{v}, \rho \rangle \in \mathcal{S}$, for some τ, ρ, and $\tau \circ \rho = \sigma$;

(S7) if $\langle \mathfrak{u}, \mathfrak{v}, \sigma \rangle \in \mathcal{S}$, $\mathfrak{v} \models \Diamond^- \psi$, $\mathfrak{u} \not\models \psi$ and $\mathfrak{u} \not\models \Diamond^- \psi$, then there is a quasiworld $\mathfrak{w} \models \psi$ such that $\langle \mathfrak{u}, \mathfrak{w}, \tau \rangle \in \mathcal{S}$, $\langle \mathfrak{w}, \mathfrak{v}, \rho \rangle \in \mathcal{S}$, for some τ, ρ, and $\tau \circ \rho = \sigma$;

(S8) if $\langle \mathfrak{u}, \mathfrak{v}, \sigma \rangle \in \mathcal{S}$, $\Diamond^+ C \in t \in T_{\mathfrak{u}}$, $t\sigma t'$, $C \notin t'$ and $\Diamond^+ C \notin t'$, then there are $\mathfrak{w}$ and $t'' \in T_{\mathfrak{w}}$ such that $C \in t''$, $\langle \mathfrak{u}, \mathfrak{w}, \tau \rangle \in \mathcal{S}$, $\langle \mathfrak{w}, \mathfrak{v}, \rho \rangle \in \mathcal{S}$, for some τ and ρ, $t\tau t'' \rho t'$, and $\tau \circ \rho = \sigma$ (if $t = t_a$, $t' = t'_a$ then one can take $t'' = t''_a$);

(S9) if $\langle \mathfrak{u}, \mathfrak{v}, \sigma \rangle \in \mathcal{S}$, $\Diamond^- C \in t \in T_{\mathfrak{v}}$, $t'\sigma t$, $C \notin t'$ and $\Diamond^- C \notin t'$, then there are $\mathfrak{w}$ and $t'' \in T_{\mathfrak{w}}$ such that $C \in t''$, $\langle \mathfrak{u}, \mathfrak{w}, \tau \rangle \in \mathcal{S}$, $\langle \mathfrak{w}, \mathfrak{v}, \rho \rangle \in \mathcal{S}$, for some τ and ρ, $t'\tau t'' \rho t$, and $\tau \circ \rho = \sigma$ (if $t = t_a$, $t' = t'_a$ then one can take $t'' = t''_a$).

The crucial step in constructing a satisfiability checking algorithm for $\mathcal{CIQ}_\Diamond$-formulas in strict linear orders is the following:

Theorem 27. *A $\mathcal{CIQ}_\Diamond$-formula φ is satisfiable in a strict linear order with ≥ 2 elements iff there exists a satisfying set for φ.*

Since the number of distinct quasiworlds for a formula φ does not exceed $\sharp(\varphi)$ and every quasiworld contains at most $\flat(\varphi)$ concept types, one can effectively check whether there exists a satisfying set for φ (e.g., simply by looking through all sets of suitable triples for φ). It follows that Theorem 27 is enough to establish the decidability of the satisfiability problem for $\mathcal{CIQ}_\Diamond$-formulas in strict linear orders. (Clearly, the satisfiability in a strict linear order with a single element is decidable.) So we focus on the proof of this theorem. One direction is easy.

Proof. ($\Rightarrow$) Suppose φ is satisfied in a $\mathcal{CIQ}_\Diamond$-model $\mathfrak{M} = \langle \langle W, < \rangle, I \rangle$ with ≥ 2 elements. Define a set $\mathcal{S}$ by putting in it all triples $\langle \mathfrak{u}, \mathfrak{v}, \sigma \rangle$ for which there are worlds $u, v \in W$ such that $u < v$, $\mathfrak{u} = \mathfrak{w}_u$, $\mathfrak{v} = \mathfrak{w}_v$ (see the proof of Theorem 16), and $t\sigma t'$ iff there is x in the domain of $\mathfrak{M}$ such that $t = t^{I(u)}(x)$ and $t' = t^{I(v)}(x)$. It is readily seen that $\mathcal{S}$ is a satisfying set for φ. □

To prove the converse we require a number of definitions. Fix a satisfying set $\mathcal{S}$ for φ. Construct a quasimodel satisfying φ by taking the limit of an inductively defined sequence of weak quasimodels over $\mathcal{S}$.

Definition 28 (Weak quasimodel). By a *weak quasimodel* (for φ) over $\mathcal{S}$ we mean a finite sequence $q = \langle \mathfrak{w}_1, \dots, \mathfrak{w}_n \rangle$ of quasiworlds for φ such that $\langle \mathfrak{w}_i, \mathfrak{w}_{i+1}, \sigma_{ii+1} \rangle \in \mathcal{S}$ for some connection σ_{ii+1} and every $i \in (0, n)$. Instead of $\mathfrak{w}_i$ we also write $q(i) = \langle T_i, T_i^o, \Phi_i \rangle$. A sequence of the form $r = \langle t_1, \dots, t_n \rangle$ such that $t_i \in T_i$ and $t_i \sigma_{ii+1} t_{i+1}$ will be called a *run* in q. As before, the run $\langle t_a^1, \dots, t_a^n \rangle$, for $t_a^i \in T_i^o$, is denoted by r_a.

It should be clear that for every $t \in T_i$, $i \in \{1, \dots, n\}$, there is a run r in q such that $r(i) = t$. It is easy to check also that if $1 \leq i < j \leq n$, then $\langle \mathfrak{w}_i, \mathfrak{w}_j, \sigma_{ij} \rangle$ is a suitable triple, where $\sigma_{ij} = \sigma_{ii+1} \circ \sigma_{i+1i+2} \circ \dots \circ \sigma_{j-1j}$.

Definition 29 (Defect). A *defect* in a weak quasimodel $q = \langle q(1), \dots, q(n) \rangle$ over $\mathcal{S}$ is

- any pair $d = \langle i, \psi \rangle$ such that $1 \leq i \leq n$, $\psi = \Diamond^+ \chi \in sub\varphi$ (or $\psi = \Diamond^- \chi \in sub\varphi$), $q(i) \models \psi$ and $q(j) \not\models \chi$ for every $j \in (i, n+1)$ (respectively, $j \in (0, i)$), and

- any triple $d = \langle i, r, C \rangle$ such that $1 \leq i \leq n$, r is a run in q, $C = \Diamond^+ D \in con\varphi$ (or $C = \Diamond^- D \in con\varphi$), $C \in r(i)$ and $D \notin r(j)$ for every $j \in (i, n+1)$ (respectively, $j \in (0, i)$).

Suppose d is a defect in a weak quasimodel $q = \langle q(1), \ldots, q(n)\rangle$ over $\mathcal{S}$. We construct a new weak quasimodel q^d which "cures" d. In accordance with the definition above, consider two cases.

Case 1: $d = \langle i, \psi\rangle$, for $\psi = \Diamond^+\chi$. Let $j \geq i$ be the maximal number for which $\langle j, \psi\rangle$ is a defect in q. If $j = n$ then, by (S2), there is a quasiworld $\mathfrak{w} \models \chi$ such that $\langle \mathfrak{w}_n, \mathfrak{w}, \sigma\rangle \in \mathcal{S}$, for some connection σ. Put $q^d = \langle q(1), \ldots, q(n), \mathfrak{w}\rangle$. When $j \neq n$, we select, according to (S6), a quasiworld $\mathfrak{w} \models \chi$ such that $\langle q(j), \mathfrak{w}, \tau\rangle \in \mathcal{S}$, $\langle \mathfrak{w}, q(j+1), \rho\rangle \in \mathcal{S}$ and $\tau \circ \rho = \sigma_{jj+1}$. Then we insert $\mathfrak{w}$ right after $q(j)$ in q thus obtaining

$$q^d = \langle q(1), \ldots, q(j), \mathfrak{w}, q(j+1), \ldots, q(n)\rangle .$$

The formula $\psi = \Diamond^-\chi$ is treated in a symmetrical way.

Case 2: $d = \langle i, r, \Diamond^+C\rangle$. Again let $j \geq i$ be the maximal number for which $\langle j, r, \Diamond^+C\rangle$ is a defect in q. If $j = n$ then, by (S4), there exists a quasiworld $\mathfrak{w}$ and a type $t \in T_{\mathfrak{w}}$ such that $\langle \mathfrak{w}_n, \mathfrak{w}, \sigma\rangle \in \mathcal{S}$, for some σ, $C \in t$, and $r(n)\sigma t$. In this case we put $q^d = \langle q(1), \ldots, q(n), \mathfrak{w}\rangle$. When $j \neq n$, we use (S8) to select a quasiworld $\mathfrak{w}$ and a type $t \in T_{\mathfrak{w}}$ such that $\langle q(j), \mathfrak{w}, \tau\rangle, \langle \mathfrak{w}, q(j+1), \rho\rangle \in \mathcal{S}$, $C \in t$, $r(j)\tau t\rho r(j+1)$ and $\tau \circ \rho = \sigma_{jj+1}$. This yields us a weak quasimodel

$$q^d = \langle q(1), \ldots, q(j), \mathfrak{w}, q(j+1), \ldots, q(n)\rangle$$

"curing" d. The case of $d = \langle i, r, \Diamond^-C\rangle$ is considered analogously.

We are in a position now to complete the proof of Theorem 27.

Proof. Suppose $\mathcal{S}$ is a satisfying set for φ and $\mathfrak{O} = \langle W, <\rangle$ a dense strict linear order without endpoints. We construct by induction a sequence of weak quasimodels q_i over $\mathcal{S}$ and a sequence of subframes $\mathfrak{O}_i = \langle W_i, <_i\rangle$ of $\mathfrak{O}$, for $i = 0, 1, \ldots$.

Step 0. Take a triple $\langle \mathfrak{u}, \mathfrak{v}, \sigma\rangle \in \mathcal{S}$ such that $\mathfrak{u} \models \varphi$ or $\mathfrak{v} \models \varphi$ (it exists by (S1)) and let $w_1 < w_2$ in $\mathfrak{O}$. Then we put $q_0 = \langle \mathfrak{w}_{w_1}, \mathfrak{w}_{w_2}\rangle$, $\mathfrak{O}_0 = \langle W_0, <_0\rangle$, where $\mathfrak{w}_{w_1} = \mathfrak{u}$, $\mathfrak{w}_{w_2} = \mathfrak{v}$, $W_0 = \{w_1, w_2\}$ and $w_1 <_0 w_2$.

Step $i+1$. Suppose we have already constructed a weak quasimodel

$$q_i = \langle \mathfrak{w}_{w_1}, \ldots, \mathfrak{w}_{w_n}\rangle \tag{4}$$

and a subframe $\mathfrak{O}_i = \langle W_i, <_i\rangle$ of $\mathfrak{O}$ such that

$$W_i = \{w_1, \ldots, w_n\},\ w_1 <_i \cdots <_i w_n.$$

If the set D_i of all defects in q_i is empty then we are done: q_i is clearly a quasimodel based on $\mathfrak{O}_i$ and satisfying φ. Otherwise we take some $d \in D_i$, construct the weak quasimodel

$$q_i^d = \left\langle \mathfrak{w}_{w_1}, \ldots, \mathfrak{w}_{w_j}, \mathfrak{w}_w, \mathfrak{w}_{w_{j+1}}, \ldots, \mathfrak{w}_{w_n}\right\rangle \tag{5}$$

for $j \in \{1, \dots, n\}$, select some $w \in W$ such that $w_j < w < w_{j+1}$ ($w_n < w$, if $j = n$, and $w < w_1$, if $j = 1$) and define $\mathfrak{O}_i^d$ to be the subframe of $\mathfrak{O}$ containing $\mathfrak{O}_i$ and w.

Define a set D_i^d of defects in q_i^d in the following way. Suppose d' is a defect in D_i different from d. If $d' = \langle k, \psi \rangle$ then we put $d' = \langle k, \psi \rangle$ in D_i^d when $k \leq j$ and d' is a defect in q_i^d; when $k > j$, we put $d' = \langle k+1, \psi \rangle$. If $d' = \langle k, r, D \rangle$ then we fix a run r' in q_i^d extending r and put $d' = \langle k, r', D \rangle$ in D_i^d when $k \leq j$ and d' is a defect in q_i^d; when $k > j$, we put $d' = \langle k+1, r', D \rangle$. Clearly, $|D_i^d| \leq |D_i| - 1$. If $D_i^d \neq \emptyset$ then we take a defect $d' \in D_i^d$, construct $q_i^{dd'}$, $\mathfrak{O}_i^{dd'}$, and so on. When all defects in D_i are cured, we obtain a weak quasimodel $q_{i+1} = \langle \mathfrak{w}_{w_1}, \dots, \mathfrak{w}_{w_m} \rangle$ and a subframe $\mathfrak{O}_{i+1} = \langle W_{i+1}, <_{i+1} \rangle$ of $\mathfrak{O}$ such that $W_{i+1} = \{w_1, \dots, w_m\}$ and $w_1 <_{i+1} \cdots <_{i+1} w_m$.

Step ω. Finally, put

$$W_\omega = \bigcup_{i<\omega} W_i, \ <_\omega = \bigcup_{i<\omega} <_i, \ \mathfrak{O}_\omega = \langle W_\omega, <_\omega \rangle, \ Q = \langle \mathfrak{w}_w : w \in W_\omega \rangle .$$

We show now that Q is a quasimodel based on $\mathfrak{O}_\omega$ and satisfying φ.

Let $u \in W_\omega$, $\mathfrak{w}_u = \langle T_u, T_u^o, \Phi_u \rangle$ and $t' \in T_u$. We are going to construct a run in Q through t'. Note first that $\mathfrak{w}_u$ belongs to a weak quasimodel q_i of the form (4), for some $i < \omega$, and there is a run r in q_i coming through t'. Define an extension of r for each act of expanding q_i.

Suppose that we "cure" a defect d in q_i and obtain q_i^d. If $d = \langle j, \psi \rangle$ or $d = \langle j, r_1, D \rangle$, for $r_1 \neq r$, then we take any run r' in q_i^d containing r and declare it to be the *extension* of r in q_i^d. If $d = \langle j, r, \Diamond^+ C \rangle$ and q_i^d is of the form (5) (so that $t' = r(k)$ for some $k \leq j$) then we define the *extension* of r in q_i^d to be the run $r(1), \dots, r(j), t, r(j+1), \dots, r(n)$, where $t \in C$ is the concept type in T_w selected in Case 2 above. For $d = \langle j, r, \Diamond^- C \rangle$ the extension of r in q_i^d is defined in a symmetrical way. Now, if r' is the extension of r in q' and r'' the extension of r' in q'' then r'' is the *extension* of r in q''. Finally, we define the *extension* of r in Q as the limit r_ω of the sequence of the extensions of r in q_{i+1}, q_{i+2}, etc.; more precisely, r_ω comes through $t \in T_w$, $w \in W_\omega$, iff the extension of r in some q_j, $j > i$, comes through t. (If the original r is r_a for some $a \in ob\varphi$, then we can always define r_ω so that it comes through all t_a^w, $w \in W_\omega$.)

The constructed extension r_ω is a run in Q coming through t'. Indeed, suppose $\Diamond^+ C \in r_\omega(w)$ for some $\Diamond^+ C \in con\varphi$ and some $w \in W_\omega$. Then the extension r' of r in q_j, for some $j \geq i$, comes through $r_\omega(w)$, say $r_\omega(w) = r'(k)$. If $\langle k, r', \Diamond^+ C \rangle$ is not a defect in r' then there is $m > k$ such that $C \in r'(m)$ and so $C \in r_\omega(v)$ for some $v >_\omega w$. And if $\langle k, r', \Diamond^+ C \rangle$ is a defect then it is cured in some extension of r', and again we must have $v >_\omega w$ with $C \in r_\omega(v)$. Conversely, assume that there is $v >_\omega w$ and $C \in r_\omega(v)$, for some $\Diamond^+ C \in con\varphi$. Consider the extension r' of r in some q_j containing both $\mathfrak{w}_w$ and $\mathfrak{w}_v$. Let $r'(k) = r_\omega(w)$ and $r'(m) = r_\omega(v)$, $k < m$. Since r' is a run in q_j and by the definition of a suitable triple, we must have $\Diamond^+ C \in r'(k) = r_\omega(w)$.

The case of $\Diamond^- C$ is considered analogously.

Thus, r_ω is a run in Q through t'. It is readily seen also that, for every $\Diamond^+\psi \in sub\varphi$ ($\Diamond^-\psi \in sub\varphi$), $Q(u) \models \Diamond^+\psi$ (respectively, $Q(u) \models \Diamond^-\psi$) iff $Q(v) \models \psi$ for some $v >_\omega u$ ($v <_\omega u$). So Q is a quasimodel based on $\mathfrak{O}_\omega$ and satisfying φ. □

This shows that the satisfiability problem for $\mathcal{CIQ}_\Diamond$-formulas in strict linear orders is decidable. To prove that it is decidable also in $\langle \mathbb{Q}, < \rangle$ we require one more definition.

Definition 30 ($\mathbb{Q}$-satisfying set). Say that a satisfying set $\mathcal{S}$ for a formula φ is *$\mathbb{Q}$-satisfying* if for every $\langle \mathfrak{u}, \mathfrak{v}, \sigma \rangle \in \mathcal{S}$ there exist $\langle \mathfrak{u}', \mathfrak{u}, \tau' \rangle \in \mathcal{S}$, $\langle \mathfrak{v}, \mathfrak{v}', \rho' \rangle \in \mathcal{S}$, and $\langle \mathfrak{u}, \mathfrak{w}, \tau \rangle \in \mathcal{S}$, $\langle \mathfrak{w}, \mathfrak{v}, \rho \rangle \in \mathcal{S}$ such that $\tau \circ \rho = \sigma$.

Theorem 31. *A $\mathcal{CIQ}_\Diamond$-formula φ is satisfiable in $\langle \mathbb{Q}, < \rangle$ iff there is a $\mathbb{Q}$-satisfying set for φ.*

Proof. ($\Rightarrow$) is established in the same way as in the proof of Theorem 27.

($\Leftarrow$) Suppose $\mathcal{S}$ is a $\mathbb{Q}$-satisfying set for φ and $\mathfrak{O} = \langle \mathbb{Q}, < \rangle$. We define a sequence of weak quasimodels q_i over $\mathcal{S}$ almost in the same way as in the proof of Theorem 27. The only difference is that now, having cured all defects at step $i+1$ and constructed a weak quasimodel

$$q'_{i+1} = \langle \mathfrak{w}_{w_1}, \ldots, \mathfrak{w}_{w_m} \rangle ,$$

we define q_{i+1} to be a weak quasimodel

$$q_{i+1} = \left\langle \mathfrak{w}_{u_1}, \mathfrak{w}_{w_1}, \mathfrak{w}_{u_2}, \mathfrak{w}_{w_2}, \ldots, \mathfrak{w}_{u_m}, \mathfrak{w}_{w_m}, \mathfrak{w}_{u_{m+1}} \right\rangle$$

in which $\langle \mathfrak{w}_{u_i}, \mathfrak{w}_{w_i}, \sigma_i \rangle \in \mathcal{S}$ and $\left\langle \mathfrak{w}_{w_m}, \mathfrak{w}_{u_{m+1}}, \sigma_{m+1} \right\rangle \in \mathcal{S}$, for some σ_i and σ_{m+1}, $i = 1, \ldots, m$, such that $u_1 < w_1 < u_2 < w_2 < \ldots < u_m < w_m < u_{n+1}$. As a result we construct a quasimodel satisfying φ and based on a subframe of $\mathfrak{O}$ which is isomorphic to $\mathfrak{O}$. □

7 OTHER TEMPORAL DESCRIPTION LOGICS

The methods of proving decidability developed above work for an arbitrary decidable description logic which is closed under the disjoint union construction in the proof of Lemma 17. Most description logics are of this sort. Of other logics, especially interesting are those which allow the construction of the concept $\{a\}$ from every object name a. Such concepts can be understood as *nominals* or *names* (see e.g. [Blackburn, 1993]. The formula $\top = \exists R.\{a\}$ is true in a model iff xRa for all objects x in its domain. It follows that logics with this construct cannot be closed under the formation of disjoint unions.

In this section we briefly explain how to modify the notions of the preceeding sections in order to cope with nominals. We will be considering two rather expressive decidable description logics, namely, $\mathcal{CQO}$ and $\mathcal{CIO}$, first introduced by De Giacomo [1995].

Let $\mathcal{CI}$ and $\mathcal{CQ}$ be the languages resulting from $\mathcal{CIQ}$ by omittingqualified number restrictions $\exists_{\geq n}$ and inversions of roles, respectively. $\mathcal{CIO}$ and $\mathcal{CQO}$ are the extensions of $\mathcal{CI}$ and $\mathcal{CQ}$ with the following concept construct: $\{a\}$ is a concept whenever a is an object name. The concept $\{a\}$ is interpreted in a model I in a straightforward manner: $\{a\}^I = \{a^I\}$. Temporal description logics $\mathcal{CIO}_{\mathcal{US}}$ and $\mathcal{CQO}_{\mathcal{US}}$ and their semantics are defined in the obvious way (we still assume that object names are rigid designators). Having concepts of the form $\{a\}$, there is no need to define as atomic formulas $a : C$ and aRb: they are equivalent to $\{a\} \to C = \top$ and $\{a\} \to \exists R.\{b\} = \top$, respectively. Now we have:

Theorem 32. *There are algorithms that are capable of deciding whether*

1. *a given $\mathcal{CIO}_{\mathcal{US}}$- or $\mathcal{CQO}_{\mathcal{US}}$-formula is satisfiable in $\langle \mathbb{Z}, < \rangle$ and in $\langle \mathbb{N}, < \rangle$,*
2. *a given $\mathcal{CIO}_{\diamond}$- or $\mathcal{CQO}_{\diamond}$-formula is satisfiable in a strict linear order as well as in $\langle \mathbb{Q}, < \rangle$.*

We will point out the most important modifications of the proof of Theorem 8. By $ob\varphi$ we will denote the set of object names a such that $\{a\} \in con\varphi$.

First we should change the definition of a quasiworld candidate: in the present context it is a pair $\langle T, \Phi \rangle$ such that the third condition of Definition 10 holds and, for every $a \in ob\varphi$, there exists precisely one $t \in T$ for which $\{a\} \in t$. Note that in the definition of a quasiworld candidate we omit the set T^o; its role is now played by the types t containing concepts of the form $\{a\}$. We denote the type t containing $\{a\}$ by t_a and define T^o to be the set of all types of the form t_a. The notion of an extended model remains the same. An extended model I realizes a quasiworld candidate iff the first condition of Definition 12 holds and, for every $a \in ob\varphi$, $t^I(a) = t_a$. De Giacomo [1995] proves the decidability of the satisfiability problem for both $\mathcal{CIO}$ and $\mathcal{CQO}$. It follows that one can effectively recognize whether a quasiworld candidate is a quasiworld.

The definition of a run also requires a modification. In Definition 14 we allowed runs r in which $r(u) = t_a^u$ and $r(v) \neq t_a^v$ for some $u \neq v$; now such runs should be forbidden (in accordance with the condition that $x \in \{a\}^{I(u)}$ iff $x \in \{a\}^{I(w)}$). More precisely, a run r still has to satisfy all the conditions of Definition 14 and also the following one: if $r(u) = t_a^v$ then $r(v) = t_a^v$, for all $u, v \in W$. The definition of a quasimodel should be clear now. The only important thing which remains to be modified is the proof of Theorem 16. Basically this reduces to the proof of an analogue of Lemma 17. Of course, we cannot claim now that $|[x]^J| = \kappa'$ for any x in the domain of J. We reformulate this lemma in the following way.

Lemma 33. *Let Q be a quasimodel for φ based on $\langle W, <\rangle$. There is a cardinal $\kappa \geq \aleph_0$ such that, for any cardinal $\kappa' \geq \kappa$, every $\mathcal{CIO}$-quasiworld ($\mathcal{CQO}$-quasiworld) $Q(w) = \mathfrak{w}$ is realized in an extended $\mathcal{CIO}$-model ($\mathcal{CQO}$-model) J in which $|[x]^J| = \kappa'$ for all x in the domain of J different from any a^J, $a \in ob\varphi$.*

Proof. The lemma is trivial if $T^o_w = T_w$, for some $w \in W$, since in this case, in any quasimodel realizing $Q(w)$, every x in the domain coincides with some $a \in ob\varphi$. (Note that in this case $T^o_{w'} = T_{w'}$, for any $w' \in W$.) So suppose this is not the case. First we consider $\mathcal{CIO}$ and $\mathcal{CQO}$ simultaneously.

For each quasiworld $Q(w) = \mathfrak{u}$ fix an extended model $I_\mathfrak{u}$ realizing $\mathfrak{u}$. Let $\Delta_\mathfrak{u}$ be the domain of $I_\mathfrak{u}$. Then we define κ to be the supremum of $\aleph_0$ and $|[x]^{I_\mathfrak{u}}|$ for all quasiworlds $Q(w) = \mathfrak{u}$ and all $x \in \Delta_\mathfrak{u}$ with $x \neq a^{I_\mathfrak{u}}$ for any $a \in ob\varphi$. We show that κ satisfies the required conditions.

Suppose $Q(w) = \mathfrak{w}$ for some $w \in W$ and $\kappa' \geq \kappa$. Take an extended model $I = \langle \Delta, R^I_0, \ldots, C^I_0, \ldots, (C\mathcal{U}D)^I, \ldots, (C'\mathcal{S}D')^I, \ldots, a^I_0, \ldots \rangle$ realizing $\mathfrak{w}$ and such that $|[x]^I| \leq \kappa$ for every $x \in \Delta$, $x \neq a^I$ for any $a \in ob\varphi$. Let $N = \{a^I : a \in ob\varphi\}$ and

$$J = \langle \Delta', R^J_0, \ldots, C^J_0, \ldots, (C\mathcal{U}D)^J, \ldots, (C'\mathcal{S}D')^J, \ldots, a^J_0, \ldots \rangle,$$

where

$$\Delta' = N \cup \{\langle x, \xi\rangle : x \in \Delta - N, \xi < \kappa'\}, \quad a^J_i = a^I_i,$$
$$C^J_i = \{\langle x, \xi\rangle : x \in (\Delta - N) \cap C^I_i, \xi < \kappa'\} \cup (C^J_i \cap N).$$

The definition of R^J_i depends on whether we deal with $\mathcal{CIO}$ or $\mathcal{CQO}$. In both cases we have for all $\xi < \kappa'$, $x, y \in \Delta - N$, $a, b \in ob\varphi$: $\langle x, \xi\rangle\, R^J_i\, \langle y, \xi\rangle$ iff $xR^I_i y$, $a^I R^J_i b^I$ iff $a^I R^I_i b^I$, and $\langle x, \xi\rangle\, R^J_i a^I$ iff $xR^I_i a^I$. In the case of $\mathcal{CIQ}$—because of the inverse role constructor—we put $a^I R^J_i \langle x, \xi\rangle$ iff $a^I R^I_i x$, for all $\xi < \kappa'$, $x \in \Delta - N$. It is readily checked that J satisfies the required conditions for $\mathcal{CIO}$. However, for $\mathcal{CQO}$ this may be not so, since a^I may have more R_i successors now. In this case it is enough to modify the definition by taking $a^I R^J_i \langle x, \xi\rangle$ iff $a^I R^I_i x$ and $\xi = 0$. □

8 OPEN PROBLEMS

This paper introduces temporal description logics as an expressive and *decidable* alternative to temporal predicate logics. We have proved the decidability of the satisfiability problem for $\mathcal{CIQ}_{\mathcal{US}}$-formulas in $\langle \mathbb{N}, <\rangle$ and $\langle \mathbb{Z}, <\rangle$, and of $\mathcal{CIQ}_\Diamond$ in strict linear orders and $\langle \mathbb{Q}, <\rangle$. It would also be of interest to find solutions to the following problems:

- Is the satisfiability problem for $\mathcal{CIQ}_{\mathcal{US}}$-formulas in strict linear orders and $\langle \mathbb{Q}, <\rangle$ decidable?

- Is the satisfiability problem for $\mathcal{CIQ}_\diamond$-formulas in $\langle \mathbb{R}, < \rangle$ decidable?
- What is the complexity of the satisfiability problems considered in this paper?

In the temporal extensions of $\mathcal{CIO}$ and $\mathcal{CQO}$ we assumed that object names (and so concepts of type $\{a\}$) are *rigid* designators: their extensions are defined globally and do not depend on the particular world. By allowing object names to be interpreted locally we obtain more expressive languages.

- Is the satisfiability problem for the resulting language decidable?

As was already noted, none of the underlying description languages considered here has the finite model property (fmp). And even if we take as the basis of our temporal logics a description logic with the fmp (say $\mathcal{ALC}$), it does not follow that the resulting temporal description logic having models with finite domains coincides with the logic whose models may have arbitrary domains (see [Wolter and Zakharyaschev, 1998]). This observation leads to the following problem:

- Are the temporal description logics considered in this paper decidable when the domains of models are assumed to be finite?

Acknowledgments. The work of the second author was supported by grant number 97–01–00975 from the Russian Fundamental Research Foundation. The authors would like to thank the anonymous referees of this paper for helpful remarks.

REFERENCES

[Artale and Franconi, 1994] A. Artale and E. Franconi. A computational account for a description logic of time and action. In *Proc. 4th Conf. on Principles of Knowledge Representation and Reasoning*, pages 3–14, Montreal, Canada, 1994. Morgan Kaufman.

[Baader and Hollunder, 1991] F. Baader and B. Hollunder. A terminological knowledge representation system with complete inference algorithms. In *Proc. workshop on Processing Declarative Knowledge, PDK-91*, pages 67–86. Lecture Notes in Artificial Intelligence, No. 567. Springer Verlag, 1991.

[Baader and Laux, 1995] F. Baader and A. Laux. Terminological logics with modal operators. In *Proc. 14th International Joint Conf. on Artificial Intelligence*, pages 808–814, Montreal, Canada, 1995. Morgan Kaufman.

[Baader and Ohlbach, 1995] F. Baader and H.J. Ohlbach. A multi-dimensional terminological knowledge representation language. *Journal of Applied Non-Classical Logic*, 5:153–197, 1995.

[Bergamaschi and Sartori, 1992] S. Bergamaschi and C. Sartori. On taxonimic reasoning in conceptual design. *ACM Trans. on Database Systems*, 17:385–422, 1992.

[Blackburn, 1993] P. Blackburn. Nominal tense logic. *Notre Dame Journal of Formal Logic*, 34:56–83, 1993.

[Borgida *et al.*, 1989] A. Borgida, R.J. Brachman, D.L McGuiness, and L. Alperin Resnick. CLASSIC: A structural data model for objects. In *Proc. ACM SIGMOD Int. Conf. on Management of Data*, pages 59–67. Portland, Oreg., 1989.

[Borgida, 1995] A. Borgida. Description logics in data management. *IEEE Trans. on Knowledge and Data Engineering*, 7:671–682, 1995.

[Brachman and Schmolze, 1985] R.J. Brachman and J.G. Schmolze. An overview of the KL-ONE knowledge representation system. *Cognitive Science*, 9:171–216, 1985.

[Catarci and Lenzerini, 1993] T. Catarci and M. Lenzerini. Representing and using interschema knowledge in cooperative information systems. *J. of Intelligent and Cooperative Information Systems*, 2:375–398, 1993.

[De Giacomo and Lenzerini, 1994] G. De Giacomo and M. Lenzerini. Boosting the correspondence between description logics and propositional dynamic logics. In *Proc. 12th Nat. Conf. on Artificial Intelligence (AAAI-94)*, pages 205–212. AAAI Press/The MIT Press, 1994.

[De Giacomo and Lenzerini, 1996] G. De Giacomo and M. Lenzerini. TBox and ABox reasoning in expressive description logics. In *Proc. 5th Conf. on Principles of Knowledge Representation and Reasoning (KR'96)*, pages 316–327, Montreal, Canada, 1996. Morgan Kaufman.

[De Giacomo, 1995] G. De Giacomo. *Decidability of Class-Based Knowledge Representation Formalisms.* PhD thesis, Univ. di Roma, 1995.

[Devanbu and Litman, 1991] P. Devanbu and D. Litman. Plan-based terminological reasoning. In *Proc. 2nd Conf. on Principles of Knowledge Representation and Reasoning*, pages 128–138, Montreal, Canada, 1991. Morgan Kaufman.

[Fagin *et al.*, 1995] R. Fagin, J. Halpern, Y. Moses, and M. Vardi. *Reasoning about Knowledge.* MIT Press, 1995.

[Finger and Gabbay, 1992] M. Finger and D. Gabbay. Adding a temporal dimension to a logic system. *Journal of Logic, Language and Information*, 2:203–233, 1992.

[Gabbay and Shehtman, 1998] D. Gabbay and V. Shehtman. Products of modal logics, part 1. *Journal of the IGPL*, 6:73–146, 1998.

[Gabbay *et al.*, 1994] D. Gabbay, I. Hodkinson, and M. Reynolds. *Temporal Logic.* Oxford University Press, 1994.

[Halpern and Shoham, 1991] J. Halpern and Y. Shoham. A propositional modal logic of time intervals. *Journal of the ACM*, 38:935–962, 1991.

[Kröger, 1990] F. Kröger. On the interpretability of arithmetic in temporal logic. *Theoretical Computer Science*, 73:47–60, 1990.

[Marx and Venema, 1997] M. Marx and Y. Venema. *Multi dimensional modal logic.* Kluwer Academic Publishers, 1997.

[Marx, 1999] M. Marx. Complexity of products of modal logics. *Journal of Logic and Computation*, 9(2):197–214, 1999.

[Reynolds, 1996] M. Reynolds. A decidable temporal logic of parallelism. Manuscript, 1996.

[Schild, 1993] K. Schild. Combining terminological logics with tense logic. In *Proc. 6th Portuguese Conf. on Artificial Intelligence*, pages 105–120, Porto, 1993.

[Schmiedel, 1990] A. Schmiedel. A temporal terminological logic. In *Proc. 9th National Conf. of the American Association for Artificial Intelligence*, pages 640–645, Boston, 1990.

[Spaan, 1993] E. Spaan. *Complexity of Modal Logics.* PhD thesis, Department of Mathematics and Computer Science, University of Amsterdam, 1993.

[Szalas and Holenderski, 1988] A. Szalas and L. Holenderski. Incompleteness of first-order temporal logic with until. *Theoretical Computer Science*, 57:317–325, 1988.

[Weida and Litman, 1992] R. Weida and D. Litman. Terminological reasoning with constraint networks and an application to plan recognition. In *Proc. 3rd Conf. on Principles of Knowledge Representation and Reasoning*, pages 282–293, Montreal, Canada, 1992. Morgan Kaufman.

[Wolter and Zakharyaschev, 1998] F. Wolter and M. Zakharyaschev. Satisfiability problem in description logics with modal operators. In *Proc. 6th Conf. on Principles of Knowledge Representation and Reasoning (KR'98)*, pages 512–523, Montreal, Canada, 1998. Morgan Kaufman.

[Wright *et al.*, 1993] G.T. Wright, E.S. Weixelbaum, G.T. Vesonder, K.E. Brown, S.R. Palmer, J.I. Berman, and H.H. Moore. A knowledge-based configurator that supports sales, engineering, and manufacturing at AT&T network systems. *AI Magazine*, 14:69–80, 1993.

INDEX